Stochastic Analysis and Applications in Physics

NATO ASI Series

Advanced Science Institutes Series

A Series presenting the results of activities sponsored by the NATO Science Committee, which aims at the dissemination of advanced scientific and technological knowledge, with a view to strengthening links between scientific communities.

The Series is published by an international board of publishers in conjunction with the NATO Scientific Affairs Division

A Life Sciences **B Physics**	Plenum Publishing Corporation London and New York
C Mathematical and Physical Sciences **D Behavioural and Social Sciences** **E Applied Sciences**	Kluwer Academic Publishers Dordrecht, Boston and London
F Computer and Systems Sciences **G Ecological Sciences** **H Cell Biology** **I Global Environmental Change**	Springer-Verlag Berlin, Heidelberg, New York, London, Paris and Tokyo

NATO-PCO-DATA BASE

The electronic index to the NATO ASI Series provides full bibliographical references (with keywords and/or abstracts) to more than 30000 contributions from international scientists published in all sections of the NATO ASI Series.
Access to the NATO-PCO-DATA BASE is possible in two ways:

– via online FILE 128 (NATO-PCO-DATA BASE) hosted by ESRIN, Via Galileo Galilei, I-00044 Frascati, Italy.

– via CD-ROM "NATO-PCO-DATA BASE" with user-friendly retrieval software in English, French and German (© WTV GmbH and DATAWARE Technologies Inc. 1989).

The CD-ROM can be ordered through any member of the Board of Publishers or through NATO-PCO, Overijse, Belgium.

Series C: Mathematical and Physical Sciences - Vol. 449

Stochastic Analysis and Applications in Physics

edited by

Ana Isabel Cardoso

TELECOM PORTUGAL,
Funchal (Madeira), Portugal

Margarida de Faria

CCM,
Universidade da Madeira,
Funchal (Madeira), Portugal

Jürgen Potthoff

Fakultät für Mathematik und Informatik,
Universität Mannheim,
Mannheim, Germany

Roland Sénéor

Centre de Physique Théorique,
Ecole Polytechnique,
Palaiseau, France

and

Ludwig Streit

CCM,
Universidade da Madeira,
Funchal (Madeira), Portugal
and
BiBoS,
Universität Bielefeld,
Bielefeld, Germany

Springer Science+Business Media, B.V.

Proceedings of the NATO Advanced Study Institute on
Stochastic Analysis and Applications in Physics
Funchal, Madeira, Portugal
August 6–19, 1993

A C.I.P. Catalogue record for this book is available from the Library of Congress.

ISBN 978-94-010-4098-3 ISBN 978-94-011-0219-3 (eBook)
DOI 10.1007/978-94-011-0219-3

Printed on acid-free paper

Originally published by Kluwer Academic Publishers in 1994
Softcover reprint of the hardcover 1st edition 1994

TABLE OF CONTENTS

Preface

Stochastic analysis and its various applications in physics have to a large extent developed symbiotically. In the past decades mathematics has provided physics with a vast and rapidly expanding array of tools and methods, while on the other hand physics has counted among the sources of direction and of structural intuition for the mathematical research in stochastics.

We hope to have captured some of the focal points of this dialogue in the NATO ASI "Stochastic Analysis and its Applications in Physics" and in the present volume.

On the mathematical side White Noise Analysis has emerged as a viable framework for stochastic and infinite dimensional analysis (*Hida, Streit*). Another growth point is the theory of stochastic partial differential equations and their applications (*Bertini et al.*, *Øksendal, Potthoff, Russo, Sénéor*). Gauge field theories have increasingly attracted the attention not only of physicists but of mathematicians as well (*Gross, Léandre, Sengupta*). On the other hand the contributions of *Lang* and of *Vilela Mendes* show the extent to which stochastic methods have found a place in the physicists' toolbox. Dirichlet forms provide a fruitful link between the mathematics of Markov processes and fields and the physics of quantum systems (*Albeverio et al.*). The deterministic-stochastic interface was addressed by *Collet* and by *Mandrekar*, Euclidean quantum mechanics by *Cruzeiro and Zambrini*, excursions of diffusions by *Truman*, and Kubo-Martin-Schwinger norms of statistical mechanics by *Streater*.

So much for a rapid synopsis of the material represented in the present volume. We owe it to the efforts and the kindness of the lecturers. Participants of the ASI had ample opportunity to profit from their openness and enthusiasm. But also the setting for such an exchange of ideas has to be right. For this thanks are due to the hospitality extended by the Centro de Formação Profissional. They also go to our untiring helpers - the staff, clerical as well as scientific, and the students of the Universidade da Madeira, who made it all possible.

Ana Isabel Cardoso, Margarida de Faria, Jürgen Potthoff, Roland Sénéor, Ludwig Streit

Funchal, Mannheim, Palaiseau 1994.

Sponsors:

NATO Scientific and Environmental Affairs Division
Fundação Luso-Americana para o Desenvolvimento
Governo Regional da Região Autónoma da Madeira
Universidade da Madeira
Direcção Regional de Emprego - Centro de Formação Profissional

LIST OF PARTICIPANTS

Sergio ALBEVERIO
Fak. f. Mathematik
Ruhr-Universität Bochum
D-44789 Bochum
GERMANY

Arvydas ASTRAUSKAS
Institute of Mathematics and Informatics
Akademijos str. 4
2600 Vilnius
LITHUANIA

Claas BECKER
Ruhr-Universität Bochum
Fakultät für Mathematik
Universitätsstr. 25
D-44789 Bochum
GERMANY

Alexander BENDIKOV
Mathematisches Institut
Universität Erlangen-Nürnberg
Bismarckstr. 1 1/2
D-91054 Erlangen
GERMANY

Fred Espen BENTH
Department of Mathematics
University of Oslo
Box 1053 Blindern
N-0314 Oslo
NORWAY

Lorenzo BERTINI
Dipartimento di Matematica
Università die Roma 'Tor Vergata'
Via della Ricerca Scientifica
I-00133 Roma
ITALY

David BETOUNES
Mathematics Department
University of Southern Mississippi
Hattiesburg, Mississippi 39406-5045
U.S.A.

Hans Jürgen BORCHERS
Institut für Theoretische Physik
Bunsenstr. 9
D-37073 Göttingen
GERMANY

Lisa BORLAND
Institute für Theoretische Physik
und Synergetic
Universität Stuttgart
Pfaffenwaldring 57/IV
D-70569 Stuttgart
GERMANY

José Manuel BRAGA
CFMC
Av. Prof. Gama Pinto 2
PL-1699 Lisboa Codex
PORTUGAL

Zdzislaw BRZEZNIAK
Fakultät für Mathematik
Ruhr-Universität Bochum
Universitätsstr. 25
D-44789 Bochum
GERMANY

Nicoletta CANCRINI
Dipartimento de Fisica
Universita "La Sapienza"
Piazza Aldo Moro 2
I-00185 Roma
ITALY

Ana Isabel CARDOSO
Calcadad de St Clara, 32
P-9000 Funchal (MADEIRA)
PORTUGAL

Pierre COLLET
Centre de Physique Theorique
Ecole Polytechnique
Plateau de Palaiseau
F-91128 Palaiseau
FRANCE

Philippe COMBE
Centre de Physique Théorique
CNRS - Luminy, Case 907
163 Avenue de Luminy
F-13228 Marseille Cedex 9
FRANCE

José Manuel N. Castanheira da COSTA
Universidade da Madeira
Matemática
Largo do Colégio
P-9000 Funchal (Madeira)
PORTUGAL

Ana Bela CRUZEIRO
CFMC
Av. Prof. Gama Pinto 2
P-1699 Lisboa Codex
PORTUGAL

Mário Dionisio CUNHA
Universidade da Madeira
Matemática
Largo do Colégio
P-9000 Funchal (Madeira)
PORTUGAL

Alexei DALETSKI
Fakultät für Mathematik
Ruhr-Universität Bochum
Universitätsstr. 150
D-44789 Bochum
GERMANY

Vladimir N. DAMGOV
Bulgarian Academy of Sciences
Space Research Institute
JK "Mladst 1A"
B1.522, Vh. 5, Ap. 98
1784 Sofia
BULGARIA

Thomas DECK
Fakultät für Mathematik
und Informatik
Universität Mannheim
D-68131 Mannheim
GERMANY

Custódia Mercês Reis Rodrigues DRUMOND
Universidade da Madeira
Matemática
Largo do Colégio
P-9000 Funchal
PORTUGAL

Margarida Maria Coelho Ribeiro de FARIA
Universidade da Madeira
Matemática
Largo do Colégio
P-9000 Funchal (Madeira)
PORTUGAL

Orlando FREITAS
Universidade da Madeira
Matemática
Largo do Colégio
P-9000 Funchal
PORTUGAL

Tadeusz GERSTENKORN
Institute of Mathematics
Lódz University
ul. s. Banacha 22
PL-90-238 Lódz
POLAND

Roman GILERAK
Institut of Theoretical Physics
University of Wroclaw
PL-50-204 Wroclaw
POLAND

José Gabriel Lira GOMES
Universidade da Madeira
Matemática
Largo do Colégio
P-9000 Funchal (Madeira)
PORTUGAL

Maribel Gomes GONÇALVES
Matemática
Universidade da Madeira
Largo do Colégio
P-9000 Funchal (Madeira)
PORTUGAL

Leonard GROSS
Department of Mathematics
Cornell University
Ithaca, NY 14853 – 7901
U.S.A.

Faruk GÜNGOR
Department of Mathematics
Faculty of Science
Istanbul Technical University
Ayazaga Kampusu
80626 Ayazaga
TURKEY

Zbigniew HABA
Institute of Theoretical Physics
University of Wroclaw
Pl. Maksa Borna 9
PL-50-240 Wroclaw
POLAND

Takeyuki HIDA
Mathematic Department
School of Sciences & Technology
Meijo University
Shiogama-guchi, Tempaku-ku
Nagoya 468
JAPAN

Helmut HÜFFEL
Institut für Theoretische Physik
Universität Wien
Boltzmanngasse 5
A-1090 Wien
AUSTRIA

Sudhir JAIN
Department of Math + Computing
Faculty of Science + Technology
University of Derby
Kedleston Road
Derby DE3 1GB
UNITED KINGDOM

Helge Kristian JENSSEN
Universitet I Oslo
Det-matematisk -
Naturvitenskapelige Fakultet
Matematisk Institut
Matematikkbygningen, Blindern
Box 1053 Blindern
N-0316 Oslo
NORWAY

Viktor K. JIRSA
University of Stuttgart
Institut for Theoretical Physics
and Synergetics
Pfaffenwaldring 57/4
D-70569 Stuttgart
GERMANY

Giovanni JONA–LASINIO
Dipartimento de Fisica
Universita "La Sapienza"
Piazza Aldo Moro 2
I-00185 Roma
ITALY

Tomasz KAPITANIAK
Division of Control and Dynamics
Technical University of Lódz
Stefanowskiego 1/15
PL-90-924 Lodz
POLAND

Witold KARWOWSKI
Institut of Theoretical Physics
Wroclaw University
Pl. Maxa Borna 9
PL-50-204 Wroclaw
POLAND

V.N. KOLOKOLTSOV
Institut für Mathematik
Ruhr-Universität Bochum
Universitätsstr. 150
D-44789 Bochum
GERMANY

Vladimir KONOTOP
Universidade da Madeira
Matemática
Largo do Colégio
P-9000 Funchal (Madeira)
PORTUGAL

Yuri G. KONDRATIEV
BiBoS
Universität Bielefeld
Universitätsstr. 25
D-33615 Bielefeld
GERMANY

Vladimir D. KOSHMANENKO
Institute of Mathematics
Repin Str. 3
Kiev 4 252601 GSP
UKRAINE

Joachim KUPSCH
Fachbereich Physik
Universität Kaiserslautern
Erwin-Schrödinger-Straße
D-67663 Kaiserslautern
GERMANY

Christian LANG
Institute für Theoretische Physik
Universität Graz
Universitätsplatz 5
A-8010 Graz
AUSTRIA

Angelika LASCHECK
Fakultät für Physik - BiBoS
Universität Bielefeld
Universitätsstr. 25
D-33615 Bielefeld
GERMANY

R. LEANDRE
Université Louis Pasteur
Department de Mathematique
7, rue René Descartes
F-67084 Strasbourg Cedex
FRANCE

Jung Soon Kim LEE
Department of Mathematics
Southern University and
A & M. College
P.O.Box 9759
Baton Rouge, Louisiana 70813
U.S.A.

Peter LEUKERT
Fakultät für Physik
Universität Bielefeld
Universitätsstr. 25
D-33615 Bielefeld
GERMANY

Jürgen LÖFFELHOLZ
Universität Leipzig
Physik
Augustusplatz 10
D-04109 Leipzig
GERMANY

Vidyadhar S. MANDREKAR
Department of Statistics and
Probabilities
Michigan State University
East Lansing, MI 48824
U.S.A.

Rui Vilela MENDES
CFMC
Av. Prof. Gama Pinto 2
P-1699 Lisboa Codex
PORTUGAL

Sandra Maria Freitas MENDONÇA
Matemática
Universidade da Madeira
Largo do Colégio
P-9000 Funchal (Madeira)
PORTUGAL

Ryszard MRUGALA
Institute of Physics
Nicholas Copernicus University
Grudziadzka 5
PL-87-100 Torun
POLAND

Peter MÜLLER
Institut für Theoretische Physik
Universität Erlangen-Nürnberg
Staudtstr. 7
D-91058 Erlangen
GERMANY

Folkert MÜLLER-HOISSEN
Institut für Theoretische Physik
Bunsenstr. 9
D-37073 Göttingen
GERMANY

Hiromichi NAKAZATO
Department of Physics
University of the Ryukyus
Okinawa 903 - 01
JAPAN

Hanna NENCKA
Matemática
Universidade da Madeira
Largo do Colégio
P-9000 Funchal (Madeira)
PORTUGAL

Bernt ØKSENDAL
Mathematisk Institut
Universitet i Oslo
Postboks 1032
Blindern 0315 Oslo 3
NORWAY

Axel PELSTER
Institut für Theoretische Physik
und Synergetic
Universität Stuttgart
Pfaffenwaldring 57
D-70569 Stuttgart
GERMANY

Eduardo Jorge Nunes PEREIRA
Centre de Quımica - Fısica Molecular
Universidade Técnica de Lisboa
Complexo I
Instituto Superior Técnico
Av. Rovisco Pais
P-1096 Lisboa Codex
PORTUGAL

António Manuel Ramalho PIRES
Estrada da Carnacha, 74
Sõ Gonçalo
P-9000 Funchal (Madeira)
PORTUGAL

António Alberto Torres Garcia PORTUGAL
Praceta Luis de Camões, no. 3, 2 andar
P-3000 Coimbra
PORTUGAL

Jürgen POTTHOFF
Lehrstuhl für Mathematik V
Universität Mannheim
A5 Seminargebäude
D-68161 Mannheim
GERMANY

Zhongmin QIAN
Department of Applied Mathematics
Shanghai Institute of Railway
Technology
1 Zhennan Rd.
Shanghai 200333
CHINA

Gert ROEPSTORFF
Institut für Theoretische Physik
RWTH Aachen
Sommerfeldstr. 26-28
52074 Aachen
GERMANY

Yuri ROZANOV
Steklov Mathematical Institute
Vavilova Str. 42
117966 Moscow GSP - 1
RUSSIA

M.G. RUDAVETS
Laboratory of Mathematical Physics
Institute of Chemical Physics
Russian Academy of Sciences
Chernogolovka
142432 Moscow Region
RUSSIA

Slawomir RUDNICKI
Institute of Theoretical Physics
and Astrophysics
University of Gdansk
Wita Stwosza 57
PL-80-952 Gdansk
POLAND

Francesco RUSSO
Université de Provence
UFR - MIM
3, Place Victor Hugo
13331 Marseille Cedex 3
FRANCE

Nuno SA
Secção de Fisica
Universidade Trás-os-Montes e Alto-Douro
Apartado 202
Vila Real
P-5001 Vila Real
PORTUGAL

Werner SACK
F.B. Mathematik
Universität Kaiserslautern
Postfach 3049
D-67653 Kaiserslautern
GERMANY

Jochen SCHMIDT
Fakultät für Mathematik
und Informatik
Universität Mannheim
D-68131 Mannheim
GERMANY

Javier Casahorran SEBASTIAN
Departamento de Fisica Teorica
Facultad de Ciencias
Universidad de Zaragoza
Zaragoza 50009
SPAIN

Roland SÉNÉOR
Centre de Physique Theorique
Ecole Polytechnique
F-91128 Palaiseau
FRANCE

Ambar SENGUPTA
Department of Mathematics
Louisiana State University
Baton Rouge LA 70803-4918
U.S.A.

António Casimiro SILVA
Universidade da Madeira
Matemática
Largo do Colégio
P-9000 Funchal (Madeira)
PORTUGAL

José Luis da SILVA
Universidade da Madeira
Matemática
Largo do Colégio
P-9000 Funchal (Madeira)
PORTUGAL

Orlando SILVA
Universidade da Madeira
Matemática
Largo do Colégio
P-9000 Funchal (Madeira)
PORTUGAL

Sankatha P. SINGH
Department of Math. & Stat.
Memorial University
St. John's, NF - A1C 5S7
CANADA

Paul B. SLATER
Community and Organization Research Institute
University of California
Santa Barbara CA 93106-2150
U.S.A.

O.G. SMOLYANOV
Faculty of Mech. and Mathematics
Moscow State University
119899 Moscow
RUSSIA

David A. STORVICK
School of Mathematics
127 Vincent Hall
206 Church Street S.E.
Minneapolis, MN - 55455
U.S.A.

Orlin STOYTCHEV
Mathematics Section
Internacional Centre for Theoretical Physics
P.O.Box 586
34100 Triestre
BULGARIA

Raymond F. STREATER
Department of Mathematics
King's College
Strand
London WC2R 2LS
UNITED KINGDOM

Ludwig STREIT
Universidade da Madeira
Colégio dos Jesuitas
Largo do Colégio
9000 Funchal (Madeira)
PORTUGAL

Veronica STROULECKAYA
Department of Mathematics
University of Nottingham
University Park
Nottingham NG72RD
UNITED KINGDOM

Serguei TCHEREMCHANTSEV
Universidade da Madeira
Matemática
Largo do Colégio
P-9000 Funchal (Madeira)
PORTUGAL

Matthias TIMPEL
Fakultät für Mathematik
und Informatik
Universität Mannheim
D-68131 Mannheim
GERMANY

Aubrey TRUMAN
Department of Mathematics
University College of Swansea
Singleton Park
Swansea SA2 8PP
UNITED KINGDOM

Hisao WATANABE
Mathematics Department
Okayama University of Science
Okayama 700
JAPAN

Werner WESTERKAMP
BiBoS
Fakultät für Physik
Universität Bielefeld
Universitätsstr. 25
D-33615 Bielefeld
GERMANY

INFINITE DIMENSIONAL DIFFUSIONS, MARKOV FIELDS, QUANTUM FIELDS AND STOCHASTIC QUANTIZATION

SERGIO ALBEVERIO[1),2)]
YURI G. KONDRATIEV[2),3)],
MICHAEL RÖCKNER[4)]

[1)] *Mathematisches Institut, Ruhr–Universität, D 44801 Bochum (Germany); SFB 237; CERFIM (Locarno);*
[2)] *BiBoS, Universität Bielefeld;*
[3)] *On leave of absence from Institute of Mathematics, Kiev (Ukraine);*
[4)] *Mathematisches Institut, Universität Bonn*

ABSTRACT. We give a survey of recent results relating infinite dimensional diffusions with stochastic partial differential equations. We also present applications to the theory of (homogeneous) Markov fields, quantum fields, stochastic quantization as well as Gibbs fields and lattice models of classical and quantum statistical mechanics.

0. Contents

A. I. Cardoso et al. (eds.), Stochastic Analysis and Applications in Physics, 1–34.

1. Introduction

In these lectures we would like to survey the theory of Dirichlet forms and associated Markov processes in its connections with recent developments, in particular those related with applications to stochastic partial differential equations, Markov fields, quantum fields and statistical mechanical models (both classical and quantum). We try to give an idea of the basic structures, methods and examples, for proofs and more details we refer to the original publications. As general introductions to the theory of Dirichlet forms we refer to the books [Fu1], [FuOT], [Si], [BH], [MR1] and the lecture notes [ABR], [R1]. We present (in Sect. 2) the theory of Dirichlet forms in its abstract version first, and then in its basic finite dimensional examples (with relations to quantum mechanics) and (Sect. 3) infinite dimensional examples (with relations to stochastic p.d.e. and Markov fields). (Other examples related with spin systems of classical and statistical mechanics are discussed in Sect. 5). In Sect. 4 we discuss some basic recent progress in the study of the uniqueness problem for Dirichlet forms, with applications to the proof of logarithmic Sobolev inequalities for infinite dimensional systems. Recent applications to stochastic quantization, Glauber dynamics and "strings" are discussed in Sect. 5.

2. Dirichlet forms

2.1 The "abstract" theory

Let $\mathcal{H}$ be a real separable Hilbert space with scalar product $(\, , \,)$. Let $\mathcal{E}$ be a positive bilinear closed symmetric form on H, with dense domain $D(\mathcal{E}) \subset \mathcal{H}$, i.e., $\mathcal{E}(u,u) \geq 0 \; \forall u \in D(\mathcal{E})$, $\mathcal{E}$ linear in both arguments, $\mathcal{E}(u,v) = \mathcal{E}(v,u) \; \forall u,v \in D(\mathcal{E})$ and $D(\mathcal{E})$, with norm given by $\mathcal{E}_1(u,v) \equiv \mathcal{E}(u,v) + (u,v)$, is complete. It is well known that to $\mathcal{E}$ there is associated a unique positive self-adjoint operator H on $\mathcal{H}$, with domain $D(H) \subset D(\mathcal{E})$, s.t. $\mathcal{E}(u,v) = (H^{\frac{1}{2}}u, H^{\frac{1}{2}}v)$ (or equivalently $\mathcal{E}(u,v) = (Hu,v) \; \forall u \in D(H), \; \forall v \in D(\mathcal{E})$). To H there is uniquely associated a strongly continuous self-adjoint contraction semigroup $T_t = e^{-tH}$, $t \geq 0$, generated by H, in $\mathcal{H}$.

Remark 1 There is a natural extension of these associations, see e.g. [MR1], to the case where the symmetry condition in $\mathcal{E}$ is dropped, if $\tilde{\mathcal{E}}$ denotes the symmetric part of $\mathcal{E}$, $\tilde{\mathcal{E}}$ is as above and moreover one has $|\mathcal{E}_1(u,v)| \leq K \mathcal{E}_1(u,u)^{\frac{1}{2}} \mathcal{E}_1(v,v)^{\frac{1}{2}}$ for some $K \in (0,\infty)$ ("(weak) sector condition") (in the symmetric case $\mathcal{E} = \tilde{\mathcal{E}}$ this holds automatically with $K=1$). In this case $\mathcal{E}$ is still associated with a linear operator H, which generates a strongly continuous contraction semigroup $T_t = e^{-tH}$ on $\mathcal{H}$ s.t. $e^{-t}T_t^{\mathbb{C}}$, $t \geq 0$ (with $T_t^{\mathbb{C}}$ the complexification of T_t) is the restriction of a holomorphic contraction semigroup on the sector $\{z \in \mathbb{C} \, \| \text{Im } z| \leq \frac{1}{K} \text{Re } z\}$.

Now let $\mathcal{H} = L^2(\mu)$, with μ a positive measure on some measurable space (E,B). Exploiting the natural order structure on $\mathcal{H}$ we can define a (symmetric) Dirichlet form $\mathcal{E}$ on $\mathcal{H}$ by asking that $\mathcal{E}$ has in addition to the properties listed above the "Dirichlet property", i.e., $u^\sharp \in D(\mathcal{E})$ and $\mathcal{E}(u^\sharp, u^\sharp) \leq \mathcal{E}(u,u) \; \forall u \in D(\mathcal{E})$ with $u^\sharp \equiv (u \vee 0) \wedge 1$. There is a 1–1 correspondence between (symmetric) Dirichlet forms and (self-adjoint) Markov (contraction) semigroups $(T_t)_{t>0}$ on $\mathcal{H}$ (where Markov means $0 \leq u \leq 1 \to 0 \leq T_t u \leq 1 \; \forall u \in \mathcal{H}$). The probabilistic interest of Markov semigroups and Dirichlet forms is that they lead to Markov processes. Let us in fact assume that $(E,\mathcal{B})$ is a Hausdorff topological space (with B both the Borel σ–algebra and the σ–algebra generated by bounded continuous functions) and μ is σ–finite on $(E,\mathcal{B})$. One says that $\mathcal{E}$ is properly associated with a Markov process $X \equiv (X_t, t \geq 0)$ on E if $E_\bullet[u(X_t)]$ is an $\mathcal{E}$–quasi–continuous μ–version of $(e^{-tH}u)(\cdot) \; \forall u \in L^2(\mu)$. $\mathcal{E}$–quasi–continuous means intuitively "continuous modulo $\mathcal{E}$–exceptional sets", where exceptionality of sets is defined purely in terms of the form $\mathcal{E}$. (The intuitive meaning of exceptionality is best understood in terms of examples, see e.g. [AHKS], [MR1]). One would now like to relate the path properties of X with analytic properties of $\mathcal{E}$. The basic theorem proven in recent years [AMR3] as the last step in a long evolution of the theory (see e.g. [Fu1], [FuOT], [MR1,2])

is the following: $\mathcal{E}$ is quasi–regular (in a sense of quasi–regularity defined purely analytically, quasi regularity being intuitively something like "being determined on compacts") iff there is a μ–tight standard process X properly associated with $\mathcal{E}$ (where μ–tight and standard can be formulated purely in probabilistic terms, thus intrinsic to the process X). In particular, in this case X is strong Markov with right continuous paths and left limits exists and are quasi–left continuous up to the life time. Moreover, for quasi regular Dirichlet forms one has the following important characterisation:
$\mathcal{E}$ local (in the sense that $\mathcal{E}(u,v) = 0$ if supp $(|(|u) \cdot \mu) \cap$ supp $(|v| \cdot \mu) = \emptyset) \leftrightarrow X$ diffusion (i.e., P_z ($t \to X_t$ continuous up to life time) $= 1$ for all $z \in E$ outside an $\mathcal{E}$–exceptional set).
Moreover, one has a unique characterization of Hunt processes in terms of properties of $\mathcal{E}$ [AMR5]. All results hold incidentally also for the non–symmetric $\mathcal{E}$ as in Remark 1, with the "Dirichlet property" defined in a suitable way. For further details on the topics in this section see the books [Fu1], [FuOT], [BH], [MR1]. A general structural result can be given whenever E is, e.g., a Souslin space (i.e., the continuous image of a separable complete metric space, e.g. a Banach space). In this case [AMR6] any quasi–regular Dirichlet form decomposes in 3 parts, $\mathcal{E} = \mathcal{E}^c + \mathcal{E}^j + \mathcal{E}^k$, where $\mathcal{E}^c$ is the "diffusion part", $\mathcal{E}^j$ is the "jump part" and $\mathcal{E}^k$ is the "killing part". More detailed results can be given under stronger assumptions on E, see [Fu1], [Sp], [ARSp]. For an important result on invariance of capacities on abstract Wiener spaces see [AFHMR].

2.2 Examples in finite dimensions

Now let E be an open subset of $\mathbb{R}^d$ and assume for simplicity supp$\mu = E$. Let $\mathcal{E} = \mathcal{E}^c$ be first defined on $C_0^\infty(E)$ by $\mathcal{E}(u,v) = \sum_{i,j=1}^d \int \frac{\partial u}{\partial x_i} \frac{\partial v}{\partial x_j} d\nu_{ij}$, $\forall u,v \in C_0^\infty(E)$, with $\nu_{ij} = \nu_{ji}$, $(\nu_{ij}(K))$ a positive definite matrix for all compacts $K \subset E$. $\mathcal{E}$ is a densely defined bilinear symmetric form in $L^2(\mu)$. From the constructive analytic point of view a basic problem is the closability of $\mathcal{E}$. One basic sufficient condition for closability is "the integration by parts" formula, i.e., the existence of a symmetric operator $\overset{\circ}{H}$ s.t. $\mathcal{E}(u,v) = (\overset{\circ}{H}\ u, v)$ $\forall u,v \in C_0^\infty(E)$. In this case the closure of the form exists and we denote it again by $\mathcal{E}$, with domain $D(\mathcal{E}) \supset C_0^\infty(E)$. If e.g. the ν_{ij} have a density ρ_{ij} with respect to Lebesgue measure dx on E and e.g. $\rho_{ij} \in L^2_{\text{loc}}$, $\frac{\partial}{\partial x_i}\rho_{ij} \in L^2_{\text{loc}}$ (with $\frac{\partial}{\partial x_i}$ in the Schwartz distributional sense) then $\overset{\circ}{H}$ exists and is in fact equal to $\overset{\circ}{H} = -\sum_{i,j=1}^d \frac{\partial}{\partial x_i}\left(\rho_{ij}\frac{\partial}{\partial x_j}\right)$ on $C_0^\infty(E)$.
But much stronger criteria are known for closability, see [AR1] (as a particular case of a situation where E is infinite dimensional). Closability criteria are also known in cases with jumps and killing, see e.g. [AS], [Sp] or [RW]. If in addition to closability of a symmetric bilinear form on a dense minimal domain, like the above $C_0^\infty(E)$, one has a suitable contraction property with respect to smooth contractions, then the

closure of the form is a Dirichlet form. E.g. in the above example, with $\nu_{ij} = \rho_{ij}dx$, for a smooth approximation $\varphi_\varepsilon(u)$ with $|\varphi'_\varepsilon(u)| \le 1$, converging to $(u \vee 0) \wedge 1$ as $\varepsilon \downarrow 0$, one has $\mathcal{E}(\varphi_\varepsilon(u), \varphi_\varepsilon(u)) = \sum_{i,j=1}^d \int_E |\varphi'_\varepsilon(u)|^2 \frac{\partial u}{\partial x_i}\frac{\partial u}{\partial x_j}\rho_{ij}\, dx \le \mathcal{E}(u,u)$, which "yields in the limit $\varepsilon \downarrow 0$" the Dirichlet property $\mathcal{E}(u^\sharp, u^\sharp) \le \mathcal{E}(u,u)$. Thus the form $\mathcal{E}$ is then a Dirichlet form. $\mathcal{E}$ being local and quasi–regular (as can be verified, see e.g. [MR1]), we know by the general theory that there is an associated diffusion process X. In the particular case $E = \mathbb{R}^d$, $\nu_{ij} = \frac{1}{2}\delta_{ij}\, dx$, X is standard Brownian motion on $\mathbb{R}^d$ and quasi–regularity is just regularity in the sense of capacity theory ($\mathcal{E}$–exceptional sets are capacity zero sets etc.).

2.3 Relation with quantum mechanics

Consider as a special case of the above Dirichlet form for $E = \mathbb{R}^d$, the form $\overset{\circ}{\mathcal{E}}_\mu$ given on the minimal domain $C_0^\infty(\mathbb{R}^d)$ by $\overset{\circ}{\mathcal{E}}_\mu\ (u,v) \equiv \int \nabla u(x)\nabla v(x)d\mu(x)$, $u,v \in C_0^\infty(\mathbb{R}^d)$ (with ∇ the gradient operator). We first assume that μ has a density $|\varphi|^2$ with respect to Lebesgue measure, where $\varphi > 0$ a.e., $\varphi \in C^2(\mathbb{R}^d)$. In this case $\overset{\circ}{\mathcal{E}}_\mu$ is closable (see above) and its closure $\mathcal{E}_\mu$ is a symmetric Dirichlet form on $L^2(\mu)$. Let H_μ be the associated Dirichlet operator. We easily see by integration by parts that $H_\mu u = -\nabla u - 2\frac{\nabla\varphi}{\varphi}\nabla u$ for $u \in C_0^\infty(\mathbb{R}^d)$. Now defining the operator $U_\varphi : L^2(\mathbb{R}^d, dx) \equiv L^2(dx) \to L^2(\mathbb{R}^d, d\mu) \equiv L^2(\mu)$ by $U_\varphi u \equiv \varphi^{-1}u$ we see that U_φ is unitary and the image H_μ^s of $\mathcal{H}_\mu$ under this unitary isomorphism of $L^2(dx)$ with $L^2(\mu)$ is $H_\mu^s \equiv U_\varphi^{-1}H_\mu U_\varphi$ and one has $H_\mu^s u = -\Delta u + Vu\ \forall u \in C_0^\infty(\mathbb{R}^d)$, with $V(x) \equiv \frac{\Delta\varphi}{\varphi}(x)$.

H_μ^s is thus a Schrödinger operator (acting in $L^2(dx)$) and H_μ is its "Dirichlet realization" (in $L^2(\mu)$). If $\varphi \in L^2(\mathbb{R}^d)$ we have $1 \in L^2(\mu)$ and $H_\mu 1 = 0$, $H_\mu^s\varphi = 0$. Thus in this case, since both H_μ^s and H_μ are positive, the infimum of the spectrum of H_μ^s is the eigenvalue 0. H_μ is sometimes called "**ground state representation**" or "**Darboux representation**" of H_μ^s. We could viceversa have started from a Schrödinger operator $-\Delta + V$ with potential V s.t. $\exists\varphi > 0$, $E > -\infty$ s.t. $(-\Delta + V)\varphi = E\varphi$. Then defining $d\mu = |\varphi|^2 dx$, we could have defined $H_\mu^s \equiv -\Delta + V - E$ as an operator in $L^2(dx)$ and we could have introduced $H_\mu \equiv U_\varphi H_\mu^s U_\varphi^{-1}$ (cf. [AHK9]).

The advantage of operating with H_μ instead of H_μ^s is that H_μ is the generator of a diffusion (whereas only for $V \ge 0$ H_μ^s can be associated with a diffusion and killing term with rate V). We also remark that starting from a Dirichlet operator H_μ we can define a corresponding operator $H_\mu^s \equiv U_\varphi^{-1}H_\mu U_\varphi$ even though φ is not C^2. For this reason H_μ^s defined in this way are sometimes called "generalized Schrödinger operators". Taking as starting point a Dirichlet generator H_μ is the so called "Dirichlet approach" to quantum mechanics (and to quantum field theory). This approach has been systematically presented in [AHKS]. For other discussions and applications see also e.g. [A], [ABM], [ABR], [AFKS], [AGKS], [AHK7], [AHK9], [AHKS2], [AJM], [AKuS], [Br], [Fu2], [SV]. It also relates to previous work by Araki, Coester–Haag on the canonical formulation of quantum field theory and to work of L. Gross on Log–Sobolev inequalities.

For any given measure μ on $\mathbb{R}^d$ the above form $\overset{\circ}{\mathcal{E}}_\mu$ in $L^2(\mu)$ is called the **classical pre–Dirichlet form given by** μ and its closure (whenever it exists) is called the **classical Dirichlet form given by** μ. This concept can be extended to finite dimensional manifolds (see e.g. [AHK6], [Fu3], [ABR]) and infinite dimensional spaces (and manifolds), see below.

A typical example of applicability of the Dirichlet approach to the definition of Hamiltonians is to take $d = 3$, $\mu = \mu_\alpha$, with $\mu_\alpha(dx) = |\varphi_\alpha|^2 dx$, $\varphi_\alpha(x) \equiv \frac{e^{\alpha|x|}}{4\pi|x|}$, $\alpha \in \mathbb{R}^d$. The classical Dirichlet form given by μ_α exists, and the corresponding generalized Schrödinger operator describes the perturbation of $-\Delta$ by a "point interaction of (renormalized) strength α at the origin". See [AGHKH], [ABR], [ABD1,2], [ABM], [Br], [SV] and references therein for the use of this model and many extensions of it in physics (solid state physics, nuclear physics, polymers physics ...). See also e.g. [AFKS], [Fu2], [A], [ABM], [AJM] for the use of the diffusion process associated with (Dirichlet) quantum mechanics.
We shall discuss below diffusion processes associated with classical Dirichlet forms in infinite dimensions, showing in particular that they solve a stochastic differential equation (this has been first discussed in [AHK5], see also for the finite dimensional situation [Fu1], [AHKS]).

3. Examples in infinite dimensions

3.1 Examples of Dirichlet forms on infinite dimensional spaces and associated diffusions

Now let us specialize the situation of 2.1 to the case where E is a locally convex Hausdorff vector space which is a Souslin space, e.g. E is a separable Banach space or a conuclear space. Let μ be a positive finite measure on E s.t. (for simplicity) $\mu(U) > 0$ for any open $U \neq \emptyset$, $U \subset E$. Let $FC_b^\infty \equiv \{f(l_1, \ldots, l_m) | m \in \mathbb{N},\ f \in C_b^\infty(\mathbb{R}^m),\ l_1, \ldots, l_m \in E'\}$ be the space of smooth bounded cylinder functions on E. Here E' denotes the dual of E. For any $k \neq 0$, $k \in E$, $u \in FC_b^\infty$, the directional derivative $\frac{\partial u}{\partial k}(z) \equiv \frac{d}{ds}u(z + sk)|_{s=0}$ in direction k is well–defined. Let us consider the bilinear form $\overset{\circ}{\mathcal{E}}_{k,\mu}(u, v) \equiv \int \frac{\partial u}{\partial k}\frac{\partial v}{\partial k} d\mu$, $u, v \in FC_b^\infty$, acting in $L^2(\mu)$.

Necessary and sufficient conditions for $\overset{\circ}{\mathcal{E}}_{k,\mu}$ to be closable are known ([AR1]). We say that k is μ–admissable if $\overset{\circ}{\mathcal{E}}_{k,\mu}$ is closable. Now let us suppose we have a rigging

$E' \subset \mathcal{H} \subset E$, where $\mathcal{H}$ is a separable Hibert space densely and continuously embeded in E and E' is the topological dual of E. Suppose there exists a complete orthonormal basis $K_0 \subset E'$ s.t. each $k_0 \in K_0$ is μ-admissable. Then $\mathring{\mathcal{E}}_\mu \equiv \sum_{k\in K_0} \mathring{\mathcal{E}}_{k,\mu}$ is closable ([AR1]) and its closure $\mathcal{E}_\mu$ is a Dirichlet form, called again the **classical Dirichlet form given by** μ.

Remark Using the chain rule we see that for any $h \in \mathcal{H}$, $h \neq 0$, $u \in FC_b^\infty$, $h \to \frac{\partial u}{\partial h}(z)$, $z \in E$, is a continuous linear functional on $\mathcal{H}$, so that, by Riesz' theorem, we can define $\nabla u(z) \in \mathcal{H}$ by $< \nabla u(z), h >_{\mathcal{H}} \equiv \frac{\partial u}{\partial h}(z)$, with $<\ ,\ >_{\mathcal{H}}$ the scalar product in $\mathcal{H}$. Then we can write $\mathring{\mathcal{E}}_\mu$ as $\mathring{\mathcal{E}}_\mu\,(u,v) = \int < \nabla u, \nabla v >_{\mathcal{H}} d\mu$.

Remark Let $E = \mathcal{S}'(I\!R^d)$, $\nu \equiv$ Gaussian white noise measure over $I\!R^d$ and consider the natural rigging of $L^2(\nu)$ (in the notation of [HKPS]) $(\mathcal{S}) \subset L^2(\nu) \equiv (L^2) \subset (\mathcal{S})^*$ Let $\Phi \in (\mathcal{S})^*_+$ then (by work of Kondratiev, Somoilenko and Yokoi, e.g. [BK]) there exists a finite positive measure μ on $\mathcal{B}(E)$ representing Φ. If $\mathring{\mathcal{E}}_\mu$ is closable in $L^2(\mu)$, then we can get $\mathcal{E}_\mu$ as the closure of $< \Phi, |\nabla u|_2^2 >$, for $u \in (\mathcal{S})$. In particular, thus the form core FC_b^∞ for $\mathcal{E}_\mu$ can be replaced by $(\mathcal{S})$ (and certain other subsets of $L^2(\mu)$: see [AHPRS1,2], [PR], [HKPS], [AR5] and references therein). (For an extension of white noise calculus to the non Gaussian case see [AKoR], [ADKS].)

If E is a Banach space or a conuclear space, one shows that the above classical Dirichlet form $\mathcal{E}_\mu$ given by μ is quasi regular [AR4], [Sch] and by the general theory it then follows that to $\mathcal{E}_\mu$ there is properly associated a μ-tight diffusion X (with infinite life time and state space E). X has laws P^z, $z \in E$, the process with law $P^\mu = \int P^z_\mu(dz)$ has μ as an invariant measure. A Fukushima decomposition holds in the sense that for any $u \in D(\mathcal{E}_\mu)$ and any quasi-continuous μ-version $\tilde{u}$, $\tilde{u}(X_t) - \tilde{u}(X_0) = M_t^{[u]} + N_t^{[u]}$, P^z-a.s., q.e. $z \in E$, with $M^{[u]}$ a martingale additive functional of finite energy and $N^{[u]}$ the continuous additive functional of zero energy which in the case $u \in D(H_\mu)$ is given by $N_t^{[u]} = \int_0^t (-H_\mu u)(X_s)\, ds$. Moreover, if μ "allows an integration by parts formula" in all directions $k \in K_0$, ($\mathring{\mathcal{E}}_\mu$ is closable and) X satisfies a stochastic equation. More precisely, we say that μ has "**logarithmic derivative** β_k" in direction $k \in K_0$ if $\exists \beta_k \in L^2(\mu)$ s.t. $\int \frac{\partial u}{\partial k}\, d\mu = -\int u\beta_k\, d\mu\ \forall u \in FC_b^\infty$. We say μ has a **logarithmic derivative** if it has a logarithmic derivative in all directions of a complete orthonormal basis $K_0 \subset \mathcal{H}$. In this case (and if in addition $< k, \cdot >_E \in L^2(\mu)$) we have that X satisfies the system of stochastic differential equations $d < k, X_t >= \beta_k(X_t)dt + w_{k,t}$, ($P^z$-a.s., q.e. $z \in E$) with $(w_{k,t}, P^z)$ a 1-dimensional Brownian motion (starting at 0, independent for $k \neq k'$). In this case $H_\mu = -\sum_{k\in K_0} \frac{\partial^2}{\partial k^2} - \beta_k(\cdot)\frac{\partial}{\partial k}$ on FC_b^∞.

Remark A characterization of "integration by parts" is given in [AKuR]. (Integration by parts formula, nowadays familiar from Malliavin Calculus, had been also discussed by several mathematicians, see e.g. [Sk], [Ko2], [DF]). In finite dimensions, i.e., when $E = \mathbb{R}^d$ the existence of β_k, $k \in K_0$, as above implies that μ is absolutely continuous with respect to Lebesgue measure, i.e., $\mu(dx) = \rho(x)dx$, $\beta_k = \partial_k \ln \rho = \partial_k^* 1$, (the latter formula makes also sense in infinite dimensions) $\varphi \in H^{1,2}_{\text{loc}}$, where $\varphi \equiv \sqrt{\rho}$ (cf. [AR1], [ARZ1,2]).

Remark

1) Sufficient conditions on (μ, E, H) are also known (see [AR2]) s.t. $\exists \beta : E \to E$ measurable s.t. ${}_{E'}< u, \beta >_E = \beta_k$ μ–a.s. and X satisfies the stochastic equation on E: $dX_t = \beta(X_t)dt + dw_t$, $X_0 = z$ (P^z a.s., q.e. $z \in E$), with w_t a Brownian motion on E over $\mathcal{H}$. In this case β is called the **logarithmic derivative** of μ.
2) For operators and processes on infinite dimensional manifolds see e.g. [AD].

3.2 Examples in quantum field theory, connections with Markov fields and stochastic p.d.e.

A. Stochastic quantization of fields

Let $E = \mathcal{S}_{-n}(\mathbb{R}^d) \subset \mathcal{S}'(\mathbb{R}^d)$, $\mathcal{H} = L^2(\mathbb{R}^d)$. (where $\mathcal{S}_{-n}$ is a Hilbert subspace of $\mathcal{S}'(\mathbb{R}^d)$, defined e.g. in [AR4]).

A.1 The free field case

In this case μ is the Gaussian measure $\mu = \mu^0 = N(0; (-\Delta_d + m^2)^{-1})$, for m a constant ($m \geq 0$ for $d \geq 3$, $m > 0$ for $d = 1,2$), with $N(0;C)$ the normal distribution with mean 0 and covariance C. By Minlos theorem μ is well defined, and it is in fact the standard Gaussian measure associated with $\mathcal{H}_{-1}$. μ is Nelson's (free Euclidean) field path space measure. The generator H_μ of the classical Dirichlet form $\mathcal{E}_\mu$ is essentially self–adjoint on FC_b^∞ (as shown in [AHK4,5]). The associated diffusion X satisfies the linear stochastic p.d.e. $dX_t = -(-\Delta_d + m^2)X_t + dw_t$, with dw_t the gaussian space–time white noise on $\mathbb{R}^{d+1}$, $X_0 = z \in E$ q.e. z, P^z a.s. (X, P^z) is the "stochastic quantization process" with invariant measure μ (Δ_d is the Laplacian on $\mathbb{R}^d$). We remark that μ is Euclidean invariant (i.e., under the transformations on $\mathcal{S}'(\mathbb{R}^d)$ induced by the Euclidean group of transformations in $\mathbb{R}^d$). In Subsect. B below we shall see how μ can be interpreted as the probability measure giving the distribution of a (Euclidean invariant) global Markov field.

A.2 The interacting case

In the case $d = 2$ one can construct non Gaussian Euclidean invariant measures $\mu = \mu^v$ whose associated classical Dirichlet form yield diffusions satisfying non linear stochastic p.d.e. These measures μ^v are obtained as $\mu^v = \text{w-}\lim_{\Lambda \uparrow \mathbb{R}^2} \text{w-}\lim_{\varepsilon \downarrow 0} \mu^v_{\Lambda,\varepsilon}$, where

Λ runs over the compact subsets of $\mathbb{R}^2$ with $\mu^v_{\Lambda,\varepsilon}(dz) \equiv \frac{M_{\Lambda,\varepsilon}(z)}{E_{\mu^0} M_{\Lambda,\varepsilon}} \mu^0(dz)$, $M_{\Lambda,\varepsilon}(z) \equiv \exp(-\int_\Lambda v_\varepsilon(z_\varepsilon)\, dx)$, with $z_\varepsilon \equiv \chi_\varepsilon * z$, χ_ε a suitable function s.t. $\chi_\varepsilon \to \delta_0$ as $\varepsilon \downarrow 0$ and v_ε a suitably regularized and renormalized "interaction density function" determined by a real–valued function v. Such measures are the "$v(\varphi)_2$–interactions" and examples have been constructed for v polynomial [GJ], exponential [AHK2] or trigonometric functions [AHK1], [AHK11]. Also in these cases the classical Dirichlet form given by μ^v is well defined, the associated diffusion X satisfies $dX_t = -(-\Delta_2 + m^2)X_t dt + \beta^v(X_t)dt + dw_t$, $< k, \beta^v(z) >= \beta^v_k(z)$, $z \in E$, $k \in C_0^\infty(\mathbb{R}^2)$ and $\beta^v_k(z)$ the $L^2(\mu)$–limit of $\lim_{\varepsilon \downarrow 0} < k, (v')_\varepsilon(z) >$. We shall see below that a uniqueness result holds for this equation (in finite volume).

B. The Euclidean case

The coordinate process $\tilde{X}$ associated with the measures $\mu = \mu^0$ resp. μ^v of Subsection A can be realized as an $\mathcal{S}'(\mathbb{R}^d)$–valued process ($d$ arbitrary in the case $\mu = \mu^0$, $d = 2$ for $\mu = \mu^v$). ${}_{\mathcal{S}} < f, \tilde{X} >_{\mathcal{S}'}$ has a natural extension to a generalized random field indexed by $\mathcal{H}_{-1}(\mathbb{R}^d)$. In particular, thinking of $x = (x^0, \hat{x}) \in \mathbb{R} \times \mathbb{R}^{d-1}$ with x^0 as "time", ${}_{\mathcal{S}} < f, \tilde{X} >_{\mathcal{S}'}$ has a natural extension to $f = \delta_t \otimes g$ with $g \in \mathcal{S}(\mathbb{R}^{d-1})$ so that then ${}_{\mathcal{S}} < f, \tilde{X} >_{\mathcal{S}'} = \tilde{X}_t(g)$. We can thus restrict μ to the σ–algebra Σ_0 generated by the "time zero fields" $X_0(g)$. Call μ_0 this restriction. μ_0 can be considered as a measure on $E = \mathcal{S}'(\mathbb{R}^{d-1})$. Then in the setting $(\mathcal{H} = L^2(\mathbb{R}^{d-1}), E)$, μ_0 defines a classical Dirichlet form. The associated Dirichlet operator is then the Hamiltonian of the field with interaction described by v, (at least on a dense domain, see below). $\tilde{X}$ has been shown to be a Euclidean invariant global Markov field in [AHKZ] in the sense that $\sigma(A) \perp \sigma(\overline{\mathbb{R}^d - A})$ conditional to $\sigma(\partial A)$ for any measurable subset A of $\mathbb{R}^d$, where $\sigma(\cdot)$ means σ–algebra generated by the fields $< \nu, \tilde{X} >$ with supp ν in A. Looked upon as a infinite dimensional process $t \to \tilde{X}_t$ (with values in $\mathcal{S}'(\mathbb{R}^{d-1})$) it is a diffusion, satisfying a pseudodifferential equation with nonlinear drift, which for $v = 0$ is simply linear and given by $d\tilde{X}_t = -\sqrt{(-\Delta_{d-1} + m^2)}\tilde{X}_t dt + dw_t$, with dw_t space–time white noise (on $\mathbb{R}^d$). For uniqueness discussions of the solution see below. To $\tilde{X}$ there is associated a quantized solution of the hyperbolic equation $(\Box + m^2)\Phi + v'(\Phi) = 0$, with $\Phi(t, \cdot) \equiv e^{itH_\mu} \tilde{X}_0(\cdot) e^{-itH_\mu}$.

Remark (i) Thus $(\tilde{X}, \mu)$ yield relativistic quantum fields ϕ. The relativistic character of ϕ is due to the Euclidean invariance of μ and its Markov property. It is possible to discuss by somewhat similar methods also hyperbolic stochastic p.d.e., see [AHRu1-2]. The regularization + renormalization procedure turns out to be necessary in order to make the nonlinearity be felt, i.e., to avoid the problem of "triviality" of the solutions. Results of this type were first derived in the above Euclidean framework in [AHK10], [AGHK].

(ii) In the extension to $d > 2$ of the above relations between homogeneous Markov fields, quantum fields and infinite dimensional diffusions in the non Gaussian interacting case (corresponding to $v \neq 0$) one meets considerable difficulties, see e.g.

[GJ]. For the time being the best results in the probabilistic spirit of this lecture have been obtained in [AHKI], [AIK1-3], [AIS], [AT] where a certain type of homogeneous global Markov nonlinear electromagnetic field has been constructed and in [AHKH1-3], [AHKHK1] where homogeneous fields indexed by hypersurfaces have been discussed. See also [AIK3,4,5], [AHKHK2], [ASch1,2], [AK2] for related work in lower dimensions.

4. The uniqueness problem

4.1 Markov uniqueness and strong uniqueness

We have seen above how to a given closable form $\overset{\circ}{\mathcal{E}}$ with a suitable contraction property ("pre Dirichlet form"), defined on a dense subset $\overset{\circ}{D}$ (e.g. $\overset{\circ}{D}= FC^{\infty}$ in the example of Section 3.1) in a Hilbert space $L^2(\mu)$ one can associate by closure a Dirichlet form $\mathcal{E} = \overline{\overset{\circ}{\mathcal{E}}}$ (e.g. $\mathcal{E} = \mathcal{E}_{\mu}$, the classical Dirichlet form given by μ). $\mathcal{E}$ is then uniquely associated with an operator L, the generator of a Markov semigroup (T_t) (see above), and the whole machinery of potential theory / probability theory works then in constructing a good process properly associated with (T_t). What about the uniqueness of the construction? A moment reflection shows that one should distinguish at least two forms of uniqueness. We can ask namely the question: given $(\overset{\circ}{\mathcal{E}}, \overset{\circ}{D})$ with $\overset{\circ}{D}\subset D(L)$ do there exist Dirichlet forms $\mathcal{E}' \neq \mathcal{E}$ s.t. $\mathcal{E}' =\overset{\circ}{\mathcal{E}}$ on $\overset{\circ}{D}$, $D(\mathcal{E}') \supset D(\mathcal{E})$ and, if L' is the Dirichlet operator to $\mathcal{E}'$, $D(L') \supset\overset{\circ}{D}$? This is equivalent with: do there exist symmetric Markov semigroups $e^{tL'} \neq e^{tL}$ s.t. $L = L'$ on $\overset{\circ}{D}$? We say that we have **Markov uniqueness** if the answer to the above question is negative.

From the point of view of functional analysis, however, another concept of uniqueness comes naturally into one's mind: suppose L is essentially self–adjoint on $\overset{\circ}{D}$, then obviously any symmetric semigroup $e^{tL'}$, with L' coinciding with L on $\overset{\circ}{D}$, has to coincide with e^{tL}, wether or not $e^{tL'}$ is Markov. Thus the essential self–adjointness of the Dirichlet operator L implies in particular Markov uniqueness.

We call for shortness the property of essential self–adjointness of L on $\overset{\circ}{D}$ "**strong uniqueness**".

Thus: strong uniqueness $\Rightarrow$ Markov uniqueness. However in general the converse is not true, i.e., one can have Markov uniqueness without having strong uniqueness. Examples have been given, in finite dimensions, by Röckner [R2] and Takeda [Ta2] (the latter example relates to our previous discussion of point interactions). In the case of forms associated with the Laplacian in a bounded region Λ of $\mathbb{R}^d$ the

different Dirichlet forms resp. Markov semigroups can be characterized by different (Dirichlet, Neumann, ...) boundary conditions. The existence on non Markov symmetric semigroups with generator coinciding with $-\Delta$ on $C_0^\infty(\Lambda)$ is proven by the existence of Krein's extension, see e.g. [Fu1], [Br].
In the case of the classical Dirichlet form $\mathcal{E}_\mu$ given by a measure $\mu = \rho dx$ on $\mathbb{R}^d$, with $\rho \neq 0$ a.e. one has, see [RZ1,2]: $\sqrt{\rho} \in \mathcal{H}^{1,2}_{\text{loc}} \Longleftrightarrow$ Markov uniqueness. A sufficient condition for strong uniqueness was given already in [Wi] by N. Wielens: if $\rho > 0$, $\sqrt{\rho} \in \text{Lip}_{\text{loc}}$ then strong uniqueness holds. We shall discuss below other conditions for strong uniqueness which can be applied also in the infinite dimensional case.
Several criteria for Markov uniqueness in the infinite dimensional case have been discussed, see [Ta1], [RZ1,2]. In particular, for the models of quantum fields discussed above given by the measures μ resp. μ^v resp. $\overset{\circ}{\mu}{}^v$ of section 3.2. Markov uniqueness (for space–time and time–zero fields) has been proven by Röckner and Zhang [RZ1,2], provided the interaction is restricted to a bounded region (space resp. space–time cut–off). A Markov uniqueness result "under change of tangent space", applicable also without such a cut–off, is discussed in [ARZ1,2] (see also [ARZ3]). For other partial results in this case see also [AK1], [AKuR].

Remark Markov uniqueness is related to "uniqueness in the sense of the martingale problem", see [AR3] for a detailed discussion of the relations.

4.2 General results on strong uniqueness

Let us consider the following concretization of the classical Dirichlet form definition from Section 3.1. For a given real separable Hilbert space $\mathcal{H}$ with the scalar product $< \cdot, \cdot >$ and norm $|\cdot|$ we fix a triplet

$$E' \equiv \mathcal{H}_+ \subset \mathcal{H} \subset \mathcal{H}_- \equiv E\,.$$

Here $\mathcal{H}_+$ and $\mathcal{H}_-$ are Hilbert spaces in duality with respect to $\mathcal{H}$ such that the natural embeddings $\mathcal{H}_+ \longrightarrow \mathcal{H}$, $\mathcal{H} \longrightarrow \mathcal{H}_-$ are dense and belong to the Hilbert–Schmidt class. The norms and scalar products on $\mathcal{H}_\pm$ will be denoted $|\cdot|_\pm$ and $(\cdot,\cdot)_\pm$ resp. In addition to the space FC_b^∞ (see Section 2.3) we will use the following spaces of smooth mappings. Denote by $C^n(\mathcal{H}_-, B)$, $n \in \mathbb{N} \cup \{0,\infty\}$ the set of mappings from $\mathcal{H}_-$ into a Banach space B that are n times continuously Fréchet differentiable. Define $C_b^n(\mathcal{H}_-, B)$, $C_{b,\text{loc}}^n(\mathcal{H}_-, B)$ as the subsets of $C^n(\mathcal{H}_-, B)$ which are characterized by the conditions of global resp. local boundedness in usual operator norms of the derivatives,

$$f^{(k)} : \mathcal{H}_- \longrightarrow \underbrace{\mathcal{L}(\mathcal{H}_-, \ldots, \mathcal{L}(\mathcal{H}_-, B)\ldots)}_{k}\,,$$

$k = 0, 1, \ldots, n$, $f \in C^n(\mathcal{H}_-, B)$. Introduce also the set $C^n_{\rm pol}(\mathcal{H}_-, B) \subset C^n(\mathcal{H}_-, B)$ of all polynomially bounded mappings which satisfy

$$\left\| f^{(k)}(x) \right\| \le C(1 + |x|_-)^p \ , \ x \in \mathcal{H}_- \ ,$$

for some $p \in I\!N$, $C > 0$.

For a given probability measure μ on $\mathcal{B}(\mathcal{H}_-)$ we suppose the existence of the directional logarithmic derivatives $\beta_\varphi \in L^2(\mu)$ for all $\varphi \in \mathcal{H}_+$. Moreover, we assume the representation

$$\beta_\varphi(x) = < \beta(x), \varphi > \ , \ \varphi \in \mathcal{H}_+ \ ,$$

for some measurable mapping $\beta : \mathcal{H}_- \longrightarrow \mathcal{H}_-$ that (see above) is called the logarithmic derivative of μ. In addition, $|\beta|_- \in L^2(\mu)$, i.e., $\int |\beta(x)|_-^2 d\mu < \infty$, will always be assumed in this subsection. As in Section 2.3 we introduce the pre Dirichlet form

$$\overset{\circ}{\mathcal{E}}_\mu (u, v) = \int < \nabla u, \nabla v > \ d\mu$$

on functions $u, v \in FC_b^\infty$. The corresponding Dirichlet operator H_μ is given by

$$H_\mu u = - \sum_{k \in K_0} \frac{\partial^2 u}{\partial k^2} - \sum_{k \in K_0} \beta_k(\cdot) \frac{\partial u}{\partial k} \equiv -\Delta u - < \beta, \nabla u >$$

on $u \in FC_b^\infty$, so that $\overset{\circ}{\mathcal{E}}_\mu (u, v) = (H_\mu u, v)_{L^2(\mu)}$. In many applications it is important to establish the essential self-adjointness of the Dirichlet operator H_μ (i.e. strong uniqueness). As we mentioned before, this means self-adjointness for the closure $\tilde{H}_\mu$ of H_μ in $L^2(\mu)$.

A. The case of a smooth logarithmic derivative

We start from the extension of the well known Kato inequality [RS] to Dirichlet operators. To this end we need to introduce a space of generalized functions on $\mathcal{H}_-$. Let us introduce the natural norm on $C_b^2 \equiv C_b^2(\mathcal{H}_-, I\!R)$

$$\|u\|_{C_b^2} = \sup_{x \in \mathcal{H}_-} \{ |u(x)| + \|u'(x)\| + \|u''(x)\| \} \ .$$

We have obviously the topological embedding $C_b^2 \subset L^2(\mu)$ so that we can introduce generalized functions on $\mathcal{H}_-$ as linear continuous functionals on C_b^2 with their action in terms of the scalar product in $L^2(\mu)$. In other words, we introduce the rigging

$$[C_b^2]' \supset L^2(\mu) \supset C_b^2 \ ,$$

where $[C_b^2]'$ is the dual space for C_b^2 with respect to $L^2(\mu)$. The condition $|\beta|_- \in L^2(\mu)$ gives $H_\mu : C_b^2 \longrightarrow L^2(\mu)$ continuously and due to the symmetry of H_μ in $L^2(\mu)$ this operator can be extended to $L^2(\mu)$ such that $H_\mu : L^2(\mu) \longrightarrow [C_b^2]'$. Now we are ready to formulate the extension of the Kato inequality to Dirichlet operators, see [Ko2], [KT2], [BK].

Theorem 1 Let $\beta \in C^2_{b,\mathrm{loc}}(\mathcal{H}_-, \mathcal{H}_-)$. If $f \in L^2(\mu)$ and $H_\mu f \in L^2(\mu)$ (the action of H_μ is meant in the sense of generalized functions) then the following inequality between generalized functions from $[C_b^2]'$ holds:

$$H_\mu |f| \le (\mathrm{sgn}\ f) H_\mu f \ .$$

As a consequence we have immediately the essential self–adjointness of H_μ on C_b^2.

Theorem 2 If $\beta \in C^2_{b,\mathrm{loc}}(\mathcal{H}_-, \mathcal{H}_-)$ then the operator H_μ is essentially self–adjoint.

Proof It is sufficient to prove that $f = 0$ follows from the equality $(H_\mu + 1)^* f = 0$, i.e., $(H_\mu + 1) f = 0$ in the sense of generalized functions. The equality $H_\mu f = -f$ implies $H_\mu f \in L^2(\mu)$ that is why the Kato inequality is applicable. We have therefore

$$H_\mu |f| \le (\mathrm{sgn}\ f) H_\mu f = -(\mathrm{sgn}\ f) f = -|f| \ ,$$

whence $(H_\mu + 1)|f| \le 0$. Noting that

$$0 \ge ((H_\mu + 1)|f|, 1)_{L^2(\mu)} = (|f|, 1)_{L^2(\mu)} = \|f\|_{L^1(\mu)}$$

we obtain $f = 0$. ∎

Remark The essential self–adjointness of H_μ on C_b^2 is still valid under weaker conditions on β. Namely, instead of $\beta \in C^2_{b,\mathrm{loc}}(\mathcal{H}_-, \mathcal{H}_-)$ and $|\beta|_- \in L^2(\mu)$, one may only assume that $\beta \in C^1_{b,\mathrm{loc}}(\mathcal{H}_-, \mathcal{H}_-)$, see [KT1].

B. An approximate criterium

The proof of the theorems mentioned above uses the smoothness of the logarithmic derivative β in an essential way. In many important applications such a smoothness does not hold (see [AKoR3,4]) and Section 5 below). Now we will consider the case when β is not smooth but admits "nice" approximations.
In the following we need two additional assumptions about the measure μ:
moment condition (MC)

$$\forall p \in I\!N \quad \int |x|_-^p \, d\mu(x) < \infty$$

integrability condition (IC)

$$\forall p \in I\!N \quad \int |x|_-^p \, |\beta(x)|_-^2 \, d\mu(x) < \infty \ .$$

The following result is proven in [AKoR4] (see also [AKoR2]).

Theorem 3 Let the measure μ satisfy the conditions (MC) and (IC). Suppose there exists a sequence $\{b_n | n \in \mathbb{N}\} \subset C^2_{\mathrm{pol}}(\mathcal{H}_-, \mathcal{H}_-)$ such that

(i) $\int |\beta(x) - b_n(x)|_-^2 d\mu(x) \longrightarrow 0, \; n \to \infty$

(ii) $\forall n \in \mathbb{N} \exists C_n \geq 0$:

$$(b_n(x), x)_- \leq C_n(1 + |x|_-^2) \; , \; x \in \mathcal{H}_-$$

(iii) $\exists C_n \geq 0$: $\forall n \in \mathbb{N} \; \forall h \in \mathcal{H}_-$

$$(b_n'(x)h, h)_- \leq C|h|_-^2 \; , \; x \in \mathcal{H}_- \; .$$

Then the operator H_μ is essentially self–adjoint.

The proof of this theorem is based on checking the conditions of Berezansky's abstract parabolic criterium (see e.g. [BK]) for the sequence of operators

$$H_n = -\Delta - < b_n, \nabla > \; , \; n \in \mathbb{N} \; .$$

More precisely, we need to construct strong solutions in $L^2(\mu)$ to the Cauchy problems

$$\begin{cases} \dfrac{d}{dt} u_n(t) + H_n u_n(t) = 0 \\ u_n(0) = f \in \mathcal{F}C_b^\infty \end{cases}$$

and show the convergence

$$\int_0^1 \|(H_\mu - H_n)u_n(t)\|_{L^2(\mu)} dt \longrightarrow 0 \; , \; n \to \infty \; .$$

To this end we use the probabilistic representation

$$u_n(t) = E\{f(\xi_{n,x}(t))\} \; ,$$

where $\xi_{n,x}(t)$ is the Brownian motion with the drift b_n and the initial condition $\xi_{n,x}(0) = x \in \mathcal{H}_-$. See [AKoR4] for details.

Remark (i) If as in [AKoR1,2] we use a sequence of bounded mappings $\{b_n | n \in \mathbb{N}\} \subset C_b^2(\mathcal{H}_-, \mathcal{H}_-)$ then we do not need the assumptions (MC) and (IC), see also [AKoR4] for a detailed version.

(ii) For the connection of essential self–adjointness of (H_μ, C_b^2) with that of (H_μ, FC_b^∞) we refer to [AKoR1].

4.3 Some dividents: Log–concave measures and logarithmic Sobolev inequalities

We will continue the considerations of the previous subsection 4.2. by establishing more spectral properties of Dirichlet operators. The property of strong uniqueness or essential self–adjointness will play an important role in our considerations. To stress this fact, we introduce the set $\mathcal{P}_{sa}(\mathcal{H}_-)$ of probability measures on $\mathcal{B}(\mathcal{H}_-)$ which is characterized by the following two conditions:

1) For any $\mu \in \mathcal{P}_{sa}(\mathcal{H}_-)$ the logarithmic derivative β exists and $|\beta|_- \in L^2(\mu)$
2) The operator H_μ (on the domain $\mathcal{F}C_b^\infty$) is essentially self–adjoint

Below we fix a measure $\mu \in \mathcal{P}_{sa}(\mathcal{H}_-)$ and a dense linear subset $\Phi \subset \mathcal{H}_+$.
Let us suppose in addition that the logarithmic derivative β of μ is differentiable in directions given by vectors in Φ, i.e.,

$$\forall \varphi, \psi \in \Phi \ \exists \ \frac{\partial}{\partial \psi} \beta_\varphi(x) \in L^2(\mu)$$

and we have the representation

$$\frac{\partial}{\partial \psi} \beta_\varphi(x) = < \beta'(x)\varphi, \psi > ,$$

where $\beta'(x) : \mathcal{H}_+ \longrightarrow \mathcal{H}_-$ are linear operators which are defined for μ–a.a. $x \in \mathcal{H}_-$ and such that $\mathcal{D}(\beta'(x)) \supset \Phi$ for μ–a.a. $x \in \mathcal{H}_-$. We set $R(x) = -\beta'(x)$. The measure μ is called **uniformly log–concave** (UCL) if $\exists \lambda > 0$: $\forall \varphi \in \Phi$ $< R(x)\varphi, \varphi > \geq \lambda |\varphi|^2$ μ–a.e. This definition was introduced in [AKoR4]. For another approach to the definition of log–concave measures see [AK1], [AnKo]. We shall say that a measure μ is Φ–ergodic iff the only measurable subsets of $\mathcal{H}_-$ which are invariant w.r.t. shifts by vectors in Φ have μ–measure 0 or 1 [AHK5], [Sk].
For the study of properties of Dirichlet operators the following identity is very useful [Ko2], [BK, Ch. 6]:

$$\int |H_\mu u|^2 d\mu = \int \left\{ \mathrm{Tr}_{\mathcal{H}} u''^2 + < Ru', u' > \right\} \, d\mu$$

($u \in FC_b^\infty$). An application of this identity to $\mu \in \mathcal{P}_{sa}(\mathcal{H}_-)$ which is (ULC) and Φ–ergodic shows that $0 \in I\!R$ is a simple eigenvalue of H_μ and there is a gap in the spectrum of the operator H_μ, moreover $H_\mu \geq \frac{1}{2}\lambda$ on the orthogonal complement to the constants in $L^2(\mu)$ [AKoR4]. We recall that μ satisfies a **log–Sobolev** inequality with the **Sobolev coefficient** $c_\mu > 0$ if for all $f \in FC_b^\infty$ with $\|f\|_{L^2(\mu)} = 1$ the following inequality is true [Gr1,2]:

$$\int |f|^2 \log |f| \, d\mu \leq c_\mu \int |\nabla f|^2 \, d\mu \, .$$

Theorem 4 [AKoR4] Suppose that the conditions (i)–(iii) of the approximate criterium from 4.2. are satisfied and, in addition, there exists a sequence of Φ–ergodic measures $\{\mu_n | n \in I\!N\}$ such that

(iv) $\forall f \in FC_b^\infty \int f\, d\mu_n \underset{n\to\infty}{\longrightarrow} \int f\, d\mu$

(v) $\forall n \in I\!N\ b_n = \beta_{\mu_n}$, where β_{μ_n} denotes the logarithmic derivative of μ_n

(vi) $\exists \lambda > 0$: $\forall n \in I\!N\ \forall \varphi \in \mathcal{H}_+$

$$< R_n(x)\varphi, \varphi > \geq \lambda |\varphi|^2 \ \mu\text{–a.a.} ,$$

where $R_n(x) = -b_n'(x)$.

Then the measure μ satisfies the log–Sobolev inequality with Sobolev coefficient $c_\mu = \lambda^{-1}$.

Due to general properties of Dirichlet operators the semigroup $T_t = e^{-tH_\mu}$, $t \in I\!R_+$, generates a semigroup of positivity preserving contractions in any space $L^p(\mu)$, $p \in [1, \infty]$. The log–Sobolev inequality gives an additional property of hypercontractivity for the semigroup: for given $p, q > 1$ and all $t \geq 0$ such that

$$e^{-\frac{t}{c_\mu}} \leq \frac{q-1}{p-1}$$

the operators T_t are contractions from $L^2(\mu)$ into $L^p(\mu)$. An important corollary of hypercontractivity is the Rothaus–Simon mass gap theorem [Ro], [Sim], [Gr1,2]. This theorem shows that $0 \in I\!R$ is a simple eigenvalue of H_μ and $H_\mu \geq c_\mu^{-1}$ on the orthogonal complement to the constants in $L^2(\mu)$. By the spectral theorem this implies exponential ergodicity of the semigroup T_t: $\forall f \in L^2(\mu)$

$$\|T_t f - E_\mu f\|_{L^2(\mu)} \leq e^{-\frac{t}{c_\mu}} \|f - E_\mu f\|_{L^2(\mu)} ,$$

where

$$E_\mu f = \int f\, d\mu .$$

Example In some applications measures on infinite dimensional spaces are used which are obtained as limits of measures on finite dimensional spaces. In particular, such a situation is typical in classical statistical physics (see Section 5). For measures of this type we have an approximation given in their definition, for which the conditions in Theorems 3,4 can be checked in many cases. Let us describe the following model example.

Let $\{L_n | n \in I\!N\}$ be a sequence of finite dimensional subspaces of Φ such that $L_n \subset L_{n+1}$, $n \in I\!N$, and $\bigcup_n L_n$ is dense in $\mathcal{H}_+$. For any $n \in I\!N$ let μ_n be a probability measure on $\mathcal{B}(L_n)$ of the form

$$d\mu_n(x_{L_n}) = \frac{1}{Z_n} e^{-U_n(x_{L_n})} dx_{L_n} ,$$

where $U_n \in C^2(L_n)$ and $x_{L_n} = P_{L_n}x$, P_{L_n} being the projector onto L_n. As usually, dx_{L_n} denotes the Lebesgue measure on L_n. Suppose that the measures μ_n converge to a probability measure μ on $\mathcal{B}(\mathcal{H}_-)$ in the following sense: $\forall f \in FC_b$

$$\int_{L_n} f(P_{L_n}x)\, d\mu_n(x_{L_n}) \longrightarrow \int_{\mathcal{H}_-} f(x)\, d\mu(x)\ ,\ n \to \infty\ .$$

Suppose also that

$$\exists \lambda > 0 : \forall n \in \mathbb{N}\ \forall \varphi \in L_n\ < U_n''(\cdot)\varphi, \varphi > \geq \lambda |\varphi|^2\ .$$

Then an application of the previous theorem shows that the measure μ satisfies the log–Sobolev inequality with the Sobolev coefficient $c_\mu = \frac{1}{\lambda}$ (cf. [AKoR4] for details).

5. Dirichlet forms and processes associated with classical and quantum Gibbs states

5.1 Stochastic dynamics for unbounded spin systems on a lattice

Denote by $\mathbb{Z}^d$ the d–dimensional integer lattice and by $\mathbb{R}^{\mathbb{Z}^d} \ni (x_k)_{k\in\mathbb{Z}^d}$ the set of all real–valued maps over $\mathbb{Z}^d$. The subset $\mathbb{R}_0^{\mathbb{Z}^d}$ of all finite sequences shall play the role of the set Φ in Subsection 4.3. The real Hilbert space $\mathcal{H} = l_2(\mathbb{Z}^d)$ is the completion of $\mathbb{R}_0^{\mathbb{Z}^d}$ with respect to the norm $|\cdot|$ generated by the scalar product

$$< \varphi, \psi >= \sum_{k\in\mathbb{Z}^d} \varphi_k \psi_k\ ,\ \varphi, \psi \in \mathbb{R}_0^{\mathbb{Z}^d}\ .$$

Let A be a bounded positive operator in $\mathcal{H}$ with the matrix $(a_{kj})_{k,j\in\mathbb{Z}^d}$ in the natural basis of $\mathcal{H}$. We suppose that $a_{kj} = a(k-j)$, where $a(\cdot) : \mathbb{Z}^d \to \mathbb{R}$ is some function on $\mathbb{Z}^d$ such that $a(-k) = a(k)$, $k \in \mathbb{Z}^d$. In addition we assume
A_1) A is strictly positive, i.e.,

$$\exists C_A > 0 : \forall h \in \mathcal{H}\ < Ah, h > \geq C_A |h|^2$$

A_2) the matrix $(a_{kj})_{k,j\in\mathbb{Z}^d}$ corresponds to a local interaction, i.e.

$$\exists N_0 \in \mathbb{N} : \forall k \in \mathbb{Z}^d,\ |k| > N_0\ a(k) = 0\ ,$$

where $|k| = \|k\|_{\mathbb{R}^d}$.

Let μ_0 be the Gaussian measure on $\mathbb{R}^{\mathbb{Z}^d}$ with zero mean and correlation operator A^{-1}. Heuristically the measure μ_0 has the representation

$$d\mu_0(x) = \frac{1}{Z} e^{-\frac{1}{2}\sum_{k,j} a(k-j)x_k x_j} \prod_{k \in \mathbb{Z}^d} dx_k$$

and describes the Gibbs state of a classical harmonic lattice system with finite radius of interaction. The rigorous definition of μ_0 can be given by its Fourier transform

$$\int_{\mathbb{R}^{\mathbb{Z}^d}} e^{i<\varphi,x>} d\mu_0(x) = e^{-\frac{1}{2}<A^{-1}\varphi,\varphi>} \quad (\varphi \in \mathbb{R}_0^{\mathbb{Z}^d}) .$$

We will consider the Gibbs state of a classical lattice system which is obtained by a perturbation of the initial one by using a one–particle potential given as the polynomial

$$P(s) = a_{2m}s^{2m} + \cdots + a_1 s + a_0 \ ; \ a_{2m} > 0 .$$

To this end for any finite set $\Lambda \subset \mathbb{Z}^d$ and $\varepsilon > 0$ we introduce the following Gibbsian modification of μ_0:

$$d\mu_\varepsilon^\Lambda(x) = \frac{1}{Z_{\Lambda,\varepsilon}} \exp\left\{-\varepsilon \sum_{k\in\Lambda} P(x_k)\right\} d\mu_0(x) ,$$

where $Z_{\Lambda,\varepsilon}$ is a normalizing factor s.t. $\mu_\varepsilon^\Lambda(\mathbb{R}^{\mathbb{Z}^d}) = 1$. It is well known that in the case of sufficiently small $\varepsilon > 0$, $\varepsilon < \varepsilon_0$, the limit $\lim_{\Lambda\to\mathbb{Z}^d} \mu_\varepsilon^\Lambda = \mu_\varepsilon$ exists in the sense of weak convergence of finite dimensional distributions (e.g. [MM]).

Let us fix some $r \in \mathbb{N}$ such that $\sum_{k\in\mathbb{Z}^d}(1+|k|)^{-r} < \infty$ and introduce the Hilbert spaces

$$\mathcal{H}_\pm = \left\{ x \in \mathbb{R}^{\mathbb{Z}^d} \middle| \ |x|_\pm^2 = \sum_{k\in\mathbb{Z}^d} x_k^2(1+|k|)^{\pm r} < \infty \right\} .$$

Then $\mu_\varepsilon(\mathcal{H}_-) = 1$.
The classical pre–Dirichlet form for the measure μ_ε has now the form

$$\overset{\circ}{\mathcal{E}}_{\mu_\varepsilon}(u,v) = \int \sum_{k\in\mathbb{Z}^d} \frac{\partial u}{\partial x_k} \frac{\partial v}{\partial x_k} d\mu_\varepsilon$$

($u, v \in \mathcal{F}C_b^\infty$), and the integration by parts formula gives the explicit form of the Dirichlet operator on FC_b^∞:

$$(H_{\mu_\varepsilon} u)(x) = -\sum_{k\in\mathbb{Z}^d} \frac{\partial^2 u}{\partial x_k^2} + \sum_{k,j\in\mathbb{Z}^d} a(k-j)x_k \frac{\partial u(x)}{\partial x_j} + \varepsilon \sum_{k\in\mathbb{Z}^d} P'(x_k)\frac{\partial u}{\partial x_k} .$$

An application of the approximate criterium from Subsection 4.2 shows that the operator H_{μ_ε} is an essentially self-adjoint operator in $L^2(\mu)$ on the domain FC_b^∞, see [AKoR3], [AKoR4].

In [AKoR3], [AKoR4] it was shown also that for sufficiently small $\varepsilon > 0$ the measure μ_ε satisfies a log–Sobolev inequality. More precisely, set

$$C_P = \inf_{s \in \mathbb{R}} P''(s)$$

and choose $\varepsilon_1 \in (0, \varepsilon_0]$ such that $\forall \varepsilon \in [0, \varepsilon_1] \quad C_A + 2\varepsilon C_P \geq 0$ (with C_A as in condition A_1). Then for any $\varepsilon \in [0, \varepsilon_1]$ the measure μ_ε satisfies a log–Sobolev inequality with a Sobolev coefficient $c_{\mu_\varepsilon} = \frac{2}{C_A}$.

An important consequence of this fact is the ergodicity of the stochastic dynamics which is associated with the measure μ_ε. Let us consider the system of Langevin's equations for the measure μ_ε:

$$\begin{cases} d\xi_k(t,x) = -\dfrac{1}{2} \displaystyle\sum_{k,j \in \mathbb{Z}^d} a(k-j)\xi_j(t,x)\, dt - \dfrac{\varepsilon}{2} P'(\xi_k(t,x))\, dt + dw_k(t) \\ \xi_k(0,x) = x_k \end{cases}$$

$x = (x_k)_{k \in \mathbb{Z}^d} \in \mathcal{H}_-$. Here $w_k(t)$, $k \in \mathbb{Z}^d$ is a family of independent standard one–dimensional Wiener process, $x \in \mathcal{H}_-$ is a given initial state. As is well known, this system has a unique strong solution

$$\xi(t,x) = (\xi_k(t,x))_{k \in \mathbb{Z}^d} \in \mathcal{H}_-$$

for any initial data $x \in \mathcal{H}_-$ (see e.g. [LR], [DR]) which defines a diffusion process with state space $\mathcal{H}_-$. We introduce the Markov semigroup P_t, $t \geq 0$, which is generated by this process as usual:

$$(P_t f)(x) = E\{f(\xi(t,x))\}\ ,$$

then $P_t : C_b(\mathcal{H}_-) \longrightarrow C_b(\mathcal{H}_-)$. This semigroup is called stochastic dynamics for the measure μ_ε. The generator of P_t coincides with the operator $\frac{1}{2}H_{\mu_\varepsilon}$ on all functions from FC_b^∞. The essential self–adjointness H_{μ_ε} on FC_b^∞ gives a unique extension of the stochastic dynamics to the space $L^2(\mu_\varepsilon)$ which is given by the semigroup $\exp(-\frac{t}{2}H_{\mu_\varepsilon})$, $t \geq 0$. Due to Subsection 4.3 we have the exponential ergodicity of this semigroup, which in particular implies the ergodicity of the process $\xi(t,x)$, $t \geq 0$, $x \in \mathcal{H}_-$:

$$\forall f \in FC_b^\infty : E\{f(\xi(t,x))\} \underset{t \to \infty}{\longrightarrow} \int f\, d\mu_\varepsilon$$

for μ_ε–a.a. $x \in \mathcal{H}_-$. This shows that we can reconstruct the measure μ_ε by using the corresponding Langevin equations. Such approach is known as stochastic quantization. For applications in different models of mathematical physics and more references see e.g. [RZ1,2], [AR4], [JLM], [IMS].

Let us remark that the results we discussed above can be extended to the case of non–polynomial anharmonic potentials, where the polynomial P in the definition of the measure μ_ε is replaced by the potential V. Another important extension involves the consideration of anharmonic systems with non–local interactions. In such systems the function $a(\cdot) : \mathbb{Z}^d \to \mathbb{R}$ has no longer a finite support on the lattice $\mathbb{Z}^d$. For the detailed study of these extensions we refer the reader to [AKoR5,6].

5.2 Stochastic dynamics for compact spin systems

Let us consider the special case of classical lattice systems when the spin space of every particle (single spin space) is a compact Riemannian manifold. In recent years there has been a growing interest in the investigation of such systems and their stochastic dynamics (see e.g. [AAAK], [DS], [SZ1,2] and references therein). The equilibrium states of these systems are described by Gibbs measures on infinite products of manifolds. The stochastic (or Glauber) dynamics corresponding to these states is given by a semigroup whose generator coincides on smooth cylinder functions with the Dirichlet operator of a Gibbs measure. Below we describe (essentially following [AKoR6]) the problem of uniqueness of the L^2–dynamics that corresponds to the question about the essential self–adjointness of Dirichlet operator considered. About constructions of Glauber dynamics in spaces of continuous functions or special spaces of smooth functions on infinite products of compact manifolds see [AAAK], [DS], [SZ1,2], [Ze].

The single spin space for our models will be the compact connected C^∞ manifold M with Riemannian metric $< \cdot,\cdot >$. $T(M)$ denotes the tangent bundle over M. We introduce the gradient operator $\nabla : C^1(M) \to C(T(M))$ and the divergence operator div on C^1–sections of $T(M)$ onto $C(M)$. We will use the normalized Riemannian measure σ on the Borel σ–algebra $\mathcal{B}$ over M.
The configuration space of the models we consider is the infinite product of manifolds $X = M^{\mathbb{Z}^d}$ endowed with the product topology. Given $\Lambda \subset \mathbb{Z}^d$

$$X \ni x \mapsto x_\Lambda = (x_k)_{k\in\Lambda} \in M^\Lambda$$

denotes the natural projection taking X onto M^Λ. For any $p \in \mathbb{Z}_+ \cup \{\infty\}$ and $\Lambda \subset \mathbb{Z}^d$, $|\Lambda| < \infty$, the symbol $C^p_\Lambda(X)$ denotes the set of functions on X of the form

$$X \ni x \mapsto f(x_\Lambda) \in \mathbb{R}$$

where f runs over the set $C^p(M^\Lambda)$ of continuous (up to the order p) differentiable functions on M^Λ. We shall also use the set of finitely based functions

$$FC^p(X) = \bigcup_{\Lambda:|\Lambda|<\infty} C^p_\Lambda(X) \; .$$

For such functions the symbols ∇_k resp. Δ_k will denote the gradient and the Laplace–Beltrami operator with respect to the variable $x_k \in M$.

For the construction of a stochastic dynamics on X we will use the so called Gibbs measures on the Borel σ–algebra $\mathcal{B}(X)$. The construction of such measures is based on a given family of potentials $\{U_\Lambda | \Lambda \subset \mathbb{Z}^d,\ |\Lambda| < \infty\}$ which describe interactions between particles of the system which are considered. About such potentials we assume the following:

1) $U_\Lambda \in C^2(M^\Lambda)$
2) the interaction in the system has a finite radius, i.e.,

$$\exists N_0 \in \mathbb{N} : \forall \Lambda \subset \mathbb{Z}^d \ , \ \text{diam}\ (\Lambda) > N_0 \text{ implies } U_\Lambda = 0 \ .$$

3) the interaction is invariant with respect to shifts on the lattice: $\forall \Lambda \subset \mathbb{Z}^d$, $|\Lambda| < \infty\ \forall k \in \mathbb{Z}^d$

$$\theta_k U_\Lambda = U_{\Lambda+k} \ ,$$

where $\theta_k f(x_\Lambda) = f(x_{\Lambda+k})$ and $\Lambda + k = \{j + k | j \in \Lambda\} \subset \mathbb{Z}^d$.

For $\Lambda \subset \mathbb{Z}^d$, $|\Lambda| < \infty$, we introduce the energy of the interaction in the volume Λ with given boundary condition $\xi \in X$ as

$$V_\Lambda(x_\Lambda|\xi) = \sum_{\Lambda' \cap \Lambda \neq \emptyset} U_{\Lambda'}(y) \ ,$$

where $y = (x_\Lambda, \xi_{\Lambda^c}) \in X$, $\Lambda^c = \mathbb{Z}^d \backslash \Lambda$. Setting

$$Z_\Lambda(\xi) = \int_{M^\Lambda} e^{-V_\Lambda(x_\Lambda|\xi)} d\sigma(x_\Lambda)$$

($d\sigma(x_\Lambda) = \times_{k \in \Lambda} d\sigma(x_k)$) we introduce the Gibbs measure in Λ with boundary condition ξ as

$$d\mu_\Lambda(x_\Lambda|\xi) = \frac{1}{Z_\Lambda(\xi)} e^{-V_\Lambda(x_\Lambda|\xi)} d\sigma(x_\Lambda) \ .$$

Due to the assumptions about the potentials these measures are well–defined for any finite volume Λ and all boundary conditions $\xi \in X$.

A probability measure μ on $\mathcal{B}(X)$ is called a **Gibbs measure** (for a given family of potentials) if for any $\Lambda \subset \mathbb{Z}^d$, $|\Lambda| < \infty$ and all $\xi \in X$ the conditional measure

$$\mu(\cdot | x_{\Lambda^c} = \xi_{\Lambda^c})$$

coincides with $\mu_\Lambda(\cdot|\xi)$ on $\mathcal{B}(M^\Lambda)$. We will denote the set of all such Gibbs measures by $\mathcal{G}$. Due to the compactness of M we have $\mathcal{G} \neq \emptyset$, but this set can have a very rich structure (see e.g. [Pr], [Ge]).

Let $\mu \in \mathcal{G}$ be a fixed Gibbs measure. The classical pre–Dirichlet form which corresponds to μ is defined for $u, v \in FC^1(X)$ by

$$\overset{\circ}{\mathcal{E}}_\mu (u,v) = \int_X \sum_{k \in \mathbb{Z}^d} < \nabla_k u, \nabla_k v > \, d\mu \ .$$

An integration by parts formula [AAAK] gives the following representation for the Dirichlet operator H_μ which is associated with $\overset{\circ}{\mathcal{E}}_\mu$:

$$(H_\mu u)(x) = - \sum_{k \in \mathbb{Z}^d} (\Delta_k u)(x) + \sum_k < \nabla_k V_k, \nabla_k u >$$

($u \in FC^2(X)$), where

$$V_k = \sum_{\Lambda : k \in \Lambda} U_\Lambda \in FC^2(X) \ .$$

Note that the Dirichlet operator H_μ is defined by the same differential expression for all Gibbs measures and this expression defines a symmetric operator in any space $L^2(X, \mu)$, $\mu \in \mathcal{G}$. The following result is proven in [AKoR5] in the case when the manifold M is the unit circle and extended to the general case in [AKoR6].

Theorem Under our assumptions about smoothness, finite radius and shift invariance of the family of potentials $\{U_\Lambda\}$ and for any Gibbs measure $\mu \in \mathcal{G}$ the Dirichlet operator H_μ defined on $FC^2(X)$ is an essentially self–adjoint operator in $L^2(X, \mu)$.

Now we can define the stochastic dynamics in any space $L^2(X, \mu)$, $\mu \in \mathcal{G}$, as the semigroup

$$T_t^\mu = e^{-t\tilde{H}_\mu} \ , \ t \geq 0 \ ,$$

where $\tilde{H}_\mu$ denotes the self–adjoint operator in $L^2(X, \mu)$ obtained as the closure of H_μ. The generator of $(T_t^\mu)_{t>0}$ is thus uniquely determined on $FC_b^2(X)$. Note, that it is easy to deduce that also H_μ defined on $FC_b^\infty(X)$ is an essentially self–adjoint operator on $L^2(X, \mu)$.
were constructed some classes of smooth functions on X which are invariant with respect to the stochastic dynamics considered. A similar result was announced in [DS] and used for the investigation of properties of the stochastic dynamics. The method used in [AAAK] is based on the decomposition of H_μ into a finite sum of differential operators and an application of a generalization of the Da Prato–Grisvard theorem [DaPG].

5.3 Euclidean Gibbs states and stochastic quantization for quantum lattice systems

Let us describe a general class of quantum lattice systems. We associate with every point $k \in \mathbb{Z}^d$ a quantum mechanical particle having (for simplicity of notations) one internal degree of freedom. To this particle there corresponds the state space $\mathcal{H}_k = L^2(\mathbb{R}, dx_k)$ and the canonical momentum and position operators

$$(p_k f)(x_k) = -i\frac{d}{dx_k}\ , \ (q_k f)(x_k) = x_k f(x_k)$$

as self-adjoint operators in $\mathcal{H}_k$ on the natural domains. To any $\Lambda \subset \mathbb{Z}^d$, $|\Lambda| < \infty$ there corresponds the state space $\mathcal{H}_\Lambda = L^2(\mathbb{R}^\Lambda, dx_\Lambda)$ and the C^*-algebra $\mathcal{A}$ of all bounded operators in $\mathcal{H}_\Lambda$ ($\equiv$ observables in the volume Λ). For $\Lambda_1 \subset \Lambda_2$ there is defined a natural norm preserving the embedding $\mathcal{A}_{\Lambda_1} \subset \mathcal{A}_{\Lambda_2}$, therefore we can introduce the C^*-algebra of local observables

$$\mathcal{A}_{\text{loc}} = \bigcup_{\Lambda:|\Lambda|<\infty} \mathcal{A}_\Lambda\ .$$

Its completion $\mathcal{A}$ is called the C^*-algebra of quasi-local observables.
In the definition of temperature or Gibbs states of the systems considered we restrict ourselves to the case of anharmonic systems with one-particle anharmonicity. For such systems the local Hamiltonian in the volume $\Lambda \subset \mathbb{Z}^d$, $|\Lambda| < \infty$, is defined as the self-adjoint operator H_Λ in $\mathcal{H}_\Lambda$ which has the form

$$H_\Lambda = -\frac{1}{2}\sum_{k\in\Lambda}\frac{\partial^2}{\partial x_k^2} + \frac{1}{2}\sum_{k,j\in\Lambda} a(k-j)x_k x_j + \sum_{k\in\Lambda} V(x_k)$$

on smooth cylinder functions. Here, as in Subsection 5.1, $(a(k-j))_{k,j\in\mathbb{Z}^d}$ is the matrix of a given strictly positive operator in $l_2(\mathbb{Z}^d)$. The one-particle potential $V = \bar{V}$ is assumed to be such that $V \in C^2(\mathbb{R})$ and

(V1) $\exists C > 0, R \geq 1 : \forall x \in \mathbb{R}$

$$|V'(s)| \leq C(1+|x|)^R$$

(V2) $\exists B \in \mathbb{R} : \forall x, y \in \mathbb{R}$

$$-(x-y)(V(x)-V(y)) \leq B(x-y)^2\ .$$

The Gibbs state of the system in the volume Λ for an inverse temperature $\beta > 0$ is given as a functional $G_{\beta,\Lambda}$ on $\mathcal{A}_\Lambda$ of the form

$$G_{\beta,\Lambda}(a) = \frac{\text{Tr}\,(Ae^{-\beta H_\Lambda})}{\text{Tr}\,(e^{-\beta H_\Lambda})} \quad (A \in \mathcal{A}_\Lambda)$$

The Gibbs state of the infinite system is given by a functional G_β on the algebra $\mathcal{A}_{\text{loc}}$ obtained by the thermodynamic limit:

$$G_\beta(A) = \lim_{\Lambda \to \mathbb{Z}^d} G_{\beta,\Lambda}(A) \quad (A \in \mathcal{A}_{\text{loc}})$$

One of these approaches to the construction of Gibbs states relies on the use of functional integration. It was suggested in [AHK3] and further developed by many authors (e.g. [KL], [GlKo], [BaKo1-3]).

In this way, the problem of constructing Gibbs states G_β is equivalent to the problem of constructing some mesure ν_β on the space of periodic trajectories on the circle S_β of length β with values in $\mathbb{R}^{\mathbb{Z}^d}$:

$$\Omega_\beta = \left\{ X(\cdot) = \left(x_k(\cdot) \right)_{k \in \mathbb{Z}^d} | x_k : S_\beta \to \mathbb{R} \ k \in \mathbb{Z}^d \right\} .$$

For $|\Lambda| < \infty$ let us consider the probability measure on Ω_β heuristically given by

$$d\nu_{\beta,\Lambda}(X(\cdot)) = \frac{1}{Z_\Lambda} \exp\{-E_{\beta,\Lambda}[X(\cdot)]\} \prod_{\substack{k \in \Lambda \\ u \in S_\beta}} dx_k(u) ,$$

$$E_{\beta,\Lambda}[X(\cdot)] = \int_{S_\beta} \{ \frac{1}{2} \sum_{k \in \Lambda} \dot{x}_k^2(u) + \frac{1}{2} \sum_{k,j \in \Lambda} a(k-j) x_k(u) x_j(u)$$
$$+ \sum_{k \in \Lambda} V(x_k(u)) \} \, du,$$

which uniquely determines the Gibbs state $G_{\beta,\Lambda}$. One can give a rigorous sense to the measure $\nu_{\beta,\Lambda}$ by the consideration of it as a perturbation of the Gaussian measure that corresponds to $V \equiv 0$ [AHK3], [GlKo]. The limit measure $\nu_\beta = \lim_{\Lambda \to \mathbb{Z}^d} \nu_{\beta,\Lambda}$ will be called the measure (or Euclidean Gibbs state) corresponding to the Gibbs state of the quantum lattice system. On an informal level such a measure is constructed using the Euclidean action $E_\beta[X(\cdot)] = E_{\beta,\mathbb{Z}^d}[X(\cdot)]$.

The stochastic quantization procedure applied to the lattice system with the Euclidean action E_β involves the following stages:

1) Constructing a random process $X_t = \{x_{k,t}(u) | k \in \mathbb{Z}^d, \ u \in S_\beta\}$, $t \geq 0$, as the solution to the Langevin equation

$$\begin{cases} \dfrac{\partial x_{k,t}(u)}{\partial t} = \dfrac{1}{2} \dfrac{\delta E_\beta(X)}{\delta x_k(\cdot)} (X_t(u)) + \dot{w}_{k,t}(u) \\ k \in \mathbb{Z}^d , \end{cases}$$

where $\dot{w}_{k,t}(u)$ is a Gaussian white noise:

$$E\{\dot{w}_{k_1,t_1}(u_1) \dot{w}_{k_2,t_2}(u_2)\} = \delta(k_1 - k_2) \delta(t_1 - t_2) \delta(s_1 - s_2) ,$$

$\frac{\delta}{\delta x_k(\cdot)}$ denotes the variational derivative.

2) Proving ergodicity of the process X_t, $t \geq 0$, i.e., the existence and uniqueness of an invariant measure $\mu_{\beta,\text{inv}}$ and the weak convergence of the distributions of the random variables X_t to $\mu_{\beta,\text{inv}}$ as $t \to \infty$.
3) Identification of $\mu_{\beta,\text{inv}}$ with the Euclidean Gibbs state ν_β, defined as the thermodynamic limit of $\nu_{\beta,\Lambda}$.

The points 1) and 2) of this scheme were realized in [AKT]. More precisely, after calculating the explicit form of the variation derivative in the Langevin equation we arrive at an infinite system of nonlinear parabolic stochastic differential equations

$$\begin{cases} \dfrac{\partial}{\partial t} x_{k,t}(u) = \dfrac{1}{2}\dfrac{\partial^2}{\partial u^2} x_{k,t}(u) - \dfrac{1}{2}\displaystyle\sum_{j\in\mathbb{Z}^d} a(k-j)x_{j,t}(u) - \dfrac{1}{2}V'(x_{k,t}(u)) - \dot{w}_{k,t}(u) \\ k \in \mathbb{Z}^d \ (t > 0,\ u \in S_\beta)\,. \end{cases}$$

It was shown that under the conditions (V1) and (V2) this system has a unique solution (mild in a commonly accepted sense) and the dependence of this solution on an initial condition $\overset{\circ}{X} = \{x_{k,0}(u) \mid k \in \mathbb{Z}^d,\ u \in S_\beta\} \in \Omega_\beta$ was studied. Moreover, under an additional condition of monotonicity on the drift, in the system discussed, it was proven that this solution is an ergodic Markov process. For a precise statement of the mentioned results and more detailed information about ergodic properties of the solution and the invariant measure we refer to [AKT]. Let us remark that the generator of the Markov process constructed coincides on FC_b^∞ with the Dirichlet operator H_{ν_β} which is generated by the Dirichlet form

$$\mathcal{E}_{\nu_\beta}(u,v) = \int_{\Omega_\beta} \sum_{k\in\mathbb{Z}^d} \left\{ \int_{S_\beta} \frac{\delta u(x(\cdot))}{\delta x_k(s)} \frac{\delta v(x(\cdot))}{\delta x_k(s)}\, ds \right\} d\nu_\beta(x(\cdot))$$

Also let us remark that the measure ν_β can be looked upon as describing a set of interacting strings (free loops) on the lattice $\mathbb{Z}^d$ and the corresponding process describes the time evolution of this system. (For another process on a loop space see also [ALR])

Acknowledgements The first two authors are very grateful to Professors Margarida De Faria, Jürgen Potthoff and Ludwig Streit for their kind invitations to a very stimulating summer school in Madeira. The second named author gratefully acknowledges financial support by the DFG. We thank Georg Schlitt for his skilful help with the setting of the manuscript.

References

[A] S. Albeverio, *Mathematical physics and stochastic analysis — a round table report*, Proc. Round Table, St. Chéron, Jan. '92, Bull Sci. Math. **117**, 125–152 (1993)

[AAAK] S. Albeverio, A.Val. Antonjuk, A.Vic. Antonjuk, Yu.G. Kondratiev, *Stochastic dynamics in some lattice spin systems*, Preprint (1993)

[ABD1] S. Albeverio, Z. Brzeźniak, L. Dabrowski, *Time dependent propagator (fundamental solution) for the Schrödinger operator with point interaction*, Preprint 1993

[ABD2] S. Albeverio, Z. Brzeźniak, L. Dabrowski, *The heat equation with point interaction in L^p spaces*, Bochum Preprint (1993)

[ABM] S. Albeverio, Ph. Blanchard, Z.M. Ma, *Feynman-Kac semigroups in terms of signed smooth measures*, pp. 1–31 in Proc. Oberwolfach Meeting "Random Partial Differential Equation", Edts. Hornung and Kotelenez, Birkhäuser (1991)

[ABR] S. Albeverio, J. Brasche, M. Röckner, *Dirichlet forms and generalized Schrödinger operators*, pp. 1-42 in Proc. Sønderborg Conf. "Schrödinger operators", Edts. H. Holden, A. Jensen, Lect. Notes Phys. **345**, Springer, Berlin (1989)

[AD] S. Albeverio, A. Daletskii *Pseudodifferential operators with symbols on solution manifolds at some infinite dimensional Hamiltonian systems*, Preprint (1993)

[ADKS] S. Albeverio, Y. Daletski, Y. Kondratiev, L. Streit, *Non gaussian infinite dimensional analysis*, BiBoS Preprint (1994)

[AFHMR] S. Albeverio, M. Fukushima, W. Hansen, Z.M. Ma. M. Röckner, *An invariance result for capacities on Wiener space*, J. Funct. Anal. **106**, 35–49 (1992)

[AFKS] S. Albeverio, M. Fukushima, W. Karwowski, L. Streit, *Capacity and quantum mechanical tunneling*, Commun. Math. Phys. **80**, 301–342 (1981)

[AGHK] S. Albeverio, G. Gallavotti, R. Høegh-Krohn, *Some results for the exponential interaction in 2 or more dimensions*, Comm. Math. Phys. **70**, 187–192 (1979)

[AGHKH] S. Albeverio, F. Gesztesy, R. Høegh-Krohn, H. Holden, *Solvable models in quantum mechanics*, Springer Verlag, Berlin (1988) (translation into Russian by Yu.A. Kuperin, K.A. Makarov, V.A. Geilerk, Mir, Moscow 1990)

[AGKS] S. Albeverio, F. Gesztesy, W. Karwowski, L. Streit, *On the connection between Schrödinger- and Dirichlet forms*, J. Math. Phys. **26**, 2546–2553 (1985)

[AHK1] S. Albeverio, R. Høegh-Krohn, *Uniqueness of the physical vacuum and the Wightman functions in the infinite volume limit for some non-polynomial interactions*, Comm. Math. Phys. **30**, 171–200 (1973)

[AHK2] S. Albeverio, R. Høegh-Krohn, *The Wightman axioms and the mass gap for strong interactions of exponential type in two-dimensional space-time*, J. Funct. Anal. **16**, 39–82 (1974)

[AHK3] S. Albeverio, R. Høegh-Krohn, *Homogeneous random fields and statistical mechanics*, J. Funct. Anal. **19**, 242–272 (1975)

[AHK4] S. Albeverio, R. Høegh-Krohn, *Quasi invariant measures, symmetric diffusion processes and quantum fields*, pp.11–59 in "Proceedings of the International Colloquium on Mathematical Methods of Quantum Field Theory", Editions du CNRS. 1976, (Colloques Internationaux du Centre National de la Recherche Scientifique, No. 248).

[AHK5] S. Albeverio, R. Høegh-Krohn, *Dirichlet forms and diffusion processes on rigged Hilbert spaces*, Z. Wahrsch. u. verw. Gebiete **40**, 1–57 (1977)

[AHK6] S. Albeverio, R. Høegh-Krohn, *Diffusion Fields, Quantum Fields and Fields with Values in Groups*, pp. 1–98 in "Stochastic Analysis and Applications", Adv. in Probability, Edt. M. Pinsky, M. Dekker, New York (1984)

[AHK7] S. Albeverio, R. Høegh-Krohn, *A remark on the connection between stochastic mechanics and the heat equation*, J. Math. Phys. **15**, 1745–1747 (1974)

[AHK8] S. Albeverio, R. Høegh-Krohn, *Hunt processes and analytic potential theory on rigged Hilbert spaces*, Ann. Inst. H. Poincaré (Probability Theory) **B13**, 269–291 (1977)

[AHK9] S. Albeverio, R. Høegh-Krohn, *The method of Dirichlet forms*, pp.250–258 in "Stochastic behaviour in classical and quantum systems", Proceedings of the Volta Memorial Conference, Como, 1977, Lect. Notes Phys. **93**, Springer-Verlag, (1979)

[AHK10] S. Albeverio, R. Høegh-Krohn, *Martingale convergence and the exponential interaction in* $\mathbb{R}^n$, pp.331–335 in "Quantum Fields-Algebras, Processes", Proc. Bielefeld Int. Colloquium, (Ed. L. Streit), Springer Verlag, Wien (1980)

[AHK11] S. Albeverio, R. Høegh-Krohn, *Uniqueness and the global Markov property for Euclidean fields, The case of trigonometric interactions*, Commun. Math. Phys. **68**, 95–128 (1979)

[AHKH1] S. Albeverio, R. Høegh-Krohn, H. Holden, *Markov cosurfaces and gauge fields*, pp.211–231 in Proc. Schladming 1984, 23. Intern. Universitätswochen für Kernphysik, Acta Phys. Austr. Suppl. XXVI (1984): "Stochastic Methods and Computer Techniques in Quantum Dynamics", Edts. H. Mitter, L. Pittner

[AHKH2] S. Albeverio, R. Høegh-Krohn, H. Holden, *Markov processes on infinite dimensional spaces, Markov fields and Markov cosurfaces*, pp.11–40 in "Stochastic Space-Time Models and Limit Theorems", Ed. L. Arnold, P. Kotelenez, D. Reidel, Dordrecht (1985)

[AHKH3] S. Albeverio, R. Høegh-Krohn, H. Holden, *Stochastic multiplicative measures, generalized Markov semigroups and group valued stochastic processes and fields*, J. Funct. Anal. **78**, 154–184 (1988)

[AHKHK1] S. Albeverio, R. Høegh-Krohn, H. Holden, T. Kolsrud, *Representation and construction of multiplicative noise*, J. Funct. Anal. **87**, 250–272 (1989),

[AHKHK2] S. Albeverio, R. Høegh-Krohn, H. Holden, T. Kolsrud, *Construction of quantised Higgs-like fields in two dimensions*, Phys. Letts. **B 222**, 263–268 (1989)

[AHKI] S. Albeverio, R. Høegh-Krohn, K. Iwata, *Covariant Markovian random fields in four space-time dimensions with nonlinear electromagnetic interaction*, pp.

69–83 in "Applications of Self-Adjoint Extensions in Quantum Physics", in Proc. Dubna Conf. 1987, Edts. P. Exner, P. Seba, Lect. Notes Phys. **324**, Springer, Berlin (1989)

[AHKS1] S. Albeverio, R. Høegh–Krohn, L. Streit, *Energy forms, Hamiltonians and distorted Brownian paths*, J. Math. Phys. **18**, 907–917 (1977)

[AHKS2] S. Albeverio, R. Høegh–Krohn, L. Streit, *Regularization of processes and Hamiltonians*, J. Math. Phys. **21**, 1636–1642 (1980)

[AHKZ] S. Albeverio, R. Høegh–Krohn, B. Zegarlinski, *Uniqueness and global Markov property for Euclidean fields: The case of general polynomial interactions*, Comm. Math. Phys. **123**, 377–424 (1989)

[AHPRS1] S. Albeverio, T. Hida, J. Potthoff, M. Röckner, L. Streit, *Dirichlet forms in terms of white noise analysis I — Construction and QFT examples*, Rev. Math. Phys. **1**, 291-312 (1990)

[AHPRS2] S. Albeverio, T. Hida, J. Potthoff, M. Röckner, L. Streit, *Dirichlet forms in terms of white noise analysis II — Closability and Diffusion Processes* Rev. Math. Phys. **1**, 313-323 (1990)

[AHRu1] S. Albeverio, Z. Haba, F. Russo, *Stationary solutions of stochastic parabolic and hyperbolic Sine–Gordon equations*, J. Phys. A **26** , L711–L718 (1993)

[AHRu2] S. Albeverio, Z. Haba, F. Russo, *Trivial solution for a nonlinear two–space dimensional wave equation perturbed by space–time white noise*, Bochum Preprint (1993)

[AIK1] S. Albeverio, K. Iwata, T. Kolsrud, *Random fields as solutions of the inhomogeneous quaternionic Cauchy-Riemann equation I. Invariance and Analytic Continuation* Comm. Math. Phys. **132**, 555–580 (1990)

[AIK2] S. Albeverio, K. Iwata, T. Kolsrud, *A model of four space–time dimensional gauge fields: reflection positivity for associated random currents*, pp. 257–269 in: Proceedings of the Liblice Conference on "Rigorous Results in Quantum Dynamics", June 1990, Edts. J. Dittrich, P. Exner, World Scientific, Singapore (1991)

[AIK3] S. Albeverio, K. Iwata, T. Kolsrud, *Homogeneous Markov generalized vector fields and quantum fields over 4-dimensional space-time*, pp. 1–18 in "Stochastic partial differential equations and applications", Proc. Trento Conf. Stochastic Partial Differential Equations, Ed. G. Da Prato, L. Tubaro (1991), Pitman Res. Notes **268**, Longman (1992)

[AIK4] S. Albeverio, K. Iwata, F. Kolsrud, *Random parallel transport on surfaces of finite type, and relations to homotopy*, Preprint (1992), pp. 50–57 in Probabilistic Methods in Mathematical Physics, Edts. F. Guerra, M.I. Loffredo, C. Marchioro, Proc. Siena, World Scientific, Singapore (1992)

[AIK5] S. Albeverio, K. Iwata, F. Kolsrud, *Moments of random fields over a family of elliptic curves and modular forms*, Preprint (1992), to appear in Proc. Locarno '91, Edts. S. Albeverio, U. Cattaneo, D. Merlini, World Scientific, Singapore (1992)

[AIS] S. Albeverio, K. Iwata, M. Schmidt, *A convergent lattice approximation for non–linear electromagnetic fields in four dimensions*, J. Math. Phys. **34**, 3327–3342 (1992)

[AJM] S. Albeverio, G.W. Johnson, Z.M. Ma, *The analytic operator–valued Feynman integral via additive functionals of Brownian motion*, Bochum Preprint (1993)

[AK1] S. Albeverio, S. Kusuoka, *Maximality of infinite dimensional Dirichlet forms and R. Høegh–Krohn's model of quantum fields*, pp. 301–330 in R. Høegh–Krohn's Memorial Volume "Ideas and methods in Quantum and Statistical Physics", Vol. 2, Edts. S. Albeverio, J.E. Fenstad, H. Holden, T. Lindstrøm, Cambridge University Press (1992).

[AK2] S. Albeverio, S. Kusuoka, *A basic estimate for two–dimensional stochastic holonomy along brownian bridges*, Preprint (1992), to appear in J. Funct. Anal.

[AKoR1] S. Albeverio, Yu.G. Kondratiev, M. Röckner, *An approximate criterium for essential self–adjointness of Dirichlet operators*, Pot. Anal. **1**, 307–317 (1992)

[AKoR2] S. Albeverio, Yu.G. Kondratiev, M. Röckner, Addendum to: *An approximate criterium for essential self–adjointness of Dirichlet operators*, Pot. Anal. **2**, 195–198 (1993)

[AKoR3] S. Albeverio, Yu.G. Kondratiev, M. Röckner, *Dirichlet operators and Gibbs measures*, Preprint (1992), to appear in Kallianpur Festschrift, Ed. G. Emch et al.

[AKoR4] S. Albeverio, Yu.G. Kondratiev, M. Röckner, *Dirichlet operators via stochastic analysis*, Preprint (1993), to appear in J. Funct. Anal. (1994)

[AKoR5] S. Albeverio, Yu.G. Kondratiev, M. Röckner, *A remark on stochastic dynamics on the infinite dimensional torus*, Preprint (1993), Publication in preparation

[AKoR6] S. Albeverio, Yu.G. Kondratiev, M. Röckner, *Stochastic dynamics for continuous spin systems on a lattice*, BiBoS Preprint (1994)

[AKoS] S. Albeverio, Yu.G. Kondratiev, L. Streit, *How to generalize white noise analysis to non–Gaussian measures*, to appear in Proc. French–German Meeting, Bielefeld 1992

[AKT] S. Albeverio, Yu.G. Kondratiev, T. Tsykalenko, *Stochastic dynamics and stochastic quantization of classical lattice systems I: Ergodicity*, Bochum Preprint (1993)

[AKuR] S. Albeverio, S. Kusuoka, M. Röckner, *On partial integration in infinite dimensional space and applications to Dirichlet forms* J. London Math. Soc. **42** (1990)

[AKuS] S. Albeverio, S. Kusuoka, L. Streit, *Convergence of Dirichlet forms and associated Schrödinger operators*, J. Funct. Anal. **68**, 130–148 (1986)

[ALR] S. Albeverio, R. Léandre, M. Röckner, *Construction of a rotational invariant diffusion on the free loop space*, C. R. Acad. Sci. Paris, t. 316, Série I, p. 287–292 (1993)

[AM1] S. Albeverio, Z.M. Ma, *Additive functionals, nowhere Radon and Kato class smooth measure associated with Dirichlet forms*, Osaka Math J. **29**, 247–265 (1992)

[AM2] S. Albeverio, Z.M. Ma, *Perturbation of Dirichlet forms - lower semiboundedness, closability and form cores*, J. Funct. Anal. **99**, 332–356 (1991)

[AMR1] S. Albeverio, Z.M. Ma, M. Röckner, *Regularization of Dirichlet spaces and applications*, CRAS, Ser. I, Paris **314**, 859–864 (1992)

[AMR2] S. Albeverio, Z.M. Ma, M. Röckner, *Non symmetric Dirichlet forms and Markov processes on general state space*, CRAS, Ser. I, Paris **314**, 77–82 (1992)

[AMR3] S. Albeverio, Z.M. Ma, M. Röckner, *Quasiregular Dirichlet forms and Markov processes*, J. Funct. Anal. **111**, 118–154 (1993)

[AMR4] S. Albeverio, Z.M. Ma, M. Röckner, *Local property of Dirichlet forms and diffusions on general state space*, Math. Ann. **296**, 677–686 (1993)

[AMR5] S. Albeverio, Z.M. Ma, M. Röckner, *Characterization of (non-symmetric) Dirichlet forms associated with Hunt processes*, Bochum Preprint (1992)

[AMR6] S. Albeverio, Z.M. Ma, M. Röckner, *A Beurling-Deny structure theorem for Dirichlet forms on general state space*, Bochum Preprint (1991) pp. 115–123 in R. Høegh–Krohn's Memorial Volume, On the Scientific Work of Raphael Høegh–Krohn, pp. in "Ideas and Methods in Mathematical Analysis, Stochastics, and Applications", Vol. 1, Edts. S. Albeverio, J.E. Fenstad, H. Holden, T. Lindstrøm, Cambridge University Press (1992)

[AnKo] A.V. Antonjuk, Yu. G. Kondratiev, *Log-Concave smooth measures on Hilbert space and some properties of corresponding Dirichlet operators*, BiBoS Preprint (1991)

[AR1] S. Albeverio, M. Röckner, *Classical Dirichlet forms on topological vector spaces-closability and a Cameron-Martin formula*, J. Funct. Anal. **88**, 395-436 (1990)

[AR2] S. Albeverio, M. Röckner, *Stochastic differential equations in infinite dimension: solutions via Dirichlet forms*, Prob. Th. Rel. Fields **89**, 347–386 (1991)

[AR3] S. Albeverio, M. Röckner, *Dirichlet form methods for uniqueness of martingale problems and applications*, Bochum Preprint (1993) (to appear in Proc. AMS Summer Inst. Stochastic Analysis, Cornell, 1993), (Edts. R. Durrett, M. Pinsky)

[AR4] S. Albeverio, M. Röckner, *Classical Dirichlet forms*, P.T.R.F (1989)

[AR5] S. Albeverio, M. Röckner, *Infinite dimensional diffusions connected with positive generalized white noise functionals*, pp. 1–21 in "White Noise Analysis", Ed. T. Hida, H. H. Kuo, J. Potthoff, L. Streit World Scient., Singapore (1990)

[AR6] S. Albeverio, M. Röckner, *Dirichlet forms, quantum fields and stochastic quantisation*, pp.1–21 in "Stochastic Analysis, path integration and dynamics", Emanations from "Summer Stochastics", Warwick 1987, Edts. D. Elworthy, J.C. Zambrini, Pitman Res. Notes, Longman, Harlow (1989).

[ARSp] S. Albeverio, M. Röckner, U. Spönemann, in preparation

[ARZ1] S. Albeverio, M. Röckner, T.S. Zhang, *Markov uniqueness and its applications to martingale problems, stochastic differential equations and stochastic quantization*, C. R. Math. Rep. Acad. Sci. Canada XV, 1–6 (1993)

[ARZ2] S. Albeverio, M. Röckner, T.S. Zhang, *Markov uniqueness for a class of infinite dimensional Dirichlet operators*, In: "Proceedings of the 9th winter school on stochastic processes and optimal control, Friedrichroda (Thuringia), March 1–7, 1992" Ed. H.J. Engelbert et al., Gordon and Breach (1993)

[ARZ3] S. Albeverio, M. Röckner, T.S. Zhang, *Girsanov transform for symmetric diffusions with infinite dimensional state space.* Ann. Probab. **21**, 961–978 (1993)

[AS] S. Albeverio, S. Song, *Closability and resolvent of Dirichlet forms perturbed by jumps*, Pot. Anal. **2**, 115–130 (1993)

[ASch1] S. Albeverio, J. Schäfer, *Rigorous approach to abelian Chern–Simons theory*, Bochum preprint 1991, pp. 143–152 in "Groups and Related Topics", Proc. German–Polish Workshop 1991, Edt. R. Gielerak et al., World Scientific, Singapore (1992)

[ASch2] S. Albeverio, J. Schäfer, *A mathematical model of abelian Chern–Simons theory*, SFB 237 - Preprint, Bochum (1991), to appear in Proc. Locarno Conf. "Stochastic Processes - Physics and Geometry II", World Scient., Singapore (1993)

[AT] S. Albeverio, H. Tamura, *On the propagator of a scalar field in the presence of a confining nonlinear electromagnetic force*, Preprint (1992)

[AZ] S. Albeverio, B. Zegarlinski, *Global Markov property in quantum field theory and statistical mechanics: a review on results and problems*, pp. 331–369 in R. Høegh-Krohn's Memorial Volume "Ideas and Methods in Quantum and Statistical Physics", Vol. 2, Edts. S. Albeverio, J.E. Fenstad, H. Holden, T. Lindstrøm, Cambridge University Press (1992)

[BaKo1] V.S. Barbulyak, Yu.G. Kondratiev, *Functional integrals and quantum lattice systems: I. Existence of Gibbs states*, Dokl. Akad. Nauk Ukraine (1991), N8, pp. 31–34 (in Russian)

[BaKo2] V.S. Barbulyak, Yu.G. Kondratiev, *Functional integrals and quantum lattice systems: II. Periodical Gibbs states*, Dokl. Akad. Nauk Ukraine (1991), N9, pp. 38–40 (in Russian)

[BaKo3] V.S. Barbulyak, Yu.G. Kondratiev, *Functional integrals and quantum lattice systems: III. Phase transitions*, Dokl. Akad. Nauk Ukraine (1991), N10, pp. 19–21 (in Russian)

[BH] N. Bouleau, F. Hirsch, *Dirichlet Forms and Analysis on Wiener Space*, de Gruyter, Berlin and New York (1991)

[BK] Yu.M. Berezansky, Yu.G. Kondratiev, *Spectral methods in infinite dimensional analysis*, Naukova Dumka; Kiev (1988), English translation to appear in Kluwer Academic Publishers (1993)

[Br] J.F. Brasche, *Schrödinger operators with singular interactions*, Bochum Preprint 1991, to appear in J. Math. Ann. Appl.

[DaPG] G. Da Prato, P. Grisvard, *Sommes d'opérateurs linéaires et équations différentielles opérationelles*, J. Math. Pures et Appl. **54**, 305–387 (1975)

[DF] Yu.L. Daletsky, S.V Fomin, *Measures and differential equations on infinite dimensional spaces*, Kluwer Academic Publishers (1992)

[DR] H. Doss, G. Royer, *Processus de diffusion associés aux mesures de Gibbs sur* $\mathbb{R}^{\mathbb{Z}^d}$, Z. Wahrsch. u. verw. Gebiete **41**, 125–158 (1979)

[DS] J.-D. Deuschel, D. Stroock, *Hypercontractivity and spectral gap of symmetric diffusion with application to the stochastic Ising model*, J. Funct. Anal. **92**, 30–48 (1990)

[Fu1] M. Fukushima, *Dirichlet Forms and Markov Processes*, North Holland, Amsterdam (1980)

[Fu2] M. Fukushima, *A note on irreducibility and ergodicity of symmetric Markov processes*, Lect. Notes. Phys. **173**, 200–207 (1982)

[Fu3] M. Fukushima, *Energy forms and diffusion processes*, pp. 65–98 in "Mathematics and Physics", Vol. 1, Ed. L. Streit, World Scientific, Singapore (1985)

[FuOT] M. Fukushima, Y. Oshima, M. Takeda, *Dirichlet forms and symmetric Markov processes*, W. De Gruyter, Berlin (1994), to appear

[Ge] H.O. Georgii, *Gibbs measures and phase transitions*, W. de Gruyter (1988)

[GlKo] S.A. Globa, Yu.G. Kondratiev, *The construction of Gibbs states of quantum lattice systems*, Sel. Math. Sov **9**, 297–307 (1990)

[Gr1] L. Gross, *Logarithmic Sobolev inequalities*, Amer. J. Math. **97**, 1061–1083 (1976)

[Gr2] L. Gross, *Logarithmic Sobolev inequalities and contractivity properties of semigroup*, Cornell University Preprint (1992)

[GJ] J. Glimm, A. Jaffe, *Quantum Physics*, Springer (1984)

[HKPS] T. Hida, H.H. Kuo, J. Potthoff, L. Streit, *White noise*, Kluwer, Dordrecht (1993)

[IMS] I.A. Ignatjuk, V.A. Malishev, V. Sidoravicius, *Convergence of the stochastic quantization method*, Prob. Theory and Math. Stat. **1**, 526–538, Mokslas, Vilnius (1990)

[JLM] G. Jona–Lasinio, P.K. Mitter, *On stochastic quantisation of field theory*, Comm. Math. Phys. **101**, 409–436 (1985)

[KL] A. Klein, L. Landau, *Stochastic processes associated with KMS states*, J. Funct. Anal. **42** (1981), pp. 368–428

[Ko1] Yu.G. Kondratiev, *Perturbations of second quantization operators*, Funct. Anal. Appl. **16**, 76–77 (1982)

[Ko2] Yu.G. Kondratiev, *Dirichlet operators and smoothness of solutions of elliptic equations*, Soviet Math. Dokl. **31**, 461–464 (1985)

[KT1] Yu.G. Kondratiev, T.V. Tsykalenko, *Dirichlet opertors and differential equations*, Selecta Math. Sov. **10**, 345–397 (1991)

[KT2] Yu.G. Kondratiev, T.V. Tsykalenko, *Infinite dimensional Dirichlet operators I: Essential self-adjointness and associated elliptic equations*, Pot. Anal. **2**, 1–22 (1993)

[LR] G. Leha, G. Ritter, *On diffusion processes and their semigroups in Hilbert spaces with an application to interacting stochastic systems*, Ann. Probab. **12**, 1077–1112 (1984)

[MM] V.A. Malishev, R.A. Minlos, *Gibbsian random fields*, Nauka, Moscow (1995)

[MR1] Z.M. Ma, M. Röckner, *An Introduction to the Theory of (Non-Symmetric) Dirichlet Forms*, Springer, Berlin (1992)

[MR2] Z.M. Ma, M. Röckner, *Markov processes associated with positivity preserving coercive forms*, Bonn SFB 256–Preprint (1994)

[Pr] C. Preston, *Random fields*, Lect. Notes Math. **534**, Springer, Berlin (1976)

[PR] J. Potthoff, M. Röckner, *On the contraction property of energy forms on infinite dimensional space*, J. Funct. Anal. **92**, 155–165 (1990)

[R1] M. Röckner, *Dirichlet forms on infinite dimensional state space and applications*, pp. 131–186 in "Stochastic Analysis and Related Topics" Edts. H. Körezlioglu, A.S. Ustumel, Birkhäuser, Boston (1992)

[R2] M. Röckner, *Potential theory on non-locally compact space via Dirichlet forms*, pp. 107–126 in "Potential Theory", Ed. M. Kishi, W. De Gruyter, Berlin, New York (1991)

[Ro] O.S. Rothaus, *Diffusion on compact Riemannian manifolds and logarithmic Sobolev inequalities*, J. Funct. Anal. **42**, 102–109 (1981)

[RW] M. Röckner, N. Wielens, *Dirichlet forms — closability and change of speed measure*, Infinite dimensional analysis and stochastic processes, Research Notes in Mathematics, Vol. 124, 119–144, Editor: S. Albeverio, Boston, London, Melbourne, Pitman 1985

[RS] M. Reed, B. Simon, *Methods of modern mathematical physics, Vol. II, Fourier analysis and self-adjointness*, Academic Press, New York (1978)

[RZ1] M. Röckner, T.-S. Zhang, *On uniqueness of generalized Schrödinger operators and applications*, J. Funct. Anal. **105**, 187–231 (1992)

[RZ2] M. Röckner, T.-S. Zhang, *Uniqueness of generalized Schrödinger operators — part II*, J. Funct. Anal. **2**, 455–467 (1994)

[Sch] B. Schmuland, *An alternative compactification for classical Dirichlet forms on topological vector spaces*, Stoch. and Stoch. Repts **33**, 75–90 (1990)

[Si] M.L. Silverstein, *Symmetric Markov Processes*, Lect. Notes Math. **426**, Springer, Berlin (1974)

[Sim] B. Simon, *A remark on Nelson's best hypercontractive estimates*, Proc. Amer. Math. Soc. **55**, 376–378 (1976)

[Sk] A.V. Skorohod, *Integration on Hilbert space*, Springer–Verlag, Berlin–Heidelberg–New York (1984)

[Sp] U. Spönemann, *An existence and a structural result for Dirichlet forms*, Bielefeld Thesis, in preparation (1994)

[SV] P. Stollmann, J. Voigt, *Perturbation of Dirichlet forms by measures*, in preparation

[SZ1] D. Stroock, B. Zegarlinski, *The equivalence of the logarithmic Sobolev inequality and Dobrushin–Shlosman mixing condition*, Comm. Math. Phys. **144**, 303–323 (1992)

[SZ2] D. Stroock, B. Zegarlinski, *The logarithmic Sobolev inequality for continuous spin systems on a lattice*, J. Funct. Anal. **104**, 299–326 (1992)

[Ta1] M. Takeda, *On the uniqueness of Markov self-adjoint extension of diffusion operators on infinite dimensional spaces*, Osaka J. Math. **22**, 733–742 (1985)

[Ta2] M. Takeda, *The maximum Markovian self-adjoint extensions of generalized Schrödinger operators*, J. Math. Soc. **44**, 113–130 (1992)

[Wi] N. Wielens, *On essential self-adjointness of generalized Schrödinger operators*, J. Funct. Anal. **61**, 98–115 (1985)

[Ze] B. Zegarlinski, *Gibbsian description and description by stochastic dynamics*, Bochum Preprint Nr. 127 (1991)

BURGERS EQUATION FORCED BY CONSERVATIVE OR NONCONSERVATIVE NOISE

L. BERTINI [2] N. CANCRINI [1] G. JONA-LASINIO [1]

1 *Dipartimento di Fisica, Università di Roma 'La Sapienza' P.le Aldo Moro 2 00185 Roma, Italy. E-mail:cancrini@sci.uniroma1.it*

2 *Dipartimento di Matematica, Università di Roma 'Tor Vergata', Via della Ricerca Scientifica 00133 Roma, Italy. E-mail: bertini@mat.utovrm.it*

INTRODUCTION

We consider the forced Burgers equation in one space dimension

$$\partial_t u_t(x) = \nu \partial_x^2 u_t(x) - u_t(x)\partial_x u_t(x) + \varepsilon f_t(x) \tag{1}$$

where $u_t(x)$ is a scalar field, $t \geq 0$ and $x \in \mathbf{R}^1$, $f_t(x)$ the forcing term, ν the viscosity and ε the strength of the noise.

The unforced Burgers equation, $\varepsilon = 0$ in (1), has been proposed as a simple model in statistical hydrodynamics describing the interaction of dissipative and non linear inertial term in the fluid motion [Bu]. It is an incomplete description of the physical system and random perturbations can be used to simulate the external noise and/or the neglected effects; they may also help to select physically relevant invariant measures. Equation (1) has also been related to a random model for growing interfaces [KPZ].

Let $\eta_t(x)$ be the white noise in space and time, i.e.

$$\mathbf{E}\left(\eta_t(x)\eta_{t'}(x')\right) = \delta(t-t')\delta(x-x') \tag{2}$$

we discuss random forcing terms either of white noise type, $f_t(x) = \eta_t(x)$, or of the form gradient of white noise, $f_t(x) = \partial_x \eta_t(x)$. We remark that, in both cases, equation (1) is Galilean invariant.

For the non forced Burgers equation $\int dx\, u_t(x)$ is conserved in time. The first choice $f_t(x) = \eta_t(x)$ breaks this conservation law whereas for the second

A. I. Cardoso et al. (eds.), Stochastic Analysis and Applications in Physics, 35–44.

choice we can rewrite (1) as

$$\partial_t u_t(x) + \partial_x \left(\frac{1}{2} u_t(x)^2 - \varepsilon \eta_t(x)\right) = \nu \partial_x^2 u_t(x) \tag{3}$$

from which it is apparent that $\int_a^b u_t(x)\, dx$ changes only through boundary terms.

With both choices the forcing term is singular and the mathematical interpretation of (1) is not apparent. For the non conservative case we here outline how an existence theorem for the Cauchy problem for equation (1) can be obtained, a detailed proof being in [BCJ]. This theorem gives an explicit expression of a strong solution (in the sense of stochastic differential equation) as a continuous process (in time) with values in the class of continuous functions (in space). A uniqueness result has been recently proven when the Burgers equation is considered in a bounded domain with Dirichlet boundary conditions [DDT].

When the stochastic forcing term is the gradient of white noise, as the perturbation is more singular if compared to the previous case, the solution is expected to be a process with values in a distribution space. Using a regularization procedure that is eventually removed we are able to construct a distribution valued process which formally solves the mild form of equation (3). However the problem of giving a rigorous meaning to equation (3) is still open.

In this paper we will refer to model A when the random perturbation of Burgers equation (1) is white noise, to model B when it is the derivative of white noise. The results for model A is a joint work of the three authors, while those for model B were obtained by L.B. and N.C. .

We write the initial condition $u_0(x)$ for the Cauchy problem for equation (1) in the form $u_0(x) = -\partial_x U_0(x)$; we assume $U_0 \in C^1(\mathbf{R})$ for model A and $U_0 \in C^0(\mathbf{R})$ for model B, the derivative has then to be interpreted in distribution sense. Furthermore, for both models, we assume U_0 has at most linear growth and u_0 has at most exponential growth. For model B the latter hypothesis means u_0 is in the dual space of test functions with a decay faster than the exponential one.

COLE-HOPF TRANSFORMATION

Our strategy to study the stochastically forced Burgers equation is the use of the Cole-Hopf transformation which reduces (1) to a linear equation with multiplicative noise. This equation is solved via a generalized Feynman-Kac formula and, performing the inverse Cole-Hopf transform, an explicit solution of (1) is constructed. The use of this technique for the stochastic equation requires some care and, as we will see, its application to model A and B presents some differences.

Let us realize the white noise $\eta_t(x)$ as the generalized derivative of the brownian sheet, i.e. $\eta_t(x) = \partial_t \partial_x W_t(x)$. The gaussian process $W_t(x)$ has correlation function

$$\mathbf{E}^w \left(W_t(x) W_{t'}(x') \right) = t \wedge t' \, C(x, x'), \qquad C(x, x') = \theta(x x') \, |x| \wedge |x'| \tag{4}$$

where $a \wedge b = \min\{a, b\}$ and θ is the indicator function of the set $[0, \infty)$.

For model A, $f_t(x) = \eta_t(x)$, the Cole-Hopf transformation $u_t(x) = -2\nu \partial_x \log \psi_t(x)$, reduces (1) to the following linear equation with multiplicative *half white* noise

$$\partial_t \psi_t(x) = \nu \partial_x^2 \psi_t(x) - \frac{\varepsilon}{2\nu} \psi_t(x) \partial_t W_t(x) \tag{5}$$

whereas for model B, $f_t(x) = \partial_x \eta_t(x)$, introducing $u_t(x) = -2\nu \partial_x \log \theta_t(x)$, we obtain a linear equation with multiplicative white noise

$$\partial_t \theta_t(x) = \nu \partial_x^2 \theta_t(x) - \frac{\varepsilon}{2\nu} \theta_t(x) \partial_t \partial_x W_t(x) \tag{6}$$

We note that equation (6), often called the stochastic heat equation, is studied not only in connection with the stochastic Burgers equation but also because it is a linearized model for the evolution a scalar field in a non stationary random medium [BC,Mo].

In order to study rigorously (5) and (6), they must be interpreted as stochastic PDE. As they contain a non trivial diffusion the stochastic differential presents the well known ambiguities. The correct choice is not a trivial point: in fact the models A and B require a different prescription.

Model A

In order to obtain, via the Hopf-Cole transformation, the solution of Burgers equation (1), the stochastic differential in (5) has to be interpreted in the Stratonovich sense. This result is reasonable since the Stratonovich differential obeys the formal rules of ordinary calculus. We thus consider

$$d\psi_t(x) = \nu\partial_x^2\psi_t(x)dt - \frac{\varepsilon}{2\nu}\psi_t(x) \circ dW_t(x) \tag{7}$$

which can be rewritten in term of Ito differential as

$$d\psi_t(x) = \left(\nu\partial_x^2\psi_t(x) + \frac{\varepsilon}{2\nu}V(x)\psi_t(x)\right)dt - \frac{\varepsilon}{2\nu}\psi_t(x)dW_t(x) \tag{8}$$

the extra term $V(x) = \frac{1}{2}C(x,x) = \frac{1}{2}|x|$ arises from the formal expression of the Stratonovich differential, see (4) above. In the next section we construct, via a generalized Feynman-Kac formula, a process which solves the mild form of (8)

$$\psi_t(x) = G_t \star \psi_0\,(x) + \frac{\varepsilon}{2\nu}\int_0^t G_{t-s} \star (V\psi_s ds - \psi_s dW_s)\;(x) \tag{9}$$

where $\psi_0(x)$ is the initial condition, related to the initial datum for the Burgers equation by $\psi_0(x) = \exp\{(2\nu)^{-1}U_0(x)\}$, $G_t(x)$ is the heat kernel (with diffusion coefficient ν) and $\star$ denotes convolution in space.

Model B

In this model the random potential is more singular and the situation is more complicated. If we follow the same steps as above, i.e. interpret (6) as a Stratonovich equation and solve it via a Feynman-Kac formula, we find a divergent expression. However after a simple renormalization - the Wick exponential - a meaningful expression is obtained. The renormalized Feynman-Kac formula solves the mild form of equation (6) when the stochastic differential is interpreted in the Ito sense

$$\theta_t(x) = G_t \star \theta_0\,(x) - \frac{\varepsilon}{2\nu}\int_0^t G_{t-s} \star \theta_s dB_s\,(x) \tag{10}$$

where θ_0 is the initial datum and $B_t(x) = \partial_x W_t(x)$ is a cylindrical Wiener process, i.e. if $f, g \in L^2(\mathbf{R})$ are test functions

$$\mathbf{E}^w\left(B_t(f)B_{t'}(g)\right) = t \wedge t' \int dx\, f(x)\, g(x) \tag{11}$$

FEYNMAN-KAC FORMULA

We introduce an auxiliary brownian motion which allows us to write a solution of (9) via a generalized Feynman-Kac formula. Let define $dP^{\beta}_{x,t} = dydP^{\beta}_{y,0;x,t}$ where $dP^{\beta}_{y,0;x,t}$ is the Wiener measure with diffusion coefficient 2ν on the paths starting at time t in x and arriving at time 0 in y. We will denote by $\mathbf{E}^{\beta}_{x,t}$ the expectation with respect to the probability measure $dP^{\beta}_{x,t}$. We stress that the brownian motion β is independent from the brownian sheet W.

Model A

A formal application of Feynman-Kac formula to equation (5), which has a time dependent potential, gives

$$\psi_t(x) = \mathbf{E}^{\beta}_{x,t}\left(\psi_0(\beta_0)e^{-\frac{\epsilon}{2\nu}\int_0^t ds\partial_s W_s(\beta_s)}\right) \tag{12}$$

however, due to the singular character of the half white noise, the integral in the exponent is meaningless. Nonetheless it is possible to prove the Feynman-Kac formula if the exponent is defined as a stochastic integral (with respect to the brownian sheet W) along the brownian curve $s \mapsto \beta_s$. Let $s_i = i\,2^{-n}t$, $i = 0, \cdots, 2^n$ be a partition of $[0,t]$ and, given a realization β_s, introduce

$$M^n_{\beta}(t) = \sum_{i=0}^{2^n-1}\left(W_{s_{i+1}}(\beta_{s_i}) - W_{s_i}(\beta_{s_i})\right) \tag{13}$$

using (4) we have

$$\lim_{n\to\infty}\mathbf{E}^w(M^n_{\beta}(t))^2 = \lim_{n\to\infty}\sum_k C(\beta_{s_k},\beta_{s_k})(s_{k+1}-s_k) = \int_0^t ds|\beta_s| \tag{14}$$

It is not difficult to verify that, for almost all β, $M^n_{\beta}(t)$ is a Cauchy sequence in $L^2(dP^w)$; we then define $M_{\beta}(t)$ as its limit. Note that β a.s. $M_{\beta}(t)$ is, under dP^w, a continuous gaussian process.

In [BCJ] the following proposition is proven.

Proposition 1 *Let* $\psi_0 = \exp\{(2\nu)^{-1}U_0(x)\}$, *and*

$$\psi_t(x) = \mathbf{E}^{\beta}_{x,t}\left(\psi_0(\beta_0)e^{-\frac{\epsilon}{2\nu}M_{\beta}(t)}\right) \tag{15}$$

then $\forall\,(t,x) \in [0,\infty)\times\mathbf{R}$

(i) *dP^w — a.s. $\psi_t(x)$ is a solution of (9).*

(ii) *$\forall p \geq 1$ $\psi_t(x) \in L^p(dP^w)$.*

(iii) *$t \mapsto \psi_t(x)$ is dP^w-a.s. locally Hölder continuous with exponent $\alpha < 1/2$. The map $x \mapsto \psi_t(x)$ is dP^w-a.s. differentiable; the derivative is given by*

$$\partial_x \psi_t(x) = G'_t \star \psi_0(x) + \frac{\varepsilon}{2\nu} \int_0^t G'_{t-s} \star (V\psi_s ds - \psi_s dW_s)(x) \tag{16}$$

where

$$G'_t \star f(x) = \int dy\, \partial_x G_t(x-y) f(y) \tag{17}$$

Furthermore the map $(t,x) \mapsto \partial_x \psi_t(x)$ is dP^w-a.s locally Hölder continuous with exponent $\alpha < 1/4$ in time and $\alpha < 1/2$ in space.

(iv) *dP^w — a.s. $\psi_t(x) > 0$*

The properties (*i*)-(*ii*) are statements about the existence of the solution of (9); they are proved through a careful application of stochastic calculus. Properties (*iii*)-(*iv*) are required to perform the Cole-Hopf transformation. The differentiability of $x \mapsto \psi_t(x)$ is proven in two steps. First, using (15), the Hölder continuity is established. Then (9) is shown to be differentiable obtaining (16). The latter step extends to the stochastic model the regularization property of the heat kernel. Using (9) and (16) the Hölder continuity of $t \mapsto \psi_t(x)$ and $t \mapsto \partial_x \psi_t(x)$ can then be proved. Finally the positivity (*iv*) follows from the Feynman-Kac characterization (15).

Model B

Due to the strong spatial divergence of the process $B_t(x)$ the stochastic curvilinear integral appearing in the Feynman-Kac formula cannot be defined as we did in the previous case. In order to define the solution of equation (10) a regularization and then a renormalization are needed.

Let $B_t^\kappa(x)$ be an ultraviolet regularization of the Wiener process B; it has correlation function given by

$$\mathbf{E}^w \left(B_t^\kappa(x) B_{t'}^\kappa(x') \right) = t \wedge t' \, \Gamma_\kappa(x - x') \tag{18}$$

where Γ_κ approximates the delta function.

When the cutoff κ is finite the stochastic curvilinear integral

$$N^{\kappa}_{\beta}(t) = \lim_{n\to\infty} \sum_{i=0}^{2^n-1} \left(B^{\kappa}_{s_{i+1}}(\beta_{s_i}) - B^{\kappa}_{s_i}(\beta_{s_i})\right) \tag{19}$$

is defined analogously to (13), but in this case its variance

$$\mathbf{E}^w(N^{\kappa}_{\beta}(t))^2 = \int_0^t ds\,\Gamma_\kappa(0) = t\,\Gamma_\kappa(0) \tag{20}$$

diverges when the regularization is removed. For this reason, to obtain a meaningful expression as $\kappa \to \infty$, the Feynman-Kac formula (15) has to be renormalized introducing the Wick exponential; we thus define

$$\theta^{\kappa}_t(x) = \mathbf{E}^{\beta}_{x,t}\left(\theta_0(\beta_0)\,\exp\left\{-\frac{\varepsilon}{2\nu} N^{\kappa}_b(t) - \frac{1}{2}\frac{\varepsilon^2}{4\nu^2}\,t\,\Gamma_\kappa(0)\right\}\right) \tag{21}$$

It is now possible to remove the regularization; the following Proposition is proven in [BC].

Proposition 2 *Let* $\theta_0(x) = \exp\{(2\nu)^{-1}U_0(x)\}$*; for any* $(t,x) \in \mathbf{R}^+ \times \mathbf{R}$, $\theta^{\kappa}_t(x)$*, defined in (21), is a Cauchy sequence in* $L^2(dP^w)$*, denoting by* $\theta_t = \theta_t(x)$ *its limit we have*

(i) For all $p \geq 1$, $\theta^{\kappa}_t(x) \to \theta_t(x)$ *in* $L^p(dP^w)$ *and* dP^w *a.s.*

(ii) For $(t,x) \in R^+ \times \mathbf{R}$, $(t,x) \mapsto \theta_t(x)$ *is* dP^w*-a.s. locally Hölder continuous. The Hölder exponent is* $\alpha < 1/2$ *in space and* $\alpha < 1/4$ *in time.*

(iii) For any $t \geq 0$*, the* $C^0(\mathbf{R})$ *valued, process* θ_t *is,* dP^w *a.s, the unique solution of (10).*

(iv) dP^w *a.s.* $\theta_t(x) > 0$

The key point in the proof of this Proposition is to establish that $\theta^{\kappa}_t(x)$ is a Cauchy sequence; it is proven using a representation of $\mathbf{E}^w\left(\theta^{\kappa}_t(x)^2\right)$ in terms of exponential moment for a local time of the brownian motion β which allows explicit computations. The results (*ii*) and (*iii*) are then proven analogously to Proposition 1. We stress however in this case the map $x \mapsto \theta_t(x)$ is not differentiable, but only Hölder continuous; in fact the process $\theta_t(x)$ has "half" the regularity of $\psi_t(x)$. The statement (*iv*) is a straightforward extension of the result in Mueller [Mu].

SOLUTION OF BURGERS EQUATION

Model A

As $\psi_t(x)$ is not twice differentiable, equation (1) is not defined in the classical sense. However the process $u_t(x)$, defined by the Cole-Hopf transformation, is a solution of its mild form, i.e.

$$u_t(x) = G_t \star u_0\,(x) - \frac{1}{2}\int_0^t ds\, G'_{t-s} \star u_s^2\,(x) + \varepsilon \int_0^t G'_{t-s} \star dW_s\,(x) \qquad (22)$$

From Proposition 1 we have the following Theorem which states the existence result for the solution of the Burgers equation with non conservative noise.

Theorem 3 *Set* $u_t(x) = -2\nu\partial_x \log \psi_t(x)$, *with* $\psi_t(x)$ *given by (15).* *Then* $u_t = u_t(x)$ *is a* $C^0(\mathbf{R})$*-valued process, adapted to the brownian sheet, such that* $\forall\,(t,x) \in [0,\infty) \times \mathbf{R}$

(*i*) $\forall p \geq 1 \;\; u_t(x) \in L^p(dP^w)$; $(t,x) \mapsto u_t(x)$ *is* dP^w*-a.s. locally Hölder continuous with exponent* $\alpha < 1/4$ *in time and* $\alpha < 1/2$ *in space.*

(*ii*) $u_t(x)$ *solves (22)* dP^w*-a.s.*

The proof of (*i*) uses Jensen inequality and the results of Proposition 1. The statement (*ii*) is proven introducing first a regularization which allows the use of stochastic calculus. One then shows the regularization can be removed.

Model B

For a fixed cutoff κ and $\theta_t^\kappa(x)$ given by (21), introduce the process

$$u_t^\kappa(x) = -2\nu\partial_x \log \theta^\kappa(x) \qquad (23)$$

it satisfies the regularized mild form of (3) with a Wick renormalization, i.e.

$$u_t^\kappa(x) = G_t \star u_0(x) - \frac{1}{2}\int_0^t ds\, G'_{t-s} \star \left((u_t^\kappa)^2 - \frac{\varepsilon^2}{2\nu}\Gamma_\kappa(0) \right)(x) + \varepsilon \int_0^t G'_{t-s} \star dB_s^k(x) \tag{24}$$

Notice that the term $\Gamma_\kappa(0)$ actually disappears due to the integration of G'.

Since $\theta_t(x)$ is not differentiable $u_t(x)$ can be defined as a distribution valued process

$$u_t(f) = 2\nu \int dx\, f'(x) \log \theta_t(x) \tag{25}$$

In the limit $k \to \infty$ the non linear term in (24) becomes then ill defined, the limiting equation may be meaningless. A reasonable expectation is that $u_t(x)$ can be interpreted as a solution of the limit equation in a weak probabilistic sense.

ACKNOWLEDGEMENTS

The participation of L.B. and N.C. to this summer school has been partially supported by a NATO-ASI grant.

G. J.-L. would like to thank Prof. L. Streit and all the organizers for the kind invitation to this NATO Advanced Institute.

REFERENCES

[BC] Bertini L., Cancrini N. : *The stochastic heat equation: Feynman-Kac formula and intermittence* Preprint (1994).

[BCJ] L. Bertini, N. Cancrini, G. Jona-Lasinio *The Stochastic Burgers Equation*, To appear in Comm. Math. Phys.

[Bu] J.M. Burgers *The Nonlinear Diffusion Equation*, Dordrecht: D. Reidel 1974.

[DDT] Da Prato G., Debussche A., Temam R. : *Stochastic Burger Equation* Preprint di Matematica n. 27 'Scuola Normale Superiore' Pisa. (1993).

[KPZ] Kardar M., Parisi G., Zhang Y.-C. : *Dynamical Scaling of Growing Interfaces* Phys. Rev. Lett. **56**, 889-892 (1986).

[Mo] Molchanov S.A. : *Ideas in Theory of Random Media.* Acta Appl. Math. **22**, 139-282 (1991) and references therein.

[Mu] Mueller C. : *On the Support of Solutions to the Heat Equation with Noise.* Stochastics and Stochastics Reports, **37**, 225-245 (1991).

STATISTICAL PROPERTIES OF PIECEWISE EXPANDING MAPS OF THE INTERVAL

P.COLLET
Centre de Physique Théorique, CNRS UPR 14
Ecole Polytechnique, F-91128 Palaiseau Cedex (France)

ABSTRACT. We give a brief introduction to the stochastic properties of expanding maps of the interval using the transfer operator method. We also describe recent results about the statistics of entrance times and times of visits to small intervals.

Introduction

We are going to consider discrete time dynamical systems. They are given by a space X (often called the phase space or the space of states of the system), and a map f of X into itself which is the time evolution. At a given time, the state of the system is completely characterized by a point $x \in X$. This point varies with time and the successive states of the system are given by iterations of f. Namely, if at time 0 the system is in the state x_0, at the successive (discrete) times 1, 2, 3, etc. the system is in the states $x_1 = f(x_0)$, $x_2 = f(x_1)$ etc. It is convenient to introduce a notation for the n th iterate:

$$x_n = f^n(x_0) = \underbrace{f \circ \cdots \circ f}_{\text{n times}}(x_0) \ .$$

We can observe immediately the analogy with stochastic processes. Indeed, let μ be a probability measure on X which is invariant by the map f, namely

$$\mu(f^{-1}(A)) = \mu(A)$$

for any measurable set A, where f^{-1} is the set theoretic inverse of f (namely, $f^{-1}(A) = \{x \,|\, f(x) \in A\}$). Then (X, μ, f) defines a (stationary) stochastic process by the above iteration formula. For stochastic processes there is no more structure introduced in the problem and one is more interested in asking questions in a measure theoretic sense. For dynamical systems, the phase space X has often a more precise structure: metric space, manifold etc. and we will concentrate below on the case of an interval. We will see that there is a rich interplay between topological/differentiable concepts and measure theoretic aspects.

The central question in the theory of dynamical systems is the study of the asymptotic time behavior of the system. The asymptotic time properties can be

A. I. Cardoso et al. (eds.), Stochastic Analysis and Applications in Physics, 45–57.

separated into two groups. The first group concerns topological or differentiable questions: bifurcations, geometry of attractors etc. The second group is concerned with statistical questions and relies on the ideas of stochastic processes once an invariant measure has been chosen. Quite often however there is a continuum of invariant probability measures which reflects the variety of the asymptotic time behavior and one has to choose one of them or to describe properties which are true for entire classes of them.

As mentioned before we will concentrate on the study of some maps of the interval $X = [0,1]$. This is because they are simpler to discuss although they have quite a rich variety of different behaviors and also one can illustrate most of the ideas which are used in the study of more complicated (higher dimensional) systems.

In order to get some simple feelings about what can happen, let us consider two simple examples.

The first one is the map $f(x) = x + \alpha \pmod 1$ where α is an irrational number. This map can also be considered as an irrational rotation on the circle. One can prove that there is a unique invariant probability measure which is the Lebesgue measure.

Our second simple example is the map $f(x) = 2x \pmod 1$ which is the doubling of the angle on the circle. Here the Lebesgue measure is also invariant but there are many others δ_0, $(\delta_{1/3} + \delta_{2/3})/2$ etc.

The main difference between these two examples is that the second one has the property of sensitive dependence on initial conditions (S.D.I.C. for short) while the first one does not have this property. S.D.I.C. reflects the instabilities of the trajectories in phase space, it involves the amplification of small errors and the loss of memory of the initial condition. S.D.I.C. can be defined as follows

$$\exists\, \delta > 0 \text{ such that, } \forall x,\ \forall \epsilon > 0,\ \exists\, y,\ |x-y| < \epsilon,\ \exists\, n(x,y), \text{ such that}$$

$$|f^{n(x,y)}(x) - f^{n(x,y)}(y)| > \delta\ .$$

In other words, close (but different) initial conditions, no matter how close will be separated by the dynamics up to a "macroscopic" scale if one waits long enough. Note that after the time $n(x,y)$ the two trajectories may come back very close again. This is in fact the main recipe for chaos, namely if one has a bounded phase space, the distance between the two trajectories cannot grow for ever, at large scales, the phase space is folded by the non linearities.

The class of dynamical systems that we are going to study are extensions of the second example above. This is the class of piecewise expanding maps of the interval. In order to define such a map we start with a finite partition of the interval, or equivalently with an increasing (finite) sequence of numbers $0 = a_0 < a_1 < \cdots < a_{r-1} < a_r = 1$. We will denote by $\mathcal{A}$ the associated partition (whose atoms are the intervals (a_s, a_{s+1}) with the usual ambiguity at the boundary).

Definition. *A map f of the unit interval is piecewise expanding (with respect to a partition $\mathcal{A}$ as above) if the restriction of f to any interval of the partition extends to a C^∞ function on the closure, and there is a positive number $\rho > 1$ such that for $0 \le s \le r-1$*

$$\left| f'_{[a_s, a_{s+1}]} \right| > \rho\ .$$

Unless otherwise stated, in all the results below we will assume that the dynamical system is given by an expanding map of the unit interval. In most cases it is in fact enough to assume that there is an integer $n > 0$ such that f^n satisfies the above hypothesis. For simplicity we will not pursue this easy generalization.

It is immediate that the map $f(x) = 2x \pmod 1$ discussed above is piecewise expanding. Using the invariance of the Lebesgue measure and the dyadic representation of the real numbers, one obtains immediately that the dynamical system generates the successive flipping of a coin, the Lebesgue measure being the fair probability distribution. One can generalize this result to all finite state Markov chains. Any finite state Markov chain can be represented by a piecewise affine map of the interval with branches of slope at least 1 (the representation is not unique). Some branches may have slope one but away from some trivially degenerate cases, there is an iterate of the map which is piecewise expanding.

Other interesting examples occur in the construction of random number generators. For example, on the interval $[0, 2147483399]$, the map

$$f(x) = 40692x \pmod{2147483399}$$

is one of the best random number generator with uniform distribution on 32 bits machines (see [L.]).

An old example with infinitely many branches is due to Gauss and is associated to the development of a number in continuous fractions. The map f of the unit interval is given by

$$f(x) = \frac{1}{x} - \left[\frac{1}{x}\right]$$

where $[\,\cdot\,]$ denotes the integer part. The successive coefficients (a_n) of the continuous fraction expansion of a number x are given by

$$a_n = \left[\frac{1}{f^n(x)}\right] .$$

We note that all the above examples have a special topological Markov property, namely for any atom I of the partition $\mathcal{A}$, the closure of $f(I)$ is the union of closures of atoms of $\mathcal{A}$. This condition is sometimes useful but we will not use it below, hence treating the general case. It makes a simple connection with the theory of sub-shifts of finite type (see [B.]).

We also mention that expanding maps of the interval appear also as sections of higher dimensional systems (see for example [A.]).

Since we will be interested in the statistical aspects of the dynamics we have to introduce an invariant probability measure. As was mentioned above there are in general many such measures and this raises the question of choosing one. The most interesting class of measures in this context will turn out to be the measures absolutely continuous with respect to the Lebesgue measure. We will see why below. But for the moment a natural question is whether any piecewise expanding map of the interval has an absolutely continuous invariant probability measure (a.c.i.p.m. for short). The answer is yes as was proven by Lasota and Yorke [L.Y.].

Theorem 1. *Any piecewise expanding map of the interval has at least one absolutely continuous invariant measure. The density is of bounded variation.*

We will not give the proof of this result but introduce the main tool which is very useful in discussing other properties of the system.

Assume that $d\mu = hdx$ is an a.c.i.p.m., then from the definition of invariance, we must have for any continuous function g the relation

$$\int g \circ f \; hdx = \int ghdx \; .$$

If one performs changes of variables in the left integral taking into account the fact that f is in general not invertible, one gets (for any h in fact)

$$\int g \circ f \; hdx = \int gP(h)dx$$

where the operator P is defined by

$$P(h)(x) = \sum_{y,\, f(y)=x} \frac{h(y)}{|f'(y)|} \; .$$

The operator P is called the transfer operator of the map f, and a function h is the density of an a.c.p.i.m. if and only if $P(h) = h$. P is somewhat analogous to (the adjoint of) the transition probability of Markov chains, its formal kernel being $\delta(x - f(y))$. We now list some useful elementary properties of the transfer operator.

Proposition 2.

1) P maps non-negative functions to non-negative functions and $|P(g)| \leq P(|g|)$.

2)

$$P^n(g)(x) = \sum_{y,\, f^n(y)=x} \frac{g(y)}{|f^{n\prime}(y)|} \; .$$

3)

$$\int g_1 \circ f \; g_2 dx = \int g_1 P(g_2) dx \; .$$

4)

$$P(g_1 \circ f \; g_2) = g_1 P(g_2)$$

Note that it follows from 1) and 3) that in the space $L^1(dx)$ the norm of P is one.

The proof of Theorem 1 relies on a main estimate for the operator P applied to functions of bounded variation. We recall the definition of the variation $\vee g$ of a function g:

$$\vee g = \sup_{0 \leq b_1 < b_2 < \cdots < b_q \leq 1} \sum_{j=0}^{q-1} |g(b_{j+1}) - g(b_j)| \; .$$

A function of bounded variation is a function whose variation is finite. The main technical lemma is the following ([L.Y.]).

Lemma 3. *If f is a piecewise expanding map of the interval, there exists positive numbers $0 < \alpha < 1$ and $\Gamma > 0$ such that for any function g of bounded variation and for any integer n*

$$\vee P^n(g) \leq \alpha^n \vee g + \Gamma \|g\|_{L^1(dx)} .$$

We refer to [L.Y.] for a proof. Once this result is established, the proof of Theorem 1 follows by a compactness argument. One can for example consider the sequence of functions

$$h_n = \frac{1}{n} \sum_{j=0}^{n-1} P^j(1) .$$

It is easy to verify that the sequence $(\vee h_n)$ is bounded, and moreover, the integral of each h_n is equal to 1 by 3) of Proposition 2. By Helly's Theorem we can extract a subsequence which converges in $L^1(dx)$ and it is easy to verify that the limit is an a.c.i.p.m.

These a.c.i.p.m. may not be unique, but they form a finite dimensional convex set. Also, if the map f has a dense orbit, the a.c.i.p.m. is unique. We refer to [H.K.] or [C.] for the proofs and other similar results.

Spectral theory of the Transfer operator

As we will see below, many dynamical properties can be read out of the spectral theory of the transfer operator P. We saw in the previous section for example that the eigenvectors with eigenvalues one correspond to the a.c.i.p.m. In order to discuss the spectrum of this operator, we first need to define a Banach space. This will be the space of functions of bounded variation equipped with the norm

$$|||g||| = \vee g + \|g\|_{L^1(dx)}$$

The key to the spectral property of P is the following result (see [H.K.] or [C.]).

Lemma 4. *The transfer operator P is quasi compact, namely there is a sequence (Q_n) of compact (here even finite rank) operators such that*

$$\lim_{n \to \infty} |||P^n - Q_n||| = 0 .$$

From the spectral theory of quasi compact operators and Proposition 1 one gets the following result (see [D.S.]).

Theorem 5. *The spectral radius of P is 1. The points of the spectrum $Sp(P)$ of P which are on the unit circle (the peripheral spectrum) consist of a finite number of eigenvalues with finite multiplicity. The rest of the spectrum lies in a disk centered at the origin with radius $r < 1$. In other words, there is a collection of finite rank projections Q_λ for $\lambda \in Sp(P)$ with $|\lambda| = 1$ and a bounded operator R such that*

a)

$$Q_\lambda Q_{\lambda'} = Q_{\lambda'} Q_\lambda = \begin{cases} 0 & \text{if } \lambda \neq \lambda' \\ Q_\lambda & \text{else.} \end{cases}$$

b)

$$RQ_\lambda = Q_\lambda R = 0$$

c)

$$|||R^n||| \leq \mathcal{O}(1) r^n$$

d)

$$P = \sum_{\lambda \in \mathrm{Sp}(P),\, |\lambda|=1} \lambda Q_\lambda + R$$

As a first consequence, we observe that since Q_1 is finite dimensional, the number of linearly independent a.c.i.p.m. is finite. It turns out that if the a.c.p.i.m. is unique, it is ergodic. We recall that this means that any integrable invariant function ($g \circ f = g$ almost surely), is almost surely constant. From now on we will assume that this is the case. We refer to [H.K.] or [C.] for a decomposition result of the general case into a finite number of ergodic components.

We now come to the mixing properties. We recall that a probability measure μ invariant by a transformation f is mixing if for any pair of functions g_1 and g_2 belonging to $L^2(d\mu)$ we have

$$\lim_{n \to \infty} \int g_1 \circ f^n g_2 d\mu = \int g_1 d\mu \int g_2 d\mu\ .$$

Note that mixing implies ergodicity, and that irrational rotations on the circle are not mixing.

There is a simple spectral criteria for an a.c.i.p.m. to be mixing.

Proposition 6. *An a.c.i.p.m. is mixing if and only if 1 is the only point in the spectrum of P of modulus one (and it has to be an isolated eigenvalue).*

Again, as in the case of ergodicity, there is in the general case a finite decomposition into mixing components (see [H.K.] or [C.]). An important consequence of mixing is that we have immediately exponential mixing (under some regularity assumptions on the functions). Indeed, combining Proposition 6 with the spectral theory of Theorem 5 we obtain the following corollary.

Corollary 7. *Let μ be a mixing a.c.i.p.m., then there is a positive number C, and a number $0 < \delta < 1$ such that for any function $g_2 \in \mathcal{B}$ and any function $g_1 \in L^1(dx)$ we have*

$$\left| \int g_1 \circ f^n g_2 d\mu - \int g_1 d\mu \int g_2 d\mu \right| \leq C\delta^n \|g_1\|_{L^1(dx)} \, |||g_2||| \, .$$

This result is often referred to as the exponential decay of correlations. Note that the exponential rate of decay depends crucially on the regularity of the function g_2. We refer to [C.I.] for more information on this problem.

From these spectral properties we can now deduce a number of interesting statistical properties. From Birkhoff's ergodic theorem, we know that if g is an integrable function for an ergodic a.c.i.p.m. $d\mu = hdx$, then the time averages of almost any trajectory converge to the integral of μ. Namely, if we define

$$S_n(x) = \sum_{j=0}^{n-1} g(f^j(x))$$

then for μ almost every x we have

$$\lim_{n\to\infty} \frac{1}{n} S_n(x) = \int g d\mu \, .$$

This is the analog of the law of large numbers in probability theory for the sequence of random variables $(g \circ f^n)$ which are however not independent. Nevertheless we can ask for the corrections to the law of large numbers, namely the analog of the central limit theorem. The first step is of course to define the asymptotic variance. This is the square of the mean square deviation $\sigma \geq 0$ formally given by

$$\sigma^2 = \lim_{n\to\infty} \frac{1}{n} \int S_n^2 d\mu \, .$$

It follows easily from the exponential mixing of Corollary 7 that if the function g is of bounded variation, the limit exists and is given by

$$\sigma^2 = \int g^2 h dx + 2 \sum_{j=1}^{\infty} \int g P^j(gh) dx \, .$$

As we observed before, the random variables $(g \circ f^n)$ are strongly coupled in general, although the exponential mixing tells us that this coupling fades away with time. However contrary to the case of independent random variables the coupling may result in sufficient cancellations to destroy the variance. For example, if the function g satisfies

$$g = u \circ f - u$$

for some square integrable function u, it is easy to verify that $\sigma^2 = 0$. The converse is in fact known to be true. Assuming now $\sigma > 0$ we can state a central limit theorem.

Theorem 8. *Assume the variance σ^2 is not zero, then there is a number $\nu > 0$ such that*

$$\sup_a \left| \mu(x | S_n / \sqrt{n\sigma} \leq a) - (2\pi)^{1/2} \int_{-\infty}^{a} e^{-s^2/2} ds \right| \leq \mathcal{O}(n^{-\nu}) .$$

We refer to [C.P.] for a proof of this strong form of the central limit theorem. The proof of the standard weaker form can be found in [H.K.] or [C.]. These references contain also some results about large deviations and references to laws of iterated logarithms and invariance principles.

Measures of chaos

We have seen above that S.D.I.C. and non-linearities are at the origin of strong chaos. However one would like to have a quantitative measurement of the intensity of chaos. A natural question in this direction is how fast do nearby trajectories separate, i.e. what is the rate of S.D.I.C. ? This can easily be settled in one dimension. Let x and $x + \epsilon$ be two nearby initial conditions (ϵ small). As long as the two trajectories stay close together, we can use the chain rule to approximate their distance, namely

$$|f^n(x) - f^n(x + \epsilon)| \approx |\epsilon| \, |f^{n\prime}(x)| = |\epsilon| \prod_{j=0}^{n-1} |f'(f^j(x))| .$$

For expanding maps, this approximation suggests an exponential growth rate for the distance between neighboring trajectories. This growth rate can therefore be conveniently written as

$$\frac{1}{n} \log |f^{n\prime}(x)| = \frac{1}{n} \sum_{j=0}^{n-1} \log |f'(f^j(x))| .$$

We now observe that the right hand side of this expression is the time average of the function $\log |f'(\cdot)|$ along the trajectory of the point x (up to time n). Therefore we can now apply Birkhoff's ergodic theorem to obtain the limit of this quantity when the time goes to infinity. In other words, for each ergodic invariant measure ν, we obtain an average growth rate of the initial error which is equal to the quantity

$$\int \log |f'| \, d\nu .$$

This number is called the Lyapunov exponent of ν. Note that it depends strongly on the measure ν, this means in fact that we may have very different exponential growth rates according to which initial condition we consider. i.e. unless the function $\log|f'|$ is constant, the growth rate is not going to be uniform. Note also that there is an interchange of limits in this result with respect to our initial goal. The Lyapunov exponent is defined using a limit of infinite time assuming that we are still in the linear regime. This is not true in general for a fixed perturbation ϵ as introduced before. For large time, one will observe an exponential growth rate close to the Lyapunov exponent but only as long as the two trajectories stay close. Once they have been macroscopically separated, the above argument cannot be applied anymore, and as explained before the non linearities take over.

There is an older measure of the loss of information by a dynamical system which is called the entropy (the Kolmogoroff-Sinai entropy). This notion is very much connected with the entropy of statistical mechanics since in a suitable sense it measures the volume of phase space which is the support an ergodic invariant measure. More precisely we start with a partition $\mathcal{A}$ and construct a sequence of successively refined partitions $(\mathcal{A}_n)$ by

$$\mathcal{A}_n = \bigvee_{j=0}^{n} f^{-j}\mathcal{A}$$

where if $\mathcal{A}$ and $\mathcal{B}$ are two partitions, $\mathcal{A} \vee \mathcal{B}$ denotes the partition which is obtained by taking the intersections of the atoms of the two partitions. A partition $\mathcal{A}$ is said to be generating if the sequence of partitions $\mathcal{A}_n$ converges to the point partition when n diverges. This is the case for the partition used in section 1 to define the expanding map f. Using a Theorem of Shannon, Mac Millan and Breiman, if μ is an ergodic invariant probability measure for f, its entropy $S(\mu)$ is a nonnegative number defined as follows. For any fixed positive number ϵ (smaller than one) there is an integer $n(\epsilon)$ such that for $n > n(\epsilon)$ we can decompose the atoms of $\mathcal{A}_n$ into two disjoint groups groups

$$\mathcal{A}_n = \mathcal{N}_n \cup \mathcal{B}_n \ ,$$

the group of nice atoms $\mathcal{N}_n$ and the group of bad atoms $\mathcal{B}_n$. The bad atoms altogether account for a small part of the total mass of the probability measure μ, namely

$$\mu\left(\cup_{I\in\mathcal{B}_n}\right) \le \epsilon \ .$$

The good atoms not only account for most of the weight but have also the property that they have about the same mass which is of the order of $\exp(-nS(\mu))$, namely

$$\forall I \in \mathcal{N}_n \qquad e^{-n(S(\mu)+\epsilon} \le \mu(I) \le e^{-n(S(\mu)-\epsilon} \ .$$

Note that since the good atoms have all about the same weight and their total mass is about one, their number is of the order of the inverse of their mass, namely $\exp(nS(\mu))$.

At this point it is natural to ask if there is any relation between these two measures of chaos, namely entropy and Lyapunov exponent. The first result is a general inequality due to Ruelle [R.] and also valid in any dimension.

Theorem 9. *For an ergodic invariant measure μ of a map of the interval we have*

$$S(\mu) \leq \lambda(\mu) .$$

The case of equality was settled by Ledrappier [Le.] for maps of the interval.

Theorem 10. *For a map of the interval an ergodic invariant measure μ satisfies the equality $S(\mu) = \lambda(\mu)$ if and only if it is absolutely continuous.*

This explains why the a.c.i.m. are singled out. They saturate Ruelle's inequality which is sometimes called the variational principle. These measures behave also in a regular way with respect to stochastic perturbations of the dynamical system. We refer to [Ke.] for detailed results and references.

Statistics of S.D.I.C. and entrance times

Problems of entrance time are at the hart of ergodic theory, and Statistical Mechanics. An important question is to estimate the time needed to observe an abnormal situation, for example how long does it take so that all the molecules of a gas will be together in half the volume of the container. If this time is larger than the age of the universe, one may decide that this abnormal situation will never occur. More precisely, assume that we have a dynamical system, then given a subset of phase space, we would like to estimate the time it takes for a typical trajectory to enter this set (assuming of course that we do not start inside). Birkhoff's ergodic theorem tells us that on the average, if μ is an ergodic invariant measure, for μ almost every initial condition, the orbit spend an average time $\mu(A)$ in the set A. This associates a time scale $1/\mu(A)$ to any measurable set of positive measure. More precisely, we can define the first entrance time of the orbit of a point x into the set A by

$$\tau_A(x) = \min\{n \geq 0\,;\ f^n(x) \in A\} .$$

From Birkhoff's ergodic theorem it follows easily that if μ is an ergodic invariant measure, then τ_A is finite μ almost surely. Similarly, we can define the process of successive visits to the set A by the sequence of random variables $\tau_A^0 = \tau_A$, and for $p \geq 1$

$$\tau_A^p(x) = \min\{n > \tau_A^{p-1}\,;\ f^n(x) \in A\} .$$

These random variables are again μ almost surely finite.

As explained before, from Birkhoff's ergodic theorem, we expect that τ_A (and τ_A^p) is of the order of $1/\mu(A)$. It is quite remarkable that even at this level of generality one can say something about the average of τ_K if the initial condition belongs to A. We refer to [Ka.] for a precise statement. We will now make some more assumptions in order to be able to derive a precise result. We will first assume that the dynamical system is an expanding map of the interval, and that the measure μ is an a.c.i.p.m. with density h, ergodic and mixing. Inspired by the limit theorems from probability theory we will also consider the case of small sets and more precisely of small intervals. We now formulate a typical result under these hypothesis (see [C.G.1]).

Theorem 11. *Let (K_j) be a sequence of intervals such that $\mu(K_j) > 0$ and $|K_j| \to 0$. Then there is a diverging sequence of positive numbers (T_j) such that for any number $t > 0$ we have*

$$\mu(\{x|\tau_{K_j}(x) > tT_j\}) = e^{-t} .$$

In other words, this means that the sequence of random variables $(\tau K_j/T_j)$ converges in law to an exponential. At this point we have only a weak relation between the times scales T_j and the natural ones $1/\mu(K_j)$, namely $\liminf_{j\to\infty} T_j\mu(K_j) > 0$. We will need a further hypothesis to prove that both time scales are of the same order. We also observe that the hypothesis about a sequence of intervals is crucial in the proof. This is probably related to the existence of other invariant measures supported for example by Cantor sets. Note also that there is no assumption on the convergence of the sequence of intervals (K_j). The sequence may freely wander around the interval $[0,1]$, this means that the result is really about small intervals and not about sequences.

As explained above, we now introduce two supplementary hypothesis which will unable us to complete the picture (see [C.G.1]).

Theorem 12. *Let (K_j) be a sequence of intervals as in the above Theorem. Assume moreover that*

1) There is a diverging sequence (m_j) of integers such that for $0 < l < m_j$, $f^l(K_j) \cap K_j = \emptyset$.

2)

$$\liminf_{j\to\infty} \inf_{x\in K_j} h(x) = \zeta > 0 .$$

Then with the notations of Theorem 1

i)

$$\lim_{j\to\infty} T_j\mu(K_j) = 1 .$$

ii) The sequence of renormalized return times $(\tau^p_{K_j}\mu(K_j))$ converges in law to a Poisson point process of parameter 1.

Condition 1) says that we do not accumulate too fast to a periodic orbit. These are places where in some sense the orbits are abnormally trapped. However periodic orbits are dense in our system, and this is why we cannot (and do not have to) impose any growth condition on the diverging sequence (m_j). This divergence must be rather slow since from the expansivity it follows that $m_j << -\log(\mu(K_j))$.

Condition 2) is a technical condition preventing a division by zero during the proof of ii) where we use conditional probabilities.

Finally we observe that conditions 1) and 2) are μ "generic" in the sense that for μ almost every x, $h(x) > 0$ and one can find a decreasing sequence of intervals (K_j) converging to x satisfying the assumption 1) of the above Theorem. This follows essentially from the fact that periodic points form a countable set hence of measure 0.

Another result concerns the statistics of close encounters of two trajectories. We recall that the basic ingredients of strong chaos are S.D.I.C. which ensured the local instability of close trajectories and non-linearities of the mapping which fold the phase space. This is indeed the only way to reconcile instability with a bounded phase space. The picture of chaos is that two nearby initial conditions give rise to trajectories which are more and more separated by the S.D.I.C. until the non linear folding takes over. After a certain number of these folding steps, the two trajectories may come back very near to one-another where the S.D.I.C. plays the main role and so on. This coming back is of course directly related to mixing. For a fixed (small) positive number ϵ, we say that the two trajectories of the initial points x and y have an ϵ close encounter at time n if

$$|f^n(x) - f^n(y)| < \epsilon \,.$$

We can define as before the first ϵ close encounter $\tau_\epsilon(x, y)$ and the process of the successive ϵ close encounters. We now have the following result.

Theorem 13. *Assume that the density h of the invariant measure μ is strictly positive. Then on $[0,1] \times [0,1]$ for the measure $\mu \times \mu$ the process of successive ϵ close encounters converges in law when ϵ tends to zero to a marked Poisson process.*

This can be interpreted as follows. On the (discrete) time line, one puts a dot when one observes an ϵ close encounter. Then one rescales the unit of time by ϵ. In the limit one gets a process which can be described as follows. First is given a Poisson process of parameter $\lambda > 0$. Every times this process puts a point on the time line (the bell rings), one chooses at random and independently of what happened before an integer $j > 0$ according to a probability distribution p_j on $\mathbb{N}$. This effect is due to the clustering of the ϵ close encounters. Indeed if two trajectories come at a distance much less than ϵ, S.D.I.C. will start to separate them exponentially fast, but it will take some time before the distance reaches ϵ. In our previous construction with the time line we will have a cluster of neighboring points which are compressed into one by the time renormalization. The marking by an integer of the Poisson process is what remains of this clustering in the limit.

The detailed proofs of these results can be found in [C.G.1] and [C.C.], see also [C.G.2] for similar results in a different setting. We will now give a very brief sketch of the proof of Theorem 11.

Let K be a subinterval of $[0,1]$, and denote by K^c its complement. The set $\{\tau_K > n\}$ is the set of initial conditions such that the corresponding orbit stays in K^c at least up to time n. In other words,

$$\{\tau_K > n\} = \prod_{j=0}^{n} \chi_{K^c}(f^n(x)) \,.$$

Using properties 3) and 4) of proposition 2 we get

$$\mu(\{\tau_K > n\}) = \int_0^1 (P\chi_{K^c})^{n+1}\, h(x) dx$$

where $P\chi_{K^c}$ is the operator acting on functions first by multiplication with χ_{K^c}, and then applying P. If $K^c = [0,1]$ this is simply the operator P and the spectral theory is given in Theorem 5. If $K \neq \emptyset$ but is nevertheless a small interval one can try to use perturbation theory to prove that the operator $P\chi_{K^c}$ has a dominant eigenvalue which is positive and of modulus smaller than one. This will essentially prove that the sequence $\mu(\{\tau_K > n\})$ decays exponentially fast with n and lead to Theorem 11 after a suitable normalization. There are however some technical difficulties and this program does not works as such. We refer to [C.G.1] for a complete proof along these lines.

References

[A.F.] R.Adler, L.Flato Geodesic flows, interval maps and symbolic dynamics. Bull. Amer. Math. Soc. **25**, 229-334 (1991).

[B] R.Bowen. *Equilibrium States and the Ergodic Theory of Anosov Diffeomorphims.* Lecture Notes in Mathematics **470**, Springer Verlag, Berlin Heidelberg New York (1975).

[C] P.Collet. Some ergodic properties of maps of he interval. To be published in the proceedings of the summer school "Dynamical Systems & Frustrated Systems", Temuco 1991.

[C.C] Z.Coelho, P.Collet. Asymptotic limit law for close approach of two trajectories in expanding maps of the circle. Prob. Theor. and Related Topics, to appear.

[C.G.1] P.Collet, A.Galves. Asymptotic distribution of entrance times for expanding maps of the interval. Preprint.

[C.G.2] P.Collet, A.Galves. Statistics of close visits to the indifferent fixed point of an interval map (with A.Galves). J. Stat. Phys. (to appear).

[C.I.] P.Collet, S.Isola. Essential spectrum in C^k of expanding Markoff maps of the interval. Commun. Math. Phys. **139**, 551 (1991).

[C.P.] Z.Coelho, W.Parry. Central limit asymptotics for subshifts of finite type. Isr. J. Math. **69**, 235 (1989).

[D.S.] N.Dunford, J.T.Schwartz. *Linear Operators I.* Interscience Publishers, New York (1963).

[H.K.] F.Hofbauer, G.Keller. Ergodic properties of invariant measures for piecewise expanding monotonic transformations. Math. Zeit. **180**, 119 (1982).

[Ka.] M.Kac. On the notion of recurrence in discrete stochastic processes. Bull. Amer. Math. Soc. **53**, 1002-1010 (1947).

[K.] G.Keller. Stochastic stability in some chaotic systems. Monat. Math. **94**, 313-333 (1982).

[L.] P.L'Ecuyer. Random number generators, good ones are hard to find. Commun. of the ACM **33**, 85 (1990).

[Le.] F.Ledrappier. Some properties of absolutely continuous invariant measures on an interval. Ergod. Th. and Dyn. Syst. **1**, 77 (1981).

[L.Y.] A.Lasota, J.Yorke. On the existence of invariant measures for piecewise monotonic transformations, *Trans. Amer. Math. Soc.* **186**, 481-488, (1973).

[R.] D.Ruelle. *Thermodynamic Formalism.* Addison Wesley, Reading (1978).

Euclidean Quantum Mechanics. An outline.

A. B. Cruzeiro[1], *Complexo II, University of Lisbon*
Av. Prof. Gama Pinto 2, 1699 Lisbon Codex, Portugal
J. C. Zambrini[1], *Complexo II, University of Lisbon*
Av. Prof. Gama Pinto 2, 1699 Lisbon Codex, Portugal

ABSTRACT. We describe a rigorous version of the formal functional calculus developed by R. Feynman in relation with his path integral formulation of Quantum Mechanics. Our approach is Euclidean but quite distinct from the usual one based on Feynman-Kac formula. A different probabilistic interpretation of the classical heat equation, suggested long ago by E. Schrödinger, that constitute the basis of our version of "Euclidean Quantum Mechanics", is used instead.

Content

1. BERNSTEIN PROCESSES ON $\mathbb{R}^N$

§1. Origin of Euclidean Quantum Mechanics

The fundamental idea is due to E. Schrödinger (1931-32) and he said about it that it constitutes:

"Une analogie avec la mécanique ondulatoire qui fut si frappante... qu'il m'est

[1] Partly Supported by JNICT.

A. I. Cardoso et al. (eds.), Stochastic Analysis and Applications in Physics, 59–97.

difficile de la croire purement accidentelle." [1]

Schrödinger's starting question was the following: suppose that a Brownian motion takes place in a region of $\mathbb{R}^N$ and that its probability density ρ is observed to be ρ_1 at time t_1 and ρ_2 at time t_2, where ρ_2 is not necessarily the solution of the Cauchy problem for the associated free heat equation with initial condition ρ_1. What is, then, the "most" probable evolution for t in I $= [t_1, t_2]$?

Using its skill in statistical mechanics (the problem does not seem a priori to have anything to do with quantum mechanics) he is able to show that, for any t in I, the answer is

$$\rho_t = \eta_t \eta_t^* \tag{1.1}$$

where

$$(1.2) \qquad \begin{cases} \dfrac{\partial \eta_t}{\partial t} = -\dfrac{1}{2}\Delta \eta_t \\ \eta_{t_2} = \eta_2 \end{cases} \qquad \text{and} \qquad \begin{cases} -\dfrac{\partial \eta_t^*}{\partial t} = -\dfrac{1}{2}\Delta \eta_t^* \\ \eta_{t_1}^* = \eta_1^* \end{cases} \qquad (1.2)^*$$

and the boundary conditions η_2, η_1^* have to be determined from the data ρ_1 and ρ_2. This requires to solve a rather complicated system of nonlinear functional equations that we call the "Schrödinger's system", to be specified soon.

A short time later, S. Bernstein [2] provided a more probabilistic interpretation of Schrödinger's problem, looking at it as the problem of the construction of a stochastic process given the joint probability density of its starting and final positions and an analogue of the usual transition probability but involving informations about the past and the future of the process. For this purpose and using a terminology which, of course, did not exist in the thirties, Bernstein suggested to replace the Markov property by the local Markov property, an intriguing observation for those familiar with the probabilistic approaches to Euclidean Quantum Field Theory in the seventies.

In 1960, A Beurling (also aware of a work of R. Fortet in this direction [3]) proved that the Schrödinger's system has, under weak restrictions one, and only one, solution. Actually, as one could easily expect, he proved a much more general result than the one we will need here, which is the following: Let h(x,t,z) be the integral kernel of e^{-tH} for H a quantum Hamiltonian such that h is strictly positive, bounded, symmetric. Then the following ("Schrödinger's") system for given strictly positive data $\rho_{-1}(x)\,dx$ and $\rho_1(z)\,dz$ has an unique solution (η_{-1}^*, η_1): [4]

$$\eta_{-1}^*(x) \int_{\mathbb{R}^N} h(x,2,z)\, \eta_1(z)\, dz = \rho_{-1}(x) \tag{1.3}$$

$$\eta_1(z)\int_{\mathbb{R}^N} \eta^*_{-1}(x)\ h(x,2,z)\,dx = \rho_1(z)$$

Notice that up to this point in time, the nature of the stochastic processes underlying Schrödinger's original problem was not investigated. In 1974, B. Jamison pushed forward this part of the problem, achieving the constructions of what he called, after Bernstein, "reciprocal processes" [5]. Nevertheless, the class of processes relevant for typical problems of quantum mechanics, for example for an Hamiltonian of the form

$$H = -\frac{\hbar^2}{2m}\,\Delta + V(x) \tag{1.4}$$

with V in the Kato class [6] (containing both the harmonic oscillator and the Coulomb potentials) was still not characterized.

One of the reasons for this was, maybe, that since long before the mid-seventies, probabilists were used to interpret a potential term like V in the Hamiltonian (1.4) as a "killing term". This interpretation dates back, at least, to M. Kac [7]. If X(.) is the underlying process (for example the familiar Wiener process), this means that the conditional expectation

$$\lim_{\Delta t\downarrow 0} \frac{1}{\Delta t} \Pr\,(\text{the process is killed during } [t, t+\Delta t] \mid X(t) = x) = V(x). \tag{1.5}$$

Since the potential V has, in quantum mechanics, exactly the same interpretation as in classical mechanics, it is clear that the probabilistic view (1.5) is more a mathematical characterization than a physical interpretation. In any case, it is completely irrelevant to quantum dynamics.

On the other hand, little doubts were possible on the nature of the process underlying an Hamiltonian like (1.4). Since at least R. Feynman's pioneering work [8], it was known by the physicists that the processes should be diffusions with diffusion matrix

$$C(x) = \frac{\hbar}{m}\,\mathbb{1} \tag{1.6}$$

where $\mathbb{1}$ is the $N \times N$ identity matrix.

Actually, according to Feynman, the diffusion matrix relevant for quantum physics was not Eq (1.6) but

$$C_Q(x) = i\,\frac{\hbar}{m}\,\mathbb{1} \tag{$\overline{1.6}$}$$

where $i = \sqrt{-1}$, and he developed his path integral approach to quantum mechanics only as a formal, but precious, analogy with the classical Brownian motion used in statistical mechanics. The reading of [9] (Cf, for example p 165 and p 177) is revealing about this. It became progressively clear [10] however that a sound probabilistic version of Feynman's point of view should start from Eq (1.6) instead from $(\overline{1.6})$.

To come back to Schrödinger's original question and with the slight advantage of posteriority, the problem can be now reformulated in the following way:

For H any Hamiltonian observable of the form (1.4) construct a family of probability measures for diffusion processes without killing, defined on a time interval I, and such that, for any time t in I, the probability density ρ_t of the processes is of the product form (1.1), where η_t and η_t^* solve two heat equations (1.2), adjoint of each other with respect to the time, and for H replacing $H_0 = -\frac{1}{2}\Delta$.

This problem was solved in 1985-1986 [11] and is the basis of this version of "Euclidean Quantum Mechanics" (EQM) that we are going to describe here.

H. Föllmer has also studied Schrödinger's initial question from the point of view of large deviations and proved that the measure $\rho_t\,dx$ is the one minimizing an entropy under the constraints $\rho_{t_1}dx$ and $\rho_{t_2}dx$, along the original line of Schrödinger's justification of the product form (1.1) [12].

On the other hand, in an apparently completely unrelated context, the need to fix two probability densities at two different times and try to produce diffusion processes associated, somehow, to a given Hamiltonian (1.4) reappeared in the beginning of the eighties when K. Yasue initiated the variational approach to Nelson's "Stochastic Mechanics" [13-14]. The present review is not about Nelson's theory, and our Euclidean quantum perspective has a mathematical structure quite distinct: Stochastic Mechanics is a direct probabilistic approach to the Schrödinger equation in $L^2(\mathbb{R}^N)$

$$\begin{cases} i\,\dfrac{\partial\psi}{\partial\tau} = H\psi \\ \psi_{t_0} = \psi_0 \end{cases} \tag{1.7}$$

and the theory inspired by Schrödinger is an indirect approach to quantum dynamics (an "analogy", in Schrödinger's words) through two heat equations. Let us remind that it is traditional, in mathematical physics, to call "Euclidean" a method involving the heat equations instead of Eq (1.7), and we will follow this tradition.

Nevertheless, it is true that a special realization of EQM provides a version of Nelson's theory completely different from the original one (whose mathematical consistency was proved by E. Carlen [15]).

We shall come back on this aspect only in our final discussion but we should already say that a certain previous familiarity with Stochastic Mechanics was very useful for the unexpected discovery of EQM.

§2. Basic aspects of the construction, properties and examples

We begin by a summary of the basic elements necessary for the construction of Bernstein diffusions, following Jamison [5], for $N = 3$.

A Bernstein transition probability

$$P(s,x,t,B,u,z) \quad , \qquad t_1 \leq s \leq t \leq u \leq t_2 \tag{2.1}$$

for $x,z \in \mathbb{R}^3$, B a Borel set in $\mathbb{R}^3$ satisfies:

(i) $B \mapsto P(s,x,t,B,u,z)$ is a probability measure on the Borel sigma algebra of $\mathbb{R}^3$.

(ii) $(x,z) \mapsto P(s,x,t,B,u,z)$ is measurable.

(iii) For any pair of Borelian sets B_1, B_2, the following 3-points counterpart of the Chapman-Kolmogorov property holds:

$$\int_{B_2} P(s,x,t,B_1,u,\xi) \,.\, P(s,x,u,d\xi,v,z) = \int_{B_1} P(s,x,t,dy,v,z) \,.\, P(t,y,u,B_2,v,z)$$

for $t_1 \leq s \leq t \leq u \leq v \leq t_2$.

What corresponds, for the Bernstein processes, to the usual construction of Markov processes from the data of an initial distribution and of a Markov transition probability is the following:

Theorem [5]

Given a Bernstein transition function $P(s,x,t,B,u,z)$ and m a probability measure on $\mathbb{R}^3 \times \mathbb{R}^3$, there is an unique stochastic process z_t such that, if $\mathcal{P}_t$ (respectively $\mathcal{F}_t$) denotes the increasing (decreasing) filtration for z_t, t in I,

(1) $E[f(z_t) \mid \mathcal{P}_s \cup \mathcal{F}_u] = E[f(z_t) \mid z_s, z_u]$ for any $s<t<u$ in I, f bounded measurable, and $E[\ldots \mid \mathcal{A}]$ denotes a conditional expectation, given the condition $\mathcal{A}$.

(2) $P(z_{t_1} \in B_1, z_{t_2} \in B_2) = m(B_1 \times B_2)$ for any Borelian B_1, B_2.

(3) $P(z_t \in B \mid z_s, z_u) = P(s, z_s, t, B, u, z_u)$ for any $t_1 \leq s \leq t \leq u \leq t_2$ and Borelian B.

(4) Choose $I = [s,u]$. Then if x_0, t_0 denotes the starting space-time point x,s,

$$P(z_s \in B_s, z_{t_1} \in B_1, \ldots, z_{t_n} \in B_n, z_u \in B_u) =$$

$$\int \ldots \int_{B_s \times B_1 \times \ldots \times B_n \times B_u} dm(x,z) \prod_{i=0}^{n-1} P(t_i, x_i, t_{i+1}, dx_{i+1}, u,z)$$

The proof of Th 3.2 is a "3-points analogue" of Kolmogorov classical construction [16]. An interesting novelty with respect to this Markovian construction is that there are two possible ways to take into consideration the third, or extra point involved in the definition of the Bernstein transition function. The first one, chosen in the formulation (4), is to fix the final position $z_u = z$ and to recognize, in the properties of the Bernstein transition probabilities defined before $P(s,x,t,B,u,z) \equiv Q_z(s,x,t,B)$, the usual properties of a Markovian transition probability [16]. It one defines an initial probability $P_s(dx)$ via the desintegration of the given joint probability m of (2), it follows from the classical theory that a Markovian process z^z exists with this transition and initial distribution. Using again the measure m to define, this time, the final distribution of the process z_u to be constructed, the existence of P, as given by (4), follows. But, and this is the whole point, we could as well have chosen to fix the initial position $z_s = x$ in the Bernstein transitions probabilities and realize that, as a function of t,B,u and z, this Bernstein transition is a backward Markovian transition Q_*^x, a concept not too familiar to theoretical physicists, but clear consequence of the presence of two adjoint heat equations in Schrödinger's framework. In this formulation, P as given in (4) is constructed from Q_*^x and the final distribution $P_u(dz)$ obtained from the joint probability m. So the probabilistic construction resulting from Schrödinger's idea is, indeed, completely time symmetric, as the roles of the initial and final probability densities $P_s(dx)$ and $P_u(dz)$.

Notice that, up to this point, there is no relation whatsoever with quantum dynamics, namely with the given Hamiltonian operator (1.4). The construction is much too general for our needs. The connection with physics is established by the following observation:

Let $H = -\frac{\hbar^2}{2m}\Delta + V$, with a potential V such that the integral kernel $(e^{-tH/\hbar})(x,y) = h(s,x,t,y) = h(x,t\text{-}s,y)$ is jointly continuous and strictly positive (Cf Kato class [6]). Define for $s \leq t \leq u$ in I,

$$p(s,x,t,y,u,z) = \frac{h(s,x,t,y)\, h(t,y,u,z)}{h(s,x,u,z)} \tag{2.2}$$

It follows immediately from the semigroup property that p(s,x,t,dy,u,z) defines a Bernstein transition probability density.

Of course, there is nothing yet, in this general framework, to rule out processes which are only local Markov but not Markov. Jamison mention an explicit example, due to Slepian, of stationary Gaussian Bernstein process which is not Markov. [5]

This is due to the lack of prescription of the joint measure m in the abovementioned framework: many m, and therefore many processes, are compatible with prescribed marginal probabilities P_s and P_u. Only one of those will be Markovian and the associated joint measure is precisely the one whose marginals provide Eq (1.3) (on I = [-1,1]), i.e., the Schrödinger system for the boundary data of the two adjoint heat equations. This means that, whatever he realized it or not, Schrödinger ruled out the relevance of non Markov processes for a probabilistic version of quantum mechanics. Other opinions about this have, however, be defended [13, 1985] and could have serious implications in quantum field theory.

The substitution of Schrödinger's special joint probability in the finite dimensional distribution given by (4) in the Theorem provides, as wanted, Markov Bernstein process z_t, t in I, such that

$$P(z_t \in B) = \int_B \eta_t \eta_t^* dx \tag{2.3}$$

where $\hbar \frac{\partial \eta_t}{\partial t} = H\eta_t$ and $- \hbar \frac{\partial \eta_t^*}{\partial t} = H\eta_t^*$ (2.4)

and appropriate, positive, boundary conditions are given for Eqs (2.4).

The relation (2.3) is the Euclidean version of Born interpretation in quantum mechanics ([11]), and the reason of the terminology we use regarding the strategy founded on it. It is the probabilistic counterpart of the famous quantum mechanical axiom according to which

$$\int_B \overline{\psi}_\tau \psi_\tau \, dx = \int_B |\psi_\tau|^2 \, dx \tag{2.5}$$

for ψ_τ solving Eq (1.7) with the Hamiltonian (1.4), represents the probability, for the particle, to be in the Borelian B at time τ. Without (2.3) the probability involved, generally for technical reasons, in probabilistic counterparts of quantum dynamics have very little to do with quantum probabilities (Cf also the conclusion).

As soon as (2.3) holds, it is clear that we are describing conservative phenomena. If B is the whole space, the total probability is conserved with time, as in quantum mechanics, and for the same reason.

It also follows from (2.3) and (2.4) that the probabilistic counterpart of the complex conjugation in L^2 is a relation of time reversal, under which η_t becomes η_t^* and reciprocally. This remark will be elaborated soon.

It should also be stressed that to each potential (or Hamiltonian) is associated in this way an infinite family of measures, indexed by boundary conditions. In the proper context of quantum physics, the relation between initial condition of Eq (1.7) and Bernstein measures will be one to one. But, without extra specifications on boundary conditions, there are much more Bernstein measures than measures "coming from quantum mechanics". Afterwards, Bernstein process (or measures) will refer to any Markovian process constructed as before and fulfilling (2.3), coming or not from quantum mechanics.

One shows easily (Cf [11], [17]) that, regarded as an usual (forward) Markov process, such a process is characterized by the following transition probability, for $s \leq t$, and B a Borelian

$$Q(s,x,t,B) = \frac{1}{\eta_s(x)} \int_B h(s,x,t,y)\ \eta_t(y)\ dy \tag{2.6}$$

and, symmetrically, as a backward Markov process, by the backward transition probability

$$Q^*(s,B,t,y) = \frac{1}{\eta_t^*(y)} \int_B \eta_s^*(x)\ h(s,x,t,y)\ dx \tag{2.6*}$$

Let us consider now an Hamiltonian slightly more general than (1.4), corresponding to the Euclidean counterpart of the physical one for a unit mass and charged particle in an electromagnetic field, namely

$$H = \frac{\hbar^2}{2} [\nabla - \frac{i\,a}{\hbar}]^2 + V \tag{2.7}$$

where $a : \mathbb{R}^3 \to \mathbb{R}^3$ is the vector potential. This Euclidean counterpart is (since $a \to i\,A$),

$$H_A = - \frac{\hbar^2}{2} (\nabla - \frac{A}{\hbar})^2 + V \tag{2.8}$$

It is not self-adjoint. As a matter of fact, its adjoint is

$$H_A^* = - \frac{\hbar^2}{2} (\nabla + \frac{A}{\hbar})^2 + V \tag{2.8*}$$

Nevertheless, nothing in the structure of EQM depends on this. Using (2.6) and (2.6)*, one computes easily the drifts and diffusion coefficients of the family of Bernstein diffusions associated with the Hamiltonian H_A. It follows that those processes solve simultaneously the forward and backward Itô's stochastic differential equations [27], t in I,

$$dz_t = \hbar^{1/2}\, dw_t + \hbar\, \frac{\nabla\eta_t}{\eta_t}(z_t)\, dt - A(z_t)\, dt \tag{2.9}$$

$$d_*z_t = \hbar^{1/2}\, d_*w_{*t} - \hbar\, \frac{\nabla\eta_t^*}{\eta_t^*}(z_t)\, dt - A(z_t)\, dt \tag{2.9*}$$

where W_{*t} denotes a Brownian motion with respect to the decreasing filtration $\mathcal{F}_t$ associated with the second (i.e. the usual) heat equation in (2.4). Also W_t is the Brownian motion familiar to theoretical physicists in the context of statistical mechanics, and associated with an increasing filtration $\mathcal{P}_t$ [16]. Afterwards, $\mathcal{F}_t$ and $\mathcal{P}_t$ will be referred to as the future and the past, respectively, for obvious reasons. Their completely symmetric use in EQM is just the modern probabilistic translation of Schrödinger's Euclidean strategy.

Before mentioning some explicit examples, let us observe that, when the Bernstein measure will come from quantum mechanics the relation with physics will follow from an analytical continuation in time of the collection of Schwinger functions

$$S_n(t_1,\ldots, t_n) = E[z_{t_1}\cdots z_{t_n}] \tag{2.10}$$

$n = 1,2,\ldots$

to the quantum mechanical analogous of these moments, namely

$$\langle Q(\tau_1) \ldots Q(\tau_n)\rangle_{L^2} \tag{2.11}$$

where $Q(\tau)$ denotes the evolution of the position observable under the Heisenberg automorphism of operator on $L^2(\mathbb{R}^3)$ and $\langle . \rangle_{L^2}$ is the expectation in this Hilbert space. Given the particular nature of Bernstein measures, one should be slightly more careful in this "Wick rotation" than in the usual one, Cf [17]. Cf also (43).

In addition one should stress that, as it is clear from the formulation of Schrödinger's starting problem, the method should hold for measures completely different from the diffusion ones associated with Hamiltonians like (1.4), as long as one is able to guess what is the right Euclidean version (in our sense) of the original quantum mechanical problem. For the case of jump processes, Cf [18].

Still in the special context of quantum mechanics, it is interesting to discuss the Hilbert space counterpart, for Schrödinger's Euclidean formulation, of the one in regular quantum mechanics. This has been done in [17]. Cf also [43].

Now let us consider a few examples in one dimension, for simplicity.

1) The first one is due to Schrödinger [1]. Let us choose $I = [s,u]$, $\hbar = m = 1$ in Eq (1.4) and $V = 0$. Let us consider the problem of a free evolution associated with the Schrödinger's system (1.3) on I, with the boundary probability densities

$$\rho_s(y) = \delta(y\text{-}x) \qquad \text{and} \qquad \rho_u(y) = h_0(s,x,u,y)$$

where h_0 denotes the integral kernel of the free heat equation. In other words, ρ_u is just the distribution following from the solution of Eq (1.2)* with the initial condition $\rho_s(y)$. The fact that $\rho_s(y)$ is singular is irrelevant to Schrödinger's argument. Then it is simple to see that the associated pair of solutions of the two free heat equations are just, for any t in I,

$$\eta_t(y) = 1 \qquad \text{and} \qquad \eta_t^*(y) = h_0(s,x,t,y)$$

since the Schrödinger's system is trivial. Using this in Eqs (2.9) and (2.9)* (here $A = 0$) and denoting respectively the forward (backward) drift by $B(B^*)$ we obtain

$$B(y,t) = 0 \qquad \text{and} \qquad B^*(y,t) = \frac{y\text{-}x}{t\text{-}s} \tag{2.12}$$

The resulting process is a familiar Brownian motion starting from x at time s. There is nothing quantum about it, and its variance displays clearly the irreversible nature of its evolution. But, as observed by Schrödinger, we could as well have chosen the symmetric situation where

$$\rho_s(y) = h_0(s,y,u,z) \quad \text{and} \quad \rho_u(y) = \delta(y\text{-}z)$$

corresponding to another trivial Schrödinger's system whose solution produces the pair

$$\eta_t(y) = h_0(t,y,u,z) \quad \text{and} \quad \eta_t^*(y) = 1$$

and therefore a Bernstein diffusion such that

$$B(y,t) = \frac{z\text{-}y}{u\text{-}t} \qquad \text{and} \qquad B^*(y,t) = 0 \tag{2.13}$$

This is a Brownian motion conditioned to end in z at time u. As the first one, it is of little direct relevance for quantum dynamics. Notice that the chosen boundary condition for these two examples are in no way time symmetric.

2) Let us consider the harmonic oscillator Hamiltonian $H = -\frac{1}{2}\Delta + \frac{\omega^2}{2}x^2$ (2.14)
The integral kernel of $e^{-(t-s)H}$ is given by the familiar Mehler formula:

$$h(s,x,t,y) = \left[\frac{\omega}{2\pi \sinh \omega(t-s)}\right]^{1/2} \times \exp\left\{-\frac{\omega}{2\sinh \omega(t-s)}\left[(x^2+y^2)\cosh \omega(t-s) - 2xy\right]\right\}$$

2.1) Consider the stationary Ornstein-Uhlembeck process with invariant measure

$$d\mu(x) = \sqrt{\tfrac{\omega}{\pi}}\, e^{-\omega x^2}\, dx \tag{2.15}$$

This corresponds to the following solutions of the Schrödinger system (1.3) on, say $I = [-1,1]$,

$$\eta^*_{-1}(x) = \left(\frac{\omega}{2\pi}\right)^{1/4} e^{-\frac{\omega}{2}x^2 + \frac{\omega}{2}}$$

and

$$\eta_1(z) = \left(\frac{\omega}{2\pi}\right)^{1/4} e^{-\frac{\omega}{2}x^2 + \frac{\omega}{2}}$$

whose forward and backward evolutions under the associated integral kernel h provide

$$\eta^*_t(y) = \left(\frac{\omega}{2\pi}\right)^{1/4} e^{-\frac{\omega}{2}y^2 - \frac{\omega}{2}t}$$

and (2.16)

$$\eta_t(y) = \left(\frac{\omega}{2\pi}\right)^{1/4} e^{-\frac{\omega}{2}y^2 + \frac{\omega}{2}t}$$

and therefore, from (2.9) and (2.9)*, the $\mathcal{P}_t$ and $\mathcal{F}_t$ stochastic differential equations

$$dz_t = dw_t - \omega z_t\, dt$$

(2.17)

$$d_* z_t = d_* w_{*t} + \omega z_t\, dt$$

Notice that the choice of I is completely arbitrary. Actually, the resulting process z_t can be extended arbitrarily far (in the past and in the future) outside its initially chosen interval of definition I. In general, as we shall see, it is not so for typical Bernstein measures. Also notice that the process is, here, time symmetric in a trivial sense.

2.2) Consider the Ornstein-Uhlembeck process starting from x_0 at time 0. Pick $I = [0,1]$. Then one verifies easily that the solution of the associated Schrödinger's system provides

$$\eta_t^*(y) = e^{-\frac{1}{2}(\text{cotg } \omega t)\, y^2 + \frac{\omega}{\sinh \omega t} x.y + \int_0^t [-\frac{\omega}{2} \text{cotg } \omega s + \frac{\omega^2}{2} \frac{x_0^2}{\sinh \omega s}]\, ds}$$
and
$$\eta_t(y) = e^{-\frac{\omega}{2} y^2 + \frac{\omega}{2} t} \tag{2.18}$$

and therefore, from (2.9) and (2.9)*, [19]

$$\begin{aligned} dz_t &= dw_t - \omega z_t\, dt \\ d_* z_t &= d_* w_{*t} + [\omega(\text{cotg } \omega t)\, z_t - \frac{\omega x_0}{\sinh \omega t}]\, dt \end{aligned} \tag{2.19}$$

This case is quite familiar and manifest a clear time irreversibility, as expected from such a strongly ergodic process. Notice that this one can only be extended towards the future. It is also of little relevance for quantum dynamics.

3) Since Euclidean quantum mechanics is time symmetric, as quantum mechanics itself, one could believe, on the basis of (2.1), for example, that it is always possible to consider exclusively pairs of identical probabilities in the Schrödinger's system. But we want, in this physical context, to preserve a one-to-one relation between solutions of the Schrödinger's equation (1.7) and Bernstein measures, and this would be much too restrictive:
Let us come back to a free case like in 1), with $I_0 = [-T_0/2, T_0/2]$ and choose two Gaussian boundary conditions

$$\begin{aligned} \rho_{-T_0/2} &= \mathcal{N}(-\tfrac{3}{2} V_0 T_0, \frac{a^2 - 9T_0^2/4}{2a}) \\ \rho_{T_0/2} &= \mathcal{N}(-\tfrac{1}{2} V_0 T_0, \frac{a^2 - T_0^2/4}{2a}) \end{aligned} \tag{2.20}$$

Where we denote by $\mathcal{N}(m, \sigma^2)$ the normal distribution of expectation m and variance σ^2, and V_0, a, are numbers.
It is not very illuminating to write down the two linear, time dependent stochastic differential equations resulting from (2.9) and (2.9)* for this free evolution. But the probability measure of the associated Bernstein diffusion is $\mathcal{N}(V_0(t-T_0), \frac{a^2 - (t-T_0)^2}{2a})$. And although it looks quite time asymmetric on I_0, it comes indeed from quantum mechanics. Actually, on $I =]T_0-a, T_0+a[$ the dynamics is time symmetric. The process cannot be extended beyond $|t-T_0| = a$. After analytical continuation in the time parameter this corresponds to a Gaussian solution of the free Schrödinger's equation, defined for any time but, of course, without well defined probability

measure.

§3. The meaning of the reversibility in time

As it should be clear from what has been said before, the idea of time reversibility is central in this strategy. But it takes various forms and is considerably more involved that its counterpart in classical mechanics.

S. Bernstein, who was the first probabilist to think about Schrödinger's idea (Cf. [2]) and to suggest a rigorous approach to construct them, regarded the data of two probabilities at two different times as a probabilistic generalization of the trajectories of classical mechanics from the point of view of the classical calculus of variation (this was so suggestive for us that we used the form of "variational process" before to stick to the one of "Bernstein processes" [11.1-2]). (Nowadays, some mathematicians refer to the resulting diffusions as "Schrödinger processes. Cf also [42]).

This provides us with a first way to understand the time reversibility involved here and essential for quantum dynamics. Many others will follow. Notice that, as observed after (2.3 - 2.5), if we consider a Bernstein process on $I=[-1,1]$, we have

$$\frac{d}{dt}\int_{\mathbb{R}^3} \eta_t\, \eta_t^*\, dx = 0 \tag{3.1}$$

where $\eta_t = e^{(t-1)H/\hbar}\,\eta_1$ and $\eta_t^* = e^{-(t+1)H}\,\eta_{-1}^*$.

In other words we have conservation of probability for the same reason as in quantum mechanics, namely the self-adjointness of the Hamiltonian H. Actually, this would remain true for the same η_t and $\eta_t^* = e^{-(t+1)H^+}\,\eta_{-1}^*$ with H^+ the adjoint of H. We have used this in the general case (2.8), (2.8)*.

Formally, Eqs (2.4) show that η_t and η_t^* are solutions of time-reversed problems.

Now, in the conditions of validity of Beurling's Theorem (Cf. (1.3)) let us permute the given boundary probabilities on I:

$$\begin{aligned}\hat{\rho}_{-1}(x) &= \rho_1(x)\\ \hat{\rho}_1(z) &= \rho_{-1}(z)\end{aligned} \tag{3.2}$$

Then we obtain another solution, and therefore another Bernstein diffusion on the same interval $I=[-1,1]$.

It is easy to check that the resulting diffusion is $\hat{z}_t = z_{-t}$. It is associated with the same dynamics (since the integral kernel h(x,2,z) used in Eq (1.3) is unchanged). This aspect is the probabilistic counterpart of the familiar fact in classical mechanics that, for conservative systems, a trajectory with reversed boundary data is another solution of the equation of motion. As we shall see in §5, a more detailed dynamical meaning can be attached to this. It will be a mathematical justification of the abovementioned Bernstein interpretation.

The process $\hat{z}_t$ will not be used in our construction, because we shall consider, like nature itself, alas, only one direction of time. Instead, we may use $\hat{z}_t$ with the usual (forward) tools of the theory of processes in order to derive the backward properties of z_t.

For example, let us consider the following regularized derivative (introduced, in the quantum context, by E. Nelson [13], but probabilists were familiar with it before):

$$Dz_t = \lim_{\Delta t \downarrow 0} E_t[\frac{z_{t+\Delta t} - z_t}{\Delta t}] \tag{3.3}$$

Then,

$$D\hat{z}_t = -D_* z_{-t}$$

when we have defined another random variable by

$$D_* z_t = \lim_{\Delta t \downarrow 0} E_t[\frac{z_t - z_{t-\Delta t}}{\Delta t}] \tag{3.4}$$

Notice that, in the first case, the conditioning was in the past and in (3.4) it is in the future.

It is immediate to check that Dz_t and $D_* z_t$ coincide respectively with the forward and backward drifts denoted by $B(z_t, t)$ and $B^*(z_t,t)$ in the examples of §2. The fact that Dz_t and $D_* z_t$ are distinct is due to the Brownian nature of the trajectory. This aspect was already present, formally, in R. Feynman's path integral approach [8-9] to quantum mechanics and will be crucial for the probabilistic representation of the commutation relation between position and momentum (Cf. §9).

We shall mention, afterwards, many consequences of this reversibility in time. Let us, for the moment, point out a promising one.

The time-reversibility of the Bernstein processes gives rise to a possible generalization of the concept of Dirichlet form, which is still under investigation. A Symmetric Dirichlet form $\mathcal{E}$ can be given by a positive definite self-adjoint operator together with a measure ρdx by $\mathcal{E}(u,v) = \int (A\, u,v)\, \rho dx$. In a similar way, we associate to a (self-adjoint) Hamiltonian H and a family of measures indexed by time $\rho_t dx = \eta_t \eta_t^* dx$ a family of Dirichlet forms:

$$\mathcal{E}_t(u,v) = \int \left(\frac{\partial}{\partial t} \log \eta_t,\ u - \frac{1}{\eta_t} H(u\eta_t)\right) v\ \rho_t dx$$

The decomposition of the density ρ_t in a product of two solutions of adjoint heat equations is of course not unique but one can show that if $\rho_t = \eta_t \eta_t^*$ is a decomposition associated to an Hamiltonian $H = -\frac{1}{2}\ \Delta + V$ and $\rho_t = \tilde{\eta}_t \tilde{\eta}_t^*$ is associated to another one $\tilde{H}$, then $\tilde{H}$ is of the form $-\frac{1}{2}(\nabla - \nabla F)^2 + (V + \frac{\partial}{\partial t} F)$ for a certain function F. This corresponds to what is called a "Gauge transformation" in physics. One can also prove that there exists only one decomposition which is associated to a self-adjoint Hamiltonian.

When ϕ is a eigenfunction for H, $H\phi = \lambda\phi$, then the solutions $\eta_t = e^{\lambda t}\phi$ and $\eta_t^* = e^{-\lambda t}\phi$ correspond to the (particular) case of a usual Dirichlet form. The associated processes are then stationary. The uniqueness decomposition result we referred to corresponds to the uniqueness of a symmetric process associated to the Dirichlet form.

§4. The Gaussian case

It is well known that when the forces are quadratic, in quantum mechanics, and the initial conditions of Eq (1.7) are Gaussian, the quantum expectations of the position and momentum $\langle Q \rangle_\psi$ and $\langle P \rangle_\psi$ move according to classical laws of motion. In the context of Feynman path integral [9] this means that the formal probability "measures" are constructed entirely in terms of classical (smooth) trajectories. The same is true in EQM.

For our needs, it is sufficient to consider the following time dependent quadratic Hamiltonian ($\hbar = 1$)

$$H = -\frac{1}{2}\Delta + \frac{1}{2} \langle a(t)x,x \rangle \tag{4.1}$$

with a(t) a symmetric positive square matrix whose dependence on t is smooth [19].

Since the relevant η_t in Eq (2.4) is of the form

$$\eta_t(x) = \exp\left(\frac{1}{2} \langle \alpha_t x, x \rangle + \langle \beta_t, x \rangle + \gamma_t\right)$$

it follows from Eq (2.9) that the forward drift is

$$\frac{\nabla\eta_t}{\eta_t}(x) = \alpha_t\, x + \beta_t \tag{4.2}$$

Moreover the coefficients α,β and γ solve the following system of equations (^{T} denotes the transposition)

$$\begin{aligned} &\dot{\alpha}_t + \alpha_t^T \alpha_t = a(t) \\ &\dot{\beta}_t + \alpha_t^T \beta_t = 0 \\ &\dot{\gamma}_t + \frac{1}{2}\,\text{Trace}\ \alpha_t + \frac{1}{2}\,|\beta_t|^2 = 0 \end{aligned} \tag{4.3}$$

The change of variable $\alpha_t = \dot{F}_t F_t^{-1}$ (for F non singular) linearizes the first ("Riccati's") equation and turn it into the Jacobi matrix equation:

$$\ddot{F}_t = a(t)\ F_t \tag{4.4}$$

The same argument with respect to the solution η_t^* of the other heat equation results into the following

Theorem [19, 20]

Any Gaussian Bernstein diffusion z_t, $t \in [-1,1]$, associated with the Hamiltonian (4.1) and with finite moments has an expectation m(t) and covariance K(t) given by

$$m(t) = F_t\, w^{-1}\, \delta^* + F_t^*\, w^{-1} \delta \tag{4.5}$$

$$K(t) = F_t^*\, w^{-1}\, F_t^T \tag{4.6}$$

where F_t and F_t^* are two nonsingular linearly independent matrix solutions of Eq (4.4), δ and δ_* are constant vectors and w is the Wronskian

$$w(F_t, F_t^*) = F_t^T\, \dot{F}_t^* - \dot{F}_t^T\, F_t^* \tag{4.7}$$

The pair of resulting stochastic differential equations is:

$$\begin{aligned} &dz_t = dw_t + (\dot{F}_t F_t^{-1}\, z_t + (F_t^{-1})^T \delta)\ dt \\ &d_* z_t = d_* w_{*t} + (\dot{F}_t^*\, F_t^{*-1}\, z_t - (F_t^{*-1})^T \delta^*)\ dt \end{aligned} \tag{4.8}$$

Here is a series of remarks regarding this Gaussian case, in one dimension (for simplicity of the notations):

α) The role of the boundary conditions of F_t, F_t^* is interesting and should be mentioned (Cf [21]).

For the needs of the semiclassical limit, it is clear that $a(t) = \nabla^2 V(z^0(t))$, where $z^0(t)$ is a classical trajectory, stationary point of the first variation functional, is sufficient.

The relevant boundary conditions manifold associated with the Jacobi equation (4.4) i.e.

$$- \frac{d^2}{dt^2} q(t) + \nabla^2 V(z^0(t))\, q(t) = 0 \tag{4.4$'$}$$

is of the form

$$\begin{cases} \dot{F}^*_{-1}\, q(-1) - F^*_{-1}\, \dot{q}(-1) = \delta^* \\ F_1\, \dot{q}(1) - \dot{F}_1\, q(1) = \delta \end{cases} \tag{4.9}$$

where $\begin{pmatrix} F^*_{-1} \\ \dot{F}^*_{-1} \end{pmatrix}$ is initial condition of (4.4)$'$, with F_t^* the corresponding solution, and $\begin{pmatrix} F_1 \\ \dot{F}_1 \end{pmatrix}$ is final boundary condition for the same Jacobi equation, with F_t the associated solution. When the Gaussian measure "comes from Quantum Mechanics", one shows that the manifold (4.9) is itself invariant under time reversal, i.e.

$$F \longleftrightarrow F^* \quad \text{and} \quad \delta \longleftrightarrow \delta^* \tag{4.10}$$

Since the Wronskian w is itself invariant under time reversal in this sense, the expressions (4.5) and (4.6) show that such Gaussian Bernstein measures are themselves invariant under time reversal although, in general, time dependent.

It is simple to reinterpret the Gaussian diffusions of §2 along this line.

For example, the elementary illustration (2.12) considered by Schrödinger corresponds to the boundary manifold on [s,u]

$$\begin{cases} q(s) = x \\ \dot{q}(u) = 0 \end{cases} \tag{4.10}$$

with $F_t = 1$ and $F^*_t = t\text{-}s$. Since (4.10) is not invariant under time reversal, this measure does not come from quantum mechanics.

In contrast, the last example of §2 corresponds to a boundary manifold invariant under time reversal since $F_t = a - (t-T_0)$, $F^*_t = a + (t-T_0)$ which are time reversed of each other on a time interval centered in T_0. Also notice that $\delta = V_0 a$ and $\delta_* = -\delta$ expresses the fact that the initial velocity V_0 changes its sign under time reversal. So the resulting Bernstein measure indeed comes from quantum mechanics.

The stationary Ornstein-Uhlenbeck process of Ex 2.1) is associated with the boundary manifold on [-T,T]:

$$\begin{cases} \omega\, e^{\omega(T)} q(-T) - e^{\omega(-T)}\, \dot{q}(-T) = 0 \\ e^{-\omega T}\, \dot{q}(T) + \omega e^{-\omega T}\, q(T) = 0 \end{cases} \tag{4.11}$$

It is invariant under time reversal. This corresponds to the familiar ground state of the harmonic oscillator. If all the measures coming from quantum mechanics were of this kind, EQM would not be of any interest.

β) F and F* being linearly independent solutions of the same Sturm-Liouville problem, they are related by

$$F^*_t = F_t\, [F^{-1}_s\, F^*_s + \int_s^t (F^{-1}_\tau)^2\, d\tau .\, w] \qquad t \in I \tag{4.12}$$

This relation is important for proving many properties of Gaussian Bernstein diffusions. For example, it is not a priori clear why K(t) of Eq (4.6) makes sense as a covariance. Since a(t) is positive, a solution of the Jacobi equations (4.4) cannot have more than one zero on [-1,1] (Sturm Theorem). If F, F* are nonzero on]-1,1[, w can be chosen positive. From Eq (4.12), $\frac{d}{dt}(F^*_t\, F^{-1}_t) = wF^{-2}_t$, so that K(t) is positive.

For α), β) and many other aspects of Gaussian Bernstein diffusions, cf [22].

γ) Of course, correlations are of the form

$$E(z_s - Ez_s)\,(z_t - Ez_t) = \begin{cases} F^*_s w^{-1} F_t & s \leq t \\ F_s\, w^{-1} F^*_t & t \leq s \end{cases} \tag{4.13}$$

δ) As in Feynman approach, there are path integrals associated naturally to

EQM (cf §6). The semiclassical approach will amount to restrict ourselves to the two first terms in the (rigorous) $\hbar$-expansion of arbitrary Bernstein diffusion

$$z_t^{\hbar} = z_t^0 + \sqrt{\hbar}\, z_t^{sc} + \hbar\, z_t^{(2)} + \dots \tag{4.14}$$

Up to the order $\hbar$ the resulting diffusions are Gaussian ones and the result of this section apply to them. Cf also [20].

ε) Most of the results of this section can be generalized in infinite dimensions. See, for example, [19], [23].

§5. Dynamical characterization

Using Itô's formula, the mean forward and backward derivatives introduced in Eqs (3.3) and (3.4) are immediately generalized to any regular function f of a Bernstein diffusion z_t.

In the rather general context of Eqs (2.9) and (2.9)*, since

$$Dz_t = \hbar \frac{\nabla \eta_t}{\eta_t} - A \tag{5.1}$$

$$D_* z_t = \hbar \frac{\nabla \eta_t^*}{\eta_t^*} - A \tag{5.2}$$

it follows from a simple computation that

$$DDz_t = \nabla V(z_t) + Dz_t \wedge \text{rot } A + \frac{\hbar}{2} \text{rot rot } A \tag{5.3}$$

and

$$D_* D_* z_t = \nabla V(z_t) + D_* z_t \wedge \text{rot } A - \frac{\hbar}{2} \text{rot rot } A \tag{5.4}$$

These are remarkable identities. If the trajectory $t \to z_t$ were smooth (i.e., for $\hbar = 0$ in (2.9)), they would both coincide with the equation of motion of a classical charged particle in an electromagnetic field, except for a change of sign in the r.h.s., usual in such Euclidean context. This is precisely the physics described by the given Hamiltonian (2.7).

This suggests to interpret Eqs (5.3) and (5.4) as a probabilistic version of Feynman's "quantization" of the classical Euler-Lagrange equations (Cf, for example, [9], Eq (7.42) when $A = 0$). We will show, in the section on functional calculus that such an interpretation is, indeed, correct.

Let us observe that Eqs. (5.3 - 5.4) although reminiscent of the one discovered by Nelson in Stochastic Mechanics ([13]) are essentially distincts in one respect. Each of them involves exclusively the information contained in one single filtration or, equivalently, in one of the two heat equations used in the construction (Cf (2.4)). This will implies that EQM has a dynamical structure considerably simpler than Nelson's theory (Cf also the conclusion) and preserves a strong analogy with Feynman's formal path integrals approach, in contrast with stochastic mechanics. On the other hand, there are indeed indubitable similarities between Nelson theory and the Euclidean framework inspired by Schrödinger.

For example, notice that neither Eq (5.3) nor (5.4) has a well defined character under time reversal. But, since they both holds, it is clear that

$$\frac{1}{2}(\mathrm{D}\mathrm{D}z_t + \mathrm{D}_*\mathrm{D}_*z_t) = \nabla V(z_t) + \frac{1}{2}(\mathrm{D}z_t + \mathrm{D}_*z_t) \wedge \text{rot } A \tag{5.4$'$}$$

is also time. Now the right hand side reduces to a regularized version of the underlying classical ("Lorentz") force (except for the sign, of Euclidean origin), with the right behaviour under time reversal. The fact that the left hand side defines a regularized acceleration as reasonable as the one introduced by Nelson in [13 (1967)] was observed by the same author [13 p. 100]. But the argument used there to prefer the other acceleration was valid exclusively in real time. The proof of existence of the Euclidean framework suggested by Schrödinger had to wait until 1985-86 [11]. Regarding Eq (5.4)′, Thieullen has obtained a weak form of this equation for a class of Bernstein but not Markovian processes [39]. For other aspects of such regularized laws of motion, cf [41].

The validity of Eqs (5.3) and (5.4) suggests the following probabilistic generalization of the concept of classical reversible system [24]:

Let us consider a "regularized" (a terminology justified by the quantum motivation) second order system

$$\mathrm{D}\mathrm{D}z = F(z) \tag{5.5}$$

We will say that the system is reversible when it has the property that if z_t is a solution then $\hat{z}_t = z_{-t}$ solves

$$\mathrm{D}_*\mathrm{D}_*\hat{z} = F(\hat{z}) \tag{5.6}$$

When $t \to z_t$ is smooth, the two operators D and D_* reduce to the ordinary derivative and this property reduces to the usual one for classical reversible systems (Cf [24]). Notice that, considered as a first order regularized system Eq (5.5) requires the introduction of the time reversal operation

$$T\begin{pmatrix} z \\ Dz \end{pmatrix} = \begin{pmatrix} z \\ -D_* z \end{pmatrix} \tag{5.7}$$

which also reduces to the classical one when the trajectories are smooth.

If we take into account the familiar classical fact that, under time reversal the magnetic field $B \equiv \text{rot } A$ changes its sign [25] we conclude from Eqs (5.3)-(5.4) that, as it should be according to Schrödinger's strategy, these two equations are just time reversed of each other and constitute the (Euclidean) quantization of the classical Lorentz law of motion.

§6. Path integrals representations

According to Feynman, the natural approach to classical mechanics, in the perspective of the quantization, is the Lagrangian one [8-9]. The same is true in EQM, as we are going to see, and this is why our framework can be regarded as a new Euclidean counterpart of Feynman path integrals method.

Let us define, for the same physical system as before, the following (forward) regularized Lagrangian [26]

$$L(z,Dz,t) = \frac{1}{2}\,|Dz|^2 + V(z) + A.\,Dz + \frac{\hbar}{2}\,\nabla.A \tag{6.1}$$

It does not look very "classical" because of the presence of the $\hbar$ term in it. But it is much better if we introduce the following (forward) action functional

$$J[z] = E_s \int_s^1 \left\{ \frac{1}{2}|Dz_t|^2 + V(z_t) \right\} dt + E_s \int_s^1 A(z_t) \circ \, dz_t - \hbar\, E_s \log \eta_1 (z_1) \tag{6.2}$$

where $\circ$ denotes the Fisk-Stratonovich integral [27].

The necessity to use the Stratonovich integrals when a magnetic field is involved was anticipated by Feynman [8, 9]. Comparing (6.2) with (6.1) we notice that the $\hbar$-dependent term is just due to the use of Itô's calculus and therefore perfectly acceptable in a quantization procedure like this one.

Introduce the following nonlinear change of variable for η_t, the underlying solution of the heat equation used in Eq (2.9), with $A = 0$ for simplicity

$$I(x,t) = -\,\hbar \log \eta_t(x) \tag{6.3}$$

Then an elementary computation shows that I solves the following quantum generalization of the classical (Euclidean) Hamilton-Jacobi equation

$$\begin{cases} -\frac{\partial}{\partial t} I + \frac{1}{2}|\nabla I|^2 - \frac{\hbar}{2} \triangle I - V = 0, & t \leq 1 \\ I(y,1) = -\hbar \log \eta_1(y) \end{cases} \tag{6.4}$$

The change of variable (6.3) appeared for the first time, and with the same aim, in the historical paper where Schrödinger introduced his famous equation (1.7) [28]. It has, more recently, proved to be useful by specialists of controlled Markov process (Cf. "logarithmic transformation" in [29]).

Coming back to classical mechanics, we observe that the key relation between the action and the Lagrangian L will survived here, in a regularized way, namely

$$DI(z_t, t) = -L(z_t, Dz_t, t) \tag{6.5}$$

and from this follows easily, using Dynkin formula [30], under weak integrability conditions, that the solution η_t involved in (2.3)-(2.4) can be represented by

$$\eta_t(y) = \exp\left\{-\frac{1}{\hbar} E_t \int_t^1 L(z_s, Dz_s, s)\, ds + E_t \log \eta_1(z_1)\right\} \tag{6.6}$$

A completely symmetric construction involving the backward filtration provides

$$\eta_t^*(y) = \exp\left\{-\frac{1}{\hbar} E_t \int_{-1}^t L(z_s, D_* z_s, s)\, ds + E_t \log \eta_{-1}^*(z_{-1})\right\} \tag{6.7}$$

These two "path integrals" representations play, in EQM, a role analogous to the one played by the path integral representation of the wave function ψ in Feynman's approach [8-9] (and of its complex conjugate $\overline{\psi}$). Like Feynman's one, it involves not more than the classical Lagrangian of the underlying system (but evaluated in regularized velocities).

η_t and η_t^* are positive however, as expected by Schrödinger in [1]. Of course, this is only an analogy with quantum mechanics according to Feynman, but a very strong one, involving well defined probability measures. A natural question, now, is to ask if the formal functional calculus developed by Feynman [9] for the need of quantum dynamics can be carried out in our Euclidean framework. The answer is positive and is the object of our second chapter.

2. FUNCTIONAL CALCULUS

§7. Introduction

In the book [9] of Feynman and Hibbs (pg. 173) the following formula can be found:

$$<\delta F[\omega]\ (\delta\omega)>_S = -\frac{i}{\hbar} <F\delta S[\omega]\ (\delta\omega)>_S \tag{7.1}$$

Here $<\ .\ >_S$ denotes the integration with respect to the "measure" "$e^{i/\hbar S[\omega]}\mathfrak{D}\omega$", $S[\omega] = \int_S (\frac{1}{2}|\dot{\omega}|^2 + V(\omega))\, d\tau$ is the action functional.

According to the authors, most of the laws of Quantum Mechanics should follow from this integration by parts formula. For example, when applied to $F \equiv 1$ and $\delta\omega$ an arbitrary "direction", the formula becomes:

$$<\ddot{\omega}>_S = -<\nabla V(\omega)>_S \tag{7.2}$$

which is interpreted as the quantized laws of motion. Taking $F[\omega] = \omega(\tau)$ and an appropriate $\delta\omega$ we obtain

$$<\omega_\ell(\tau)\frac{(\omega(\tau)-\omega(\tau-\Delta\tau))_m}{\Delta\tau}>_S - <\frac{(\omega(\tau+\Delta\tau)-\omega(\tau))_m}{\Delta\tau}\omega_\ell(\tau)>_S = i\hbar\delta_{\ell m} \tag{7.3}$$

This is the space-time version of the Heisenberg commutation relations between position and momentum operators. Notice that the question of ordering of operators seems to be reinterpreted here as a question of conditioning in the past or in the future.

In [26] we have used Malliavin Calculus to give a mathematical meaning to the integration by parts formula (7.1) and its consequences in the framework of Euclidean Quantum Mechanics.

§8. Malliavin Calculus

Let $\mu = \mu^\hbar$ be the Wiener measure over the space of paths

$$\Omega_0 = C([0;1];\mathbb{R}^3) \cap \{\omega(0) = 0\}$$

characterized by:

$$E_\mu \exp i <\ell,\omega> = \exp\left(-\frac{\hbar}{2}\,||\ell||^2_{H^1}\right)$$

$\forall \ell \in \Omega_0^* \subset H^1$, where $H^1 = \{\phi \in \Omega_0:\ \dot{\phi}$ exists and $\int |\dot{\phi}|^2\, d\tau < +\infty\}$ is called the Cameron-Martin space. Then (Ω_0, H^1, μ) is known as the classical Wiener space.

In Malliavin calculus (cf. [36]) the derivative of a Wiener functional

$$E_{\rho\mu}(\nabla F \mid B)_{H^1} = E_{\rho\mu}(F\,[\delta B - (\nabla \log \rho \mid B)_{H^1}]) \tag{8.4}$$

§9. Functional calculus for the Bernstein processes

Let $H = -\frac{\hbar^2}{2}\Delta + V$ be our starting Hamiltonian observable. In order to write formula (8.4) for the Bernstein diffusions associated to H, we have to determine ρ, i.e., the law of the processes on the space of (continuous) paths. Together with the assumption on V already mentioned (belonging to the Kato class), we use here some extra conditions on the potential, namely that:

$$\sup_x E_x(\exp \alpha \int_0^1 |V|\,(\hbar^{1/2}\omega(t))\,dt) < +\infty \qquad \text{and}$$

$$\sup_x E_x \int_0^1 ||\nabla V||^p\,(\hbar^{1/2}\omega(t))\,dt < +\infty \tag{9.1}$$

for some $\alpha > 1$, $p > 1$. Nevertheless these assumptions could be weakened. For a regular choice of boundary conditions η^*_{-1} and η_1 for the two adjoint heat equations (bounded and positive, in particular) we can write Girsanov formula for the forward description of the Bernstein process z_t. We have:

$$\rho(\hbar^{1/2}\omega) = \exp\left(\int_0^1 \hbar^{1/2}\frac{\nabla\eta_t}{\eta_t}\,d\omega(t) - \frac{1}{2\hbar}\int_0^1 ||\frac{\nabla\eta_t}{\eta_t}||^2\,dt\right)$$

On the other hand, by Itô's formula [27]:

$$\log \eta_1(z_1) - \log \eta_0(z_0) = \int_0^1 \hbar^{1/2}\frac{\nabla\eta_t}{\eta_t}\,d\omega(t) - \hbar\int_0^1 \left(\frac{1}{2}\frac{\Delta\eta_t}{\eta_t} + \frac{1}{2}||\frac{\nabla\eta_t}{\eta_t}||^2 + \frac{1}{\hbar}\frac{\dot\eta}{\eta}\right)dt$$

From the two last equalities and the heat equation solved by η_t, namely $\dot\eta_t = -\hbar\Delta\eta_t + \frac{1}{\hbar}V\eta_t$, we derive the following expression for the density ρ of the process with respect to the Wiener measure $\mu = \mu^\hbar$ with initial distribution given by the law of z_0:

$$\rho(z) = \frac{\eta_1(z_1)}{\eta_0(z_0)} \exp -\frac{1}{\hbar}\int_0^1 V(z_t)\,dt \tag{9.2}$$

If we consider the backwards description of the process, then the density of its law w.r.t. the law of the $\mathcal{F}_t$ Wiener measure (corresponding to the Brownian motion w_* on [-1,0]) can be expressed by the following analogous (time reversed)

$F:\Omega_0 \to \mathfrak{G}$ (where $\mathfrak{G}$ is some abstract Hilbert space) in the direction of $\phi \in H^1$ is defined by:

$$\mathbb{D}_\phi F[\omega] = \lim_{\substack{\varepsilon \to 0 \\ \text{(a.e.)}}} \frac{1}{\varepsilon} (F[\omega+\varepsilon\phi] - F[\omega]) \tag{8.1}$$

The corresponding gradient is the linear operator $\nabla F[\omega] \in \mathcal{L}(H^1;\mathfrak{G})$ given by $\nabla F[\omega](\phi) = \mathbb{D}_\phi F[\omega]$.

Let $e_k(\tau) = \frac{\sin (k-1)\pi\tau}{(k-1)\pi} \cdot \sqrt{2}$, $k>1$, and $e_1(\tau) = \tau$, be an orthonormal basis of the space H^1. The scalar product $(\phi \mid e_k)_{H^1} = \int \dot{\phi}\, \dot{e}_k \, d\tau$ extends to Ω_0 and we denote this extension in the same way, namely $(\omega \mid e_k) = \int \dot{e}_k \, d\omega$ (i.e. the Itô's integral of $\dot{e}_k$). When ∇F is an Hilbert-Schmidt operator, i.e.,

$$||\nabla F||^2_{H.S.}[\cdot] = \sum_{k \geq 1} ||\nabla F[\cdot](e_k)||^2_{\mathfrak{G}} < +\infty,$$

one can define $\nabla^2 F$. Iterating this procedure, one defines derivatives of Wiener functionals of all orders as well as the corresponding Sobolev spaces:

$$W^p_r(\Omega_0,\mu) = \{F \in L^p_\mu(\Omega_0) \colon E_\mu \, ||\nabla^i F||^p_{H.S.} < +\infty \quad , \ \forall 1 \leq i \leq r\},$$

where $||\nabla^i F||^2_{H.S.}[\cdot] = \sum_{k_1,\ldots,k_i} ||\nabla^i F[\cdot](e_{k_1},\ldots,e_{k_i})||^2$.

The "divergence" is defined as the adjoint of the gradient operator in L^2_μ; namely, when $B:\Omega_0 \to H^1$ is in L^2_μ, its divergence, denoted here by δB, is the real-valued functional in L^2_μ, when it exists, satisfying, for all $F \in W^2_1(\Omega_0,\mu)$:

$$E_\mu(\nabla F \mid B)_{H^1} = E_\mu(F\delta B) \tag{8.2}$$

When $B[\omega]$ is $\mathcal{F}_t$-adapted, one can show by using Cameron-Martin-Girsanov's theorem that the divergence coincides with the Itô's integral, namely:

$$\delta B[\omega] = \frac{1}{\hbar} \int_0^1 \frac{d}{dt} B[\omega](t) \, d\omega(t) \tag{8.3}$$

If, instead of considering the Wiener measure, we consider a measure which has a density with respect to μ (this is a more natural choice in the quantum context) we obtain a generalization of the integration by parts formula (8.2). We have, for $\rho>0$, $\rho \in W^{1+\varepsilon}_1(\Omega_0)$, and $B \in W^{2+2/\varepsilon}_1(\Omega_0;H^1)$ the following equality, valid for every $F \in W^{2+2/\varepsilon}_1(\Omega_0)$:

formula:

$$\rho_*(z) = \frac{\eta^*_{-1}(z_{-1})}{\eta^*_0(z_0)} \exp - \frac{1}{\hbar} \int_{-1}^{0} V(z_t)\, dt \tag{9.3}$$

After verifying that ρ satisfies the conditions required for the Malliavin integration by parts formula (8.4) we compute $\nabla \log \rho$:

$$(\nabla \log\rho \mid \Phi)_1\, [z] = \frac{\nabla\eta_1}{\eta_1}(z_1)\ \Phi[z](1) - \frac{\nabla\eta_0}{\eta_0}(z_0)\Phi[z](0) - \frac{1}{\hbar} \int_0^1 \nabla V(z_t)\ \Phi[z](t)\, dt$$

Furthermore, if we consider a Wiener functional F which is $\mathcal{P}_s$-adapted, with $\Phi[z](t) = 0$ for $t \in [0,s]$, and using (8.3):

$$\begin{aligned} E(F\delta\phi) &= E(F \int_s^1 \frac{d}{dt}\ \Phi[z](t)\ \frac{1}{\hbar}\ dz(t)) \\ &= E(FE_s \int_s^1 \frac{d}{dt}\ \Phi[z](t)\ \frac{1}{\hbar}\ Dz(t)) \\ &= E(\frac{1}{\hbar}\ F(\Phi[z](1)Dz(1) - \Phi[z](s)Dz(s)) - \frac{1}{\hbar}\ F \int_s^1 \Phi[z](t)\ DDz(t)\ dt) \end{aligned}$$

Therefore the following theorem holds:

Theorem [26]:

Let V belong to the Kato class and satisfies the assumption (9.1). Let $s \in [0,1]$. For every $\mathcal{P}_t$-adapted functional $\Phi \in \bigcap_{p>1} W^p_1(\Omega_0;H^1)$ such that $\Phi[z](t) = 0$ for $t \in [0,s]$ and for every $F \in \bigcap_{p>1} W^p_1(\Omega_0)$ we have:

$$E(\nabla F \mid \Phi) = E(F \int_s^1 \Phi[z](t)\ \frac{1}{\hbar}\ (-DDz_t + \nabla V(z_t))\ dt) + \Phi[z](1)(\frac{1}{\hbar}\ Dz(1) - \frac{\nabla\eta_1}{\eta_1}(z_1))) \tag{9.4}$$

Remarks:

1) Starting from (9.3) we can as well obtain a corresponding "backward" theorem, i.e., an integration by parts formula w.r.t. the future filtration $\mathcal{F}_t$.

2) We interpret formula (9.4) as a regularized, rigorous version of Feynman's one (7.1). Indeed, it corresponds to the regularized notion of action functional as defined by (6.2) and the extra terms are just boundary conditions.

Let us now consider some applications of the theorem, along the line of those suggested by Feynman. Taking $F \equiv 1$ and Φ "arbitrary" we obtain:

$$E\int_s^1 \Phi[z](t)\,\tfrac{1}{\hbar}\,(-DDz_t + \nabla V(z_t))\,dt) + E(\Phi[z](1)(\tfrac{1}{\hbar}\,Dz(1) - \tfrac{\nabla\eta_1}{\eta_1}(z_1))) = 0$$

from which we deduce the forward laws of motion:
$DDz_t = \nabla V(z_t)$ a.e., with the boundary conditions z_0 given, and $Dz(1) = \hbar\,\frac{\nabla\eta_1}{\eta_1}\,(z_1)$. By a symmetric deduction we obtain as well the backwards laws of motion $D_*D_*z_t = \nabla V(z_t)$ a.e.

Notice that the laws of motion obtained here are valid almost everywhere and not only in the mean. So, they are strong (Euclidean) version of Feynman's Eq (7.2). It is also worthwhile to observe that one of the advantages of such variational characterization (besides its elegance; cf also §10) of the laws of motion lies in the fact that it provides the boundary conditions associated to them, in contrast with the characterization of §5.

As an example, let us take the free case ($V = 0$) of a Brownian motion w_t from a point x at time $t = -1$. Then $DDw_t = 0$ and $Dw_1 = 0$; on the other hand $D_*D_*w_t = 0$ and $D_*w_t = \frac{w(t)-x}{t+1}$. Notice that the backwards velocity is singular at the initial time $t = -1$, an expression of the physical fact that, since the position of the particle is exactly known at time -1, its momentum cannot be defined (Heisenberg principle).

Another application of the theorem is the derivation of the commutation relations. Let $F[z] = z(t)$, $t \in]0,1[$ and

$$\Phi[z](\tau) = \begin{cases} 0 & ,\ \tau < t - \Delta t \\ 1 + \frac{\tau - t}{\Delta t} & ,\ t - \Delta t \le \tau \le t \\ 1 & ,\ \tau > t \end{cases}$$

We obtain (cf. [26] for the details):

$$\hbar = E(z(t)\ (D_*z_t + \int_t^1 \nabla V(z(\tau))\,d\tau)) + \text{b.c.}$$

With the same F and

$$\Phi[z](\tau) = \begin{cases} 0 & ,\ \tau < t \\ \frac{\tau - t}{\Delta t} & ,\ t \le \tau \le t + \Delta t \\ 1 & ,\ \tau > t + \Delta t \end{cases}$$

we find $0 = E(z(t)\ (Dz(t) + \int_t^1 \nabla V(z(\tau))\,d\tau)) + \text{b.c.}$ From the two equalities we derive

$$E(z^i(t)D_*z^j(t) - z^i(t)\,Dz^j(t)) = \hbar\delta_{ij} \tag{9.5}$$

which is the rigorous version of Feynman's formula (7.3). For more, cf [44].

We also deduce the relation, quite familiar to probabilists:

$$\lim_{\Delta t \to 0} E\left(\frac{1}{\Delta t}(z_{t+\Delta t} - z_t)^2\right) = \hbar$$

which is the well defined counterpart of Feynman's requirement $(\overline{1.6})$.

§10. Variational principle

One can use Malliavin calculus to prove that, as suggested by Feynman in real time, Bernstein processes are "extremals" of the action functional introduced in paragraph 6. More precisely, we say that a process z is extremal for a Wiener functional F on the set of diffusions of type $d\hat{z}_t = \hbar^{1/2} dw_t + b(\hat{z}_t, t)\, dt$ having an absolutely continuous law w.r.t. $\mu^{\hbar}$ if, for every $\mathcal{P}_t$-adapted "variation" $\Phi: \Omega_0 \to H^1$ we have $(\nabla F | \Phi) = 0$ a.e. Then the following result holds:

Theorem [26]. A necessary and sufficient condition for a Bernstein diffusion z to be extremal of the action functional (6.2) (when $A = 0$) on the set of diffusions $\hat{z}$ with diffusion matrix $\hbar\, \mathbb{1}$, having an absolutely continuous law w.r.t. $\mu^{\hbar}$ and such that $\hat{z}_0$ is given, and $D\hat{z}(1) = \hbar\, \frac{\nabla \eta_1}{\eta_1}(\hat{z}_1)$, is that the following Euclidean law of motion holds for $t \in [0,1]$:

$$DDz(t) = \nabla V(z_t) \qquad \text{a.e.}$$

A symmetric result can be derived for the backwards description of the Bernstein diffusion and with respect to the time reversed action functional

$$\mathcal{J}_*[z] = E_s \int_{-1}^{s} \left(\tfrac{1}{2}\, |D_* z_t|^2 + V(z_t)\right) dt - \hbar\, E_s \log \eta^*_{-1}(z_{-1})$$

The corresponding Euler-Lagrange equation is indeed $D_* D_* z(t) = \nabla V(z_t)$ a.e.

Let us observe that, as it should be true for an action functional, the addition to a given Lagrangian L of a term $DF(z_t,t)$, in the forward case, or $D_* F(z_t,t)$ in the backward one, for F any regular function such that these additional terms make sense, does not change the laws of motion.

3. Euclidean Quantum Field Theory

In the perspective of Euclidean Quantum Mechanics, the extension to infinite dimensions (Field Theory) is based on the construction of Bernstein processes taking values on some infinite dimensional spaces (of distributions).

Here we shall describe the construction for two dimensional quantum fields, one dimension for the time and another for the space. We start from a regularized process, where the relevant fields are continuous and enable us to use Malliavin calculus techniques. The physically relevant processes and results follow from removing this regularization. They have been presented in [23].

§11. The regularized infinite dimensional Bernstein processes

Let A_ε, $\varepsilon>0$, denote the (pseudo) differential operator $A_\varepsilon = (-\Delta + m^2)^{1/2} + \varepsilon/m(-\Delta)$ acting on functions defined on the (space) interval $[-\ell,\ell]$ and with Dirichlet boundary conditions for simplicity. Let $\mathcal{H}_\varepsilon$ be the Hilbert space with scalar product

$$(h,g)_\varepsilon = \int_{-\ell}^{\ell} (A_\varepsilon h)\,(x)\,g(x)\,dx \tag{11.1}$$

For $\varepsilon>0$ the norm is equivalent to the Sobolev H^1 norm over $[-\ell,\ell]$. Let μ_ε be the Gaussian measure characterized by

$$E_{\mu_\varepsilon} \exp i\,(h,\varphi)_\varepsilon = \exp -\tfrac{1}{4}\,\|h\|_\varepsilon^2 \tag{11.2}$$

Then $(\Omega, \mathcal{H}^\varepsilon, \mu_\varepsilon)$, where $\Omega = \{\varphi \in C\,([-\ell,\ell]) : \varphi(-\ell) = \varphi(\ell) = 0\}$ is an abstract Wiener space in the sense of Gross.

Notice that a physically relevant measure in 2-dimensional Euclidean Quantum Field Theory is the measure μ_ε for $\varepsilon = 0$. The construction for this case is obtained by a limiting procedure (cf. next paragraph) and the case $\varepsilon>0$ can be therefore seen as a regularization of the physical one.

The measure μ_ε is the invariant measure for the Ornstein-Uhlenbeck process X_t^ε, given by:

$$\begin{aligned} dX_t^\varepsilon &= A_\varepsilon^{1/2} dW_t^\varepsilon - A_\varepsilon X_t^\varepsilon\,dt \\ X_0^\varepsilon &= \varphi \in \Omega \end{aligned} \tag{11.3}$$

where $E[W_t^\varepsilon(x)\;W_{t'}^\varepsilon(y)] = \min(t,t')\;A_\varepsilon^{-1}\,(x,y)$ characterizes a Brownian motion with values in Ω.

We can make the ε-dependance explicit by: $W^\varepsilon = A_\varepsilon^{-1/2} W$, where $E[W_t(x) W_{t'}(y)] = \min(t,t')\;\;\delta(x\text{-}y)$; so, we can regard the paths of the

ε-independent Wiener process W as a model of the probability space defining the expectation values.

The generator of the O.U. process X_t^ε is given by $\mathcal{L}^\varepsilon F = -\frac{1}{2}\delta^\varepsilon \nabla^\varepsilon F$, where ∇^ε is the Malliavin derivative on the Wiener space $(\Omega, \mathcal{H}^\varepsilon, \mu_\varepsilon)$, namely the linear operator

$$(\nabla^\varepsilon F(\varphi), h)_\varepsilon = \lim_{\substack{\lambda \to 0 \\ (a.e.)}} \frac{1}{\lambda}(F(\varphi + \lambda h) - F(\varphi)) \qquad , h \in \mathcal{H}^\varepsilon,$$

and δ^ε the corresponding divergence (cf. paragraph 8). When applied to cylindric Wiener functionals, the ones used for the finite-dimensional distribution of the processes, the operator $\mathcal{L}^\varepsilon$ can also be characterized by

$$\mathcal{L}^\varepsilon F(\varphi) = \frac{1}{2} \text{Trace}\,(A_\varepsilon^2 (\nabla^\varepsilon)^2 F)(\varphi) - (A_\varepsilon \varphi, \nabla^\varepsilon F(\varphi))_\varepsilon$$

In order to construct Bernstein processes we shall need the following Feynman-Kac formula on Wiener spaces:

Theorem [37,23]

Let $V^\varepsilon \geq -K_\varepsilon$, where K_ε is a positive constant, be a functional belonging to the Sobolev space $W_4^p(\Omega, \mu_\varepsilon)$ for some $2 \leq p < +\infty$. Then the operator $H_\varepsilon = \mathcal{L}^\varepsilon + V^\varepsilon$ generates a strongly continuous contraction semigroup $e^{-t(H_\varepsilon + K_\varepsilon)}$ such that the following representation holds, for $\mathcal{O}_\varepsilon \in W_2^p(\Omega, \mu_\varepsilon)$, $t>0$:

$$e^{-tH_\varepsilon}\, \mathcal{O}_\varepsilon(\varphi) = E_\varphi[\mathcal{O}_\varepsilon(X_t^\varepsilon) \exp\,(-\int_0^t V^\varepsilon(X_s^\varepsilon)\, ds)] \tag{11.4}$$

μ_ε- a.e.

This Feynman-Kac formula apparently breaks the time symmetry of our construction; actually this is true but the theory uses as well the corresponding backward (time reversed) formula.

Let us consider the variable time t in the interval [0,T] and introduce our familiar notation $\eta_t^\varepsilon = e^{-(T-t)H_\varepsilon}\, \mathcal{O}_\varepsilon$, $\eta_t^{*\varepsilon} = e^{-(t+T)H_\varepsilon}\, \mathcal{O}_\varepsilon^*$ where $H^\varepsilon = \mathcal{L}^\varepsilon + V^\varepsilon$ is, from now on, the Hamiltonian. Then, by using Girsanov transformation theorem starting from the O.U. process X_t^ε, we construct Bernstein processes associated to H^ε by solving (in the weak sense) the corresponding stochastic differential equations. More precisely, we have the following result:

Theorem [37,23]

Let $V_\varepsilon \geq -K_\varepsilon$, $V_\varepsilon \in W_4^p(\Omega, \mu_\varepsilon)$ for some $2 \leq p < +\infty$, and let $\mathcal{O}_\varepsilon, \mathcal{O}_\varepsilon^* > 0$ be bounded functionals in $W_2^p(\Omega,\mu_\varepsilon)$ such that $E_{\mu_\varepsilon} \log^2 \theta_\varepsilon < +\infty$, $E_{\mu_\varepsilon} \log^2 \theta_\varepsilon^* < +\infty$. Assume that the normalization condition $E_{\mu_\varepsilon}(\theta_\varepsilon^* \, e^{-2TH_\varepsilon} \theta_\varepsilon) = 1$ is fulfilled. Then there exists a Bernstein process z_t^ε, $t \in [-T,T]$, with state space Ω, which is weak solution of the following stochastic differential equation:

$$dz_t^\varepsilon = A_\varepsilon^{1/2} dW_t^\varepsilon - (A_\varepsilon z_t^\varepsilon - A_\varepsilon \nabla^\varepsilon \log \eta_t^\varepsilon(z_t^\varepsilon))\, dt$$

$$P(z_0^\varepsilon \in \Gamma) = \int_\Gamma \theta_\varepsilon^*(\varphi)\, e^{-2TH_\varepsilon}\, \theta_\varepsilon(\varphi)\, d\mu_\varepsilon(\varphi) \tag{11.5}$$

Furthermore, for $t \in [-T,T]$ the following Euclidean Born interpretation holds

$$P(z_t^\varepsilon \in \Gamma) = \int_\Gamma (\eta_t^{*\varepsilon})\, \eta_t^\varepsilon \, d\mu_\varepsilon \tag{11.6}$$

(to compare with Eq (2.3), Chapt 1).

As in finite dimensions, the process should also be described by its backwards stochastic differential equation, namely:

$$d_* z_t^\varepsilon = A_\varepsilon^{1/2} d_* W_{*t}^\varepsilon + (A_\varepsilon z_t^\varepsilon - A_\varepsilon \nabla^\varepsilon \log \eta_t^{*\varepsilon}(z_t^\varepsilon))\, dt,$$

where W_{*t}^ε is a backward, Ω-valued, Bronwian motion.

One can develop a functional calculus for these infinite dimensional Bernstein processes by using Malliavin calculus for two parameter processes (cf., for example, [38]) and working on the space $C([0;T];\Omega) \simeq C([0,T] \times [-\ell,\ell];\mathbb{R})$ endowed with a measure which is the law of the O.U. process X_t^ε on this path space. In particular, an integration by parts formula can be derived in this framework (cf. [37] and [23]).

Let us describe here the dynamical characterization of the processes z_t^ε. The derivatives D^ε and D_*^ε of a Wiener functional (defined on (Ω,μ_ε)) are defined, in analogy with the finite-dimensional derivatives of §5 Chapt 1 by

$$D^\varepsilon F(z_t^\varepsilon) = \lim_{\lambda \to 0} \frac{1}{\lambda} E_t^\varepsilon(F(z_{t+\lambda}^\varepsilon) - F(z_t^\varepsilon))$$

$$D_*^\varepsilon F(z_t^\varepsilon) = \lim_{\lambda \to 0} \frac{1}{\lambda} E_t^\varepsilon(F(z_t^\varepsilon) - F(z_{t-\lambda}^\varepsilon)) \tag{11.7}$$

Then the Bernstein processes are solutions of the following regularized Euler-Lagrange equations:

$$D^\varepsilon D^\varepsilon z_t^\varepsilon = A_\varepsilon^2 z_t^\varepsilon + A_\varepsilon \nabla^\varepsilon V^\varepsilon(z_t^\varepsilon) \qquad \text{a.s.} \tag{11.8}$$

with the boundary condition $D^\varepsilon z_T^\varepsilon = - A_\varepsilon z_T^\varepsilon + \nabla^\varepsilon \log \eta_T^\varepsilon(z_T^\varepsilon)$ a.s. They are as well solutions of corresponding backward Euler-Lagrange equations. These laws of motion are, in analogy with the finite-dimensional framework, associated to the following (forward) regularized action functional:

$$\mathfrak{J}^\varepsilon(z^\varepsilon) = E_s \int_s^T L^\varepsilon(z^\varepsilon, D^\varepsilon z^\varepsilon)\, dt - E_s\, [\log \eta_T^\varepsilon\, (z_T^\varepsilon)]$$

where $L^\varepsilon(z^\varepsilon, Dz^\varepsilon) = \frac{1}{2} \|A_\varepsilon^{-1/2}(D^\varepsilon z_t^\varepsilon + A_\varepsilon z_t^\varepsilon)\|_\varepsilon^2 + V^\varepsilon(z_t^\varepsilon)$ is the Lagrangian.

§12. The Quantum Fields

As we mentioned before, a physically relevant case in Euclidean Quantum Field Theory corresponds to the case $\varepsilon \to 0$. The Gaussian measure μ_ε does have a weak limit μ_0, namely:

$$E_{\mu_0} \exp i(\varphi, h)_0 = \exp - \frac{1}{4} \|h\|_0^2 \tag{12.1}$$

where $\|h\|_0^2 = \int_{-\ell}^{\ell} A_0 h(x).h(x)\, dx \qquad$ for $A_0 = (-\Delta + m^2)^{1/2}$.

The main mathematical difficulty now is that the support of the measure μ_0 is no longer a space of continuous paths but a space of distributions.

We remark that the regularized O.U. process X_t^ε can start from $\varphi \in \operatorname{supp} \mu_0$. Let us write $X_t^\varepsilon = Z_t^\varepsilon + Y_t^\varepsilon = e^{-tA_\varepsilon}\varphi + Y_t^\varepsilon$. Then, when ψ^ε is either Z^ε or Y^ε, we define the "Wick power of ψ^ε" by the formal power series

$$:\exp(\psi^\varepsilon, f): = \exp(-\tfrac{1}{2} E[(\psi^\varepsilon, f)^2]) \exp(\psi^\varepsilon, f) = \sum \frac{1}{n!} :(\psi^\varepsilon, f)^n: \tag{12.2}$$

We can define as well

$$:X^n: = :(Z^\varepsilon + Y^\varepsilon)^n: = \sum \binom{n}{m} :(Z^\varepsilon)^{n-m}: :(Y^\varepsilon)^m: \tag{12.3}$$

Then $:\psi^\varepsilon(x)^n:$ is a random field with a.s. continuous paths and we can define the so-called "polynomial interaction" by

$$V_n^\varepsilon(t) \equiv V_n^\varepsilon(t, X_t^\varepsilon) = \int_{-\ell}^{\ell} :(X_t^\varepsilon(x))^n:\, dx \tag{12.4}$$

We will also consider the "exponential interaction":

$$V_E^\varepsilon(t) = \sum_{n\geq 0} \frac{\alpha^n}{n!} V_n^\varepsilon(t) \tag{12.5}$$

which is the integral over $[-\ell,\ell]$ of : $\exp X_t^\varepsilon$: and, therefore, is positive.

The conditions to apply Feynman-Kac's formula (11.4) are fulfilled and this allows us to define the semigroup e^{-tH_ε}. Now one can prove (cf. [23]) that $V_n^\varepsilon \underset{\varepsilon\to 0}{\to} V$ in $L^2_{\mu_0}$ and also that $V_E^\varepsilon \underset{\varepsilon\to 0}{\to} V_E$ in $L^2_{\mu_0}$ for $\alpha^2<2\pi$. The corresponding semigroups also have a strong limit in $L^2_{\mu_0}$ when $\varepsilon\to 0$, and a Feynman-Kac's formula holds for these potentials in the limiting case.

We obtain, also by limiting procedure, a Bernstein process which is a weak solution of the stochastic differential equation on $\mathcal{S}'(\mathbb{R})$:

$$\begin{aligned} &dz_t^0 = A_0^{1/2}\, dW_t^0 - (A_0 z_t^0 - A_0\, \nabla^0 \log \eta_t^0(z_t^0))\, dt \\ \text{with} \\ &P(z_t^0 \in \Gamma) = \int_\Gamma \mathcal{O}^*\, e^{-2TH}\, \mathcal{O}\, d\mu_0 \end{aligned} \tag{12.6}$$

where $\eta_t^0 = e^{-(T-t)H}\mathcal{O}$, $H = -\mathcal{L}^0 + V$ and V is the polynomial or the exponential interaction.

Taking the limit when $\varepsilon\to 0$ in equations (11.8) is a more subtle question, since the sample paths z_t^ε become singular. To introduce a reasonable regularized derivative for the generalized process z_t^0, we take any test function $h \in L^2(dx)$ and define

$$D^0(z_t^0,h) = -(A^0\, z_t^0,h) + (A_0\nabla^0 \log \eta_t^0,h) \tag{12.7}$$

Again by interpreting the equality in a distributional sense we will have the following forward equation of motion:

$$D^0D^0z_t^0 = -A_0^2\, z_t^0 - \nabla V(z_t^0) \tag{12.8}$$

Since $A_0^2 = -\Delta + m^2$, this is a regularized version of the (Euclidean) Klein-Gordon equation.

Formally, the associated Lagrangian will be given by:

$$L(z^0,Dz^0) = \frac{1}{2}\, \|Dz^0 + A^0z^0\|_{L^2}^2 + V(z^0) \tag{12.9}$$

which is actually a well-defined object. It gives rise to the same law of motion as the formal following one: (Cf. §10)

$$L(z^0,Dz^0) = \frac{1}{2} ||Dz^0||^2_{L^2} + \frac{1}{2} ||A^0z^0||^2_{L^2}+V(z^0)$$

which is a regularized version of the classical Lagrangian involved in usual 2-dimensional Quantum Field Theory, namely:

$$\int_{-\ell}^{\ell} (\frac{1}{2} | \frac{\partial}{\partial t} z|^2 + \frac{1}{2} | \frac{\partial}{\partial x} z|^2 + \frac{1}{2} m^2 z^2 + \mathcal{V}(z)) \, dx$$

for some potential $\mathcal{V}$.

Conclusion

We have seen that, starting from this "analogie avec la mécanique ondulatoire ... si frappante..." mentioned by Schrödinger (1931-32) in [1], it is possible to lay the foundations of what seems to be, indeed, a promising Euclidean framework.

A priori, one could consider that this is of no use. After all, Feynman-Kac formula seems to be sufficient for many physical purposes and, on the other hand, Nelson's stochastic mechanics is an approach in terms of diffusion processes directly in real time, i.e. associated directly to the Schrödinger equation (1.7). So why to follow another Euclidean detour in the first place?

Regarding the first point, we are convinced that Schrödinger's Euclidean analogy is considerably deeper, in quantum physics, than Feynman-Kac formula.

Of course, Feynman-Kac formula retains a important role in Euclidean Quantum Mechanics. This should be clear from the construction of Bernstein diffusions and also from the description of the functional calculus. Feynman-Kac formula remains here a key result of perturbation theory. However, for those theoretical readers more interested by structural aspects of theoretical physics than by technical or computational ones, one should underline that the way probabilistic concepts enter in Feynman-Kac formula has little to do with quantum physics. As a matter of fact, these probabilities are not "quantum" in any way. They describe irreversible phenomena of a physical nature completely different from the quantum phenomena. In contrast, the strategy suggested by Schrödinger, i.e. EQM, introduces probabilities precisely where they enter in quantum physics, i.e. in the Born interpretation. The fundamental time symmetry of the physical theory is preserved as well as a number of structural aspects of quantum theory, going far beyond the Schrödinger's equation itself. Of course, this remains to be explored in details. What we call Euclidean Quantum Theory is this program of exploration.

Regarding the second point, it is a fact that along the years, the real time approach of Nelson ("Stochastic mechanics") has accumulated the clues that it was not as straightforward as anticipated.

Let us recall that Nelson theory associates to a solution of the Schrödinger equation (1.7), for an Hamiltonian H like (1.4), a well defined diffusion process X_τ with diffusion coefficient (1.6) and compatible with Born interpretation (2.5). In other words, for example,

$$P(X_\tau \in d^3x) = |\psi(x,\tau)|^2\, d^3x \tag{1}$$

Moreover, the (forward) drift of the associated diffusion is of the form

$$b(x,\tau) = \hbar \operatorname{Re}\left(\frac{\nabla\psi}{\psi}(x,\tau)\right) + \hbar \operatorname{Im}\left(\frac{\nabla\psi}{\psi}(x,\tau)\right) \tag{2}$$

An important element of Nelson's original interpretation of this result [13] was that the following formula holds for any such process (and the chosen Hamiltonian)

$$\frac{1}{2}\,(DD_*X_\tau + D_*DX_\tau) = -\nabla V(X_\tau) \tag{3}$$

Notice the "mixture of the filtrations" $\mathcal{P}_\tau$ and $\mathcal{F}_\tau$ involved in the left hand side, and the fact that, here, it is not true that $DD_*X_\tau = D_*DX_\tau$.

The mathematical existence of the diffusions X_τ has been proved only in 1984 by Carlen [15] in functional analytical terms.

As said in §1 Chap 1, Schrödinger's strategy can be adapted to Nelson's theory. This has been done originally in [11, JMP (86)]. Of course, Schrödinger's method is Euclidean and the constraint (1) is in real time. To realize it, however write the given ψ_τ as

$$\psi_\tau(x) = e^{(R+iS)/\hbar} \equiv \rho^{1/2}(x,t)\, e^{\frac{i}{\hbar}S(x,t)} \tag{4}$$

Then, on shows that Nelson's process is a particular Bernstein diffusion but that the scalar potential involved in the two relevant adjoint heat equations (2.4) is not the physical potential V but a modified potential of the form ([11])

$$\mathcal{V}(x,t) = \hbar^2 \frac{\Delta \rho^{1/2}}{\rho^{1/2}}(x,t) - V(x) \tag{5}$$

Mathematically speaking, it means that, in spite of the appealing Eq (3), the potential driving Nelson's processes is not V but a complicated nonlocal one depending, in particular, of the initial condition of the Schrödinger's equation. Probabilistically, for a given V, the diffusions of Nelson will have very little in common.

Physically speaking, the presence of $\mathcal{V}$ behind the stage is also a rather bad news. Modified potentials like (5) are familiar since, at least Bohm [31] (cf. also [32]) and do not seem to have been given, as yet, a physically conclusive interpretation. (But cf [33]). On the other hand, (5) shows immediately that all the puzzling consequences of quantum nonlocality will be present in stochastic mechanics, an aspect apparently not so trivial without using Schrödinger's strategy (Cf, for example [13 (1985), §23]).

Nelson has been introduced recently as "the founder of stochastic mechanics and its severest critic" [34]. In spite of the strong reservations expressed by Nelson regarding his theory, this one seems to be attractive to many since it is still periodically and sometimes painfully rediscovered (Cf, notably, [35], to compare with [11] PRA (1986), JMP (1986)).

Euclidean quantum mechanics is a fresh probabilistic approach which contains, in particular, probability measures analogue to the (nonexistent) ones relevant to physics, along the lines suggested long ago by Schrödinger and later by Feynman. It has the character of a program. For those probabilists who do not care especially about its initial motivation, it may present an interest since it forces us to look for probabilistic structures generalizing those of classical mechanics, i.e., quite distincts from those known today. Cf, for example [40].

For those who, like us, are motivated by quantum physics, it may give a new hope to develop the systematic analogy between quantum phenomena and probabilistic ones, whose existence was suggested by Feynman but never really proved.

Of course, radical theoretical physicists will not regard Euclidean Quantum Mechanics as an interesting theory before it is proved that this Euclidean detour enables us to describe new experimental facts (in "real time", of course) that regular quantum theory was not able to predict up to now. We believe that they are right to think so. Only if this happens, Schrödinger's initial guess that this striking analogy cannot be accidental will be proved right. This would be an impressive tribute to the father of wave mechanics, who never accepted the orthodox interpretation of his own equation.

References

[1] E. Schrödinger, Ann. Inst. Henri Poincaré, 2, 269 (1932).

[2] S. Bernstein, "Sur les liaisons entre les grandeurs aléatoires", Verh. des Intern. Mathematikerkongr, Zurich, Band 1 (1932)

[3] R. Fortet, J. Math. Pures et Appl. Vol. IX, 83 (1940).

[4] A. Beurling, Annals of Mathematics, vol. 72, nº 1, July, 189 (1960).

[5] B. Jamison, Z. Wahrscheinlichkeitstheoric V. Gebiete, 30, 65 (1974).

[6] T. Kato, Ist. J. Math., vol. 13, 135 (1973).

[7] M. Kac, "On some connections between probability theory and differential and integral equations", Proc. Second Berkeley Sympos. Math. Statist. and Probab., Univ. of California Press, Berkeley, 189 (1951).

[8] R. Feynman, Rev. Mod. Phys., 20, 367 (1948).

[9] R. Feynman and A.R. Hibbs, "Quantum Mechanics and Path integrals", McGraw-Hill, N.Y. (1965).

[10] R. H. Cameron, J. Math. Phys. 39, 126 (1960).

[11] J.C. Zambrini, Phys. Rev. A, vol. 33, nº 3, 1532 (1986);
J.C. Zambrini, J. Math. Phys., 27(9), 2307 (1986);
J.C. Zambrini, Phys. Rev. A, vol. 35, nº 9, 3631 (1987).

[12] H. Föllmer, "Random fields and diffusion processes" in "École d'été de probabilités de St-Flour XV-XVII", Ed. P.L. Hennequin, Lect. Notes in Mathematics nº 1362, 101 (1988).

[13] E. Nelson, "Dynamical Theories of Brownian Motion", Princeton Univ. Press (1967);
E. Nelson, "Quantum Fluctuations", Princeton Univ. Press (1985).

[14] K. Yasue, J. Math. Phys., 22, 1010 (1981);
K. Yasue, J. Funct. Anal., 41, 327 (1981).

[15] E. Carlen, Commun. Math. Phys., 94, 293 (1984).

[16] A.D. Wentzell, "A course in the theory of stochastic processes", McGraw--Hill (1981).

[17] S. Albeverio, K, Yasue, J.C. Zambrini, Ann. Inst. H. Poincaré, Phys. Th. vol. 49, n$^{\underline{o}}$3, 259 (1989).

[18] B. Djehiche, J. Math. Phys. 33(9), 3050 (1992); Potential analysis 2,349 (1993).

[19] A.B. Cruzeiro, J.C. Zambrini, "Ornstein-Uhlenbeck processes as Bernstein processes", in Barcelona Seminar on Stochastic Analysis, Edit D. Nualart and Marta Sanz Solé, Birkhäuser, Prog. in Prob. Series, n$^{\underline{o}}$ 32, p 40 (1993).

[20] T. Kolsrud and J.C. Zambrini, J. Math. Phys., 33(4), April 92, 1301.

[21] J.C. Zambrini, "Time reversal and Gaussian measures in Quantum Physics", in 3rd Max Born Symposium: Stochasticity and Quantum Chaos", Edit. Z. Haba et al. (Sept. 93), to appear.

[22] A. Brandão, in preparation.

[23] A.B. Cruzeiro, J. Haba, J.C.Zambrini, "Bernstein diffusions and Euclidean Quantum Field Theory", in Monte Verità Meeting, Edit. E. Bolthausen and F. Russo, Birkäuser, to appear.

[24] J. Moser, "Stable and random motions in dynamical systems", Annals of Mathematics Studies, Princeton Univ. Press (1973).

[25] L.D. Landau, E.M. Lifshitz, "The classical theory of fields", Pergamon Press, Oxford (1975).

[26] A.B. Cruzeiro and J.C. Zambrini, "Malliavin Calculus and Euclidean Quantum Mechanics. I Functional Calculus", J. Funct. Anal., vol. 96, n$^{\underline{o}}$ 1, Feb 1991, 62.

[27] N. Ikeda and S. Watanabe, "Stochastic differential equations and diffusion processes", 2nd ed., North-Holland, Amsterdam (1989).

[28] E. Schrödinger, Annalen der Physik (4), vol. 79 (1926).

[29] W.H. Fleming, "Controlled Markov processes and viscosity solution of Nonlinear evolution equations", Lezioni Fermiane, Pisa (1986).

[30] E.B. Dynkin, "Markov processes", English translation in 2 volumes, Springer, Berlin (1965).

[31] D. Bohm, Phys. Rev. 85, 166, 180 (1952); W. Pauli in "L. de Broglie, Physicien et Penseur", Ed. Albin Michel, Paris (1953).

[32] D. Bohm and J.P. Vigier, Phys. Rev. 96, 208 (1954).

[33] P. Garbaczewski and J.P. Vigier, Phys.Lett. A 167, 445 (1992).

[34] E. Nelson, "Taking Formalism Seriously", The Mathematical Intelligencer, vol. 15, n$^{\underline{o}}$ 3, 8 (1993).

[35] M. Nagasawa, "Principle of superposition and interference of diffusion processes", in Seminaire de Probabilités XXVII, Lect. Notes in Math., n$^{\underline{o}}$ 1557, Ed. J. Azéma, P.A. Meyer and M. Yor (1993).

[36] P. Malliavin, "Stochastic Calculus of variations and hypoelliptic operators", in Proc. of Int. Symp. Stoch. Diff. Eq., Kyoto, 1976 (K. Itô, ed), Tokyo, (1978).

[37] A.B. Cruzeiro and J.C. Zambrini, "Malliavin Calculus and Euclidean Quantum Mechanics, II" (in preparation).

[38] D. Nualart, M. Sanz, Z. Wahrscheinl. verw. Gebiete 70, 573, (1985).

[39] M. Thieullen, Prob. Th. Relat. Fields, 97, 231-257 (1993).

[40] A. Wakolbinger, J. Math. Phys. 30(12), 2943 (1989); T. Misawa, J. Math. Phys. 34, 775 (1993).

[41] B.C. Lévy and A.J. Krener, J. Math. Phys. 34(5), 1846 (1993).

[42] T. Mikani, Comm. Math. Phys. 135, 21 (1990).

[43] T. Kolsrud and J.C. Zambrini in "Stochastic Analysis and Applications", Ed. A.B. Cruzeiro and J.c. Zambrini, Progress in Probability vol.26, 123, Birkhäuser, Boston (1991).

[44] A.B. Cruzeiro and J.C. Zambrini, same Birkhäuser vol. as [43], 83 (1991).

[45] S. Albeverio, K. Hoegh/Krohn and L. Streit, J. Math. Phys. 18, 907 (1977).

ANALYSIS ON LOOP GROUPS

LEONARD GROSS[1]
Department of Mathematics
Cornell University
Ithaca, NY 14853

ABSTRACT. This paper surveys some recent results on Schrödinger operators over loop groups.

1. Introduction.

These lecture notes are intended to survey some recent work on analysis over infinite dimensional manifolds of the form $C(S^1, G)$ wherein G is a compact finite dimensional Lie group. Spaces of functions from S^1 into a finite dimensional Riemannian manifold M have for a long time provided interesting examples of infinite dimensional manifolds. These spaces have been intensely investigated because of the insight they give into the finite dimensional manifold itself as well as for other reasons (see e.g. [PS]). Until recent years analysis over these infinite dimensional manifolds was limited to the kind of problems that did not require integration over the infinite dimensional manifold itself. There was (and is) no precise analog of a Riemann volume element on these infinite dimensional manifolds. Stimulated in part by quantum field theory and in part by Malliavin's probabilistic approach to hypoellipticity, there has developed over the past twenty years a technology aimed in large part at dealing with problems of elliptic differential operators over nonlinear infinite dimensional manifolds. The analysis of such operators (e.g., Laplacians and Ornstein–Uhlenbeck operators) over infinite dimensional linear spaces (possessing for example a Gaussian measure) has quite a long history. But the earliest efforts to formulate a general theory of infinite dimensional Riemannian manifolds which permit a reasonable integration theory appear to be that of Kuo [Ku 1, 2, 3] and Eells and Elworthy [E, EE, Ew]. Their manifolds were modeled on abstract Wiener spaces [Gr 1]. M. A. Piech [P] initiated the exploration of a Hodge decomposition theorem in this context. However the important case of a loop space, $C(S^1, M)$ does not seem to be subsumed under the structures of these early pioneering works.

For loop spaces the natural measure seems to be pinned Brownian motion measure. We will describe this in more detail later for the special case of interest to us — that in which the target manifold is a compact Lie group. There seems to

[1] This work was partially supported by NSF Grant DMS-9200774.

A. I. Cardoso et al. (eds.), Stochastic Analysis and Applications in Physics, 99–118.

be a natural infinite dimensional analog of the Laplacian (actually an analog of the Ornstein–Uhlenbeck operator) on these loop spaces. Whereas, for a finite dimensional Riemannian manifold, the relations between spectral properties of the Laplacian and the topology of the manifold have been extensively developed the corresponding issues in infinite dimensions are in a very primitive state. One of the oldest of these connections is the Hodge–deRham theorem, whose infinite dimensional analog this author along with many others take as a first long range goal for loop spaces. We will recall this theorem now and then discuss some of the infinite dimensional questions which it leads to.

Let M denote a compact finite dimensional Riemannian manifold. Denote by d the exterior derivative operator on differential forms over M. Specifically we interpret d as an operator in the Hilbert space of square integrable forms over M (with respect to Riemann–Lebesgue measure) on a domain which makes it a closed operator. Write d^* for its Hilbert space adjoint. The Laplacian is given by

$$-\Delta = d^*d + dd^*.$$

Theorem 0. (Hodge–deRham)

a) *Image of d is closed.*

b) $\ker d = \ker \Delta \oplus$ *Image d*

c) *dimension* $\ker(\Delta \mid k$ *forms*$) = \dim H^k(M, \mathbb{R})$ *(real cohomology in degree k),* $k = 0, 1, \ldots, \dim M$.

The key step in the proof of a) and b) is the proof that if zero is in the spectrum of Δ then it is an isolated point of the spectrum. Since $-\Delta \geq 0$ this amounts to showing that

$$\inf\{(\text{spectrum} - \Delta) - \{0\}\} > 0.$$

This is the spectral gap problem for $-\Delta$. The existence of a spectral gap follows from the fact that $-\Delta$ has compact resolvent.

The long range goal of this work is to prove a version of the Hodge–deRham theorem for an interesting class of infinite dimensional manifolds — the loop spaces $M = C(S^1; G)$ — wherein G is a finite dimensional compact manifold.

In order to find an analog of the Hodge–deRham theorem for loop spaces $M = C(S^1; G)$ with G a compact manifold the following questions must be answered.

Question 1. What are the "right" k forms over M?

Question 2. What is d on these k forms?

Question 3. What is a useful substitute for Riemann–Lebesgue measure on M?

Question 4. Does d^* exist as a densely defined operator?

Question 5. If $-\Delta = d^*d + dd^*$ does $-\Delta$ have a spectral gap above zero on k–forms?

Question 6. How does $\dim \ker \Delta$ relate to the topology of M or G?

A few answers. 1–6. If M is a linear space, specifically an abstract Wiener space, all six questions have been answered completely by Shigekawa [Sh 3]. This definitive paper is a good guide for some of the algebraic and analytic questions that come up in the nonlinear case.

There are many papers that have addressed related questions in case M is linear. I will focus here just on work in the nonlinear case.

For Questions 1 and 2 there is a paper by Airault and Malliavin [AM] which defines a class of k–forms on based loop space, defines d, and proves a Stokes theorem. An alternative approach to k–forms on loop spaces is given in Jones and Leandre [JL].

For Question 3 the mathematical community deems pinned Brownian motion measure as a natural choice in the case of based loops and modifications of this in the case of unbased loops. This measure has been intensely studied in various other contexts.

For Question 4 the existence of d^*, even on 0–forms, resisted analysis in the nonlinear case until the recent breakthrough by Bruce Driver [D 1,2] on quasiinvariance. His work was followed by important extensions by E. Hsu [Hs].

For Question 6 there is a work by S. Kusuoka [Kyoto 1990] that gives results on the infinite dimensional deRham theorem in degrees zero and one. Part c) of Theorem 0 is thus considered by him apart from parts a) and b).

Concerning Question 5 a spectral gap has not been proven in any nonlinear case. Unlike the finite dimensional compact case, compact resolvent for the infinite dimensional Laplacian Δ cannot be expected to hold. Even in the flat case [Sh 3] compact resolvent fails. It would not be surprising if a spectral gap holds while the rest of the spectrum is continuous. With a view toward proving a logarithmic Sobolev inequality (which in turn implies a spectral gap via the Rothaus–Simon theorem [Gr 9]) E. Getzler [Ge 1] computed the quantities necessary to apply the powerful method of Bakry–Emery. Getzler showed that the method failed in the simplest case — that in which G is a compact semi–simple Lie group. This author [Gr 6] showed that a logarithmic Sobolev inequality holds in this case after adding on to $-\Delta \mid$ 0–forms a suitable potential, V. Unfortunately this does not imply a spectral gap because $-\Delta + V$ is not a Dirichlet form operator. However if, perchance, $-\Delta + V$ has a unique and a fortiori strictly positive lowest eigenstate ψ then the Darboux transforms of $-\Delta + V$ by ψ (alias h–transform and alias ground state transform) is essentially a Dirichlet form operator which one could hope satisfies a logarithmic Sobolev inequality and has therefore a spectral gap. It would not be unaesthetic in this case to change from pinned Brownian motion measure ν on the based loop space to the almost equally natural measure $\psi^2 d\nu$. The goal of the work reported on in this lecture is a small step toward an answer to Question 5. It is to prove uniqueness of the lowest eigenstate for $-\Delta + V$ when

V is reasonable. Existence has previously been shown [Gr 6]. We will be interested only in the case G = compact Lie group.

2. Pinned Brownian motion on based loops.

Henceforth we will denote by G a connected Lie group and by $\mathcal{G}$ its Lie algebra which we identify with the tangent space to G at the identity. We will always assume that $\mathcal{G}$ has an $\mathrm{Ad}\, G$ invariant inner product. This imposes a severe restriction on G. If $G = \mathbb{R}^k$ then $\mathrm{Ad}\, G$ acts trivially, so any inner product on $\mathbb{R}^k$ (which of course can be identified with $\mathcal{G}$) is $\mathrm{Ad}\, G$ invariant. If, on the other hand, G is compact then by averaging over G any inner product on $\mathcal{G}$ one obtains an $\mathrm{Ad}\, G$ invariant inner product. Combining these two cases we see that if G = compact $\times \mathbb{R}^k$ then $\mathcal{G}$ has an $\mathrm{Ad}\, G$ invariant inner product. It is a fact [He] that the only connected groups with an $\mathrm{Ad}\, G$ invariant inner product on its Lie algebra are those of the form compact $\times \mathbb{R}^k$. These groups are said to be of compact type. Our results are therefore limited to this case. G valued Wiener space is defined by

$$W_G = \{g \in C([0,1], G) : g(0) = e\}$$

where e is the identity element of G. W_G is clearly a group under pointwise multiplication:

$$(g_1 g_2)(s) = g_1(s) g_2(s).$$

Write

$$\varphi(g) = g(1).$$

Then $\varphi : W_G \to G$ is a homomorphism of the group W_G onto G. Let

$$\mathcal{L} = \ker \varphi = \{g \in W_G : g(1) = e\}$$

$\mathcal{L}$ is called the *based loop group* of G. $\mathcal{L}$ is clearly a normal subgroup of W_G. Moreover W_G is a topological space under the topology of uniform convergence. This topology can be metrized by any metric of the form

$$dist(g_1, g_2) = \sup\{\rho(g_1(s), g_2(s)) : 0 \le s \le 1\}$$

where ρ is a left invariant metric on G. Of course $\mathcal{L}$ is closed in this topology.

Now $\mathcal{G}$ is a Euclidean space and the notion of $\mathcal{G}$ valued Brownian motion is therefore well defined. We may realize it as a measure w (Wiener measure) on the $\mathcal{G}$ valued Wiener space

$$W_{\mathcal{G}} := \{b \in C([0,1], \mathcal{G}) : b(0) = 0\}.$$

w is determined by the condition that for any orthonormal basis $e_1, \ldots, e_r$ of $\mathcal{G}$ the coordinate functions $s \mapsto \langle b(s), e_i \rangle$ are mutually independent real valued Wiener processes.

Consider the Stratonovich stochastic differential equation

$$dg(s) = db(s) \circ g(s) \quad g(0) = e. \tag{2.1}$$

Here, as in the remainder of these notes, we will use a notation as if the group G were a matrix group. Thus in the important and prototypical case wherein $G = SU(2)$ we may take $\mathcal{G} = \mathrm{su}(2) := \{2 \times 2$ complex matrices which are skew Hermitian and of trace zero$\}$. Since solutions to Stratonovich stochastic differential equations behave in many ways like solutions to ordinary differential equations we may expect that the solution to (2.1) lies in G for each s. In fact we take it as known to the reader that the SDE (2.1) determines an a.e. defined map $I : b \mapsto g$ (the Ito map) from $W_{\mathcal{G}}$ into W_G. (See e.g. [IW].) The map I pushes the measure w forward to a probability measure P on the Borel sets of W_G

$$P = I_* w.$$

The coordinate process $s \mapsto g(s)$ is a G valued process generally known as the G valued Wiener process (or G valued Brownian motion starting at the identity element e). This process has the following properties.

a) g has left independent increments. That is, if $0 = s_0 < s_1 < s_2 < \cdots < s_n \leq 1$ then $\{g(s_i)g(s_{i-1})^{-1}\}_{i=1}^n$ constitutes an independent set of G valued random variables on (W_G, P).

b) Let $e_1, \ldots, e_r$ be an orthonormal basis of $\mathcal{G}$. For any element ξ in $\mathcal{G}$ denote by $\widetilde{\xi}$ the right invariant vector field on G which extends ξ. Thus each $\widetilde{e}_j$ is a first order differential operator on G. The Laplacian on G is defined by

$$\Delta = \sum_{j=1}^{r} \widetilde{e}_j^{\,2}.$$

The reader can verify easily that this operator is basis independent. The heat equation, $\partial u/\partial t = (1/2)\Delta u$, has a fundamental solution $p_t(x)$ satisfying

(i) $p_t(x) > 0$ for $t > 0$ and $x \in G$.

(ii) $p_t \in C^\infty(G)$ for $t > 0$.

(iii) $\int_G p_t(x)dx = 1$ where dx is Haar measure.

(iv) $\exp[t\Delta/2] =$ convolution by p_t for $t > 0$.

In item (iv) Δ is to be interpreted as a self-adjoint operator in $L^2(G, dx)$ with core $C_c^\infty(G)$. Then

$$P(g(t) \in A \mid g(s) = x) = \int_A p_{t-s}(yx^{-1})dy, \quad 0 \leq s < t \leq 1. \tag{2.2}$$

Thus the coordinate process for P is a G valued Markov process with transition semigroup $\exp[t\Delta/2]$.

The properties a) and b) determine P uniquely. Furthermore the endpoint distribution is

$$\varphi_* P = p_1(x)dx.$$

The *finite energy subgroup* K of W_G is the set of absolutely continuous functions k in W_G such that

$$\int_0^1 |k(s)^{-1}dk(s)/ds|_{\mathcal{G}}^2 ds < \infty. \tag{2.3}$$

We also write

$$K_0 = K \cap \mathcal{L}.$$

Before stating the significant property of K and K_0 let me point out what these spaces and measures reduce to in case $G = R^k$. In this case G may be identified with its Lie algebra and the Ito map is just the identity because the SDE (2.1) reduces to $dg = db$, so $g(s) = b(s)$. $W_G = W_{\mathcal{G}}$ is now a Banach space in the sup norm and $P = w$. The finite energy subgroup K is the Cameron–Martin subspace of Wiener space. The significance of the CM subspace lies in the fact that the translated measure $w_k(A) := w(k + A)$ is absolutely continuous with respect to w if and only if k lies in K [CM, Ma, S2]. The ramifications of this for differential operators over W_G will be discussed in the next section.

Returning now to the general case of a group with an AdG invariant inner product on its Lie algebra we have the following extension of the Cameron–Martin–Maruyama theorem.

Theorem A. (Albeverio and Hoegh–Krohn 1975) *For any element k in K the translated measure $P_k(A) := P(kA)$ is absolutely continuous with respect to P.*

For a proof see [AH, MM, Sh 4].

We will end this section with the corresponding theorem for the based loop space $\mathcal{L}$.

Pinned Brownian motion measure is the conditioned measure

$$\mu_x = P(\ \ | \, g(1) = x).$$

In general such conditioned measures are defined for almost every x (w.r.t. $p_1(x)dx$ in this case). But because of the regularity inherent in measures P induced by solutions to stochastic differential equations such as (2.1) the measures μ_x are well defined for all x in G. This can be regarded as a general consequence of Malliavin's powerful techniques for handling regularity problems of solutions to stochastic differential equations. See e.g. [AM]. See also [Gr 6] for more detailed discussion of the "Brownian bridge" measures μ_x. We will be interested in the measure μ_e. Since $\mathcal{L}$ is a closed subset of W_G it is a Borel set. Moreover $\mu_e(\mathcal{L}) = 1$. Henceforth we write

$$\mu = \mu_e \mid \text{Borel sets in } \mathcal{L}.$$

This is the measure on $\mathcal{L}$ that we will use in the rest of these notes.

Theorem B. (M. and P. Malliavin) *For each element k in K_0 the measure $\mu_k(A) := \mu(kA)$ is absolutely continuous with respect to μ.*

Proof: See [MM].

3. Differential operators over loop groups.

Let

$$H = \{h \in C([0,1], \mathcal{G}) : h \text{ is absolutely continuous,}$$
$$|h|^2 := \int_0^1 |\dot{h}(s)|^2_{\mathcal{G}} ds < \infty, \text{ and } h(0) = 0\}.$$

Here $\dot{h}(s) = dh(s)/ds$. For any function h in the Hilbert space H write

$$(e^h)(s) = \exp h(s), \quad 0 \le s \le 1.$$

K is a topological group in a metric determined by (2.3) [AH]. Moreover e^h is in K and the map $h \to e^h$ maps injectively a neighborhood of 0 in H onto a neighborhood of the identity in the topological group K. (See e.g. [Gr 5, Lemma 2.1].) Furthermore each element h in H determines a one parameter group $t \mapsto e^{th}$ in K. Thus we may think of H as the tangent space of K at the identity element of K (which is, by the way, the constant function equal to e).

An element of H determines a right invariant vector field ∂_h not only on K but also on W_G given, as usual, by

$$(3.1) \qquad (\partial_h F)(g) = dF(e^{th}g)/dt\big|_{t=0}, \quad h \in H, \ g \in W_G.$$

Here $F : W_G \to \mathbb{R}$ is assumed for the moment to be smooth in the sense that the derivative in (3.1) exists. For example if $0 < s_1 < s_2 < \cdots < s_n \le 1$ and u is in $C^\infty(G^n)$ and is bounded and has bounded first derivatives then $\partial_h F$ clearly exists if F is defined by

$$(3.2) \qquad F(g) = u(g(s_1), \ldots, g(s_n)), \quad g \in W_G.$$

Correspondingly, if for some g in W_G, $\partial_h F(g)$ exists for all h in H and is a continuous linear functional of h then the *gradient* of F at g is defined by the equation

$$(3.3) \qquad ((\nabla F)(g), h)_H = (\partial_h F)(g), \quad h \in H.$$

The gradient clearly exists for functions of the form (3.2).

Now let

$$H_0 = \{h \in H : h(1) = 0\}.$$

Then $(e^h)(1) = \exp h(1) = e$ for h in H_0. So e^h is in K_0. Just as we identified H with the tangent space to K at the identity we may identify H_0 with the tangent space to K_0 at the identity. Moreover each element h in H_0 determines a right invariant vector field ∂_h on $\mathcal{L}$ as well as on K_0 by the equation

$$(\partial_h f)(\gamma) = df(e^{th}\gamma)/dt\big|_{t=0}, \quad h \in H_0,\ \gamma \in \mathcal{L}.$$

where $f : \mathcal{L} \to \mathbb{R}$ is "smooth" in some sense that we will not specify here. Of course $\partial_h f(\gamma)$ will exist for quite a large class of functions on $\mathcal{L}$ including C^∞ cylinder functions such as (3.2). This time the gradient of a function on $\mathcal{L}$ will take its values in H_0. The gradient, denoted $\nabla_0 f$ is defined by

$$((\nabla_0 f)(\gamma), h)_H = (\partial_h f)(\gamma), \quad h \in H_0,\ \gamma \in \mathcal{L}$$

and of course $(\nabla_0 f)(\gamma) \in H_0$ by definition.

Exercise 3.1. Let $G = SO(3)$. Because the fundamental group of G is Z_2, $\mathcal{L}$ is a disjoint union of two closed subsets, the two homotopy classes. Define $f : \mathcal{L} \to \mathbb{R}$ by

$$f(\gamma) = \begin{cases} 1 & \text{if } \gamma \text{ is homotopic to the identity map} \\ 0 & \text{otherwise.} \end{cases}$$

So f is not constant. Compute $\nabla_0 f(\gamma)$ wherever it exists.

We have so far defined the operators ∂_h, ∇ and ∇_0 in a pointwise sense. But we are really interested in interpreting these operators in $L^2(W_G, P)$ or $L^2(\mathcal{L}, \mu)$. Naturally we want closed versions of these operators if possible. Since the C^∞ cyliner functions of the form (3.2) are dense in $L^2(W_G, P)$ (resp., $L^2(\mathcal{L}, \mu)$) it is clear that all of these operators are densely defined. Concerning the issue of closedness and closure consider first the momentum operator $P = i^{-1}d/dx$ in $L^2(\mathbb{R}, dx)$. There are (at least) three standard ways to define the domain $\mathcal{D}$ of this operator and its action on that domain, depending on how one defines the derivative.

i) (Strong L^2 derivative) Let f be in $\mathcal{D}_s$ if f is in $L^2(R, dx)$ and

$$f' := L^2 \lim_{h \to 0} h^{-1}(f(x+h) - f(x)) \tag{3.4}$$

exists. Put $P_s f = i^{-1} f'$ for f in $\mathcal{D}_s$.

ii) (Weak L^2 derivative) Let f be in $\mathcal{D}_w$ if f is in $L^2(\mathbb{R}, dx)$ and there exists a function f' in $L^2(\mathbb{R}, dx)$ such that

$$\int_{-\infty}^{\infty} f'(x)\varphi(x)dx = -\int_{-\infty}^{\infty} f(x)(d\varphi(x)/dx)dx \quad \forall \varphi \in C_c^\infty(R).$$

Put $P_w f = i^{-1} f'$ for f in $\mathcal{D}_w$.

iii) (Closure on $C_c^\infty(\mathbb{R})$) Let P_c be the closure in $L^2(\mathbb{R}, dx)$ of the operator $i^{-1}d/dx \mid C_c^\infty(\mathbb{R})$ and let $\mathcal{D}_c$ be its domain.

Now all three of these operators are equal and in particular $\mathcal{D}_s = \mathcal{D}_w = \mathcal{D}_c$. For example one can prove $\mathcal{D}_s = \mathcal{D}_w$ easily with the help of the Fourier transform. In particular all of these operators are closed.

However if one replaces Lebesgue measure on $\mathbb{R}$ by Gauss measure the situation is just a little different in ways that illuminate and carry over to the infinite dimensional case. Let $d\nu(x) = (2\pi)^{-1/2}e^{-x^2/2}$ on $\mathbb{R}$. Let $f(x) = (1+x^4)^{-1}\exp(x^2/4)$. One can verify by a straightforward computation that f is in $L^2(\mathbb{R},\nu)$, that f' is also in $L^2(\mathbb{R},\nu)$ and that if φ is in $C_c^\infty(\mathbb{R})$ with $\varphi(x) = 1$ for $|x| \le 1$ then the sequence $f_n(x) := f(x)\varphi(x/n)$ converges to f in $L^2(\nu)$ norm while each f_n is itself in $C_c^\infty(\mathbb{R})$. Moreover the derivatives f_n' converge to f' in $L^2(\nu)$ norm. Thus if the operator d/dx is to be realized as a closed operator in $L^2(\mathbb{R},\nu)$ and contain the rather minimal set $C_c^\infty(\mathbb{R})$ in its domain then the function f must be in its domain also. But it is clear that for any $h \neq 0$ the function $x \mapsto f(x+h)$ is not in $L^2(\mathbb{R},\nu)$. Hence the strong derivative (3.4), with Lebesgue measure replaced by Gauss measure ν, does not exist for this function f. Now, if φ and ψ are in $C_c^\infty(\mathbb{R})$ then an integration by parts gives

$$(\varphi',\psi)_{L^2(\nu)} = (\varphi,(-d/dx+x)\psi)_{L^2(\nu)}. \tag{3.5}$$

Hence $(d/dx) \mid C_c^\infty(\mathbb{R})$ is contained in the L^2 adjoint of $(-d/dx+x) \mid C_c^\infty(\mathbb{R})$, which adjoint of course is a closed operator. This adjoint is by definition the weak derivative. In fact the closure of $((d/dx) \mid C_0^\infty(\mathbb{R}))$ is exactly equal to this weak derivative operator, as can be proved with the help of a Hermite expansion in a manner that goes over to infinite dimensions, but which we will not discuss here. Thus in the Gaussian versions of i), ii) and iii) above the versions ii) and iii) are equal while i) is not closed. These domain issues have been systematically studied by I. Shigekawa and S. Watanabe in infinite dimensions [Sh1, W, IW].

Since a densely defined operator in a Hilbert space has a closed extension if and only if its adjoint is densely defined the key point in proving closability of the operators ∂_h above is the following integration by parts formula, which in the linear case, $G = \mathbb{R}$, was first proven by Cameron [C].

CAMERON'S INTEGRATION BY PARTS FORMULA

$$\int_{W_G} (\partial_h F_1)F_2\,dP = \int_{W_G} F_1(-\partial_h + j_h)F_2\,dP \quad h \in H \tag{3.6}$$

where

$$j_h = \int_0^1 \langle \dot{h}(s), db(s)\rangle. \tag{3.7}$$

I have written (3.6) with the appearance of the general group G (of compact type) rather than merely $\mathbb{R}$ because it holds also in the general case as well and is a consequence of Theorem A of the previous section. For a discussion of the functions F_1 and F_2 for which (3.6) holds and a derivation of (3.6) from Theorem A see [Gr 7, Section 3]. At any rate it is clear that both sides of (3.6) are meaningful for bounded C^∞ cylinder functions (3.2) with bounded first derivatives.

A historical note: Cameron [C] deduced (3.6) from the Cameron–Martin translation theorem [CM] which these authors proved in case $h \in H$ and $\dot{h}$ is of bounded variation. (Here $G = \mathbb{R}$.) As a result Cameron stated his integration by parts formula only for these h. But for functions of the form (3.2) both sides of (3.6) are clearly continuous in h in the H topology and hence (3.6) extends immediately to all $h \in H$. Maruyama [Ma] extended the Cameron–Martin translation theorem to the general case — $h \in H$ soon after. This integration by parts formula is the basis for perturbation theory in the functional integral approach to quantum field theory and has therefore been ubiquitous in the physics literature since at least the late 40's. The degree of informality in its use and its non–Wiener space context may make it difficult to recognize for some. But the central role of H in the integration by parts formula was brought out clearly by Segal [S 1,2] in the late 50's and the full integration by parts formula was used by the author in 1967 [Gr 2] to prove regularity theorems for potential theory over abstract Wiener spaces.

It is Cameron's integration by parts formula that forces one to consider differentiations only in the directions H rather than some larger "tangent space" to W_G. For example, if v is in $C([0,1];\mathcal{G})$ and $v(0) = 0$ then, putting $(e^v)(s) = \exp v(s)$, it is clear that $t \mapsto e^{tv}$ is a one parameter subgroup of W_G and that $(\partial_v F)(g) := dF(e^{tv}g)/dt\big|_{t=0}$ exists for many functions F, e.g., those of the form (3.2). It would be reasonable to declare $\{\partial_v : v \text{ continuous}, v(0) = 0\}$ to be the Lie algebra of W_G. So, if say, $G = \mathbb{R}$ then we would simply be declaring the Banach space $W_{\mathbb{R}}$ to be its own tangent space. But if v is not in H then the random variable j_v (see (3.7)) fails to exist, the Cameron–Martin translation theorem fails to hold, ∂_v^* fails to be densely defined, and ∂_v fails to have a closed version in $L^2(W_{\mathbb{R}}, P)$. Thus in proving his integration by parts formula Cameron was the first to point out that it is of significant interest to differentiate a function f on the Banach space $W_{\mathbb{R}}$ ($\equiv$ Wiener space) only in a specific, dense set of directions, H, rather than all of the directions available. This derivative in H directions, the Cameron derivative, and the corresponding Cameron gradient (3.3), as distinguished from the gradient in all $W_{\mathbb{R}}$ directions (i.e., the Frechet or Gateaux derivatives) has played a central role in analysis over Wiener spaces ever since. The Cameron gradient was called the H–derivative in [Gr 2] to emphasize the distinguished role of H.

Since a fairly self contained description of the operators ∂_h, ∇ and ∇_0 is given in [Gr 7, Section 3] we will not specify further the exact domains of these operators. But we may and will take them to be closed. Thus $\partial_h^*\partial_h$ is a nonnegative self–adjoint operator on $L^2(W_G, P)$ as is $\nabla^*\nabla$. Here ∇ is a closed densely defined operator $\nabla : L^2(W_G, P) \to L^2(W_G, P) \otimes H$. Similarly $\nabla_0^*\nabla_0$ is a nonnegative self-adjoint operator on $L^2(\mathcal{L}, \mu)$. Of course we have

$$\nabla_0^*\nabla_0 = \sum_n \partial_{h_n}^* \partial_{h_n}$$

where $\{h_n\}_{n=1}^\infty$ is an orthonormal basis of H_0. The measure preserving Ito map I described in Section 2 sets up an isomorphism of $L^2(W_G, P)$ with $L^2(W_{\mathcal{G}}, w)$ and under this isomorphism $(1/2)\nabla^*\nabla$ is carried into the very well understood number operator, whose spectrum is $\{0, 1, 2, \dots\}$. Although P is a thinly disguised version of the Gaussian measure w, its conditioned measure μ on $\mathcal{L}$ is in no sense Gaussian if G is noncommutative. Correspondingly, the operator $\nabla_0^*\nabla_0$ is very poorly understood. It is this operator which we have discussed in the Introduction (on 0–forms). The little that is known about it will be discussed in the next section.

Exercise 3.2. Let $G = SO(3)$ again. Show that dimension kernel $\nabla_0^*\nabla_0 \geq 2$. Hint: You may assume that the function f described in Exercise 3.1 is in the domain of $\nabla_0^*\nabla_0$.

4. Statement of results.

The object of our study is the operator $\nabla_0^*\nabla_0$ on $L^2(\mathcal{L}, \mu)$. I have not been able so far to prove a spectral gap for this operator. But the following theorems amount to a small step in this direction. We continue to take $G =$ compact group times $\mathbb{R}^k$.

Theorem 4.1. *If G is simply connected then*

$$\textit{dimension of kernel } \nabla_0^*\nabla_0 = 1.$$

Corollary 4.2. *If G is not simply connected then*

$$\dim\ker(\nabla_0^*\nabla_0) = \textit{cardinality of fundamental group of } G.$$

Corollary 4.3. *If G is simply connected and $V : \mathcal{L} \to \mathbb{R}$ is measurable with V and e^{-V} in $\cap_{p<\infty} L^p(\mathcal{L}, \mu)$ then for any real number $\lambda > 0$ the operator*

$$\nabla_0^*\nabla_0 + \lambda|b(1)|^2 + V$$

is bounded below and the bottom of its spectrum is an eigenvalue λ_0 of multiplicity one. Moreover λ_0 belongs to a strictly positive eigenfunction.

Theorem 4.4. *Assume that G is simply connected. Then the left action of K_0 on $\mathcal{L}$ is ergodic. That is, if $f : \mathcal{L} \to \mathbb{R}$ is measurable and if for each $k \in K_0$*

$$f(k\gamma) = f(\gamma) \text{ for } \mu \text{ a.e. } \gamma \tag{4.1}$$

then there is a real number c such that $f(\gamma) = c$ for μ a.e. γ.

Before proceeding further let us remark on the relations between these results.

Remark 4.5. Theorems 4.1 and 4.4 are equivalent in the sense that each one follows from the other quite easily. For example if we know Theorem 4.1 holds and if f satisfies (4.1) and is in $L^2(\mathcal{L}, \mu)$ then for any h in H_0, $f(\gamma) = f(e^{th}\gamma)$ a.e., so $0 = d/dt\big|_{t=0} f(e^{th}\gamma) = \partial_h f(\gamma)$. Hence $\nabla_0 f = 0$. So $f \in \ker \nabla_0^* \nabla_0 = \{\text{constants}\}$. This argument is a little informal because we have ignored the problem of strong derivatives discussed in the preceding section. But it is not hard to make it rigorous. This shows that Theorem 4.1 implies Theorem 4.4. The converse is of similar simplicity. See [Gr 7, Lemma 5.2] for details.

Remark 4.6. Corollary 4.2 is a rather easy consequence of Theorem 4.1. If $\widehat{G}$ denotes the universal covering group of G then a homotopy class of paths in G lifts up to the set of continuous paths in $\widehat{G}$ which begin at the identity and end at some fixed point x in $\widehat{G}$. Theorem 4.1 can be applied to this set of paths in $\widehat{G}$ (even though they don't end at e) to conclude that $\ker \nabla_0^* \nabla_0 \big|$(single homotopy class) has dimension one. This proves Corollary 4.2. See [Gr 7, Section 9] for details.

Remark 4.7. Corollary 4.3 has two parts — existence and uniqueness (of lowest eigenstates). There are two sets of techniques, one for existence and one for uniqueness. Both techniques live in the context of positivity preserving semigroups, which are always generated by such Schrödinger operators. For existence one needs a regularity condition in the form of a logarithmic Sobolev inequality (see e.g. [Gr 3, 4, 6, 9]) while for uniqueness one needs precisely Theorem 4.1 (see e.g. [Gr 7, Section 10]). Once one has these two ingredients the existence and uniqueness follow by techniques which were developed by several authors in the context of constructive quantum field theory in the early 1970's and late 1960's. These techniques are infinite dimensional versions of the Perron–Frobenius theorem, which ensures the existence of maximum eigenvectors of matrices with positive entries.

In the rest of these notes we will focus just on Theorems 4.1 and 4.4 and on the following finite dimensional theorem. It seems to this author remarkable that the next theorem, which involves only analysis over the finite dimensional group G and which could therefore have been discovered many years ago was not actually discovered until forced by the infinite dimensional problems which we discussed above. Denote by T the space of finite rank tensors over $\mathcal{G}$. T is an algebra with respect to tensor multiplication. Let J be the two sided ideal in T generated by $\{\xi \otimes \eta - \eta \otimes \xi - [\xi, \eta]; \xi, \eta \in \mathcal{G}\}$. Denote by T^* the completion of T in the norm $\|\sum_{k=0}^m \beta_k\|^2 = \sum (k!)^{-1} |\beta_k|^2_{\mathcal{G}^{\otimes k}}$ with $\beta_k \in \mathcal{G}^{\otimes k}$ and write $J^\perp$ for the annihilator of

J in T^* for the pairing $\langle\alpha,\beta\rangle = \sum_{n=0}^{\infty}\langle\alpha_n,\beta_n\rangle$. For any vector ξ in $\mathcal{G}$ and v in T define $R_\xi v = v\otimes\xi$. Then since $R_\xi J\subset J$ we have $R_\xi^* J^\perp \subset J^\perp$. Let $A_\xi = R_\xi^* \mid J^\perp$. Finally let $\Omega = 1\oplus 0\oplus 0\oplus\cdots$ be the unit zero rank tensor. One verifies easily that $\Omega\in J^\perp$.

Let $d\nu(x) = p_1(x)dx$ be the heat kernel measure on G at time $t=1$. (See Section 2.)

Theorem 4.8. *Assume G is simply connected and of compact type. There exists a unique surjective isometry*

$$U : L^2(G,\nu)\to J^\perp \tag{4.2}$$

such that

a) $U(1)=\Omega$

b) $U\widetilde{\xi}U^{-1} = -A_\xi,\ \ \xi\in\mathcal{G}$

where $\widetilde{\xi}$ denotes the closed, right invariant first order differential operator in $L^2(G,\nu)$ associated to ξ.

Example 4.9. Let $G=\mathbb{R}^k$. Since G is commutative J is the ideal generated by 2–tensors of the form $\xi\otimes\eta-\eta\otimes\xi$, $\xi,\eta\in\mathcal{G}$. In particular J contains all skew symmetric 2–tensors. Hence the second order component of any tensor in $J^\perp$ is necessarily symmetric. Moreover since all tensors of the form $(\xi\otimes\eta-\eta\otimes\xi)\otimes\zeta$ with ξ,η,ζ in $\mathcal{G}$ are in J it follows that for any tensor u in $J^\perp$ its 3rd order component u_3 is symmetric in its first and second factors. Similarly, putting ζ on the other side we see that u_3 is also symmetric in its second and third factors. Thus u_3 is symmetric. Similarly all the components of u are symmetric. Thus

$$J^\perp = \{\text{symmetric tensors in } T^*\}.$$

Now the heat kernel measure on $\mathbb{R}^k$ is just Gauss measure:

$$d\nu(x) = (2\pi)^{-k/2}e^{-|x|^2/2}dx.$$

Moreover the right invariant first order differential operators on $\mathbb{R}^k$ are just the operators

$$\widetilde{\xi} = \xi\cdot\nabla.$$

Choosing ξ to be a standard basis element of R^k we see that Theorem 4.8 reduces, when $G=\mathbb{R}^k$, to the statement that there exists a unique surjective isometry

$$U : L^2(\mathbb{R}^k,\text{Gauss})\to\{\text{sym.tensors in } T^*\}$$

such that

a) $U(1)=\Omega$

b) $U(\partial/\partial x_j)U^{-1} = -A_{e_j}$

This is precisely the finite dimensional version of the Wiener–Ito–Kakutani–Segal isomorphism [S 1]!

5. Sketch of proofs.

Let us think in terms of proving Theorem 4.4 (ergodicity of the left action of K_0 on $(\mathcal{L}, \mu)$). However the differential operators that enter into Theorem 4.1 will play a significant role in the proof. Throughout this section we take G to be simply connected. Since many portions of the proof "factor" commensurately with the product $G =$ compact group times $\mathbb{R}^k$ and since the case $G = \mathbb{R}^k$ is already well known (see e.g. Example 4.9) we will assume henceforth that G is both compact and simply connected. Now $(\mathcal{L}, \mu)$ is in this case in no reasonable sense a Gaussian measure space. On the other hand μ is defined (by conditioning) in terms of the measure P on W_G, which via the Ito map I of Section 2 is a Gaussian measure, w, on $W_{\mathcal{G}}$, just thinly disguised. Of course one must use, somehow, the definition of μ and thereby reduce the problem concerning μ to a problem concerning P.

Step 1. Reduction of the problem of ergodicity over the non Gaussian measure space $(\mathcal{L}, \mu)$ to a problem over the Gaussian measure space (W_G, P).

As we noted in Section 2 the map $\varphi : g \to g(1)$ is a homomorphism from the group W_G onto G with kernel $\mathcal{L}$. Thus $W_G/\mathcal{L} = G$. Of course this does not mean that $W_G = \mathcal{L} \times G$ (it does in the trivial case, i.e., $G = R^k$ which we are not considering). However W_G is always fibered over G with fiber $\mathcal{L}$ and so for some small neighborhood V of e in G, $\varphi^{-1}(V)$ is indeed a product: $\varphi^{-1}(V) = \mathcal{L} \times V$. Hence if $f : \mathcal{L} \to \mathbb{R}$ is K_0 invariant then we can extend it to a K_0 invariant function $F : W_G \to \mathbb{R}$ by putting

$$F(\gamma \times v) = f(\gamma) \text{ on } \varphi^{-1}(V)$$

and $F(g) = 0$ off the tubular neighborhood $\varphi^{-1}(V)$. The reader is advised to draw a picture in order to see that f is constant if and only if $F(g)$ depends only on $\varphi(g)$ (i.e., only on $g(1)$). See [Gr 7, Lemma 9.1] and [Gr 6, Lemma 4.3] for details of the construction of this tubular neighborhood and measure theoretic details. Now to prove Theorem 4.4 it suffices to consider only bounded functions f, in which case F is also bounded. Hence it suffices to prove the following theorem.

Theorem 5.1. *Any K_0 invariant function F in $L^2(W_G, P)$ depends only on the endpoint of g. That is, if for each element $k \in K_0$*

$$F(kg) = F(g) \text{ for } P \text{ a.e. } g \in W_G$$

then there exists a function $u : G \to \mathbb{R}$ such that

$$F(g) = u(g(1)) \text{ for } P \text{ a.e. } g \in W_G. \tag{5.1}$$

The strategy we will use for proving Theorem 5.1 captures the spirit of two of the standard proofs of ergodicity of the irrational flow on the torus. In one proof one takes a function F in L^2 (torus) which is invariant under the irrational flow and expands it in a double Fourier series. Invariance easily implies that all the Fourier coefficients are zero except for the constant term. So F is constant a.e. In the second proof one convolutes F with a C^∞ approximate δ function. The resulting continuous (in fact C^∞) function is constant on the dense orbits, hence is constant. Hence F is an L^2 limit of constants, hence is equal a.e. to a constant. Both of these ideas will be visible below, the first in Step 2, the second in Step 3. In our Gaussian setting, however, it should come as no surprise that we will use Hermite expansions (i.e., Ito expansions) instead of group representations. We will first characterize the multiple Ito expansions of K_0 invariant functions.

Step 2. Characterization of the Ito expansion of K_0 invariant functions in $L^2(W_G, P)$.

Theorem 5.2. *For any function F in $L^2(W_G, P)$ write*

$$F = \sum_{n=0}^{\infty} \int_{\Delta_n} \langle v_n(s_1, \dots, s_n), db(s_1) \otimes \cdots \otimes db(s_n) \rangle_{\mathcal{G}^{\otimes n}} \quad \textit{(Ito)} \tag{5.2}$$

for the Ito expansion of F. Here $\Delta_n = \{(s_1, \dots, s_n) : 0 < s_1 < \cdots < s_n < 1\}$ and $v_n \in L^2(\Delta_n, \mathcal{G}^{\otimes n})$. Then F is K_0 invariant if and only if

a) *each v_n is constant a.e. on Δ_n, say $v_n(\cdot) = (-1)^n \alpha_n \in \mathcal{G}^{\otimes n}$ a.e. on Δ_n and*

b) *the element $\alpha = \alpha_0 + \alpha_1 + \alpha_2 + \cdots$ (which is automatically in T^*) is in $J^\perp$.*

Sketch of proof of Theorem 5.2. Just as in the discussion of Remark 4.5, K_0 invariance of F is equivalent to the set of equations

$$\partial_h F = 0 \text{ for all } h \in H_0. \tag{5.3}$$

Now a symmetric n tensor over the real Hilbert space $L^2([0,1]; \mathcal{G})$ can be identified with a certain function u_n in $L^2([0,1]^n; \mathcal{G}^{\otimes n})$ satisfying some symmetry properties and in fact the restriction v_n of u_n to Δ_n completely determines u_n. In this sense the sequence $v := (v_0, v_1, v_2, \dots)$ of Ito expansion coefficients in (5.2) determines an element of the symmetric Fock space over $L^2([0,1]; \mathcal{G})$ in an isometric way. Moreover if G were simply $\mathbb{R}^k$ then the differential operators ∂_h would act on this Fock space simply as annihilation operators A_h, as already mentioned in Example 4.9. But in the noncommutative case — which is really the only case of interest to us — it turns out that the action of ∂_h on Fock space is given by the operator

$A_h - d\Gamma(adh)$. See [Gr 7, Section 4] for details. In particular the equations (5.3) can now be transferred to explicit and reasonably simple equations on the coefficients v_n to deduce equivalence of (5.3) with the conditions a) and b) of Theorem 5.2. See [Gr 7, Section 5] for details.

Step 3. Regularization and reduction to the C^∞ category. We now know, from Theorem 5.2, that any K_0 invariant function F in $L^2(W_G, P)$ has the representation

$$F = \sum_{n=0}^{\infty} (-1)^n \int_{\Delta_n} \langle \alpha_n, db(s_1) \otimes \cdots \otimes db(s_n) \rangle_{\mathcal{G}^{\otimes n}} \quad \text{(Ito)} \tag{5.4}$$

for some $\alpha \equiv (\alpha_0, \alpha_1, \alpha_2, \dots)$ in $J^\perp$. We will use this representation to regularize the problem, temporarily, so as to remove the measure zero subtleties that dominate the ergodicity problem. Let b^ε be a C^∞ regularization of Brownian paths: one can think of convoluting b by a C^∞ approximate δ function. Define g^ε by the equation

$$dg^\varepsilon(s)/ds = (db^\varepsilon(s)/ds) g^\varepsilon(s), \; g^\varepsilon(0) = e. \tag{5.5}$$

Define F^ε by replacing b in (5.4) by b^ε. Since (5.5) sets up a one to one correspondence between the smooth $\mathcal{G}$ valued paths b^ε and the smooth G valued paths g^ε we may think of F^ε as a function of g^ε instead of b^ε. But the same arguments used to show that any F given by (5.4) is K_0 invariant also show that $F^\varepsilon(g^\varepsilon)$ is $C^\infty \cap K_0$ invariant. Hence, even more easily than in the dense orbit argument for the irrational flow on the torus, one can see that $F^\varepsilon(g^\varepsilon)$ depends only on the endpoint of g^ε. That is, there is a function $u : G \to \mathbb{R}$ such that

$$F^\varepsilon(g^\varepsilon) = u(g^\varepsilon(1)).$$

Moreover it happens that u does not depend on ε but only on α. Thus we may hope now to let $\varepsilon \downarrow 0$ to obtain $F(g) = u(g(1))$, which would finally prove Theorem 5.1. Unfortunately the series (5.4), with b replaced by b^ε, does not converge, as $\varepsilon \downarrow 0$, to the function F defined by the series (5.4) of multiple Ito integrals but instead converges to this series with Ito integrals replaced by Stratonovich integrals. One has thereby inadvertently "reverse Wick ordered" the original function F — a terribly singular operation. In order to compensate for this erroneous limit one must start with a different α — one which corresponds to a highly regular function $\widehat{F}$ — so that in the limit described above one gets the desired function F represented as $u(g(1))$. To carry this out one must first show that $J^\perp$ contains a dense set of sufficiently regular elements α (analytic vectors for the annihilation operators) and then apply some of the machinery of Hu–Meyer, Johnson and Kallianpur, and Sugita which deals with the connection between multiple Ito integrals and multiple Stratonovich integrals. See [Gr 7, Sections 6, 7, 8] for details.

Finally let us see how Theorems 5.1 and 5.2 yield Theorem 4.8. By these two theorems the set of K_0 invariant functions F in $L^2(W_G, P)$ are in one to one correspondence with functions $u : G \to \mathbb{R}$ (see equation (5.1)) and at the same time in one to one correspondence with elements α in $J^\perp$ (see equation (5.4)). Taken together we therefore have a one to one correspondence $\alpha \leftrightarrow u$ which, because the heat kernel $p_1(x)$ gives the distribution of $g(1)$ in G, is an isometry, as required in Theorem 4.8. The condition b) in Theorem 4.8 is simply a consequence of the correspondence $\partial_h \sim A_h - d\Gamma(ad\, h)$ mentioned above. For details see [Gr 7].

6. Epilogue.

Theorem 4.8 is finite dimensional in the sense that it concerns functions over the finite dimensional Lie group G rather than over the infinite dimensional groups W_G or $\mathcal{L}$. Moreover it is non probabilistic in the sense that the measure ν on the left side of (4.2) is determined by a partial differential equation (the heat equation) while the right side is defined by some multilinear algebra. Yet the proof that I have sketched above is infinite dimensional as well as probabilistic in the sense that it analyzes functions over W_G and uses Brownian motion (in $\mathcal{G}$ and G) extensively. This theorem really ought to have a direct analytical proof. A largely analytic proof has recently been found by Omar Hijab [Hi]. Hijab found an explicit formula for the map $u \to \alpha$ from $L^2(G, \nu)$ into $J^\perp$ discussed at the end of Section 5 and showed that it is isometric. His formula avoids the above indirect construction which uses the G valued process as an intermediate step. There is also another benefit to his approach. The reader may be familiar with the fact that the Hermite polynomials H_n arise naturally when applying the creation operators $(\partial/\partial x_j)^*$ to the "vacuum state" 1 in $L^2(R^k$,Gauss). Specifically $H_n(x_j) = [(\partial/\partial x_j)^*]^n 1$. We pointed out in [Gr 8] that our analysis of $L^2(G$,heat kernel) (G compact) leads similarly to certain functions on G which are analogs of the Hermite polynomials. For example $(\widetilde{\xi}^*)^n 1$ would be such a function, where $\widetilde{\xi}$ is a right invariant vector field on G. Hijab has shown that these functions are given by the precise analog of a well known formula for Hermite polynomials. One need only replace $e^{-x^2/2}$ by the heat kernel, $p_1(x)$, on G!

Unfortunately the surjectivity of the map $L^2(G, \nu) \to J^\perp$ given by Hijab's formula still depends on a little bit of the infinite dimensional analysis in [Gr 7]. Even taking this into account Hijab's theorem already allows one to bypass the most difficult estimates in [Gr 7] associated to the replacement of multiple Stratonovich integrals by multiple Ito integrals and can therefore be used to give a considerable simplification of the main results in [Gr 7] — namely Theorems 4.1 and 4.4. above.

But in addition B. Driver has recently found a way to eliminate completely the G valued Brownian motion from the proof of Theorem 4.8. His methods build on those of Hijab and those of Brian Hall [H]. Brian Hall has found a noncommutative analog of the Segal–Bargmann transform. This transform is a natural isomorphism between (complex) $L^2(\mathbb{R}^k$, Gauss) and the space of holomorphic functions in $L^2(\mathbb{C}^k$,

Gauss). There are thus three distinct looking spaces that are naturally isomorphic, the third being {symmetric tensors over $\mathbb{C}^k$}. (Complexify Example 4.9.) In Brian Hall's work he replaces $\mathbb{R}^k$ by a compact group G, he replaces $\mathbb{C}^k$ by the complexification of G and he replaces $e^{-|x|^2/2}$ by the appropriate heat kernel, both in order to define the spaces as well as to define the transform (a kind of modified Laplace transform). This transform seems to capture in a clean analytical way some of the spirit of "Wick ordering". See [H].

Finally it should be mentioned that Theorem 4.8 extends easily to quotient spaces. If G is a simply connected compact group and K is a connected closed subgroup with Lie algebra $\mathcal{K} \subset \mathcal{G}$ then the isomorphism U carries the subspace $L^2(G, p_1(x)dx)^K$ of right K invariant functions in $L^2(G, p_1(x)dx)$ onto the subspace of $J^\perp$ which is orthogonal to the right ideal $\mathcal{K}T$. Since one has a natural isometry between $L^2(G/K, q_1(y, eK)dy)$ and $L^2(G, p_1(x)ds)^K$ where $q_t(y, z)$ is the heat kernel density on G/K for the natural Laplacian on G/K, one gets a unique isometry $V : L^2(G/K, q_1(y, eK)dy) \to (J + \mathcal{K}T)^\perp$ with analogous properties to a) and b) of Theorem 4.8.

References.

[AH] S. Albeverio and R. Hoegh–Krohn, *The energy representation of Sobolev Lie groups*, Compositio Math. **36** (1978), 37–52.

[AM] H. Airault and P. Malliavin, *Integration geometrique sur l'espace de Wiener*, Bull. Sc. math 2^e serie **112** (1988), 3–52.

[C] R. H. Cameron, *The first variation of an indefinite Wiener integral*, Proc. Amer. Math. Soc. **2** (1951), 914–924.

[CM] R. H. Cameron and W. T. Martin, *On transformations of Wiener integrals under translations*, Ann. of Math. **45** (1944), 386–396.

[D1] B. Driver, *A Cameron–Martin type quasi–invariance theorem for Brownian motion on a compact Riemannian manifold*, J. of Funct. Anal. **110** (1992), 272–376.

[D2] ———, *A Cameron–Martin type quasi–invariant theorem for pinned Brownian motion on a compact Riemannian manifold*, Trans. Amer. Math. Soc., in press.

[E] J. Eells, *Integration on Banach manifolds*, Proc. 13th Biennial Seminar of the Canadian Mathematical Congress, Halifax (1971), 41–49.

[EE] J. Eells and K. D. Elworthy, *Wiener integration on certain manifolds*, "Some Problems in Non–Linear Analysis", Edizioni Cremonese, Rome, C.I.M.E. IV (1971), 67–94.

[Ew] K. D. Elworthy, *Stochastic differential equations on manifolds*, Cambridge Univ. Press, London Math. Soc. Lecture Note Series **70** (1982).

[Ge1] E. Getzler, *Dirichlet forms on loop spaces*, Bull. Sc. math. 2^e serie **113** (1989), 151–174.

[Gr1] L. Gross, *Abstract Wiener spaces*, U. of Calif. Press, Proc. Fifth Berkeley Symposium on Math. Stat. and Prob. (1965), 31–42.

[Gr2] ———, *Potential theory on Hilbert space*, J. Funct. Anal. **1** (1967), 123–181.

[Gr3] ———, *Existence and uniqueness of physical ground states*, J. of Funct. Anal. **10** (1972), 52–109.

[Gr4] ———, *Logarithmic Sobolev inequalities*, Amer. J. Math. **97** (1975), 1061–1083.

[Gr5] ———, *Logarithmic Sobolev inequalities on Lie groups*, Ill. J. of Math. **36** (1992), 447–490.

[Gr6] ———, *Logarithmic Sobolev inequalities on loop groups*, J. of Funct. Anal. **102** (1991), 268–313.

[Gr7] ———, *Uniqueness of ground states for Schrödinger operators over loop groups*, J. of Funct. Anal. **112** (1993), 373–441.

[Gr8] ———, *The homogeneous chaos over compact Lie groups*, in "Stochastic Processes" (S. Cambanis, et al., eds.), Springer–Verlag, New York, 1993.

[Gr9] ———, *Logarithmic Sobolev inequalities and contractivity properties of semigroups*, in "Dirichlet Forms" Lecture Notes in Mathematics #1563 (G. Dell'Antonio, U. Mosco, eds.), Springer–Verlag, New York, 1993.

[H] B. Hall, *The Segal–Bargmann "Coherent State" transform for compact Lie groups*, J. Funct. Anal. (to appear).

[He] S. Helgason, *Differential Geometry and Symmetric Spaces*, Academic Press, San Diego, 1962.

[Hi] O. Hijab, *Hermite functions on compact Lie groups*, J. Funct. Anal. (to appear).

[Hs] E. P. Hsu, *Quasiinvariance of the Wiener measure and integration by parts in the path space over a compact Riemannian manifold*, J. of Funct. Anal. (to appear).

[IW] N. Ikeda and S. Watanabe, *Stochastic Differential Equations and Diffusion Processes*, North–Holland, New York, 1981.

[JL] J. D. S. Jones and R. Leandre, *L^p Chen forms on loop spaces*, in "Stochastic Analysis" (Barlow and Bingham, eds.), Cambridge University Press, 1991, pp. 103–162.

[K3] S. Kusuoka, *DeRham cohomology of Wiener-Riemannian manifolds*, Preprint RIMS, Kyoto University, August 1990.

[Ku1] H. H. Kuo, *Integration theory on infinite-dimensional manifolds*, Trans. Amer. Math. Soc. **159** (1971), 57–78.

[Ku2] ———, *Diffusion and Brownian motion on infinite-dimensional manifolds*, Trans. Amer. Math. Soc. **169** (1972), 439–457.

[Ku3] ———, *Ornstein–Uhlenbeck process on a Riemann–Wiener manifold*, Kyoto 1976, K. Itô, Ed., Proc. Symp. on Stoch. Diff. Eqs., 187–193.

[MM] M–P. Malliavin and P. Malliavin, *Integration on loop groups I. Quasi invariant measures*, J. of Funct. Anal. **93** (1990), 207–237.

[Ma] G. Maruyama, *Notes on Wiener integrals*, Kodai Math. Sem. Reports, No. 3, June 1950, 41–44.

[P] M. A. Piech, *The exterior algebra for Wiemann manifolds*, J. of Funct. Anal. **28** (1978), 279–308.

[PS] A. Pressley and G. Segal, *Loop Groups*, Oxford University Press, Oxford, 1986.

[S1] I. E. Segal, *Tensor algebras over Hilbert spaces*, Trans. Amer. Math. Soc. **81** (1956), 106–134.

[S2] ———, *Distributions in Hilbert space and canonical systems of operators*, Trans. Amer. Math. Soc. **88** (1958), 12–41.

[Sh1] I. Shigekawa, *Absolute continuity of probability laws of Wiener functionals*, Proc. of the Japan Academy **54** (1978), 230–233.

[Sh2] ———, *Derivatives of Wiener functionals and absolute continuity of induced measures*, J. Math. Kyoto Univ. **20** (1980), 263–289.

[Sh3] ———, *DeRham–Hodge–Kodaira's decomposition on an abstract Wiener space*, J. Math. Kyoto Univ. **26** (1986), 191–202.

[Sh4] ———, *Transformations of the Brownian Motion on the Lie Group*, in "Proceedings, Taniguchi International Symposium on Stochastic Analysis, Katata and Kyoto, 1982" (Kiyosi Itô, ed.), North–Holland, Amsterdam, 1984, pp. 409–422.

[W] S. Watanabe, *Lectures on Stochastic Differential Equations and Malliavin Calculus*, Tata Institute of Fundamental Research, Springer–Verlag, Berlin–Heidelberg–New York, 1984.

[Wi] N. Wiener, *The homogeneous chaos*, Amer. J. Math. **60** (1938), 897–936.

WHITE NOISE ANALYSIS AND APPLICATIONS

TAKEYUKI HIDA
Department of Mathematics
Meijo University
Tenpaku, Nagoya 468, Japan

1 Introduction

This article presents some of the recent development in white noise analysis as an infinite dimensional calculus.

Significant progress has been made since the time when the notion of a generalised white noise functional was introduced in 1975, and the subject of white noise analysis has undergone a vast development. At present it seems to be a good time to revisit the idea of having introduced such a general notion and to think of further progress on white noise analysis in search of new problems in this line.

This article starts, therefore, with a brief observation of the basic idea of how to carry on stochastic analysis. With this idea we may overview the recent progress of the analysis. Then, we shall see further development on the work and new problems naturally proposed for random fields that are expressed as generalized white noise functionals.

The idea is illustrated as follows. When we are given a functional of random variables, it would be fine if it can be expressed as a function of a system of basic random variable. The most basic system of random functions is an I.I.D. (independent identically distributed) variables; among others an independent system of Gaussian variables:

$$\{X_n; n \in \mathbf{Z}\} .$$

A continuous analogue of an I.I.D. system is a generalized stochastic process with *independent values at every point.* Assuming the distribution to be Gaussian, we are given a *white noise*, which is the time derivative $\dot{B}(t)$ of a Brownian motion $B(t)$:

$$\{\dot{B}(t); t \in R\} .$$

Either X_n or $\dot{B}(t)$ may be thought of as a member of the variable system for random functions to be discussed.

A. I. Cardoso et al. (eds.), Stochastic Analysis and Applications in Physics, 119–131.

Once a system of variables is given, we then come to choose a class of basic functions of the given variables. In the continuous parameter case, we introduce such functionals as

$$\varphi(\dot{B}(t), t \in R) .$$

We have first to take polynomials in the $\dot{B}(t), t \in R$, which are considered to be most basic. This was the original idea of introducting the notion of generalized white noise functionals. With great contributions by many authors, the most suitable class of generalized functionals has been defined (see [9], Chapt. 4) and it is now denoted by $(S)^*$.

We were then led to a stochastic process $X(t)$ which lives in the space $(S)^*$. One may ask what is a generalization of $X(t)$ when t is generalized. Maybe t is multi-dimensional vector in R^d (note: $\{t\}$is still one-dimensional) or t is a manifold of dimension $< d$.

Suppose the latter be chosen. Then we are given a random field $X(C)$ indexed by a manifold C in R^d. The calculus is still in line with our white noise analysis, if $X(C)$ is assumed to be in $(S)^*$.

A generalization of the stochastic differential equation for $X(t)$ is now a generalized *stochastic variational equation* for $X(C)$.

The purpose of the present article is now clearly stated. It is to give a proposed approach to a stochastic variational equation for a random field $X(C)$. First some necessary background for a multiparameter white noise shall be prepared, then we shall discuss various cases where actual solutions are obtained. This is of course far from a general theory, but gives a proposal for a white noise approach to this question.

2 Background

First an R^d-parameter white noise is introduced. Take a nuclear space E of smooth functions on R^d. Then, on the dual space E^* is a probability measure μ which is uniquely determined by the characteristic functional $C(\xi)$ given by

$$C(\xi) = \exp[-\frac{1}{2}\|\xi\|^2] , \qquad \|\cdot\| \text{ being the } L^2(R^d) - \text{norm}, \tag{2.1}$$

in such a way that

$$C(\xi) = \int_{E^*} \exp[i < x, \xi >] d\mu(x) . \tag{2.2}$$

Ordinary (complex-valued) white noise functionals can be realized as members of the complex Hilbert space $(L^2) = L^2(E^*, \mu)$.

The analysis on (L^2) has now been well established, and we were led to introduce a class $(S)^*$ of *generalized* white noise functionals in order to carry on stochastic analysis. Indeed, our favourite class of generalized white noise functionals is taken to be $(S)^*$, which is introduced by the following Gel'fand triple:

$$(S) \subset (L^2) \subset (S)^* , \tag{2.3}$$

where (L^2) is the Hilbert space of ordinary white noise functionals, and where (S) is a test functional space being nuclear and $(S)^*$ is the dual space of (S).

This choic of $(S)^*$ is fitting for our white noise analysis, including the stochastic variational calculus mentioned in the introduction. Significant characteristics of the space $(S)^*$ are now in order.

1. As was illustrated in the Introduction each $\dot{B}(t)$ may be thought of as an *atomic or elementary random element* for the expression of random functions. It seems quite reasonable first to take polynomials in the $\dot{B}(t)$, which are most basic functions once variables are given. Such polynomials are members of $(S)^*$ and are dense there.

2. The space $(S)^*$ contains, in addition to the polynomials, enough functionals, like Gauss kernels and delta functions of a Brownian motion, in order to carry on the variational calculus and on the so-called stochastic *causal* calculus, where the time evolution is involved explicitly.

3. While, $(S)^*$ is minimal in a sense that important examples are included, and the stochastic harmonic analysis, which is the main topic of this paper, can be developed exactly within this space. In particular, on $(S)^*$ the Lévy Laplacian is acting.

4. There is a transformation which is called the S-transform of $(S)^*$-functionals. It carries $(S)^*$ to the space of functionals $U(\xi)$ called *U-functionals*, of a smooth function $\xi \in E$. Our analysis can therefore appeal to the classical theory of functional analysis.

5. Good characterization of $(S)^*$-functionals is given by Potthoff-Streit in terms of U-functional (see next section).

Another notion that must be prepared is the *infinite dimensional rotation group* O_∞ introduced by H. Yoshizawa in 1969 [18]. It plays important roles in the white noise calculus, because, for one thing, the white noise measure μ is kept invariant under the action of the rotation group. Based on this group the so-called *harmonic analysis,* in particular, variation of $X(C)$ can be obtained by the action of the certain subgroups of O_∞. This will be prescribed in the next section.

Then, we will discuss several interesting applications. Among others, having been motivated by equations of Schrödinger type in relativistic quantum field theory, we shall deal with some class of stochastic variational equations in §4.

3 Tools for the analysis

We introduce a rotation group of the space E. Let g be a linear homeomorphism of E such that $\|g\xi\| = \|\xi\|$ holds for every $\xi \in E$. Such a transformation g is called *a rotation* of E. Obviously the collection of those g's forms a group and is topologized by the compact-open topology in the usual manner. The group is denoted by $O(E)$ and is called the *rotation group of* E. It is often called an *infinite dimensional rotation group* and denoted simply by O_∞, if no need to specify E. The adjoint g^* of g is a linear transformation acting on the space E^*. It is well known that

$$g^*\mu = \mu \qquad \text{holds for any} \quad g^* \quad \text{with} \quad g \in O_\infty . \tag{3.1}$$

This property is a most fundamental property of O_∞.

Next we shall establish an integral representation of the functionals in $(L^2) = L^2(E^*, \mu)$. It is done by the S-Transform:

$$(S\varphi)(\xi) = \int_{E^*} \varphi(x+\xi)d\mu(x), \qquad \varphi \in (S)^* . \tag{3.2}$$

An alternative expression of the S-transform is

$$(S\varphi)(\xi) = \exp[-\frac{1}{2}\|\xi\|^2] \cdot \int_{E^*} \exp[< x, \xi >]\varphi(x)d\mu(x) , \tag{3.3}$$

where the integral looks like the Laplace transform. The S-transform actually gives an integral representation of (L^2)-functionals.

What is important, regarding the S-transform, is a characterization of U-functionals. Given a functional $U(\xi)$ on E. Set

$$f(\lambda, \xi, \eta) = U(\lambda\xi + \eta) . \tag{3.4}$$

Proposition 3.1. (J. Potthoff and L. Streit). *A functional $U(\xi)$ on E is a U-functional if and only if*

i) *$f(\lambda, \xi, \eta)$ is extended to an entire analytic function $f(z, \xi, \eta)$ over the complex plane, and*

ii) *$f(z, \xi, 0)$ is of exponential type of order 2 for ξ in a bounded region in E_n for some $n > 0$.*

For proof, we refer to [9, Chapt. 4].

Next topics to be prepared in this section are the creation and annihilation operators. Let $\varphi(x)$ be an $(S)^*$-functional and let $U(\xi)$ be the U-functional associated with φ. Define a differential operator ∂_t by the formula

$$\partial_t\varphi = S^{-1}\frac{\delta}{\delta\xi(t)}U(\xi) , \qquad \frac{\delta}{\delta\xi(t)} \quad : \quad \text{the Fréchet derivative} . \tag{3.5}$$

The ∂_t is a continuous derivation acting on (S) and it plays the role of *annihilation.* While its adjoint ∂_t^* defines a *creation.*

There is a very important Laplacian operator called thte *Lévy Laplacian* and denoted by Δ_L. Its counterpart, denoted by $\tilde{\Delta}_L$, acting on U-functionals is given by using the Gâteaux derivative D_ξ in the direction ξ: Let $\{\xi_n\}$ be an equally dense (see [1] 3ème Part. Chap. III) complete orthonormal system in $L^2(I^d), I = [0,1]$. Set

$$\tilde{\Delta}_L = \lim_{N\to\infty} \frac{1}{N} \sum_{n=1}^{N} D_{\xi_n}^2 \, . \tag{3.6}$$

Then. we are given the Lèvy Laplacian Δ_L (see [9] Chapt. 6):

$$\Delta_L = S^{-1}\tilde{\Delta}_L S \ , \qquad S \quad \text{being the} S-\text{transform} \ . \tag{3.7}$$

It is convenient to use Kuo's notation in the calculation:

$$\Delta_L = \int \partial_t^2 (dt)^2 \ , \tag{3.8}$$

which is somewhat formal, however a reasonable and correct interpretation can be given. An advantage of the expression (3.8) is that the parameter set can be taken to be R^d or any domain instead of I^d.

It is known that the Lèvy Laplacian has close connection with the Lèvy group $\mathcal{G}$ (see [1] 3ème Part.) which is an important subgroup of O_∞ involving simultaneous permutations of infinitely many coordinates, so that they can *never* be approximated by the finite dimensional rotations. The group $\mathcal{G}$ is a discrete group.

The group $\mathcal{W}$ defined by

$$\mathcal{W} = \mathcal{G} \vee G_\infty$$

is called the *windmill subgroup* of O_∞.

The Lévy Laplacian acts, in a sense, effectively on the space $(S)^*$ of generalized white noise functionals and it annihilates all the functionals in (L^2), on which the infinite dimensional Laplace-Beltrami operator $\Delta_\infty = \int \partial_t^* \partial_t dt$ acts effectively. Moreover, with the notation

$$U_g\varphi(x) = \varphi(g^* x) \ , \qquad \varphi \in (S)^*, \qquad g \in O_\infty \ , \tag{3.9}$$

the following assertion can be proved (cf. [11]).

Proposition 3.2. i) *Lévy Laplacian commutes with* $U_g, g \in \mathcal{W}$, *while,* Δ_∞ *does not.*

ii) *Gauss kernels*: $\exp[c\int_T x(t)^2 dt] :, c \neq \frac{1}{2}$, *are eigen functions of* Δ_L *with eigenvalue* $\frac{2c}{1-2c}|T|$.

The importance of the Lévy Laplacian is thus emphasized.

There is another important subgroup of O_∞. There is a rotation that comes from a diffeomorphism $\beta(u)$ of $\bar{R}^d$:

$$g_\beta \ ; \quad \xi(u) \longrightarrow \xi(\beta(u))\sqrt{|\beta'(u)|} \ , \qquad \beta'(u) : \quad \text{Jacobian} \ . \tag{3.10}$$

If $\{g_t\}$ is a one-parameter subgroup of O_∞; each member g_t comes from a diffeomorphisms β_t with the property $\beta_t \circ \beta_s = \beta_{t+s}$, then we have

$$g_t g_s = g_{t+s} \ .$$

Further if g_t is continuous in t, then $\{g_t\}$ is called a *whisker*. Whiskers will serve to deform a manifold C in R^d, when variational calculus is performed.

Thus one can see one of the characteristics of our analysis; the use of the rotation group O_∞.

4 Random fields $X(C)$

Consider a random field $\{X(C); C \in \mathbf{C}\}$ indexed by C which runs through a certain class of manifolds in R^d. As is explained before, it is assumed that $X(C)$ is a random variable in $(S)^*$; namely $X(C) = X(C,x) \in (S)^*$.

We are interested in the so-called causal calculus, so that we further assume that $X(C,x)$ is $\mathcal{B}(C)$ measurable, where $\mathcal{B}(C)$ is the σ-field generated by $\{< x, \xi > ; supp(\xi) \subset (C)\}$, (C) is defined below.

When the variation $\delta X(C)$ is taken, we need to have white noise with parameter set C Hence the choice of the class $\mathbf{C}$ has to be specified. Our favorite $\mathbf{C}$ for the questions in this article, is

$$\mathbf{C} = \{C; C \text{manifold diffeomorphic to the sphere} S^{d-1}\} \ . \tag{4.1}$$

Often a subclass of $\mathbf{C}$ is chosen depending on the fields in question. It is proved that a white noise with parameter set C is well-defined, because a Gel'fand triple

$$E(C) \subset H(C) \subset E^*(C) \ , \qquad\qquad H(C) = L^2(C, d\sigma(s)) \ , \tag{4.2}$$

is formed, where $d\sigma(s)$ is the surface element over C, and the white noise measure μ_C can be introduced on $E(C)^*$. This μ_C may be regarded as a marginal probability distribution derived from μ. The details of how to restrict μ has been illustrated in [7] and omitted here.

Another kind of restriction of μ may be considered. For C in $\mathbf{C}$, let (C) denote the domain enclosed by C. Then, by restricting ξ to (C), denote it by $\xi_{(C)}$, the restricted characteristic functional, and hence the marginal distribution of μ, denote it by $\mu_{(C)}$, is defined. Now we claim (also, see [7]).

Theorem 4.1. *The family $\{\mu_{(C)}; C \in \mathbf{C}\}$ is consistent in the sense that for any $(C_1), (C_2)$ and (C_3) such that $(C_3) \subset (C_1) \cap (C_2)$, the measure $\mu_{(C_3)}$ is a marginal distribution of $\mu_{(C_1)}$ and of $\mu_{(C_2)}$.*

We are now ready to discuss random fields $\{X(C); C \in \mathbf{C}$ in line with the white noise analysis.

Our general assumption is that $X(C)$ is a functional of the form

$$X(C) = X(C, x) = \Phi(\int_{(C)} F(C, x(u), u)du), \quad x \in E^* \text{ ,and } X(C) \in (S)^* \text{ .} \tag{4.3}$$

[Idea of Stochastic Variation]. We now remind the *stochastic infinitesimal equation* proposed by P. Lévy (in 1953) for a stochastic process $\{X(t);$ $t \in R\}$:

$$\delta X(t) = \Phi(X(s), s \le t, Y(t), t, dt), \qquad Y(t) : \text{innovation .} \tag{4.4}$$

The Φ and $Y(t)$ determine the probabilistic structure of the $X(t)$ in question.

A generalization to a field $X(C)$ would be

$$\delta X(C) = \Phi(X(C'), (C') \subset (C), \quad Y(s), u(s) \in C, C, \delta C), \tag{4.5}$$
$$Y(s) = Y(u(s)) \ : \text{innovation.}$$

This is a guiding formula for our variational calculus. With this, the assumption (4.3) would be acceptable.

What we are going to discuss is far from the general theory of variational calculus, however we shall review some interesting cases which can be discussed within our framework.

i) It is often useful to apply the S-transform to $X(C)$ to let the given equation turn into a variational equation for a non-random functional of $\xi \in E$. The following result gives us some idea.

Theorem 4.2. (P. Lévy [1], and [14]) *Let μ be a vector valued function expressing C, and let $U(\mu) = U(\xi, \mu)$ be the S-transform of $X(C)$. Let the variation of $U(\mu)$ be expressed in the form*

$$\delta U(\eta) = \int_C f(\eta, U, s)\delta\eta(s)d\sigma(s) , \quad d\sigma : \textit{surface element} , \tag{4.6}$$

with the initial condition $U_0 = U(\eta_0)$. Assume that f is continuous in the three variables and satisfy the following condition: Let ξ be fixed arbitrarily. For any U, V and η such that $|U - U_0| < c, |V - V_0| < c, \eta = \eta_0 + \lambda\xi$ the inequalities

$$\int_C f(\eta, U, s)^2 d\sigma(s) \le M^2 \qquad \textit{and}$$

$$\int_C |f(\eta, U, s) - f(\eta, V, s)| d\sigma(s) \le K^2 |U - V|$$

hold with absolute constants M and K.

Then, the equation (4.6) has the unique solution $U(\eta)$, and hence $X(C)$ is uniquely determined.

A proof of the theorem is given by using the classical theory of functional analysis. Note that there is not one-to-one correspondence between C and η. Namely, when we take variation, there is a freedom to have diffeomorphisms under which C is kept invariant, but η is transformed to other function. This remark is useful when the whitening of the given field is discussed.

ii) There is another approach to stochastic variational equations which is worth to be mentioned (see [6]). Namely, the variation of $X(C)$ is assumed to be obtained by the action of the conformal group $C(d)$ which is an important subgroup of O_∞. It is well known that a white noise enjoys the conformal invariant property, and therefore it is quite reasonable for us to restrict the operators acting on $X(C)$ to the class of those which come from the conformal group $C(d)$. Also it should be noted that the group $C(d)$ is viewed as a subgroup of O_∞ which is of different type from the Lévy group and even different from the limit of finite dimensional rotation groups.

As is discussed in [6], one can find a stochastic variational equation by using the infinitesimal generators α_j of the actions U_t^j that come from one-parameter subgroups $\{g_t^j\}$ of $C(d)$:

$$U_t^j \varphi(x) = \varphi((g_t^j)^* x), \qquad \varphi \in (S)^*, \qquad j = 1, 2, \ldots, (d+1)(d+2)/2 \, . \tag{4.7}$$

Thus the stochastic variational equation is expressed in terms of the generators α_j that describes a variation δC.

iii) It is noted that the irreducible unitary representation theory of Lie groups helps to get the innovation in examples like ii).

5 Operators of exponential type

I. A successful approach has been done for an operator of the following form

$$\exp[\int_{-\infty}^t F(t-u)\partial_u^* \, du] \, , \qquad F\text{smooth and in} L^2([0,\infty)) \, , \tag{5.1}$$

of exponential type when F. Oosawa has discussed an equation (we call it the *Oosawa equation*) that describes the movement of a primitive animal, a paramecium.

Equations involving operators of the form (5.1) can be discussed within our framework by transforming the given equation to those in terms of U-functionals; in particular, for the Oosawa equation we can see asymmetry in time evolution for the movement, because of the effect of the operators of the form (5.1). Unfortunately the theory goes to somewhat different line so is not discussed here.

II. We are interested in more general exponential type operators which can be dealt with as an application of our analysis on $(S)^*$. In fact, we shall come to the stochastic variational equations of the Schrödinger type, having been motivated by the equations in the relativistic quantum field theory, in particular by Tomonaga-Schwinger-Dyson's equation (good literatures are [2], [3] and [12]).

Let the class **C** of manifolds be taken as before. Also, the $X(C)$ under consideration is again assumed to be a member of $(S)^*$. Taking the Potthoff-Streit characterization theorem for $(S)^*$-functionals (*cf* Proposition 3.1.) into account, we consider the following class **H** of operators acting on $(S)^*$:

$$\mathbf{H} = \{\text{polynomials in the } \partial_t \text{ and the } \partial_t^*,\ t \in R^d, \text{ of degree} \leq 2\}\ .$$

Take an operator $H(C)$ in **H** of a particular type, given by

$$\begin{aligned} H(C) \ &= \ \int_{(C)} P(\partial_u)du\ ,\ P : \text{polynomial of degree} \leq 2, \qquad (5.2) \\ \text{i.e.,} \qquad & P(\partial_u) = \alpha(u)\partial_u^2 + \beta(u)\partial_u + \nu(u), \qquad \alpha, \beta, \nu \text{ smooth}\ , \end{aligned}$$

where (C) is the open domain enclosed by C. Then,

$$X(C) = \exp[H(C)] \cdot X_0, \qquad X_0 \in (S)\ , \tag{5.3}$$

is defined and satisfies

$$\delta X(C) = \delta H(C) \cdot X(C) = \int_C P(\partial_s)ds \cdot X(C)\ . \tag{5.4}$$

Let X_0 be taken to be $\exp[< x, \eta >]$, and set

$$Y(C) = \exp[H(C)]X_0\ . \tag{5.5}$$

Then, we have

$$Y(C) = \exp[\int_{(C)} P(\eta(u))du] \cdot X_0\ . \tag{5.6}$$

Similar formula can be shown for an integral of $P(\partial_u^*)$ over (C), call it $H(C)^*$. Namely, for X_0 given above, set

$$Z(C) = \exp[\int_{(C)} P(\partial_u^*)du] \cdot X_0\ . \tag{5.7}$$

Then, we have $Z(C) = S^{-1}\{\exp[\int_{(C)} P(\xi(u))du + < \xi, \eta > + \frac{1}{2}\|\eta\|^2]\}$, where S is the S-transform. Or, we may write

$$Z(C) = S^{-1}\{\exp[\int_{(C)} P(\xi(u))du]\}\diamond X_0, \qquad \diamond : \text{the Wick product}\ .$$

Now take C_0 and C_1 such that $(C_0) \subset (C_1)$. Assume that $supp(\eta) \subset (C_0)$. Hence $Z(C_0)$ is $\mathcal{B}(C_0)$-measurable. Then, for $X = \exp[< x, \xi >]$ with $supp(\xi) \subset (C_0)$, we have

$$< Z(C_1) - Z(C_0), \quad X >_\mu = 0\ . \tag{5.8}$$

This is proved as follows. $Z(C_1) = \exp[\int_{(C_1)-(C_0)} P(\partial_u^*)du] \cdot Z(C_0)$, and $\exp[\int_{(C_1)-(C_0)} P(\partial_u)du] \cdot X = X$ by (5.6). Hence (5.8) is proved.

The fact that $Z(C_0)$ is $\mathcal{B}(C_0)$-measurable and the equation (5.8) shows that $Z(C_0)$ *is like the conditional expectation* of $Z(C_1)$ relative to $\mathcal{B}(C_0)$. Such an observation would be helpful for understanding Dirac's idea [2].

It is interesting to take $H(C)$ involving both ∂_u and ∂_u^*, and in addition its formal expression is like self-adjoint (i.e. symmetric in ∂_u and ∂_u^*). A general form of $H(C)$ that is indexed by C and is quadratic with the property mentioned above, may be expressible as

$$\begin{aligned} H(C) &= \int\int_{(C)^2} F_1(u,v)\{\partial_u\partial_v + \partial_u^*\partial_v^*\}dudv + \int\int_{(C)^2} F_2(u,v)\partial_u\partial_v^* dudv \\ &\quad + \int\int_{(C)^2} F_3(\mathrm{u},\mathrm{v})\partial_u^*\partial_v dudv \ . \end{aligned} \tag{5.9}$$

The kernels $F_i(u,v)$, $i = 1,2,3$, in (5.9) are all assumed to be symmetric functions (could be generalized functions). The regularity or irregularity of those kernels F_i shall be specified depending on the problems in question.

To fix the idea, we shall consider the following case:

$$H_1(C) = \int\int_{(C)^2} F_1(u,v)\{\partial_u\partial_v + \partial_u^*\partial_v^*\}dudv, \qquad F_1 \ : \text{symmetric} \ .$$

We are led to define a random field $\{X(C)\}$ by the following formula, where the Wick ordering : : , must be employed.

$$X(C) =: \exp[iH(C)] : X_0 \ , \qquad H = H_1 \ . \tag{5.10}$$

In what follows the variation $\delta X(C)$ of $X(C)$ is formed by such a δC as $(C + \delta C) \subset (C)$. Then, we have

Proposition 5.1. *The random field* $X(C)$ *given by* (5.10) *satisfies the following stochastic variational equation*

$$\frac{\delta X(C)}{\delta C}(p) = i : \{\partial_p \int_C F_1(p,u)\partial_u du + \partial_p^* \int_C F_1(p,u)\partial_u^* du\} \ H_1(C) : X(C) \ , \qquad p \in C \ . \tag{5.11}$$

Proof is in the paper [7], so is omitted.

[Remark] The equation (5.9) is rephrased as

$$\frac{\delta X(C)}{\delta C}(s) = \frac{\delta H(C)}{\delta C}(s) \cdot X(C) \ , \qquad (C) \supset (C_0) \ , \tag{5.12}$$

by using an operator $\delta H/\delta C$ derived from $H(C)$.

Theorem 5.1. *Let* $H(C)$ *be one of the operators that appear in the sum* (5.9). *A random field* $X(C)$ *given by* (5.10) *satisfies the stochastic variational equation* (5.12).

Conversely, the solution to the equation (5.12) *with an initial condition*

$$X(C_0) = X_0, \qquad X_0 = X(C_0) \,, \tag{5.13}$$

is given by $X(C) =: \exp[iH(C)] : X_0 \,, \qquad (C) \supset (C_0)$.

So far we have discussed variational equations for a given $X(C)$. The converse problem is more difficult. Namely, the problem to solve a stochastic variational equation. For this purpose it is an idea to deform C to have $C + \delta C$ to get $\delta X(C)$ only by a one-parameter family $\{\beta_t\}$ of diffeomorphisms that can define a whisker $\{g_t\}$. To fix the idea we consider an equation

$$\frac{\delta X(C)}{\delta C}(u) = H(u, C) \cdot X(C) \tag{5.14}$$

with some initial data. Then, imitating the classical theory of functional analysis, we claim

Theorem 5.2. *Denote by* $\delta_i, i = 1, 2$, *the variations of* C *by the one-parameter family of diffeomorphisms of* R^d. *Then an integrability condition of the equation* (5.14) *is*

$$\frac{\delta_1 \delta_2 X(C)}{\delta C \delta C_1} = \frac{\delta_2 \delta_1 X(C)}{\delta C_1 \delta C} \,. \tag{5.15}$$

Proof. The S-transform is applied to $X(C)$, then the classical theory of functional analysis, in particular integrability condition for functional differential equations can be applied. Good references for details are [1] 2ème Part.§ 4 and [15].

[Concluding remarks] Because of the importance of the Lévy Laplacian, it would be fine if we could take care of an operator of the form

$$\Delta_L(C) = \int_{(C)} \partial_t^2 (dt)^2$$

and to discuss the operator

$$H(C) = \Delta_L(C) + \Delta_L(C)^* \,.$$

Another remark is concerned with the *derivation* property of Δ_L under the Wick product. Based on the elementary system of variables $: x(t)^2 : , t \in R^d$, one may develop a new calculus.

Example. The Gauss kernel.

Acknowledgement The author wishes to express his deep thanks to the organizer Professor L. Streit for his kind invitation to this conference.

Also, he is grateful to Ms Hannelore Litschewsky for typing the final version of this paper in a beautiful shape.

References

[1] P. Lévy, *Problèmes concrets d'analyse fonctionnelle.* Gauthier-Villars, 1951.

[2] P.A.M. Dirac, The Lagrangian in quantum mechanics. Phys. Z. USSR 3 (1933), 64 - 72.

[3] F.J. Dyson, The radiation theories of Tomonaga, Schwinger, and Feynman. Physical Review 75 (1949), 486 - 502.

[4] W. Heisenberg, Die Grenzen der Anwendbarkeit der bisherigen Quantentheorie. Z. f. Physik 110 (1938), 251 - 266.

[5] T. Hida, A note on generalized Gaussian random fields. J. Multivariate analysis, 27 (1988), 255 - 260.

[6] ——, White noise analysis and Gaussian random fields. Proc. 24th Karpacz Winter School on Theoretical Physics. World Scientific Pub. 1989, 277 - 289.

[7] ——, Random fields as generalized white noise functionals. Acta Appl. Math. to appear.

[8] T. Hida, H.-H. Kuo and N. Obata, Transformations for white noise functionals, J. Functional Anal. 111 (1993), 259 - 277.

[9] T. Hida, H.-H. Kuo, J. Potthoff and L. Streit, *White noise - An infinite dimensional calculus.* Kluwer Academic Publisher, 1993.

[10] T. Hida, K.-S. Lee and S.-S. Lee, Conformal invariance of white noise. Nagoya Math. J. 98 (1985), 87 - 98.

[11] T. Hida, N. Obata and K. Saitô, Infinite dimensional rotations and Laplacians in terms of white noise calculus. Nagoya Math. J. 128 (1992), 65 - 93.

[12] K.-S. Lee, White noise approach to Gaussian random fields. Nagoya Math. J. 119 (1990), 93 - 106.

[13] Si Si, Gaussian processes and conditional expectations. BiBoS Notes 292/87, Univ. Bielefeld, 1987.

[14] ———, Variational calculus for Lévy's Brownian motion. Proc. Gaussian Random Fields, 1990 at Nagoya, World Scientific, 1991, 364 - 373.

[15] ———, Stochastic variational calculus in white noise analysis, Preprint.

[16] S. Tomonaga, On a relativistically invariant formulation of the quantum theory of wave fields. Progress of Theoretical Physics 1, No. 2 (1946), 27 - 42.

[17] ———, On infinite field reactions in Quantum Field Theory, Physical Review 74 (1948), 224, with a note by J.R. Oppenheimer.

[18] H. Yoshizawa, Rotation group of Hilbert space and its application to Brownian motion. Proc. International Conf. on Functional Analysis and Related Topics, Tokyo, 1969, 414 – 423.

COMPUTER STOCHASTICS IN SCALAR QUANTUM FIELD THEORY

C.B. LANG
Inst. f. theoret. Physik
Universität Graz
Universitätsplatz 5
A-8010 Graz, AUSTRIA

Abstract. This is a series of lectures on Monte Carlo results on the non-perturbative, lattice formulation approach to quantum field theory. Emphasis is put on 4D scalar quantum field theory. I discuss real space renormalization group, fixed point properties and logarithmic corrections, partition function zeroes, the triviality bound on the Higgs mass, finite size effects, Goldstone bosons and chiral perturbation theory, and the determination of scattering phase shifts for some scalar models.

Key words: Scalar quantum field theory, Φ^4 model, Monte Carlo methods

1. Lattice Quantum Field Theory

In this series of lectures I attempt to cover some numerical applications to stochastic analysis in relativistic quantum field theory (QFT). Following a more general introduction I then focus on the non-perturbative quantization of *scalar* QFT. This emphasis is dictated by the limited time and place available as well as by the progress in the subject. Scalar QFT meanwhile is the most advanced of the interesting models, both in theoretical understanding as well as in numerical stochastic calculations.

If you are interested in particular topics of lattice field (including gauge field) theory, a good starting point is to look up the proceedings of the annual workshops in lattice field theory. For the recent years they have been published in Nucl. Phys. B (Proc. Suppl.); up to now there are volumes 9(1989), 17(1990), 20(1991), 26(1992), 30(1993). In every one of these volumes there are also reviews, although mainly targeted at the devotees of the field.

There are only few books giving surveys. Almost all important results of the first, hot days of Monte Carlo computations are given in the reprint collection [116] and the monograph [24]. The more rigorously oriented person should maybe start with [120]. Recently two volumes on the field have been published, one is the monograph [118] and the other contains a collection of review article written by experts on various subtopics [25].

In the first part I attempt an overview on lattice field theory and the related computational issues. The role of the Φ^4 model in particle physics is discussed and the lattice formulation for that model is introduced. In the second chapter some rigorous statements on the model are reviewed and then lattice calculation results concerning the fixed point structure and the issue of triviality are presented. In the

A. I. Cardoso et al. (eds.), Stochastic Analysis and Applications in Physics, 133–191.

subsequent chapter further results on the so-called triviality bound are discussed; also the particle spectrum and in particular the finite size effects produced by the massless Goldstone states are reviewed. In the last section I present recent results that use the scalar model as a testing ground for new methods to determine scattering phase shifts non-perturbatively.

1.1. Regularization and Quantization

1.1.1. *Space-time Lattice*

Quantum field theory usually is discussed within the perturbative framework. Even there it is obvious that there are divergences in elementary contributions. In a functional integration (Feynman path integral) formulation one finds that the integration is not well-defined. The functional integral has to be regularized in order to be able to identify the singularities.

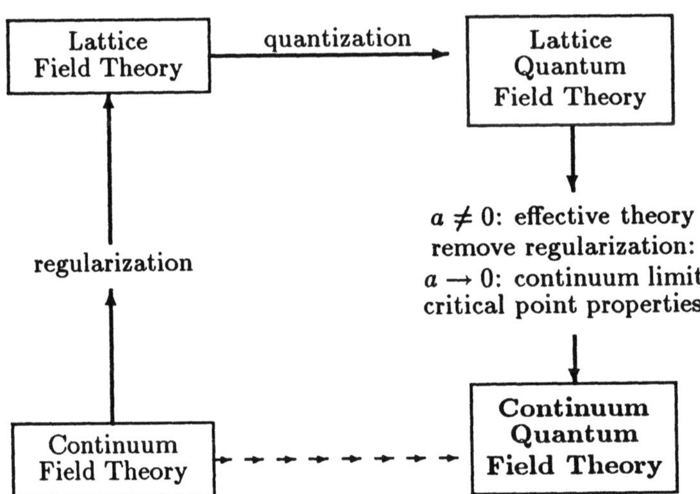

Replacing the space-time continuum by a lattice provides a suitable regularization as well as a computational tool [142]. We can define the continuum QFT as the limit of the lattice QFT for vanishing lattice spacings. Important symmetries of the Lagrangian (like e.g. gauge invariance) may be sustained in the lattice formulation. On the other hand, obvious symmetries of space-time like translational and rotational invariance are replaced by their lattice subgroups and eventual recovery in the continuum limit has to be established.

We will discuss euclidean QFT. The regularization replaces the continuous, $O(4)$ symmetric space-time arena by a (in practical calculations) finite lattice with symmetry $O(4;\mathbb{Z})$.

lattice	Λ e.g. $\mathbb{Z}_N^D \subset \mathbb{Z}^D$
field	$\Phi : \Lambda \mapsto \mathbb{X}$ (e.g. $\mathbb{C}$) $\in \Gamma$
configuration	$\Gamma \in \mathbb{X}^\Lambda$
action	$S_\Lambda : \Gamma \mapsto \mathbb{R}$
partition function	$Z_\Lambda = \int d\mu_\Lambda(\Phi) \exp[S_\Lambda(g, \Phi)]$
free energy	$F_\Lambda = \ln Z_\Lambda$
Gibbs factor	$\rho_\Lambda(\Phi) = Z_\Lambda^{-1} \exp[S_\Lambda(g, \Phi)]$
expectation values	$\langle A \rangle_\Lambda = \int d\mu_\Lambda(\Phi) \rho_\Lambda(\Phi) A(\Phi)$

Quantization thus amounts to determination of expectation values in a 4D statistical system, like e.g. the Ising model (cf. for instance [117]). With g we denote the coupling constant (or a vector of constants) parametrizing the action.

1.1.2. *Scaling*

The focus of interest in that process are two limits:

Thermodynamic limit: The infinite volume limit $|\Lambda| \to \infty$.

Continuum limit: Is obtained at a critical point $g \to g_c$ in coupling constant space, where the lattice spacing constant $a(g) \to 0$; the lattice spacing acts as a regularization parameter.

The second limit needs some discussion. The physical quantities like mass or length of the continuum theory should approach constant values, however on the lattice they are expressed in multiples of the lattice spacing a. These lattice quantities are dimensionless and are determined from the connected correlation functions (propagators) of some operators

$$\begin{aligned} \langle A(0)A(x)\rangle_c &\propto \exp(-x/\xi) = \exp(-ax/a\xi) \\ &= \exp(-m_{\mathrm{phys}}ax) , \quad \text{i.e. } m_{\mathrm{phys}}a = 1/\xi . \end{aligned} \tag{1.1}$$

These correlation functions describe propagation in euclidean time and the operators may be representing n-particle states. The exponential decay should hold for sensible theories [120].

Only at a critical point the correlation length ξ diverges[1] and allows to let $a \to 0$ while keeping a fixed value for the physical mass parameter. One has to fix the scale $a(g)$ by trading one physical parameter. Once the scale is set, all other parameters may then be expressed in terms of this scale parameter $a(g)$. However, all other physical parameters should scale correspondingly, only then can we identify the correct continuum theory.

In practical situations we distinguish between two qualities of scaling behaviour.

Scaling: The dimensionless ratios like $\frac{m_1(g)a(g)}{m_2(g)a(g)} = \frac{m_1(g)}{m_2(g)}$ scale towards the physical ratio $\frac{m_1}{m_2}$ as $a \to 0$

Asymptotic scaling: The scaling behaviour $a(g) \propto f(g)$ as determined from the renormalization group β-function.

The relative importance of irrelevant operators determines the quality of scaling. Whereas scaling may be observed quite early one has to be quite close to the critical point in order to observe asymptotic scaling.

1.1.3. *Continuum Limit*

Let us collect some points important in the context of the continuum limit.

- The lattice action should obey Osterwalder-Seiler positivity when approaching the continuum limit in order to guarantee existence of a transfer matrix formalism and positive norm Hilbert states [115].
- The continuum limit is obtained at a critical point (2nd order phase transition) of the lattice system: $g \to g^*$, where the dimensionless correlation length $\xi \to \infty$.
- All physical quantities are multiples of the lattice spacing and should scale correspondingly, $a(g) \to 0$ for $g \to g^*$.
- Lattice translation invariance mod(a) should become continuous translation invariance.
- The space-time symmetry (cubic group for cubic lattices) should become the continuous $O(4)$ (rotational) invariance. This may be observed in off-shell observables (like the 1-particle propagator) as well as in on-shell ones (like differential scattering cross-sections).
- Fields, couplings and masses in the lattice action are bare quantities and are renormalized in the quantization in continuum limit. Renormalized quantities will have a subscript r henceforth.
- Because of renormalization, the continuum properties of a quantized theory may be quite different from those of the bare theory. In statistical physics this observation is linked to the notion of universality: Theories with different micro-structure may show universal critical (macroscopic) properties. Since the critical properties are just the physical properties of our quantized continuum theory this limits the class of liable continuum QFTs.

All these points should be observed in a sensible investigation of the continuum limit. Most of these properties can be studied only in the quantized theory in a

[1] At the critical point the correlation function decay polynomially $\propto |x|^{d-2+\eta}$

non-perturbative environment. In existing Monte Carlo calculations several of these features have been checked and some of these checks will be discussed in this series of lectures.

1.2. Particle Physics and the Φ^4 Model

1.2.1. *Lattice Fields*

Typically we have to deal with the following types of lattice fields.

Scalar bosons: Occur in the action usually with local terms like Φ_x, Φ_x^2, Φ_x^4 and nearest neighbour terms like $\Phi_x U_{x,\mu} \Phi_{x+\mu}$; the n.n. term is a lattice discretization of the continuum derivative term. Monte Carlo integration methods for these fields include the standard Metropolis or heat bath updating as well as cluster- and multigrid methods. The computational effort grows with $O(L^4)$ up to $O(L^6)$ depending on the algorithm (L denotes the linear extension of the 4D lattice),

Gauge (vector) bosons: Action terms include the Wilson plaquette interaction TrU_p, where the plaquette variable is the path ordered product of link fields $U_{x,\mu}$ around a plaquette. The gauge field coupling $\beta \propto \frac{1}{g^2}$ where g is the usual perturbative coupling constant. Monte Carlo updating algorithms include Metropolis, heat bath and overrelaxation; some progress has been made in designing multigrid algorithms. The computational effort grows with $O(L^5)$ up to $O(L^6)$.

Fermions: Occur in the action with terms like $\overline{\Psi}_x \Psi_x$ and $\overline{\Psi}_x Q_{xy} \Psi_y$ which includes Dirac γ-matrices and nearest neighbour coupling κ. Updating algorithms are an order of magnitude more demanding with regard to computational resources; they include molecular dynamics updating, Langevin and "hybrid" Monte Carlo methods as well as some recent attempts towards multigrid methods. The central numerical problem is the inversion of the lattice Dirac operator Q, thus the total effort grows with $O(L^9)$ up to $O(L^{16})$ depending on the algorithm.

For the formulation of the models on finite volumes one usually assumes periodic boundary conditions, $\Phi(x + L\hat{\mu}) = \Phi(x)$ (for fermions: antiperiodic b.c.). The implicit assumption is that the thermodynamic limit should be independent on the choice of boundary conditions. Periodic b.c. seem to produce relatively small finite size effects.

1.2.2. *Scalar Model*

The standard model of electroweak and strong interactions describes extremely successfully nowadays results in particles physics in an energy domain from a few MeV up to more than 100 GeV. Its central symmetry is $SU(2) \times U(1) \times SU(3)$. The $SU(3)$ is that of QCD, represented by quarks, gluons and their interaction. The non-perturbative lattice study of that model is a central topic in lattice calculations [29], but I will not try to cover it here. The symmetry groups $SU(2) \times U(1)$ are those of quantum flavour dynamics and are represented by the fundamental leptons, 4 gauge bosons and the dublet $\Phi = (\Phi_1 + i\Phi_2, \Phi_3 + i\Phi_4)$ of complex Higgs fields. In particular the action for the Higgs fields has a symmetry $SU(2)_L \times SU(2)_R$ isomorphic to $O(4)$. Thus we may represent the Higgs fields by a vector of 4 real scalar fields, too.

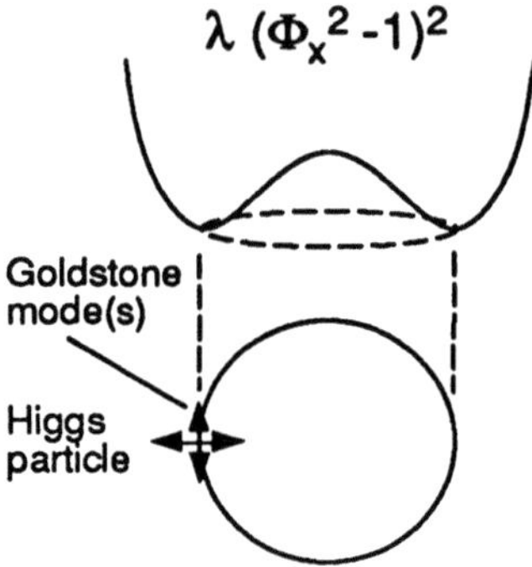

Fig. 1. Standard model symmetry breaking scenario.

In the standard scenario one assumes that there is a symmetry breaking potential like the well-known Mexican hat potential $\lambda(\Phi^2 - 1)^2$. In the Goldstone realization the vacuum chooses spontaneously one direction (out of the infinite set of inequivalent representations, cf. fig. 1), thus one is left with one massive component (the Higgs particle) and three massless components (the Goldstone modes), which, in the so-called Higgs mechanism, become the longitudinal modes of three of the (originally massless) vector bosons. The three vector bosons acquire mass in that process. For sake of the later discussion let us distinguish the two processes:

Symmetry breaking: The $O(4)$ symmetry is spontaneously broken and in the new vacuum we find one massive and three massless modes.

Mass generation: Three massless gauge bosons buddy with the three Goldstone modes and become massive gauge bosons, Z and $W^{\pm}$.

Both mechanisms are non-perturbative as they involve infinitely many degrees of freedom. The results mentioned are based on tree-level perturbation theory and some additional assumption about the vacuum structure. One of the basic questions to be answered in the non-perturbative lattice simulations is, whether these results survive quantization and whether they remain valid at arbitrary Higgs coupling strength.

The essential non-perturbative ingredient is the $O(4)$ Φ^4 model. This is the main motivation for studying that theory non-perturbatively in some detail.

1.2.3. *The Φ^4 Model*

The continuum action

$$S_{cont} = \int dx[\frac{1}{2}(\partial_\mu \Phi_x)^2 + \frac{1}{2}m_{cont}^2 \Phi_x^2 + \frac{1}{4!}\lambda_{cont}(\Phi_x^2)^2 + j_{cont}\Phi_x] \tag{1.2}$$

has three parameters, the bare mass, the 4-point coupling, and the external field. Discretization is not unique and a particular simple form of a lattice action is obtained by replacing the derivative term by a nearest neighbour difference. With suitable redefinitions of the couplings we get

$$S_{latt} = \sum_{\Lambda} [-2\kappa \sum_{\mu} \Phi_x \Phi_{x+\mu} + \quad \lambda(\Phi_x^2 - 1)^2 + J\Phi_x] \,. \tag{1.3}$$

We have several important cases.

– $\Phi_x \in \mathbb{Z}_2$: the Ising model,

– $\Phi_x \in \mathbb{R}$: the simple 1-component Φ^4 model,

– $\Phi_x \equiv \Phi_x^\alpha \in O(4)$: the 4-component Φ^4 model, i.e. the Higgs model.

For notational simplicity we sometimes use field variables

$$\varphi \equiv \sqrt{2\kappa}\Phi \,, \quad j \equiv J/\sqrt{2\kappa} \,. \tag{1.4}$$

In the form (1.3) we identify immediately two limits of the system. For $\lambda \to 0$ we approach a gaussian model. For $\lambda \to \infty$ only configurations with $|\Phi_x| = 1$ contribute and we have a nonlinear sigma model (for the 1-component version: Ising model). The coupling κ, λ and j in the lattice action are bare quantities, given at the generic lattice scale a. Values measured in the quantized system like e.g. the mass or the 4-point coupling are renormalized quantities and will be denoted by a subscript: m_r, λ_r.

At the critical point of the model, $\kappa = \kappa_c$, critical exponents characterize the singular behaviour of various quantities [125, 104]. Introducing the reduced coupling

$$t \equiv \frac{\kappa_c - \kappa}{\kappa_c} \tag{1.5}$$

we summarize in table I the critical behaviour and definitions of the critical indices. In sect. 2 we discuss the issue of triviality of the model. If it is indeed trivial (as all evidence points to) the critical exponents are those of mean field theory and there are non-leading logarithmic corrections. The exponents of these corrections are given in table I and depend on N of the group $O(N)$ [14, 83] ($N = 1$ for the one-component model).

The two definitions given for the renormalized charge agree in the continuum limit (sometimes one uses the continuum definition action with an extra factor 6: $\lambda_r = 3(m_r/\langle\varphi_r\rangle)^2$). Here Z denotes the field renormalization constant (and will be discussed in sect.3 in more detail) and not the partition function (sorry, but thats how it is). Only two of the critical exponents are independent (cf. e.g. [76]). The mean field values are for the scaling exponent of the correlation length $\nu = \frac{1}{2}$ and for the anomalous dimension $\eta = 0$. From the scaling relations we find $\alpha = 2 - d\nu = 0$, $\delta = \frac{d+2-\eta}{d-2+\eta} = 3$, $\gamma = \nu(2-\eta) = 1$ and $\beta = \gamma/(\delta - 1) = \frac{1}{2}$.

1.3. Monte Carlo Methods

Quantization amounts to determination of expectation values

$$\langle A \rangle = \frac{1}{Z} \sum_{\Phi \in \Gamma} A(\Phi) \exp[S(\Phi)] \,. \tag{1.6}$$

TABLE I
Summary of the expected critical behaviour in the N-component model in terms of the reduced critical coupling $t \equiv (\kappa_c - \kappa)/\kappa_c$ and the external field j.

Order parameter			
$\langle \Phi \rangle$	$\sim$	$\lvert t\rvert^{\beta} \lvert \ln \lvert t\rvert \rvert^{\frac{3}{N+8}}$	for $j = 0$
	$\sim$	$j^{1/\delta} \lvert \ln \lvert j\rvert \rvert^{\frac{1}{3}}$	for $t = 0$
Susceptibility			
$\chi = \langle \Phi(p=0)^2 \rangle_c = \frac{1}{\lvert\Lambda\rvert} \sum_x \langle \Phi_0 \Phi_x \rangle_c$	$\sim$	$\lvert t\rvert^{-\gamma} \lvert \ln \lvert t\rvert \rvert^{\frac{N+2}{N+8}}$	
Internal (link) energy			
$\langle E \rangle = \sum_{x,\mu} \langle \Phi_x \Phi_{x+\mu} \rangle$			
Specific heat			
$C_V = -\kappa^2 \frac{\partial \langle E \rangle}{\partial \kappa} = \kappa^2 (\langle E^2 \rangle - \langle E \rangle^2)$	$\sim$	$\lvert t\rvert^{-\alpha} \lvert \ln \lvert t\rvert \rvert^{\frac{4-N}{N+8}}$	
Correlation length			
$\langle \Phi_0 \Phi_x \rangle_c \propto \exp(-m_r \lvert x\rvert)$ for $t \neq 0$			
$\propto 1/\lvert x\rvert^{d-2+\eta}$ for $t = 0$			
$m_r = 1/\xi$	$\sim$	$\lvert t\rvert^{\nu} \lvert \ln \lvert t\rvert \rvert^{\frac{-N-2}{2N+16}}$	
Renormalized 4-pt coupling			
$\lambda_r = -\frac{\langle \Phi(p=0)^4 \rangle_c}{\langle \Phi(p=0)^2 \rangle_c^2}$			
or $= \frac{1}{2}(m_r/\langle \varphi_r \rangle)^2$	$\sim$	$\lvert \ln \lvert t\rvert \rvert^{-1}$	
where $\langle \varphi_r \rangle = \langle \varphi \rangle / \sqrt{Z}$			

In Monte Carlo integration one samples the field configurations with an equilibrium probability distribution $P(\Phi) \simeq \exp[S(\Phi)]$. This allows to determine expectation values according

$$\langle A \rangle = \lim_{T \to \infty} \frac{1}{T} \int_0^T d\tau \, A(\Phi(\tau)) = \lim_{N \to \infty} \frac{1}{N} \sum_{n=1}^{N} A(\Phi_n). \tag{1.7}$$

The stochastic variable is the field $\Phi(\tau)$ evolving with the stochastic (computer) time τ. It is determined from a homogeneous Markov process with transition probability T,

$$P(\Phi_n = \varphi_n | \Phi_{n-1} = \varphi_{n-1}) = T_1(\varphi_n | \varphi_{n-1}) \tag{1.8}$$

i.e. T_k depends only on the numbers of time steps k but not on the history. If $T_k(\varphi'|\varphi) > 0$ for all pairs (φ', φ) and the process is aperiodic then it is ergodic. A Monte Carlo integration algorithm usually constructs such a Markov sequence of configurations, similar to a sequence of snap shot of some realistic ferromagnet. Thus the reason for calling the method a simulation. In realistic simulations ergodicity and the related problem of relaxation are important questions. In particular if there are topologically different sectors in configuration space some Monte Carlo updating algorithms may have problems connecting them.

If implemented correctly the Monte Carlo integration leads to correct results with a purely statistical error that decreases with the number of iterations like $1/\sqrt{N}$. The results are for finite size lattices, however. Control of both, the statistical error and the finite size effects, belongs to the central issues of this field.

One starts with an (arbitrary) configuration. The Markov sequence of configurations

$$\varphi_0 \rightarrow \varphi_1 \rightarrow \varphi_2 \rightarrow \dots \tag{1.9}$$

produces samples of the corresponding probability distributions, which, depending on certain sufficient conditions, approach the equilibrium distribution,

$$P^{(0)} \rightarrow P^{(1)} \rightarrow P^{(2)} \rightarrow \dots \rightarrow P \text{ (equilibrium distribution)}, \tag{1.10}$$

where $P^{(n)} = T^n P^{(0)}$. In a calculation one therefore starts to measure observables only after a sufficient number of "equilibrating" Monte Carlo steps. Subsequent updates then produce configurations according to the required distribution.

Meanwhile we have a collection of updating algorithms [145]. Some of them proceed by changing the configurations locally, i.e. just one variable at a time, others try to implement collective updates of nonlocal nature.

Local updating algorithms

- **Metropolis**: The simplest and oldest [107] algorithm; we discuss it below.
- **Heat bath**: Like an iterated Metropolis updating, optimizing the local acceptance rate.
- **Overrelaxation**: A sometimes very effective generalization of the heat bath algorithm.
- **Microcanonical**: Reformulates the updating in terms of a deterministic, discrete Hamiltonian evolution in a space with doubled number of variables: (Φ, Π). A variant method introduces a microcanonical demon.
- **Hybrid Monte Carlo**: Combines a number of microcanonical updates (a "trajectory") with a final Monte Carlo acceptance step.
- **Langevin**: Uses the stochastic differential equation for the construction of configurations; equivalence to the microcanonical and the hybrid method may be demonstrated (in certain limits).
- **Multicanonical**: Or multimagnetical or entropy sampling, samples an ensemble with modified energy density, particularly well suited for systems with degenerate ground states (like at 1st order phase transitions or in spin-glasses) [5, 6].

Non-local updating algorithms

- **Multigrid**: Employs scale changing transformations to identify possible collective updating modes, essentially a simultaneous updating on several length scales.
- **Cluster**: Identifies the system-specific dynamical nonlocal degrees of freedom; this method is very efficient for certain models and will be discussed below in more detail.

1.3.1. *Metropolis Algorithm*

Considering a stochastic process $X(\tau)$, detailed balance is a sufficient condition that the probability distribution approaches the equilibrium distribution $\exp[S(x)]$,

$$P_1(x|y)e^{S(y)} = P_1(y|x)e^{S(x)} \Rightarrow \lim_{\tau\to\infty} P(X(\tau)=x) \simeq e^{S(x)} . \tag{1.11}$$

Summation over x shows that the limiting distribution is an eigenvector of P_1. The Metropolis algorithm [107], starting at some value x (i.e. some initial configuration), consists of the steps:

1. Choose some y according to some *a priori* transition probability $P_T(y|x)$.
2. Accept the new value y with the probability

$$P_A(y|x) = \min\left(1, \frac{P_T(x|y)\exp[S(y)]}{P_T(y|x)\exp[S(x)]}\right) \tag{1.12}$$

$$= \min(1, \exp[S(y) - S(x)]) \text{ for } P_T(x|y) = P_T(y|x) . \tag{1.13}$$

   It is straightforward to see, that the resulting $P_1 = P_T P_A$ fulfills detailed balance. In particular for symmetric P_T the information necessary to decide on acceptance or rejection comes only from the change of the action ΔS with regard to the change of the variable. If this change is local, i.e. just at some field variable Φ_x, the change of action may be determined from the field values in the local neighbourhood.
3. Repeat these steps from the begin.

One MC iteration is one sweep through the lattice, that is one attempt to update for each field variable on the lattice. Typical calculations perform $O(10^6)$ MC iterations for given lattice size and couplings constant values.

The exponential autocorrelation time controls the statistical dependence of subsequent configurations and may be defined

$$|P^{(n+1)} - P| \leq \exp(-1/\tau_{exp})|P^{(n)} - P| \tag{1.14}$$

or from the autocorrelation of some observable A,

$$\Gamma_A(n) = \langle A(\varphi_0)A(\varphi_n)\rangle - \langle A\rangle^2 \simeq \exp(-n/\tau_{exp,A}) , \quad \tau_{exp,A} \equiv \sup_A \tau_{exp,A} , \tag{1.15}$$

or from the integrated autocorrelation time

$$\tau_{int,A} = \frac{1}{2} + \sum_{n>0} \frac{\Gamma_A(n)}{\Gamma_A(0)} . \tag{1.16}$$

The latter quantity controls directly the statistical error in estimates of $\langle A\rangle$; a run of n measurements contains only $n/2\tau_{int,A}$ effectively independent measurements. This relaxation behaviour affects particularly systems with large correlation lengths ξ (near critical points), since the autocorrelation length grows like

$$\tau_{exp} \simeq \min(L,\xi)^{z}\ , \quad \tau_{int,A} \simeq \min(L,\xi)^{z_A} \tag{1.17}$$

where z and z_A denote the dynamical critical exponents. They depend on the updating algorithm and have the value 2 for most local algorithms like the Metropolis algorithm (for overrelaxation algorithms this value may be smaller: ~ 1 [145]). The computational effort for the simulation of a system of linear extension L near a critical point[2], where $\xi \simeq L$, grows like L^{d+z_A}. This phenomenon is called *critical slowing down*. For more detailed discussions cf. [123, 124, 145].

In the MC simulation the statistical error of the measured quantities shrinks with the inverse square root of the number of statistically independent configurations. Improvement of the performance therefore can be achieved only by improving the updating algorithm by making them computationally simple (and quick), by reducing the autocorrelation time and by reducing the dynamical critical exponent z. This is the motivation for the introduction of multigrid- and cluster updating algorithms.

1.3.2. *Cluster Algorithm*

Fortuin and Kastelyn [81, 82] pointed out that the Ising model has an alternative formulation as a bond percolation model. This leads to clusters of connected bonds [23], which are the nonlocal dynamical objects of the Ising model.

Swendsen and Wang [131] suggested a very efficient updating algorithm based on alternating between the spin and the cluster representation. For a given spin configuration one first identifies Ising clusters built of neighboured sites with parallel spins. Some of the bonds are then deleted with probability $\exp(-2\kappa)$, the resulting bond clusters are the dynamical FK-CK-SW-Clusters [81, 23, 131]. With a probability 0.5 one now can flip simultaneously all spins in a cluster, repeating this for all clusters. This results in a new spin configuration and the updating step is completed (cf. fig. 2).

It turns out that this algorithm gives $z \simeq 0.3$ for the 2D and $z \simeq 1$ for the 4D Ising model. Wolff proposed a modification, where one randomly chooses a site and identifies the cluster connected to it like in the SW-algorithm; this cluster is then flipped unconditionally. This results in $z \simeq 0$ for the 4D system!

The cluster algorithm originally was designed for systems like the Ising or the Potts model, where the spin variables take values in a discrete group. Wolff [144] observed, that one may extend its scope of application to continuous groups, that have embedded Ising variables (the group manifold may be split into disjoint parts, cf. [124]), like e.g. the group $O(N)$. This allowed to utilize the updating method for N-component Φ^4 theories. The cluster method has provided a major breakthrough, the efficiency of the algorithm has brought forward results for scalar models with highest computational precision.

[2] For systems with several coexisting phases the autocorrelation length grows even exponentially with $\simeq \exp(cL^{d-1})$; multicanonical algorithms address this tunneling problem.

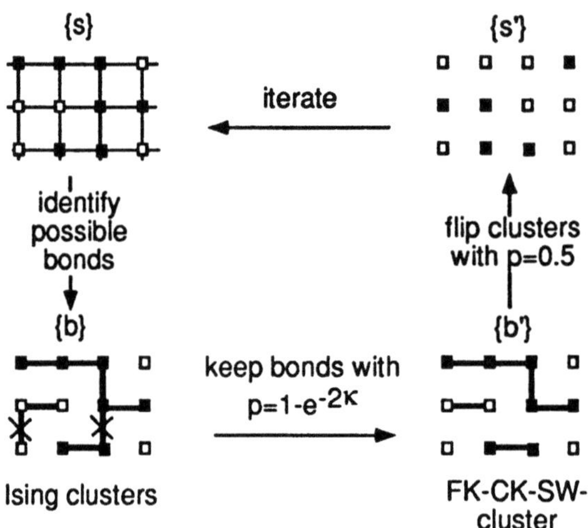

Fig. 2. The Swendsen-Wang cluster updating algorithm: $\{s\} \to \{b\} \to \{b'\} \to \{s'\}$.

The cluster representation allows the construction of observables with reduced variance, so-called *improved estimators* [127]. The propagator between spins at two sites, e.g., may be obtained by just counting how often both spins belong to the some cluster:

$$\langle \phi_x \phi_y \rangle = \langle \sum_{\text{all clusters c}} \chi_c(x)\chi_c(y) \rangle \quad \text{where} \quad \chi_c(x) = \begin{cases} 1 & \text{if } x \in c \\ 0 & \text{if } x \notin c \end{cases} . \tag{1.18}$$

For a given set of configurations this allows determination of the propagators for larger distances with much better statistical quality than achieved from the direct measurement.

The clusters are the dynamical objects of the critical system, a realization of Fisher's droplets. One therefore would expect relations between the geometrical properties and some critical indices. Due to scale invariance at a critical point clusters should look alike on different length scales. In quantum mechanics this is expressed by the fractal structure of the quantum paths, which are of Brownian nature. We might expect such a fractal structure also in the field configurations of quantum field theory (cf. fig. 3).

The relation between such a fractal structure and the odd critical index may be argued as follows [126]. Under a scale transformation we have scaling of the reduced critical temperature t (cf. table I) in the form

$$\xi \to b\xi \ , \quad t \to b^{-1/\nu} t \tag{1.19}$$

where ν denotes the scaling index of the correlation length. In a volume L^d the total magnetization scales according to table I

$$M(L) \propto L^d t^\beta \quad \to \tag{1.20}$$

$$M(bL) \propto (bL)^d b^{-\beta/\nu} t^\beta = b^{d-\beta/\nu} M(L) . \tag{1.21}$$

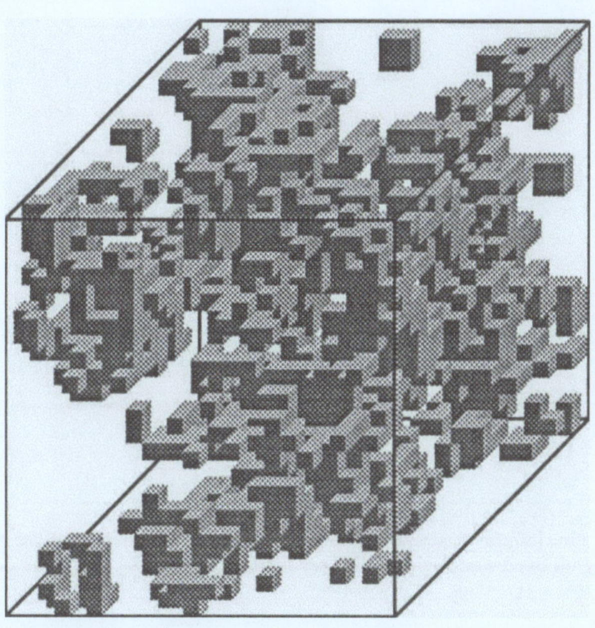

Fig. 3. **A 3D cross section through a large cluster of a 4D Ising system near the critical point**

We therefore expect finite size scaling of $M(L) \simeq L^{d-\beta/\nu}$, an anomalous dimension. This scaling behaviour should be followed by the clusters, as they are relevant for the critical behaviour. They should therefore exhibit a fractal dimension $d_f = d - \beta/\nu = d/(1+1/\delta)$. In 4D we expect mean field critical exponents $\delta = \frac{1}{3}$ and thus $d_f = 3$.

In a computer study on a finite lattice the clusters are of limited extension. Small clusters are essentially one-dimensional lattice animals, whereas large clusters will feel the boundaries and fill the space and then behave like d-dimensional objects. Scaling behaviour we can expect only at intermediate size clusters. However, the larger the lattice, the larger the domain of intermediate size. In a study of the 4D Φ^4 theory with $O(4)$ symmetry [80] the relation between mass s (number of sites belonging to the cluster) and size R (the square root of the mean square radius between points of the cluster) was studied.

In fig. 4 results of simulations for 4D lattices of various size demonstrate the existence of a scaling region. Its slope is the fractal exponent in the relation

$$s \propto R^{d_f} \tag{1.22}$$

with a value $d_f \approx 3.05$ indeed compatible with the expectations. An earlier study for the 3D Ising model [138] had produced a value $d_f = 2.51$ also compatible with the expected relation to the corresponding critical exponents in $d = 3$.

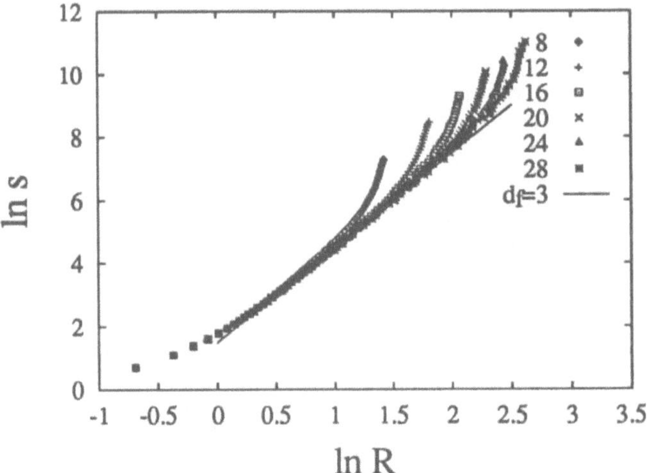

Fig. 4. In this log-log plot [80] of observed combinations (s, R) of clusters at the critical point the intermediate scaling region becomes obviously larger for larger lattices. Its slope defines the fractal dimension.

2. Triviality of Φ^4

2.1. Rigorous Results

The claim "This theory is trivial" should be formulated more explicitly: Defining the continuum theory as the limit $a \to 0$ of the lattice theory, then all n-point functions may be expressed through products of 2-point functions (a theory of generalized free fields), which even may be those of a free particle (a theory of free fields).

Let me summarize here the main results. For more details one should refer to a review like [51, 20].

$D < 4$: The continuum limit defines an interacting field theory. The renormalized 4-point coupling λ_r is bounded ($0 \leq \lambda_r \leq const$) and perturbation theory applies (cf. [49, 50, 15] and further references in [20]).

$D > 4$: The continuum limit defines a trivial theory with mean field behaviour; all connected $2n$-point functions (Wightman-distributions) vanish. Most likely this is a theory of free fields (cf. [1, 40, 16]).

$D = 4$: No rigorous proof of triviality exists, however there is a collection of results which leave little room for another result.

- Triviality has been conjectured by Wilson [141, 143] in the framework of perturbative renormalization group and, under this assumption, many scaling results have been derived (cf. [14]).
- The renormalized 4-point coupling λ_r is bounded ($0 \leq \lambda_r \leq const$); for the single component theory the critical exponents are bounded from below by gaussian (mean field) values [51].

- Considering the equivalence to random walks [132, 133] led to exact inequalities like $\lambda_r \leq 3p_{is}$ (where p_{is} denotes the intersection probability of 4D random walks) [1, 2, 3, 40, 28].
- Rigorous control of block spin transformations at the critical point (the massless theory) proves triviality for ***small enough*** λ_r [48]; close to the critical point there are logarithmic corrections [59, 60].

If the continuum limit is trivial and therefore defined by a fixed point of gaussian nature then its nature does not change in the complete $SU(2)$ gauge Higgs model [62] – although another additional FP at some different value cannot be excluded. ***Assuming triviality*** one can find regions, where the high temperature series (expansion in $1/\kappa$) still gives reliable results and the (renormalization group) scaling behaviour is already valid. This overlap allows to continue n-point functions from the symmetric (hot) phase through the phase transition [102]. Triviality implies a bound on the renormalized coupling and on the Higgs mass, even if the model is just an effective theory [19, 26]. We will discuss these issues below in more detail.

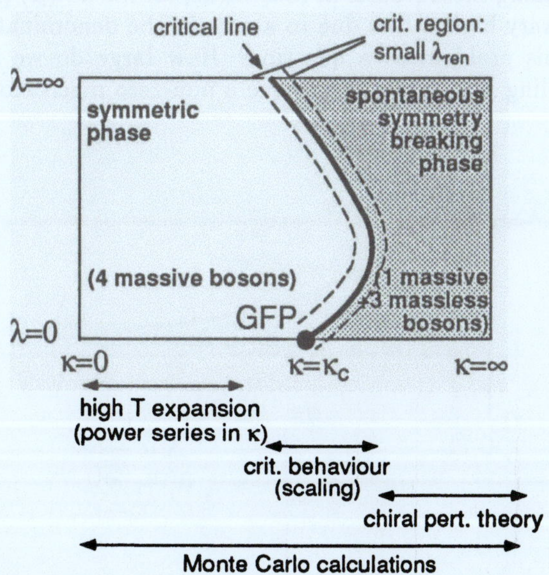

Fig. 5. Phase diagram of the 1- or (in parentheses) 4-component Φ^4 model

The phase diagram of the Φ^4 model in the (κ, λ) plane is sketched in fig.5. For the 1-component model one has a massive Higgs boson in both phases. At the phase transition the dimensionless mass vanishes due to $a \to 0$. The 4-component model has four degenerate massive bosons in the symmetric phase. In the broken phase one finds one massive boson and three massless Goldstone modes (although $a > 0$). The figure also gives regions of applicability of the standard perturbative and non-perturbative methods. Monte Carlo calculations cover the whole domain of interest but they are limited to finite volume. On the other hand, the analytic methods are limited to certain domains and have to assume dominance of the gaussian fixed point (GFP). A thorough analytic understanding, however, helps to analyze the Monte Carlo results in a reliable way, as I try to demonstrate below.

Before we turn to Monte Carlo methods and results let us briefly review the results of renormalization group. Perturbation theory gives the β-function that describes the rescaling properties of the coupling with respect to a change of the scale μ. For the 1-component model the one-loop result is

$$\left(\frac{\partial \lambda}{\partial \ln \mu}\right) = \beta(\lambda) \equiv \frac{3}{16\pi^2}\lambda^2 + O(\lambda^3)\,. \tag{2.1}$$

Solution of the differential equation gives

$$\lambda(\mu) = \frac{\lambda(\Lambda)}{1 - \lambda(\Lambda)\frac{3}{16\pi^2}\ln(\mu/\Lambda)} \quad \text{or} \quad \lambda(\Lambda) = \frac{\lambda(\mu)}{1 + \lambda(\mu)\frac{3}{16\pi^2}\ln(\mu/\Lambda)}\,. \tag{2.2}$$

The cutoff Λ corresponds to the inverse lattice spacing $1/a$, thus $\lambda(\Lambda) = \lambda_{bare}$. For the physical scale μ one could choose some low-energy mass (like the Higgs mass), thus $\lambda(\mu) = \lambda_r$. Taking these results at face value, for fixed $\lambda(\mu)$ the cutoff Λ cannot be pushed to arbitrary high values due to a zero in the denominator (Landau pole). Or, to rephrase this problem as a question: How large do we have to make the microscopical coupling λ_{bare} in order to have a non-zero macroscopical coupling λ_r?

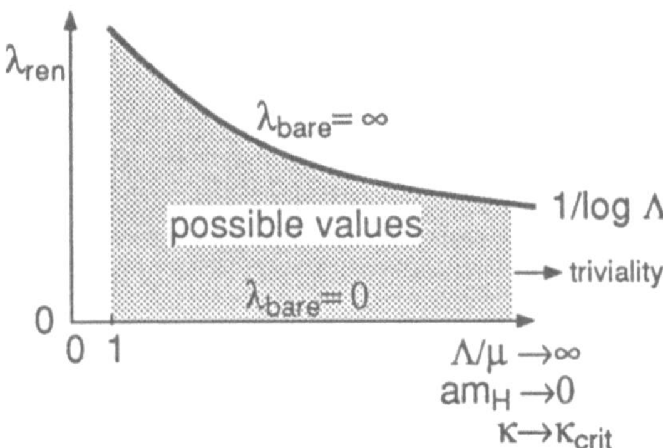

Fig. 6. The renormalization of the quartic coupling as expected from renormalization group arguments.

In fig.6 this situation is summarized. Trajectories of fixed bare coupling are bounded from above by the $\lambda_{bare} = \infty$ values of λ_r. Holding λ_r fixed but attempting to raise the cutoff (approach the critical point, i.e. the continuum limit) is not possible - one gets stuck at the upper bound at some value of Λ (or, equivalently, at some non-zero value of $a(\kappa) > 0$. The bound forces the renormalized coupling to approach zero in the continuum limit. Only at finite cutoff non-vanishing values of λ_r are permitted. Thus even a trivial theory may have an effective interaction below the cutoff.

This discussion is based on the one-loop perturbative result and the renormalization group equations. It was one of the challenges of the Monte Carlo approach to confirm (or not confirm) this scenario. Although the asymptotic shape of the upper bound is given, the multiplicative factor is of non-perturbative nature. Also, the shape at small values of Λ is regularization dependent. The corresponding amplitudes are universal up to regularization dependent corrections $O(\mu^2/\Lambda^2)$.

In a sensible regularization the cutoff should be larger than the masses of the physical states, thus $\Lambda/\mu > 1$. If the situation is like discussed above, this provides an upper bound on λ_r. Perturbation theory in the electro-weak gauge coupling g relates the renormalized coupling with the ratio between the Higgs mass m_H and the mass of the heavy vector boson W,

$$\lambda_r = \frac{1}{2}R^2 , \quad R = \frac{m_H}{F} = \frac{g}{2}\frac{m_H}{m_W} , \quad F \equiv \langle\varphi\rangle/\sqrt{Z} . \tag{2.3}$$

(As mentioned, sometimes the continuum notation with an extra factor 6 is used: $\lambda_r = 3R^2$, cf. table I.) Given the experimental values ($F \simeq 246$ GeV, or $m_W = 80.6(4)$ GeV, $g^2 \simeq 0.4$) the non-perturbative upper bound on the renormalized 4-point coupling provides an upper bound on the Higgs mass [19, 26], unrelated to the usual unitarity-bound.

What evidence can we expect from Monte Carlo simulations? There are three groups of results

(a) Phase structure and critical exponents, derived from bulk quantities like the internal energy, susceptibility or specific heat; detailed investigation of the finite size properties of the partition function allows quantitative statements on finite size scaling, the thermodynamic limit and logarithmic corrections.

(b) Monte Carlo renormalization group (MCRG) methods allow to determine the flow of coupling and the fixed point structure, including the critical exponents.

(c) Propagator measurements lead to the mass spectrum and values of the renormalized couplings. These results are of particular interest concerning the triviality bound on the Higgs mass and its regularization dependence. The energy spectrum also allows the determination of phase shifts.

In the subsequent discussion of this section I first present more qualitative results from real space renormalization group, then a high precision determination of the partition function and the logarithmic corrections. In sect. 3 the non-perturbative results on the bound are discussed.

2.2. Real Space Renormalization Group

In perturbative renormalization group the number of couplings considered is very small, although the number of couplings generated in the transformation may be infinite. The Monte Carlo renormalization group method, also called real space renormalization group (RSRG) method since it works in position space, allows to deal with a larger coupling space, although one has to truncate as well.

The *Scaling Hypothesis* suggests, that the singularities of the thermodynamic quantities are due to the divergence of the correlation length [37, 4, 22]. At the critical point the system should be scale invariant and under scale changing transformations

$$Z(\Lambda) \xrightarrow{T(\Lambda,\Lambda')} Z'(\Lambda') \tag{2.4}$$

the physical properties (i.e. the properties at long distance in terms of lattice spacing) and thus the partition function should be invariant. Here the transformation T is element of a semigroup. On a lattice it is realized by the Wilson-Kadanoff *block-spin transformation* [143], where the local degrees of freedom are "integrated out". We have

scale transformation	Λ	$\rightarrow$	Λ'
linear factor b	$\vert\Lambda\vert/\vert\Lambda'\vert$	$=$	b^D
invariance	$Z(\Lambda)$	$=$	$Z'(\Lambda')$
field variables	$\Phi \in \Gamma$	$\rightarrow$	$\Phi' \in \Gamma'$
correlation length	$n_\xi a$	$\rightarrow$	$n'_\xi a' = (n_\xi/b)ba$

For lattice systems T transforms an ensemble of configurations on Λ into an ensemble of configurations on Λ'. The correlation length changes by the scale factor b.

Given that the original ensemble was distributed according to a Gibbs measure for an action $S(\kappa, \Phi)$ let us for the moment assume that the blocked ensemble also follows such a measure, but for a different action $S(\kappa', \Phi')$ (cf. the discussion further down). Let us formulate the BST (block spin transformation) $\Phi \rightarrow \Phi'$ by introducing a normalized transformation probability P with the properties

(a) normalization: $\int d\mu_{\Lambda'}(\Phi')P(\Phi', \Phi) = 1$

(b) positivity: $P(\Phi', \Phi) \geq 0$; with (a) this implies that for all Φ there is a Φ' such that $P(\Phi', \Phi) \neq 0$, i.e. there is a non-vanishing probability of mapping into each sector of Γ'.

(c) symmetry: all symmetries of the action S should be respected.

We then may describe the transformation by

$$\begin{aligned} Z(\Lambda) &= \int d\mu_\Lambda(\Phi) \exp[S(\kappa, \Phi)] = \int d\mu_\Lambda(\Phi) d\mu_{\Lambda'}(\Phi')P(\Phi', \Phi) \exp[S(\kappa, \Phi)] \\ &\equiv \int d\mu_{\Lambda'}(\Phi') \exp[S(\kappa', \Phi')] = Z'(\Lambda') \end{aligned} \tag{2.5}$$

with

$$\exp[S(\kappa', \Phi')] = \int d\mu_\Lambda(\Phi)P(\Phi', \Phi) \exp[S(\kappa, \Phi)] \, . \tag{2.6}$$

The BST is largely arbitrary [17] and may be chosen optimal from the point of view of simulational simplicity. One has considered both, deterministic and probabilistic forms,

$$\begin{aligned} P(\Phi', \Phi) &= \prod_{x'} \delta(\Phi'_{x'}, f_{x'}(\Phi)) && \text{(deterministic)}, \\ &= \prod_{x'} \exp(\Phi'_{x'} \sum_\alpha \rho_\alpha f^\alpha_{x'}(\Phi)) && \text{(probabilistic)}. \end{aligned} \tag{2.7}$$

A deterministic transformation with a linear relation between the spins and the blockspins makes the new spin at x' an average over the old spins within some block

$B_{x'}$,

$$f_{x'}(\Phi) \propto \sum_{x \in B_{x'}} \Phi_x \, , \tag{2.8}$$

whereas a nonlinear form could be chosen to keep certain expectation values (in the example below: $\langle \Phi^2 \rangle$) fixed during the transformation,

$$f_{x'}(\Phi) = \text{sign}[\sum_{x \in B_{x'}} \Phi_x] \frac{1}{16} \sum_{x \in B_{x'}} \Phi_x^2 \, . \tag{2.9}$$

For Ising spin systems this choice gives the majority rule. Probabilistic transformations have adjustable parameters ρ_α that may be used to optimize certain features.

Under repeated transformations the (dimensionless) correlation length shrinks and the equivalent action changes.

$$n_\xi \to n'_\xi \to n''_\xi \to \; \ldots \; n^*_\xi = \begin{cases} 0 & \text{trivial FP} \\ \infty & \text{critical FP} \end{cases} , \tag{2.10}$$

$$\kappa \to \kappa' \to \kappa'' \to \; \ldots \; \kappa^* \quad \text{(fixed point)} \, . \tag{2.11}$$

The couplings renormalize according

$$\kappa' = \kappa'(\kappa) \tag{2.12}$$

which serves as the definition of the BST in action space, and, assuming simple FP (fixed point) properties, $\kappa^* = \kappa'(\kappa^*)$. If we start at a critical point, the correlation length is infinite all the way: One stays on the critical hypersurface. If the correlation length was finite it will approach zero in the iteration of the BST, corresponding to the approach to a trivial (hot or cold) FP.

At the FP we may linearize (2.12) to obtain

$$\kappa'_\alpha(\kappa) = \kappa^*_\alpha + T^*_{\alpha\beta}(\kappa - \kappa^*)_\beta + o(||\kappa - \kappa^*||) \quad \text{with} \quad T_{\alpha\beta} = \left(\frac{\partial \kappa'_\alpha}{\partial \kappa_\beta} \right)_{\kappa^*} \tag{2.13}$$

or, for the n-th step

$$T^{(n+1,n)}_{\alpha\beta} = \frac{\partial \kappa^{(n+1)}_\alpha}{\partial \kappa^{(n)}_\beta} \longrightarrow T^*_{\alpha\beta} \, . \tag{2.14}$$

The eigenvalues $\{\lambda\}$ of the FP values of the linearized BST decide on the number of relevant ($\lambda > 1$), marginal ($\lambda = 1$), or irrelevant ($\lambda < 1$) operators (couplings) at the critical FP. (There may also be redundant operators. They occur e.g. if the set of couplings is overcomplete [109, 58]). Fig. 7 shows an example for the RG flow. Starting at the FP with a small perturbation, the flow will lead away into the direction of relevant operators. This defines the renormalized trajectory, along which there are no contributions of irrelevant operators, i.e. no corrections to scaling.

Usually almost all couplings (corresponding operators in the action) are irrelevant. The attraction domain of a FP is high-dimensional and only a few (two or three) coupling are relevant. This behaviour is the reason for universality classes: The same critical surface is domain of attraction for many different initial actions.

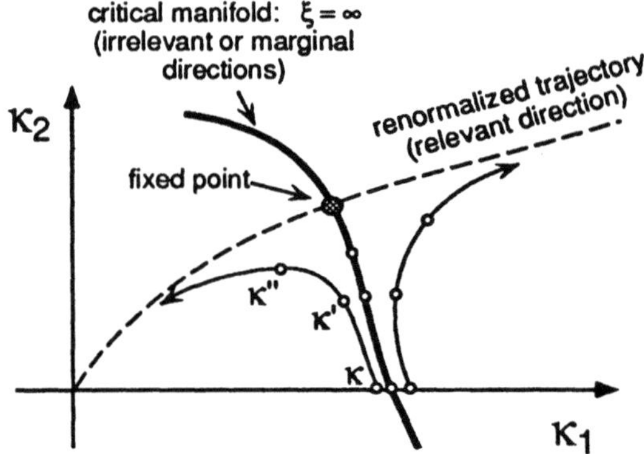

Fig. 7. Schematic display of the renormalization flow in coupling space; starting from the vertical axis at $\kappa_2 = 0$

Although the position of the critical points may be different, the critical behaviour is not, as it is always given by the properties of T^*.

The linearized BST T^* factorizes into even and odd subspaces, depending on the symmetry properties of the corresponding terms in the action under reflection $\Phi \leftrightarrow -\Phi$. The relevant eigenvalues are related to the corresponding critical exponent. This may be seen from the scaling behaviour of e.g. the correlation length. Assuming that its critical behaviour is given by some critical index ν we find

$$\begin{aligned} \xi(\kappa) &= b\xi'(\kappa'(\kappa)) \rightarrow \\ (\kappa-\kappa^*)^{-\nu} &= b\left(\kappa'(\kappa)-\kappa^*\right)^{-\nu} = b\lambda^{-\nu}(\kappa-\kappa^*)^{-\nu} \end{aligned} \tag{2.15}$$

which is consistent only for

$$\nu = \frac{\ln b}{\ln \lambda} \,. \tag{2.16}$$

If there are marginal eigenvalues there is a line of FPs in the critical surface.

The position of the FP on the critical surface may depend on details of the BST [130]. In particular it may be moved into directions of redundant operators [38]. In practice the amount of redundancy is often not clear. Improvement may be desirable for both, the FP action (cf. [58]) and the renormalized trajectory [130, 47, 92]. Recent attempts to optimize the BST such that the FP action becomes particularly simple and as local as possible seem to be promising for asymptotically free theories [69].

Before I continue, let me mention possible difficulties and caveats.

- The transformation may introduce spurious singularities as discussed by Griffiths and Pearce [56, 57].
- Some BSTs at first order transitions may produce non-Gibbsian measures ([136, 137], cf. however [106]).
- The FP may describe a nonlocal action (no exponential damping of nonlocal terms) or it may be at infinity in the parameter space of the action.
- Depending on the lattice geometry and dimensions only a limited range of blocking factors is available.

– Like in perturbative renormalization group the number of couplings generated in the transformation may be infinite. In practical calculations one therefore has to rely on truncations.

For the following discussion we write the action as a sum of operators,

$$S = \sum_\alpha \kappa_\alpha S_\alpha \,. \tag{2.17}$$

Expectation values and correlators are related to derivatives of the free energy

$$\frac{\partial F}{\partial \kappa_\alpha} = \langle S_\alpha \rangle \,, \quad \frac{\partial^2 F}{\partial \kappa_\alpha \partial \kappa_\beta} = \frac{\partial \langle S_\alpha \rangle}{\partial \kappa_\beta} = \langle S_\alpha S_\beta \rangle - \langle S_\alpha \rangle \langle S_\beta \rangle \,. \tag{2.18}$$

BSTs provide different possibilities to obtain information on the critical properties.

Critical exponents: The cross-correlations between observables measured on the original and the blocked configurations allow to get estimators for the truncated T-matrix [105, 128].

$$\frac{\partial \langle S_\alpha^{(n+1)} \rangle}{\partial \kappa_\beta^{(n)}} = \sum_\gamma \frac{\partial \langle S_\alpha^{(n+1)} \rangle}{\partial \kappa_\gamma^{(n+1)}} \frac{\partial \kappa_\gamma^{(n+1)}}{\partial \kappa_\beta^{(n)}} = \sum_\gamma \frac{\partial \langle S_\alpha^{(n+1)} \rangle}{\partial \kappa_\gamma^{(n+1)}} T_{\gamma\beta}^{(n+1,n)} \,, \tag{2.19}$$

$$\frac{\partial \langle S_\alpha^{(n)} \rangle}{\partial \kappa_\beta^{(m)}} = \langle S_\alpha^{(n)} S_\beta^{(m)} \rangle - \langle S_\alpha^{(n)} \rangle \langle S_\beta^{(m)} \rangle \,. \tag{2.20}$$

Here T is obtained from (2.19) by inversion. Due to the truncation there will be inherent errors (see however [121]). This approach has been very successful for 2D and 3D spin models where it has provided the maybe best estimates of critical exponents [18].

Flow in the space of observables: Repeated BST leads one to the renormalized trajectory. Matching of observables allows to relate couplings of different simulations.

Flow in the space of couplings: This would allow to identify FPs and universality class in a very intuitive and direct way.

Unlike in perturbation theory, in non-perturbative RSRG calculations it is no straightforward task to measure renormalized couplings. It is simple to construct the blocked configurations (in position space), but one has no direct means to know about the corresponding action. There is the problem to learn the values of (many, in the thermodynamic case infinitely many) couplings of the blocked action, which is not known directly. A possibility is by operator matching: Matching the observables determined by Monte Carlo simulations on lattices of the size of the blocked lattice [122, 75, 63, 12]. However, for this one has to combine statistically independently determined expectation values which needs high statistics. Another, very efficient method has been suggested by Swendsen [129] and amounts to determine conditional expectation values in the background of given configurations.

MCRG calculation for the one component Φ^4 model have been performed on lattices up to size 16^4 and the results were compatible with the expected scenario: one relevant even and one relevant odd direction in operator space. The result for the even critical exponent was good ($\nu = 0.53(4)$ from e.g. [11, 90], or from [21]) but the statistical accuracy not sufficient to identify the expected next-to-leading marginal eigenvalue. No convincing signal for logarithmic corrections could be determined in those calculations.

However, from the flow structure of the renormalized couplings one could try to find out, whether different values of the bare quartic coupling are in one universality class. This was done in [90] in a space 24 different couplings corresponding to 6 quadratic (local and distance 1, 2, 3, and 4 spin-spin products) and 18 quartic (local and neighbouring 2-, 3- and 4-spin products) operators. The simulation was done on a 16^4 lattice with most of the initial couplings set to zero. Only the nearest-neighbour coupling κ and the bare 4-point coupling λ were non-zero. For given fixed values of λ (0.2, 0.4 and 2) the model was simulated for the critical value of κ and BSTs were performed down to lattices of size 8^4 and 4^4. The renormalized couplings in the 24D space were determined with Swendsen's method [129].

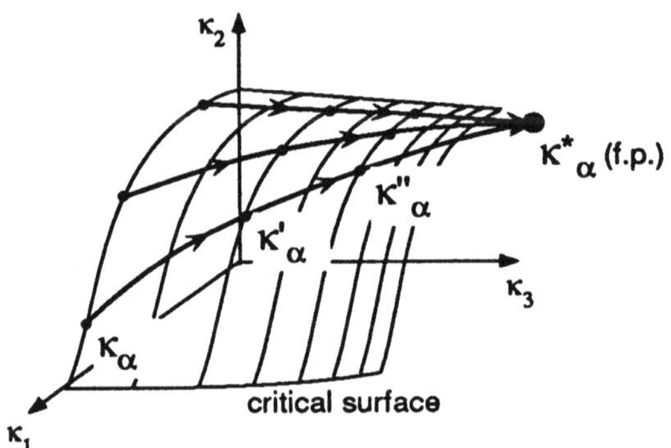

Fig. 8. Flow structure in coupling space; starting on the critical surface in the (κ_1, κ_2) plane (corresponding to the bare quartic coupling and n.n. coupling) the RG flow moves towards a common FP, independent of the bare quartic coupling: it is irrelevant.

The observed flow may be summarized as in fig. 8. From different starting values of the bare quartic coupling on the critical surface the flow is towards a common FP. In the study different BSTs were considered. As the FP position may depend on the particular BST one could expect different flow structure, which indeed was observed [21, 89, 90]. The deterministic nonlinear type of BST discussed above produced a flow towards the FP characteristic for the Ising system. The overall result demonstrated, that the bare quartic coupling is not relevant for the FP and thus for the continuum theory – in agreement with the notion of triviality.

2.2.1. *Critical Indices from Bulk Quantities*

Bulk quantities like the internal energy per link $\langle E \rangle$ or the magnetization $\langle \Phi \rangle$ are the most direct way to the determination of scaling behaviour. However, they are affected by finite size effects and therefore the critical exponents may be masked behind the smooth crossover between the phases typical for finite systems. Finite size scaling [10, 4] relates finite size properties with the critical exponents in the thermodynamic limit. However, usually one needs very precise determination of the quantities involved. The histogram method, an old idea [30] reshaped into the multi-histogram approach by Ferrenberg and Swendsen [31, 32, 33], allows to combine measurements at different couplings in an efficient way.

We rewrite the partition function in terms of the density of states (for simplicity we assume that the set of energy values is discrete, otherwise we have to rely on some binning of the allowed range).

$$\begin{aligned} Z(\kappa) &= \int d\mu_\Lambda(\Phi) \exp[\kappa S(\Phi)] = \int d\mu_\Lambda(\Phi) \sum_E \delta(S-E) \exp[\kappa S(\Phi)] \\ &= \sum_E \rho(E) \exp(\kappa E) = \sum_E p(E|\kappa) \,. \end{aligned} \tag{2.21}$$

The probability for a configuration having the total action E for coupling κ is $p(E|\kappa)$ and is estimated by the corresponding histogram values $h(\kappa, E)$ obtained in the Monte Carlo simulation of the system. Simulations at different values of the coupling provide different estimators for the spectral density $\rho(E)$:

$$\begin{aligned} \kappa_0 &\rightarrow \rho_0(E) = h(\kappa_0, E) \exp(-\kappa_0 E) \,, \\ \kappa_1 &\rightarrow \rho_1(E) = h(\kappa_1, E) \exp(-\kappa_1 E) \,, \qquad \text{etc.} \end{aligned} \tag{2.22}$$

However, the statistical accuracy for these different estimators will be best for those values of E close to the corresponding expectation value (i.e. near the peak of the corresponding histogram). A suitable combination of histograms [31, 32, 33], however, gives a reliable estimator over a broader range of energies (and therefore couplings). From this estimator for the spectral density one may construct energy histograms at arbitrary values of the coupling, as in fig. 9.

This technique may be generalized to other observables and terms in the action. Taking into account energy and magnetization we have the partition function

$$Z(\kappa, h) = \sum_{E,M} \rho(E, M) \exp(\kappa E + hM) \tag{2.23}$$

allowing the precise determination of expectation values of arbitrary functions of E and M,

$$\langle f(E, M) \rangle = \frac{1}{Z} \sum_{E,M} \rho(E, M) f(E, M) \exp(\kappa E + hM) \,. \tag{2.24}$$

An example is the specific heat C_V; in fig. 10 both, the values determined at specific couplings and the continuous curve from the multi-histogram analysis is shown. Each curve contains all the information of the corresponding individual couplings (for that lattice size).

The representation (2.23) also allows for analytic continuation to complex values of κ and h since only the exponential has to be continued. One therefore can determine the zeroes of Z in the complex κ and h planes. In the actual calculation

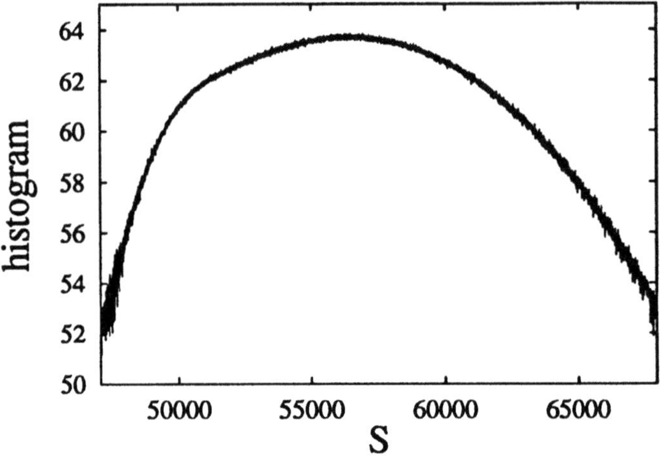

Fig. 9. The logarithm of the probability distribution (histogram) as reconstructed from the multihistogram: $\ln[\rho(E)e^{\kappa_0 E}]$ vs. E, for $\kappa_0 = 0.15055$ and lattice size 16^4.

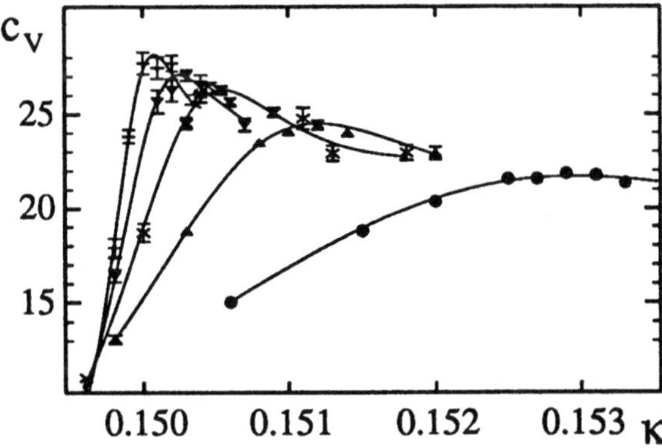

Fig. 10. The specific heat C_V for the one-component Φ^4 model determined for lattice sizes 8^4, 12^4, 16^4, 20^4, and 24^4 (from [84]).

this can be done reliably only close to the values of the couplings, where the original histograms have been determined, otherwise the statistical errors of the estimated spectral density blow up and lead to large uncertainties. Experience shows, that extrapolation is feasible in an elliptical domain with the real positions of measurements between the foci of the principal axis.

The study of zeroes of the partition function has been started by Lee and Yang [146, 147]; zeroes in the complex external field h are called Lee-Yang zeroes. They have to lie at purely imaginary values of h (on the unit circle in the fugacity $z = e^{-2h}$ plane) [146, 147]. Zeroes in the even coupling (the temperature) have been studied in particular by Fisher [36] and are thus usually called Fisher zeroes. On a finite lattice the partition function is a finite-degree polynomial in $e^{-4\kappa}$ and z.

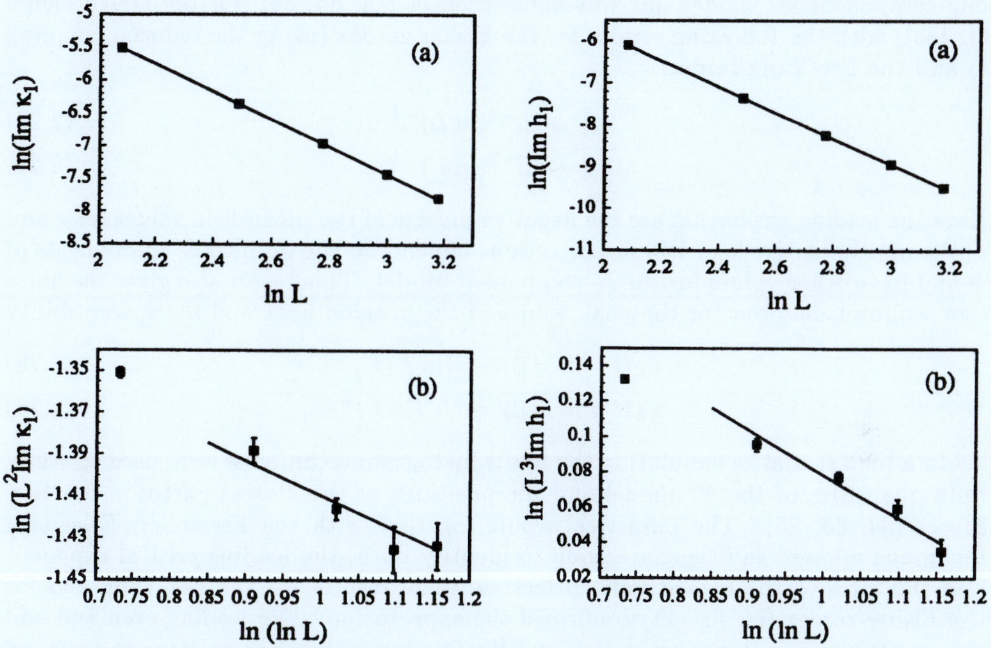

Fig. 11. **Left:** Finite size scaling behaviour of the imaginary part of the closest Fisher-zero κ_1 for lattice size L = 8, 12, 16, 20 and 24 at $h = 0$. (a) The leading exponent is related to the slope of the straight line and a fit gives -2.088(6) slightly differing from the mean field expectation $-1/\nu = -2$. This difference is due to the presence of logarithmic corrections as demonstrated in (b) where the leading behaviour has been cancelled by considering $L^2 \mathrm{Im}\kappa_1$. **Right:** Finite size scaling behaviour of the imaginary part of the closest Lee-Yang-zero h_1 for lattice size L = 8, 12, 16, 20 and 24 and $\kappa = 0.149703$. (a) The leading exponent is related to the slope of the straight line and a fit gives -3.083(4) slightly differing from the mean field expectation $-1/\delta = -3$. This difference is due to the presence of logarithmic corrections as demonstrated in (b) where the leading behaviour has been cancelled by considering $L^3 \mathrm{Im}h_1$ (figures from [86]).

Expressing, e.g. Z through a product of zeroes we find, that the specific heat – the second derivative of its logarithm – is peaked for those real values of κ that are closest to a nearby complex zero:

$$Z \propto \prod_i (\kappa - \kappa_i) \quad \rightarrow \quad C_V(\kappa) \propto \sum_i \frac{1}{(\kappa - \kappa_i)^2} \,. \tag{2.25}$$

In the thermodynamic limit the zeroes accumulate and pinch the real κ-axis (and, for critical coupling, the real point $h = 0$): They build the critical singularity. The Λ-dependence of the positions of the zeroes closest to the real coupling axis can be related to the thermodynamic critical exponents (finite size scaling) and the logarithmic corrections.

Assuming that the gaussian FP controls the leading critical behaviour, perturbative renormalization group methods may be applied to predict the critical properties [14] and the finite size scaling behaviour of the partition function zeroes [13]. For the one-component Φ^4 model this was done recently [84, 86, 85] (for the $O(N)$ model cf. [83]) with the following results for the Fisher zeroes (using the reduced coupling t) and the Lee-Yang zeroes:

$$|t_j| \sim L^{-2}(\ln L)^{-\frac{1}{6}} , \tag{2.26}$$

$$|h_j| \sim L^{-3}(\ln L)^{-\frac{1}{4}} . \tag{2.27}$$

Here the leading exponents are the negative inverse of the mean field values for ν and δ, the multiplicative logarithmic corrections depend on the number of components of Φ and have other values for the N-component model. This result also gives the finite size scaling behaviour for the peak values of the specific heat and the susceptibility

$$c_L(t = h = 0) \sim (\ln L)^{\frac{1}{3}} , \tag{2.28}$$

$$\chi_L(t = h = 0) \sim L^2(\ln L)^{\frac{1}{2}} . \tag{2.29}$$

In a high statistics simulation the multi-histogram techniques were used to study bulk quantities of the Φ^4 model and the positions of the closest partition function zeroes [84, 86, 85]. The numerical result together with the Ferrenberg-Swendsen technique allowed sufficient precision to identify both, the leading critical exponent and the form of the logarithmic correction. The results for the Lee-Yang and for the Fisher-zeroes (cf. fig. 11) confirmed the expectation. The leading even and odd exponents agree with the mean field result. The logarithmic corrections are in good agreement with the expressions (2.26) and (2.27) derived on ground of perturbative renormalization group.

3. Consequences of Triviality

Let us now concentrate on results derivable under the impact of triviality like the bound on the renormalized coupling and the Higgs mass. We therefore discuss in this section in particular the $O(4)$ model because of its close relationship to the unified theory of electro-weak interaction. The notation refers to the lattice action (1.3) given above.

3.1. Results from Non-Monte Carlo Techniques

Using a local potential approximation and Wilson recursion relation technique for renormalization group calculations Hasenfratz and Nager [68] obtained analytic and non-perturbative results on the bound for the one-component model. The effect of the inherent approximations is hard to judge in that approach. A real breakthrough, however, were the results of Lüscher and Weisz [100, 101, 102] and for this reason I want to briefly summarize them.

The authors study the $N(\geq 1)$-component model on a hypercubic lattice. Starting from the symmetric phase they proceed as follows (cf. fig. 5):

Symmetric phase: The high temperature (small κ) expansion for the renormalized mass m_r, coupling λ_r and the critical value κ_c is determined up to $O(\kappa^{14})$. The values seem to be reliable up to relatively small values of $m_r \simeq 0.5$, already within the scaling region.

Scaling (critical) region: Here one uses the results of perturbation theory in the renormalized quantities: m_r and $\lambda_r(\kappa, \lambda)$ for fixed bare λ, using 3-loop β-functions. One has to assume that the gaussian FP defines the universality class. The a priori unknown integration constants like C_1 in

$$\kappa < \kappa_c : am_r(\lambda) = C_1(\lambda)(\beta_1 \lambda_r)^{-\beta_2/\beta_1^2} \exp\left(-\frac{1}{\beta_1 \lambda_r}\right) \{1 + O(\lambda_r)\} \tag{3.1}$$

contain the non-perturbative information and are taken from the κ-expansion values in the overlap region. Corrections to scaling are $O(a^2(\ln a)^c)$. The scaling form is used to continued through the phase transition to the phase with broken symmetry in a region $m_r \leq 0.5$.

Broken symmetry phase: The Integration constants below and above the critical point are related through the massless theory at the critical point,

$$\kappa > \kappa_c : C_1'(\lambda) = e^{\frac{1}{6}} C_1(\lambda) . \tag{3.2}$$

They are renormalization prescription dependent.

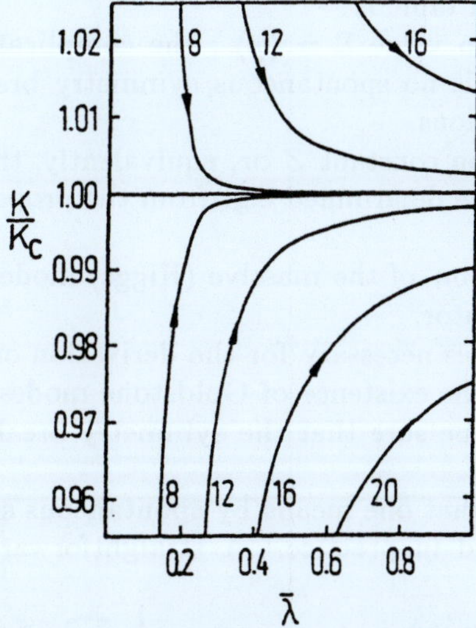

Fig. 12. Quantitative plot of renormalization group trajectories for the $O(4)$ model (from [102]). The abscissa variable $\bar{\lambda}$ corresponds to values of the bare coupling $0 \leq \lambda \leq \infty$. The range of κ-values roughly covers the scaling region $|m_r| \leq 0.5$. The curves correspond to constant renormalized 4-point coupling – they approach the critical line, but they never touch it: In the continuum limit one is forced to vanishing values of λ_r.

This program exploits standard analytic techniques in a very efficient way and worked nicely [100, 101, 102]. In fig. 12 trajectories of constant renormalized 4-point coupling are shown: They never cross the critical line, therefore only a non-interacting continuum limit is feasible.

We summarize the restrictions or assumptions for this approach:

- the κ-series is truncated;
- triviality (gaussian FP) is assumed;
- validity of perturbation theory in the renormalized couplings (renormalization group at the 3-loop level) is assumed.

Direct Monte Carlo simulations can check these assumptions and confirm the validity of the approximations. They also can determine further results concerning, for instance, the regularization dependence. A consistent picture gives weight to both, the analytical and the numerical approach.

3.2. Monte Carlo Simulations

First results go back to 1984 [140, 135] but stayed widely unnoticed. Only with the very high statistics studies of various groups [65, 108, 87, 88, 7, 74] since 1987 the results became convincing enough. These studies worked with lattice sizes up to 16^4. Most results have been obtained in the broken phase, both for $j = 0$ and $j > 0$. Typically several 100.000 sweeps and measurements per point have been performed, using first the Metropolis, later the cluster algorithm. In summary thousands of hours supercomputer CPU-time (like e.g. Cray-YMP or equivalent) have been spent for these analyses. Some studies were concerned with the one-component model but, again, most concentrated on the $O(4)$ model.

For the determination of the renormalized coupling in the phase of broken symmetry we need (cf. (2.3) and table I)

- The vacuum expectation value $\Sigma = \langle\varphi\rangle$; the complication is, that for $j = 0$ in finite volumes there is no spontaneous symmetry breaking and one has to consider suitable definitions.
- The field renormalization constant Z or, equivalently, the *pion decay constant* $F \equiv \Sigma/\sqrt{Z}$. This may be determined e.g. from the propagators of the massless Goldstone modes.
- The renormalized mass m_r of the massive (Higgs) mode, determined from the measured Higgs propagator.

In addition to these quantities necessary for the derivation of λ_r we may also want to identify clear signals for the existence of Goldstone modes, like e.g. propagators at $j = 0$ and $j > 0$, just to be sure that the symmetry breaking mechanism is the conjectured one.

The problem to define, what one means by spontaneous symmetry on finite lattices, turned out to be hard in practice. A straightforward definition originally employed was

$$\Sigma = \langle|\overline{\varphi}^\alpha|\rangle \quad \text{where} \quad \overline{\varphi}^\alpha = \frac{1}{|\Lambda|}\sum_x \varphi_x^\alpha, \tag{3.3}$$

and m_r, F and Z were determined from the propagators of the field components parallel and perpendicular to the value $\overline{\varphi}^\alpha$ of the corresponding configuration, which meandered around in the $O(N-1)$ subspace. That seemed to work nicely. In fig. 13 we show results for the Higgs mass determined that way on lattices of various size. Further results showed that the field renormalization constant stays close to the value 0.97 from the PT up to $m_r \approx 1$.

A better understanding of the infinite volume limit of the *ad hoc* ansatz (3.3) was important to achieve systematically improved quality of the results for Σ. Now,

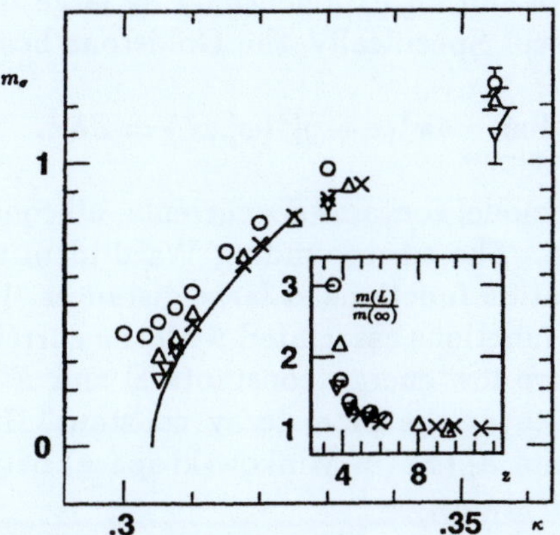

Fig. 13. **The dimensionless renormalized Higgs mass $a(\kappa)m_r$ determined on lattices of various size and extrapolated to infinite lattice size (from [113]).**

for $j \neq 0$ the symmetry is broken explicitly, all quantities are well defined and can be measured without reference to spontaneous symmetry breaking. For small values of the external field the would-be Goldstone states are still very light and therefore they feel the finiteness of the volume best. We have a situation of two different kinds of long correlation lengths (fig. 14). At the 2nd order PT $a(\kappa)m_r \to 0$, because $a(\kappa) \to 0$ for $\kappa \to \kappa_c$, since the theory is critical. Along the 1st order PT $a(\kappa)m_G(j) \to 0$ because the Goldstone boson mass $m_G(j) \to 0$ for $j \to 0$. In the discussion below we will denote masses of the massive mode m_σ and of the Goldstone mode m_π in analogy to the nonlinear σ-model.

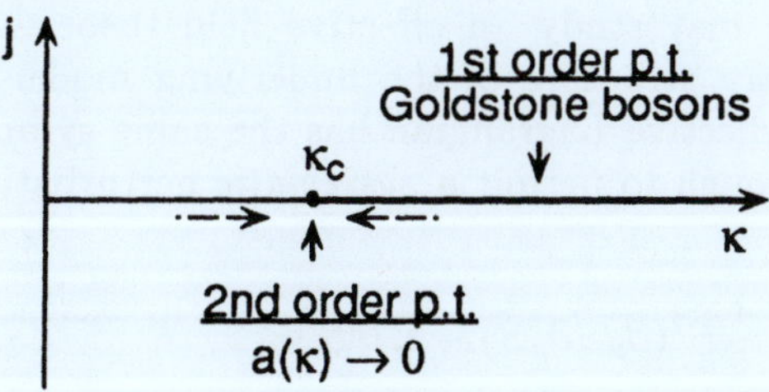

Fig. 14. **Phase structure in the (κ, j)-plane.**

Let us recall again the important parameters of the model. The field Φ^α has four real components. In the symmetric (high temperature) phase the particle spectrum has 4 degenerate states. In the low temperature phase $\kappa > \kappa_c$, at infinite volume $O(4)$ symmetry is spontaneously broken. In this phase

$$\langle \varphi_x^0 \rangle = \Sigma \,, \quad \langle \varphi_x^i \rangle = 0 \quad (i = 1,2,3) \,. \tag{3.4}$$

Here the spectrum of the theory contains three massless Goldstone bosons corresponding to the excitations in the O(4) directions $\alpha = i = 1,2,3$. Therefore, the correlation functions do not fall off exponentially at large distances, but with an inverse power of the distance. Specifically, the Goldstone boson two-point function satisfies

$$\lim_{|x-y|\to\infty} 4\pi^2|x-y|^2\langle\varphi_x^i\varphi_y^j\rangle = Z\delta^{ij}. \tag{3.5}$$

For $O(4)$ symmetry the model contains six currents, all conserved at $j = 0$, whose charges generate the group. The corresponding Ward identities strongly constrain the behaviour of the correlation functions at large distances. In fact, the asymptotic behaviour of all the Green functions associated with the currents and with the fields φ_x^α is determined by the two low energy constants Σ and F [41]. In analogy with QCD, F is often referred to as the pion decay constant. It specifies the matrix elements of the axial currents $A_\mu^j(x)$ (in Minkowski space) between the ground state and single Goldstone boson states,

$$\langle 0|A_\mu^j(0)|\pi^k(p)\rangle = i\delta^{jk}p_\mu F\ . \tag{3.6}$$

The low energy constants Z, Σ and F are related through

$$F = \Sigma/\sqrt{Z}\ . \tag{3.7}$$

For $j > 0$ the Goldstone boson mass m_π has a non-vanishing value which in the lowest order of chiral perturbation theory is

$$m_\pi^2 = j\frac{\Sigma}{F^2}\ . \tag{3.8}$$

Whenever (almost) massless states dominate the system the low energy (long distance) structure is of a universal character determined by the symmetries of the dynamics in terms of a few low energy constants and by the geometry of the finite system. One therefore may study an effective field theory with only Goldstone modes and the symmetry structure of the underlying model $O(N) \to O(N-1)$ [95, 43, 67, 111]. The effective Lagrangian has the same symmetry as the original system but is simple enough to permit a *systematic* perturbative expansion. It has the form

$$\begin{aligned} L_{\text{eff}} &= \frac{1}{2}F^2(\partial_\mu\vec{S}\partial_\mu\vec{S}) - \Sigma\vec{H}\vec{S} \\ &\quad +\text{further terms of higher order in } \vec{H} \text{ and } \vec{S}\ , \end{aligned} \tag{3.9}$$

$$Z(\vec{H}) = \int d\mu(\vec{S})\exp(\int dx L_{\text{eff}})\ . \tag{3.10}$$

In higher orders of this perturbative Lagrangian some further low energy constants Λ_Σ, Λ_F and Λ_M are required [67]. They are the scale parameters determining the logarithmic dependence of Σ, F and m_π on j. In systems of finite size L this expansion amounts to an expansion in powers of L^{-2}. It is reliable because the interaction between the Goldstone modes is weak at low momenta.

The first application of this idea [42, 95] has been made in the context of the study of finite size effects in QCD with light quarks, whose low energy properties are determined by the chiral SU(2)⊗SU(2)≃O(4) symmetry [41]. Therefore one refers to this approach as "chiral perturbation theory". A comprehensive description can be found in [43, 94, 67].

The effective theory and its expansion relates values of quantities determined at finite volumes and $j > 0$ with those at $j = 0$ at infinite volume,

$$\Sigma(j,V), F(j,V), Z(j,V) \xrightarrow{\text{Chiral Pert. Th.}} \Sigma, F, Z. \tag{3.11}$$

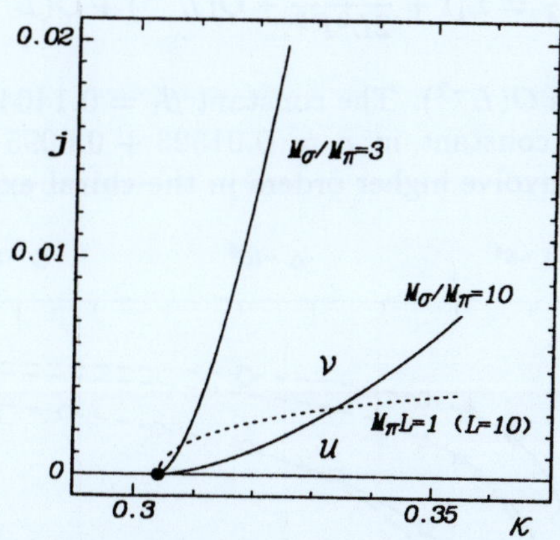

Fig. 15. The typical domains of application for two type of expansions indicated by lines of constant m_σ/m_π in the (κ, j) plane. The dashed curve corresponds to $m_\pi L = 1$ on a 10^4 lattice. The regions below (above) the dashed curve are the $\mathcal{U}$ ($\mathcal{V}$) domains for $L = 10$ (from [64]).

There are two types of expansions, valid in different domains with a non-vanishing overlap (cf. fig. 15).

($\mathcal{U}$) $m_\pi \lesssim 1/L$: This domain is characterized by very small symmetry breaking, where the symmetry is restored. The finite size effects are large here. The expansion is in powers of L^{-2} keeping the total magnetic energy

$$u_0 = \Sigma j L^4 , \tag{3.12}$$

fixed. Thus j is treated as a quantity of the order $O(L^{-4})$. In $\mathcal{U}$ the correlation length ξ_π grows as L^2 for $L \to \infty$.

($\mathcal{V}$) $m_\pi \gtrsim 1/L$: In this region j is still small, although larger than in $\mathcal{U}$. Finite size effects are smaller than in $\mathcal{U}$. A typical range observed is $m_\sigma/m_\pi \gtrsim 5 - 10$. Here the symmetry is not restored. The expansion in powers of L^{-2} in that domain keeps fixed

$$v_0 = \Sigma j L^2/F^2 = m_\pi^2 L^2 . \tag{3.13}$$

For large u_0 the results of the expansion $\mathcal{U}$ should smoothly go over into those of the expansion $\mathcal{V}$. There is an overlap of the regions of validity of both expansions around $m_\pi L \simeq 1$ [64]. In fig. 15 the dashed curve indicates where the $\mathcal{U}$ and $\mathcal{V}$ regions in the κ-j plane meet for $L = 10$.

We do not want to go into details of the various expansions. Just in order to catch the spirit of the approach let us give an example for one result of the $\mathcal{U}$-expansion.

$$\langle\varphi_x^0\rangle_{V,j} = \frac{u}{jL^4}\frac{I_2(u)}{I_1(u)} + 2\rho_2\frac{\Sigma^2}{F^4}j + O(L^{-6}), \tag{3.14}$$

where the second term is $O(L^{-4})$ and

$$\frac{u}{jL^4} = \Sigma[1 + \frac{3\beta_1}{2L^2F^2} + O(L^{-4}) + O(L^{-6})]. \tag{3.15}$$

The second term here is $O(L^{-2})$. The constant $\beta_1 = 0.14046$ depends on the lattice geometry, as does the constant in $\rho = 0.01523 + 0.0095\,O(\ln L\Lambda_\Sigma)$. The terms $O(L^{-4})$ are small and involve higher orders in the chiral expansion.

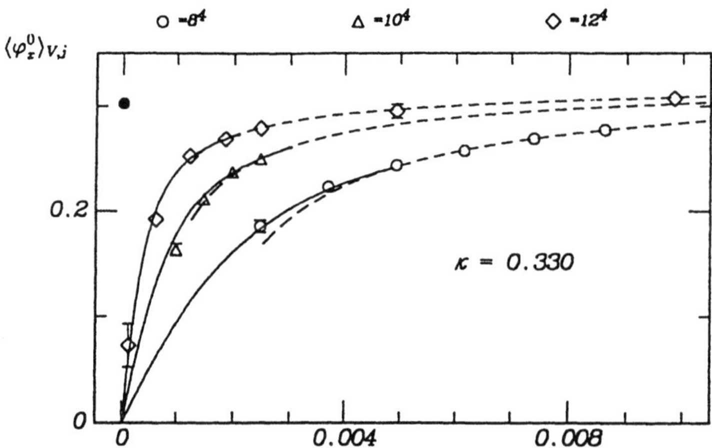

Fig. 16. Example for the type of data obtained for $\langle\varphi_x^0\rangle_{V,j}$ (from [64]). The solid curves correspond to a fit according (3.14) up to $O(L^{-2})$, the resulting value of the only parameter Σ is indicated by the full dot at $j = 0$.

Eq. (3.14) gives the functional form of the j and L dependence in terms of the thermodynamic quantities Σ and F. Fig. 16 gives an example of the type of data and fit quality obtained. The leading term in (3.14) provides the value of Σ. The non-leading term in principle can give F, but is statistically not sufficiently reliable. For this parameter one better uses the t-dependence of the Goldstone propagator.

The values $\Sigma(\kappa)$ obtained in this way prove to be very precise, as demonstrated in fig. 17. One should keep in mind, that these *are the thermodynamic values*! The fit to the scaling behaviour gives [64]

$$\Sigma^2 = 0.672(6)(-t)|\ln(-t)|^{\frac{1}{2}} \text{ and } \kappa_c = 0.3036(10). \tag{3.16}$$

In table II the values of Σ obtained by the discussed chiral perturbation expansion method [64] is compared with the original "brute force" ansatz (3.3) and one finds surprising agreement - although the new method definitely is of higher quality.

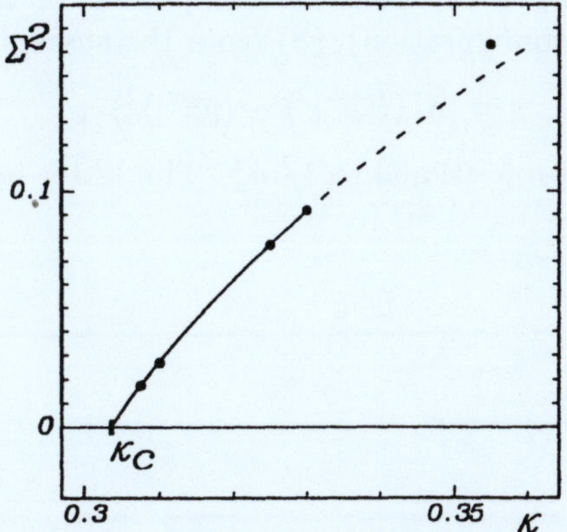

Fig. 17. The values of Σ obtained from the fit of $j > 0$, finite volume data to the chiral perturbation expression; the curve shows the fit to the expected scaling behaviour (from [64]).

TABLE II
Results on various quantities of the $O(4)$ Φ^4 model as obtained with the action (1.3) in various studies.

κ	Σ	$\langle\lvert\overline{\varphi}^\alpha\rvert\rangle$	F	Z	m_σ	m_σ/F	Ref.
0.3130(7)	0.196(1)		0.199(5)	0.976(6)	0.50	2.517(65)	[102]
0.3101(5)	0.166(6)		0.168(5)	0.974(7)	0.40	2.380(69)	[102]
0.3077(3)	0.130(3)		0.132(3)	0.973(5)	0.30	2.280(58)	[102]
0.3058(2)	0.093(2)		0.094(2)	0.972(5)	0.20	2.129(46)	[102]
0.3046(1)	0.051(1)		0.052(1)	0.971(5)	0.10	1.941(34)	[102]
0.355	0.4036(9)	0.402	0.4109(13)	0.965(2)	1.09(10)	2.65(25)	[64]
0.330	0.3027(11)	0.301	0.3075(16)	0.969(3)	0.92(7)	2.99(24)	[64]
0.325	0.2769(12)	0.276	0.2807(19)	0.973(5)	0.81(2)	2.89(9)	[64]
0.310	0.1643(63)	0.163	0.1668(73)	0.97(1)	0.39(2)	2.34(22)	[64]
0.3075	0.132(13)	0.128	0.135(14)	0.96(1)	0.29(1)	2.15(30)	[64]
0.3100	0.15549(10)		0.1584(7)	0.964(7)	0.4073(7)	2.572(15)	[54]
0.3080	0.12827(6)		0.1301(3)	0.972(3)	0.3223(12)	2.477(14)	[54]
0.3060	0.09045(10)		0.0915(2)	0.977(3)	0.2072(7)	2.264(14)	[54]

Let us summarize the different ways to determine the low energy parameters.

Determination of Σ: (a) from the $j > 0$ data with help of the discussed chiral perturbation expansion in j and L^{-2}; (b) at j=0 from the distribution of the "local" mean magnetization $\langle|\overline{\varphi}^\alpha|\rangle$ defined in (3.3) – for a detailed discussion of this constraint effective potential approach cf. [52, 55, 53, 54].

Determination of Z: from the Goldstone propagator (a) at $j > 0$ from chiral perturbation expansion formulae [64] and (b) at $j = 0$ using the field components transversal to the direction defined by the magnetization $\langle|\overline{\varphi}^\alpha|\rangle$.

Determination of m_σ : (a) from the component of the field parallel to the direction of $\overline{\varphi}^\alpha$ (for each configuration); (b) from the susceptibility

$$\chi = |\Lambda| \left(\langle|\overline{\varphi}^\alpha|^2\rangle - \langle|\overline{\varphi}^\alpha|\rangle^2\right) , \tag{3.17}$$

which is in leading order proportional to $1/m_\sigma^2$. The latter method appears to work very efficiently [54].

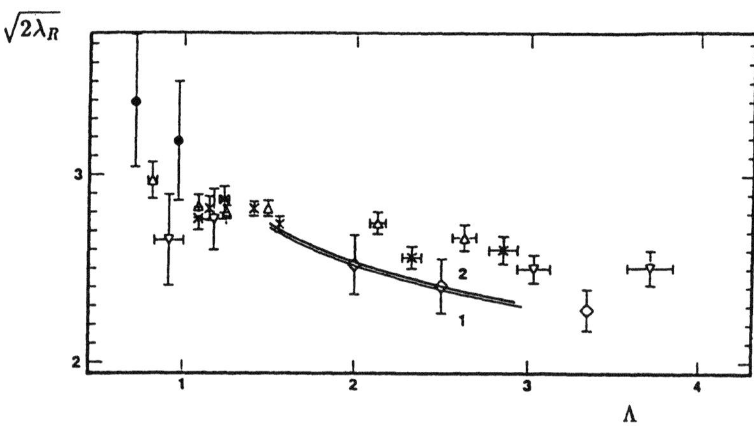

Fig. 18. Comparison of results for the renormalized coupling for different values of the cutoff $\Lambda = 1/(am_\sigma)$ (from [113] in 1990).

In table II we compare the results derived over the years from various groups. One observes increasing quality of the numbers. Comparing the Monte Carlo results for the renormalized coupling at values $m_\sigma < 0.5$ with those of [100, 101, 102] one finds good agreement (within the statistical errors). This was also confirmed by further Monte Carlo results of other groups [87, 88, 7]. The resulting estimates for the upper bound on the renormalized coupling are shown in fig. 18 and fig. 19. One clearly notices the progress in the statistical accuracy.

If we assume that the cutoff (given in units of m_σ in the figure) should not be below ~ 2 then we read off an upper bound of $R \lesssim 2.6(3)$. With (2.3) this translates into $m_H \lesssim 670(80)$ GeV. This bound comes from the field theoretic properties of the Φ^4 model as a direct consequence of triviality and should not be mixed up with the so-called unitarity bound on the Higgs mass. The bound expresses the fact, that for this part of the standard model we cannot perform the continuum limit. Or, put differently, for a given Higgs mass there is a limited domain of validity of the regularized Φ^4 theory. If the Higgs mass comes out to be much smaller, e.g. of the order of 200 GeV, the bound allows a cutoff of the order of the Planck length. Up to the allowed cutoff we still can consider the theory a model with an effective interaction. However, we conclude that the theory cannot be a truly fundamental one (we cannot remove the cutoff completely) and we therefore may expect constituent structure.

All these results have been obtained away from the critical point, therefore one does have contributions from non-leading operators. These corrections to scaling are

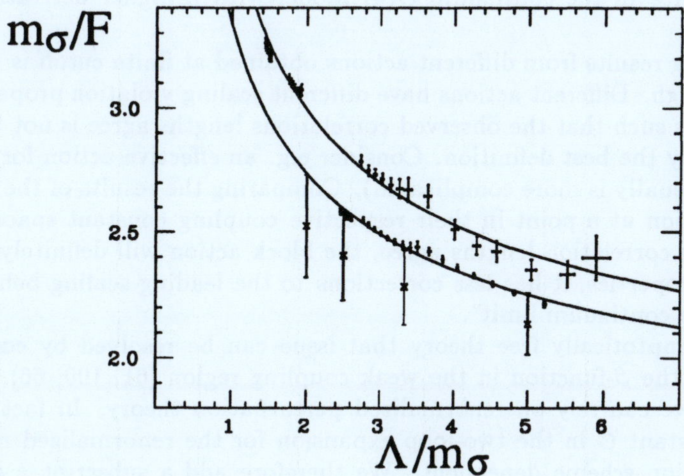

Fig. 19. Results for the renormalized coupling for different values of the cutoff $\Lambda = 1/(am_\sigma)$. The 4 points with larger error bars are from [102] and correspond to twice the error quoted, following their discussion of uncertainties. The upper group of points corresponds to a different action as will be discussed below (from [54] in 1992).

regularization dependent and $O(a^2)$. The cutoff then represents new, possibly more fundamental physics which would replace the standard-model at higher energies. However, its influence is effective on all scales and manifests itself in the cutoff dependence of all observables. In describing the experiments through a lattice model, the higher in energy we go the stronger we expect such cutoff effects to be. More and more terms would have to be included in the effective action in order to explain experiments with sufficient accuracy. Within the scaling region one expects to find these dependencies to be only weak, but this knowledge is mainly based on rough estimates and there are only few explicit studies [7, 8, 70, 53, 54].

Value and shape of the bound depend on the adopted regularization scheme, i.e. on:

- The lattice geometry; an investigation of the scalar field theory on various types of lattices is required [7, 110, 112, 70]. Methods for comparing the results obtained by means of different regularization schemes are being developed [66, 102, 91]; they require calculations of high precision in a region of the coupling parameter space where the universal asymptotic scaling associated with the ("trivial") gaussian fixed point sets in.
- Additional terms in the action; an example is the introduction of a straight distance 2 coupling in order to remove the $O(p^4)$ dependence in the lattice perturbative propagator (Symanzik improvement).
- Other terms of $O(1/\Lambda^2)$ in the action; Heller et al. [72, 73, 74] have studied these effects in the $O(N)$-model by systematically adding operators of increasing order in $1/\Lambda$, both, in continuum and on the lattice, in particular on an F_4-lattice [110].
- Other regularization schemes: Recent work demonstrates the regularization

dependence in the continuum $O(4)$ model with a higher derivative regulator [79, 78].

Comparing results from different actions obtained at finite cutoff is not straightforward, though. Different actions have different scaling violation properties. Choosing couplings such that the observed correlations lengths agree is not the only and not necessarily the best definition. Consider e.g. an effective action for block variables (which usually is more complicated). Comparing the results of the original and the block action at a point in their respective coupling constant spaces where the dimensionless correlation lengths agree, the block action will definitely have better continuum properties, it has less corrections to the leading scaling behaviour. It is "closer to the continuum limit".

In an asymptotically free theory that issue can be resolved by comparing the behaviour of the β-function in the weak coupling region [61, 100, 66]. In our case in principle we can rely on renormalized perturbation theory. In fact the proportionality constant C in the two-loop expansion for the renormalized mass in (3.1) is regularization scheme dependent. We therefore add a subscript s denoting the regularization scheme (lattice action or geometry). Asymptotically, at small enough small λ_r, $a_s/a_{s'} = C_s/C_{s'}$ can be determined by lattice perturbation theory.

For a comparison of different effective theories at finite cutoff and not too small λ_r we can no longer rely on that approach. It would be appropriate to choose another physical observable $\mathcal{O}$, that may serve as a means to measure the distance from the continuum limit independently. Comparing results due to different actions S_s (where the index distinguishes between the different actions and regularization schemes) at the same values of $\mathcal{O}$ then provides an estimate for the regularization dependence. This observable could be for example a $\pi\pi$-scattering cross-section, the σ-decay width or the violation of euclidean rotational invariance of some quantities [102, 8, 70, 54, 93]. Different operators $\mathcal{O}$ will usually have different cutoff dependence and their choice needs further motivation: There is no "best" observable. Quantities like the ratio R should be considered versus such a "distance" $\mathcal{O}$ to the continuum limit. Of course we could still use the inverse correlation length $m_{\text{phys}}a_s$, but in order to compare different schemes we have to rescale a_s at each point such that $\mathcal{O}$ agrees.

In a Monte Carlo study of the 1-component Φ^4 model [93] the violation of euclidean rotational invariance (in real space and in momentum space) of the Higgs-propagator was used as measure for the "distance to the continuum limit". Four kinds of lattice actions, respectively lattice geometries, were considered.

C : The standard action on the hypercubic lattice (1.2).

C' : An action including diagonal nearest neighbour terms with the same coupling [7].

F_4 : The nearest neighbour action on an F_4-lattice [8, 9]. It's embedding into a hypercubic lattice may be imagined by removing all odd sites of the grid. The symmetries for this case forbid terms like $\sum_\mu p_\mu^4$ in the inverse momentum space propagator and guarantee Lorentz-invariance to a higher order in the momentum cutoff $O(\Lambda^{-4})$ than the first two hypercubic lattice schemes which had violating terms $O(\Lambda^{-2})$.

S : The Symanzik improved action which includes anti-ferromagnetic couplings to the next-to-nearest neighbours along the 4 euclidean axis-directions. These additional terms are incorporated in order to remove the undesired $O(\Lambda^{-2})$ contributions. The remaining corrections of $O(\Lambda^{-4})$ in the lattice dispersion relation guarantee tree-level improvement [134].

All four lattices may be embedded into the regular hypercubic one.

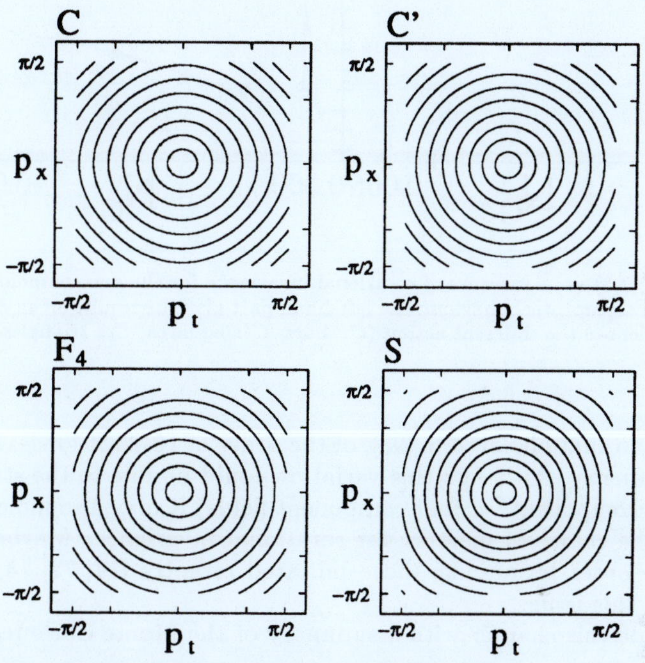

Fig. 20. Fig. from [93], showing the amount of rotational symmetry in the inverse boson propagator in momentum space, determined for 4 different actions.

Fig. 20 compares the rotational invariance properties for the boson propagator in momentum space. The deviation from the exact continuum symmetry may be quantified by considered the coefficients in an expansion in spherical harmonics. Fig. 21 exhibits the final results for $R = m_\sigma/F$ (proportional to the square root of the renormalized coupling) as functions of a measure for the deviation from rotational symmetry. As discussed above, this may be a more appropriate measure than the correlation length itself. One find a sizable difference, about 30-40%, between the values for the actions $\boldsymbol{C}$ and $\boldsymbol{C'}$ (which violate rotational invariance to a larger extent) on one hand and $\boldsymbol{F_4}$ and $\boldsymbol{S}$ on the other (which have no $O(p^4)$ violation in the tree-level propagator).

In the results for the $O(4)$ model for $\boldsymbol{F_4}$ [8] and $\boldsymbol{S}$ [54] the authors estimated the regularization dependence by comparing the results at given value of another measure of rotational symmetry, derived from a perturbation expansion of the Goldstone-Goldstone scattering cross-section. There the difference between $\boldsymbol{F_4}$, $\boldsymbol{S}$ compared to

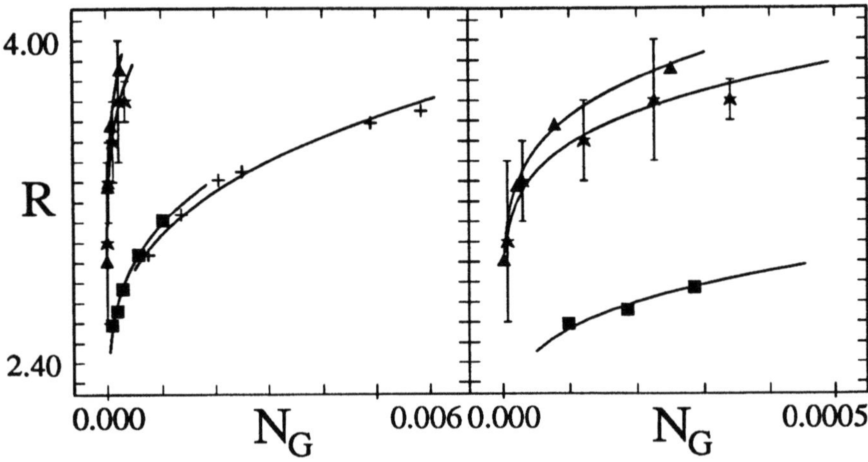

Fig. 21. The ratio (2.3) vs. a measure of rotational invariance for the one-component model; the curves are fits to the asymptotic behaviour; the left-hand part give the results on an extended scale. The symbols used denote the different action (C: bars, C': squares, F_4: triangles, S: asterisks) (fig. from [93]).

C was smaller than that observed in [93], of the order of 10-20 % for 1-10% violation of rotational symmetry. Although this variation could be due to the study of different models (one component vs. four component fields) it may also indicate a strong dependence on the choice of measure for rotational invariance. Maybe systematic studies of higher order terms, like those initiated recently [74, 72, 73, 79, 78] are helpful to clarify this issue.

Let us conclude this chapter with a summary of the Monte Carlo results for the $O(4)$ Φ^4 model.

- The observed scaling behaviour is consistent with the triviality conjecture.
- One clearly identifies Goldstone bosons and the massive Higgs mode.
- One finds good agreement with analytically derived results in regions, where the comparison is possible ($am_H < 0.5$).
- Finite size scaling works.
- Chiral perturbation theory provides an excellent description of $j > 0$ finite volume results and allows a reliable extrapolation to the thermodynamic situation of spontaneous symmetry breaking.
- Allowing for a cutoff $> 2m_H$ (or "continuum-like behaviour" up to a few percent) one obtains an upper bound for the renormalized coupling, that translates (using the experimental values for the gauge coupling and the W-mass) into the bound on the Higgs mass [71],

$$m_H < 710(60)\text{GeV}. \tag{3.18}$$

- Some (relatively weak) regularization dependence of this bound (and other quantities) has been observed.

- Further work includes fermions (Yukawa models), but no drastic change of the bound has been found.

The recent review of Heller [71] summarizes the situation and gives further relevant references.

4. Scattering: Phase Shifts and Resonances

Most observables that have been considered in lattice calculations so far are *masses* (from propagators) and *coupling constants*. Most experimental measurements in high energy physics, however, lead to cross sections and their analyses to scattering phase shifts. Indeed, resonances are described by the slope and position of the values of the phase shift passing through $\pi/2$ and the width is a quantity derived under certain additional assumptions (like the Breit-Wigner shape). In fact, considering the $O(4)$ Φ^4 model we are in a situation, where the Higgs boson is a resonance (coupling to Goldstone states) and, strictly speaking, no asymptotic state of the theory. For the determination of the Higgs mass it is therefore not sufficient to just determine the single particle propagator over larger and larger distances. In that limit one will only learn about the lowest lying energy level in that channel, which is not the Higgs mass but a 2-particle threshold.

In finite systems the energy spectrum is discrete and may be related to the phase shifts in the infinite volume system [96, 97, 103, 99, 98]. The leading volume dependence is connected to the scattering lengths and early calculations in the 4D Ising model determined the s-wave scattering length [108, 39]. Meanwhile there are several studies:

- In a pioneering work Lüscher and Wolff obtained phase shifts in the 2D $O(3)$ nonlinear σ-model [103]. There the result is known analytically from conformal field theory and one found agreement with the non-perturbative calculations.
- Below I discuss another 2D study [45, 46, 44] with the aim to determine phase shifts in a resonating situation and to find the resonance parameters in a non-perturbative way.
- Fiebig et al. [34, 35] investigate the 3D QED and determined phase shifts for the scattering of mesonic bound states (in the quenched situation, i.e. without dynamical fermions in the vacuum).
- In the 4D $O(4)$ Φ^4 model there have been studies in the symmetric phase [114] and in the broken phase with an additional external symmetry breaking field [148, 149, 139]. The latter model is a test case for resonating phase shifts as well, since one has a situation with light particles (the Goldstone bosons have non-zero mass due to the external field) and one heavy resonance (the Higgs particle decays into the Goldstone states and is no asymptotic state of the theory).

Only the efficient algorithms available for bosonic systems allow the high precision that is necessary to determine the energy levels and phase shifts, as discussed below. For this reason the scalar field theories are the leaders in that approach – the reason to discuss approach and result in this last section of this review. Also, in that process a lot can be learned on the representation of physical states an the lattice.

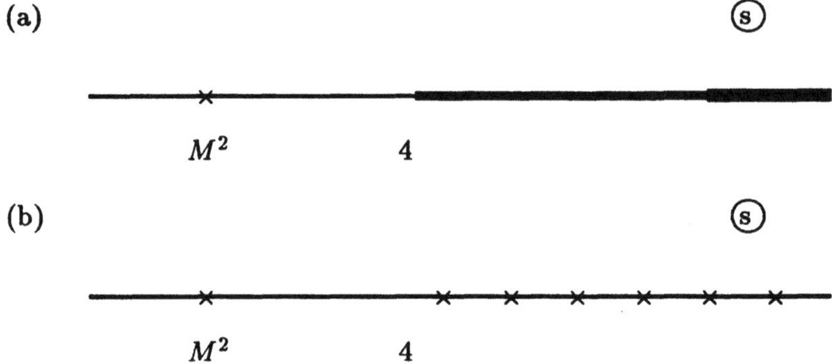

Fig. 22. The spectral density in the $s = W^2$-plane; (a) infinite volume, (b) finite volume.

4.1. Energy Eigenspectrum of Finite Systems

Consider the propagation of a state $|N\rangle$ in a channel with 1-particle (mass M) and 2-particle states (mass $m = 1$, i.e. we choose the light state as the mass unit for simplicity of notation, where not stated otherwise). In infinite volume the analytical structure of the propagator in the relativistic variable s is as shown in fig. 22(a), with an elastic cut starting at $4m^2 \equiv 4$, higher inelastic thresholds and (for $M < 2m$) a bound state below the elastic threshold. The dispersion relation representation gives

$$T(s) = \text{pole} + \frac{1}{\pi} \int_4^\infty ds' \frac{\rho(s')}{(s'-s)} \tag{4.1}$$

or, Fourier transformed to euclidean time t,

$$G(t) = c_0 e^{-Mt} + \frac{1}{\pi} \int_4^\infty dW' e^{-W't} \rho(W') \tag{4.2}$$

where W denotes the energy and ρ is the spectral density. If the single particle state is above threshold it describes a resonance and is no longer part of the discrete spectrum and no longer an asymptotic state.

In finite volume the continuous spectrum is replaced by a discrete set (cf. fig. 22(b)),

$$G(t) = c_0 e^{-Mt} + \sum_i c_i e^{-W_i t}, \tag{4.3}$$

with the eigenspectrum $\{W_i\}$. A possible resonance is represented by the form of the discrete spectrum and not by a single energy level like a bound state.

How can we determine the eigenspectrum in realistic lattice calculations? The answer is given by standard methods of quantum mechanics. One has to find a (sufficiently) complete set of operators N_n with the correct quantum numbers (symmetry properties) of that channel. Each $N_n(t), n = 1, 2...$ is defined on a single time slice, and t denotes the separation of the time slices. In the Monte Carlo simulation one then measures the connected correlator,

$$C_{nm}(t) = \langle N_n{}^*(0) N_m(t)\rangle_c \equiv \langle N_n{}^*(0)N_m(t)\rangle - \langle N_n{}^*\rangle\langle N_m\rangle \,. \tag{4.4}$$

The transfer matrix formalism (see e.g. [77]) yields the spectral decomposition

$$C_{nm}(t) = \sum_l \langle N_n^* | l\rangle e^{-W_l t} \langle l | N_m\rangle_c \tag{4.5}$$

$$= \sum_l v_n^{(l)*} v_m^{(l)} e^{-W_l t} \,. \tag{4.6}$$

For simplicity we assume non-degenerate W_l, ordered increasingly. The amplitudes $v_n^{(l)} = \langle l | N_n\rangle$ are the projections of the states $|N_n\rangle$ (generated by the operators N_n out of the vacuum) on the energy eigenstates $\langle l|$ of the scattering problem.

In actual calculations one has to truncate the correlation matrix, since one can consider only a finite number r of operators. The generalized eigenvalue problem

$$C(t)\, \zeta^{(k)}(t) = \lambda^{(k)}(t,t_0)\, C(t_0)\, \zeta^{(k)}(t) \,, \quad k = 1, 2, .., r \,. \tag{4.7}$$

allows one to find the energy eigenvalues W_l efficiently even for comparatively small values of t. In (4.7) we assume $t_0 < t$, e.g. $t_0 = 1$. We require the N_n , $n = 1, 2, .., r$ to be linearly independent, thus $C(t_0)$ is regular and the generalized eigenvalue problem is well defined. Indeed, it may be transformed to a standard eigenvalue problem for the matrix

$$C^{-\frac{1}{2}}(t_0)\, C(t)\, C^{-\frac{1}{2}}(t_0), \tag{4.8}$$

with identical eigenvalues $\lambda^{(k)}(t,t_0)$ and eigenvectors $\mathbf{u}^{(k)}(t) = C^{\frac{1}{2}}(t_0)\zeta^{(k)}(t)$. The solution for the eigenvalues is

$$\lambda^{(k)}(t,t_0) = e^{-(t-t_0)W_k} \,. \tag{4.9}$$

Some approximations have to be done in the numerical calculations, among them the restriction to a finite set of lattice operators and therefore to a finite set of eigenvalues. However, as has been pointed out in [103], the correction due to the truncation is of $O(\exp[-(t-t_0)W_{r+1}])$ by a perturbation calculation. The range of t-values is dictated by statistical errors of the values of the correlation function and is usually between 1 and 8.

4.2. Spectrum and Phase Shifts

Consider quantum mechanics in a $(d-1)$-dimensional spatial box, with periodic (torus) boundary conditions with cubic symmetry. Assume that the interaction region is localized and smaller than the extension of the box. Solving the static Schrödinger equation outside the interaction region ($V = 0$, i.e. the Helmholtz equation) and expanding the solution into cubic ($SO(d-1,\mathbb{Z})$) spherical harmonics one finds a relation between the discrete values of the momenta and the phase shift at these points [97, 103, 99]. As a simple example we consider the scalar A_1^+ sector and neglect higher angular momenta (> 3). We find

$$\exp(2i\delta_0(k)) = \frac{m_{00}(k)+i}{m_{00}(k)-i}, \tag{4.10}$$

$$m_{00}(k) = \mathcal{Z}_{00}(1;q^2)/q\pi^{\frac{3}{2}} \quad \text{where} \quad q = \frac{kL}{2\pi}, \tag{4.11}$$

i.e. $m_{00}(k) = \cot\delta_0(k)$, the usual K-matrix. The function $\mathcal{Z}_{00}$ is a generalized zeta-function, and one has

$$\begin{aligned} m_{00}(k) &= \cot q\pi && \text{for d=2}, \\ &= \frac{1}{4\pi^2 q}\sum_{\vec{n}\in Z^3}\frac{1}{\vec{n}^2-q^2} && \text{for d=4}. \end{aligned} \tag{4.12}$$

This gives a relation

$$\delta(k_n) = n\pi - \Phi(k_n) \tag{4.13}$$

between the quantized values of k and the phase shift. For $d = 4$ the functional form of $\Phi(k)$ is determined numerically (cf. [149]). For $d = 2$ one has the simple form $\Phi(k_n) = k_n L/2$ and thus

$$2\delta(k_n) + k_n L = 2n\pi \quad (n \in \mathbb{Z}), \tag{4.14}$$

and this quantization condition may be easily visualized for the cases without and with interaction (fig. 23). Because of the periodic b.c. of the wave functions one has $\Psi(0) = \Psi(L)$, $\dot{\Psi}(0) = \dot{\Psi}(L)$. Due to the (localized) interaction a non-zero phase shift affects the quantization of the momenta.

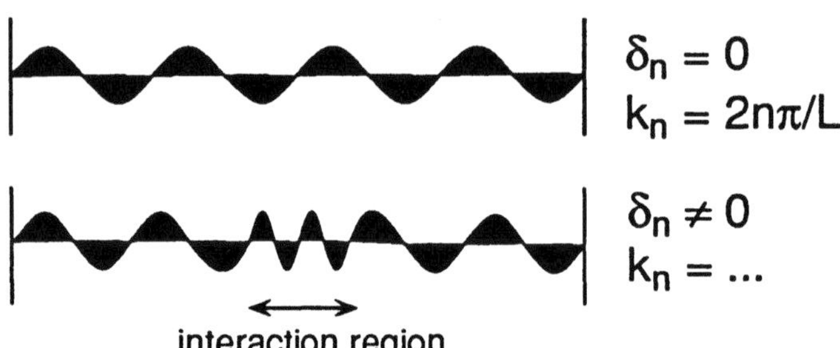

Fig. 23. The quantization in d=1 (a) without and (b) with interaction phase shift.

These arguments and the equation (4.12) may be applied to a quantum field theory problem under the following restrictions and assumptions.

- The interaction region is localized $< L$ and, in particular, the size of the single particle states is smaller than the lattice: $1/m \ll L$. This is due to that one wants a pair of two free particles as asymptotic states.
- The polarization effects are controllable. Of course there will be interaction, e.g. self interaction of the light particles around the torus world. In fact this is perturbatively the leading contribution. This effect can be understood and accounted for.

- On is confined to the elastic regime $4 < s < 9$ (for a 3-particle inelastic threshold, or 16 for a 4-particle threshold, depending on the quantum number of the channel).
- One understands the lattice artifacts (they are $O(a^2)$). This is because we work close to but never at the phase transition. These terms are due to the grained structure of the lattice and leading contributions may be studied already in the non-interacting gaussian model.

For the determination of the energy spectrum one should consider correlation functions of a sufficiently large number of observables with the correct quantum numbers, capable to represent the eigenspace of scattering states [103].

To summarize, for a given set of couplings the determination of the phase shifts in the scattering sector (at total momentum zero) consists of the following steps.

1. For given lattice size L determine the single particle state; check for polarization effects and get the mass m.
2. Determine the correlation matrix $C(t)$ for various values of t, solve the eigenvalue problem for this matrix, and obtain the spectrum from the exponential decay (4.9). The dimension r of C cannot be chosen too large, because then the inversion becomes numerically unstable. We denote the energy eigenvalues by $W_n(L)$.
3. From the values $W_n(L)$ one finds k_n with help of the dispersion relation
$$W_n = 2\sqrt{m^2 + k_n^2}. \tag{4.15}$$
This uses the fact that the 2-particle energy is the sum of the (back to back) outgoing single particle energies.
4. From the values k_n one obtains $\delta(k_n)$ from (4.13).

For each lattice size L one obtains in this way several (typically the lowest 1–4) energy eigenvalues and thus pairs k_n, $\delta(k_n)$. The procedure is then repeated for other lattice sizes. Eventually one gets more and more values k and δ, filling the domain $0 < k < \sqrt{5/4}$ (or $\sqrt{3}$ if the 4-particle threshold is the first inelastic one). In fig. 24 the spectrum of free 2-particle states (vanishing phase shift) for the 4D system is shown.

4.3. Models and Results

Here I want to discuss two studies of scalar models and their results, concentrating on common features.

4.3.1. *2D Resonance Model*

The authors [45, 46, 44] study a model, which describes two light particles φ that couple to a heavier particle η giving rise to resonating behaviour. For this one couples two Ising fields through a 3-point term,

$$S = S_{\text{Ising}}(\varphi, \kappa_\varphi) + S_{\text{Ising}}(\eta, \kappa_\eta) + \frac{g}{2} \sum_{x \in \Lambda, \mu=1,2} \eta_x \varphi_x (\varphi_{x-\hat{\mu}} + \varphi_{x+\hat{\mu}}) \,. \tag{4.16}$$

The values of the fields are restricted to $\{+1, -1\}$ and S_{Ising} denotes the usual action of the Ising model. The 3-point term is introduced in a nonlocal manner,

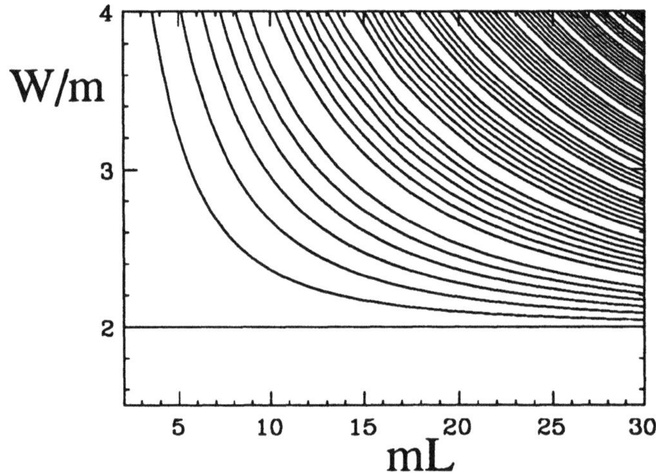

Fig. 24. The free 2-particle energy spectrum ($\delta = 0$) according to (4.13); the density increases with the volume (from [148]).

because $\varphi_x^2 \equiv 1$. The system is studied in the symmetric phase with regard to both nearest-neighbour coupling κ_φ and κ_η.

The case $g = 0$ corresponds to the situation of two independent Ising models, each with a 2nd order phase transition at $\kappa_c = \frac{1}{2}\ln(1+\sqrt{2}) \simeq 0.44068$. In the scaling limit the model describes an interacting boson with mass $m = -\log(\tanh\kappa) - 2\kappa$ [77]. Reformulating the Ising model at the phase transition as a theory of non-interacting fermions, it was shown that the scattering matrix assumes the value -1 independent of the momentum [119].

In the coupled case ($g > 0$), if we identify η and φ with particle states, the term proportional to g gives rise to transitions like $\eta \longrightarrow \varphi\varphi$ rendering η a resonance in the $\varphi\varphi$ channel, when kinematically allowed. The phase diagram of this model has interesting structure with limiting cases including random external fields and random bond models. Its structure is discussed in more detail in [46].

The values of the couplings $\kappa_\varphi, \kappa_\eta$ are adjusted such that $m_\varphi = 0.19$ and $m_\eta = 0.50$. The system is studied for the case of two independent Ising models $g = 0$ as well as for the values g=0.02, 0.04. The lattice sizes $L \times T$ considered were in the range $L = 12\ldots60$ and $T = 60\ldots100$. Cluster updating was used, with typically 200 K measurements per point. In the 2-particle channel 4–6 operators N_n were considered and 2–5 lowest eigenvalues determined from diagonalizing the correlation function at distances $\Delta t = 2\ldots8$.

4.3.2. $O(4)$ Φ^4 *Model*

Zimmermann et al. [148, 149, 139] study the $\lambda \longrightarrow \infty$ limit of the $O(4)$ model, i.e. the nonlinear σ-model, with a small external field. As discussed, this is a theory with three degenerate light pions $\vec{\pi}$ (would-be-Goldstone bosons) and a heavy σ (the Higgs boson).

The bare coupling parameters κ and j were adjusted to have renormalized masses $m_\pi = 0.23$ and $m_\sigma = 0.70$. The system was simulated with the cluster algorithm on lattices of size $L^3 \times T$ with $L = 12 \ldots 32$ and $T = 32 \ldots 40$ (40-50 K measurements per point). The authors considered 6 operators in the 2-particle channel and determined typically 2–5 lowest energy eigenvalues from the correlator at distances $\Delta t = 2 \ldots 8$.

4.3.3. *Single Particle States*

For the determination of the energy spectrum a precise knowledge of the single particle mass and related finite size effects is important. The operator of a φ state with momentum

$$p_\nu = 2\pi\nu/L\ , \quad \nu = -L/2+1, \ldots, L/2 \tag{4.17}$$

may be represented through

$$\varphi(p_\nu, t) = \frac{1}{L^{d-1}} \sum_x \varphi(x,t) \exp\left(ip_\nu x\right) . \tag{4.18}$$

Its connected correlation function over temporal distance t decays exponentially $\propto \exp\left(-E_\nu |t|\right)$ defining the single particle energy E_ν; in particular we have $E_{\nu=0} = m_\varphi$.

The observed mass, as compared to the "real" mass at vanishing lattice spacing and infinite volume, incorporates contributions from polarization due to self interaction around the torus (decreasing exponentially with L) and lattice artifacts due to the finite ultra-violet cutoff (polynomial in the lattice constant a).

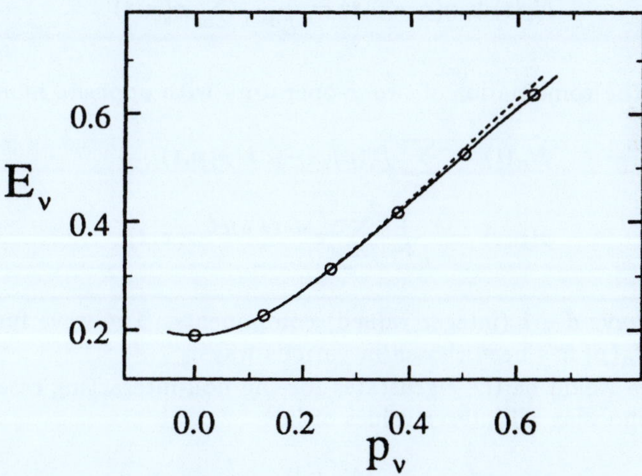

Fig. 25. Results (E_ν, p_ν) as obtained from the analysis of the single particle propagator in the 2D Ising model ($g = 0$, $L = 50$, from [46]). The dashed curve represents the continuum d.r. (4.19) and the full curve the lattice d.r. (4.20).

Fig. 25 is a plot of E_ν vs. p_ν, obtained for the pure Ising model ($g = 0$) and $L = 50$. Comparing the values (E_ν, p_ν) with the continuum spectral relation ,

$$E_\nu = \sqrt{m^2 + p_\nu^2} \tag{4.19}$$

one observes deviations $O((ap)^2)$, as expected [103]. Using the dispersion relation for the lattice propagator of a gaussian particle with mass m,

$$E_\nu = \operatorname{arcosh}(1 - \cos p_\nu + \cosh m) \ , \tag{4.20}$$

one finds excellent agreement with the measured values. Thus this particular single particle state shows little deviation from the free lattice form. These results for the 2D Ising model holds for the situations $g \neq 0$ as well and was confirmed in the 4D study [148, 149, 139], too.

Let us, for the moment, stay with the 2D Ising model. It may be considered the limit of a Φ^4 theory with infinite bare 4-point coupling λ. We are in the symmetric phase, thus we expect the leading interaction to be due to an effective 4-point term. The finiteness of the spatial volume allows self-interaction with one particle running around the torus. This leads to a finite volume correction to the particle mass proportional to $\lambda_r e^{-mL}/\sqrt{L}$. In the 2D model excellent agreement with this behaviour was observed. This mass shift on finite lattices has also been confirmed in Monte Carlo simulations of the Ising model in 4 dimensions [108].

4.3.4. *Scattering Sector*

In the 2-particle channel one considers operators with total momentum zero and quantum numbers of the η, a set as complete as possible. These operators may be constructed from the $p = 0$ single η-operator,

$$N_1(t) \equiv \eta(p = 0, t) = \frac{1}{L^{d-1}} \sum_x \eta(x, t) \tag{4.21}$$

as well as from the combination of two φ-operators with opposite momentum,

$$N_n(t) = \sum_p f_n(p) \varphi(-p, t) \varphi(p, t) \ , \tag{4.22}$$

$$\varphi(p, t) = \frac{1}{L^{d-1}} \sum_x e^{2i\pi(x.p)/L} \varphi(x, t) \ , \tag{4.23}$$

where p and x have $d - 1$ (integer valued) components. The wave function in momentum space $f_n(p)$ has been chosen in various ways.

- Plane waves would be the eigenstates for the non-interacting case. This corresponds to choosing all integer values of $|p|^2$,

$$\begin{aligned} f_n(p) &= \delta(n, |p|) \quad \text{for } d = 2 \ , \\ f_n(p) &= \delta(n, |p|^2) \quad \text{for } d = 4 \ . \end{aligned} \tag{4.24}$$

- However, one could also choose special forms like the periodic singular solutions of the Helmholtz equation

$$f_n(p) = \frac{1}{p^2 - q_n^2} \tag{4.25}$$

  as has been tried in [149].

The question here is which choice gives the best representation of the scattering eigenstates. Any "complete" set would do in principle, but in practice one is limited to a small number of operators and thus there may be "bad" and "better" choices. In the calculations discussed here no clear priority for one of the discussed forms was found. Both worked equally good; thus one favours the simpler form (4.24).

These operators and products thereof are now measured for each configuration and the resulting correlation function (4.4) is computed for various values of t, diagonalized and the energy spectrum determined from the exponential decay (4.9).

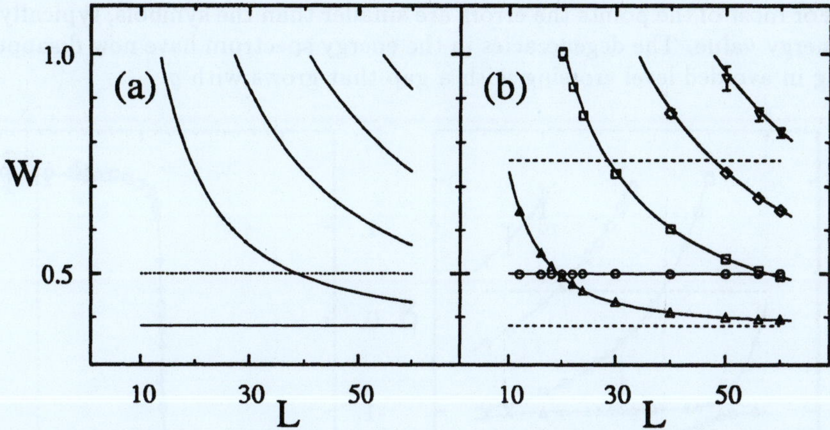

Fig. 26. **The lowest energy levels for (a) the non-interacting gaussian model and (b) the Ising model; the decoupled η-state is also shown. The full lines denote curves due to phase shifts (a) $\delta = 0$ and (b) $\delta = -\frac{\pi}{2}$ (fig. from [46]).**

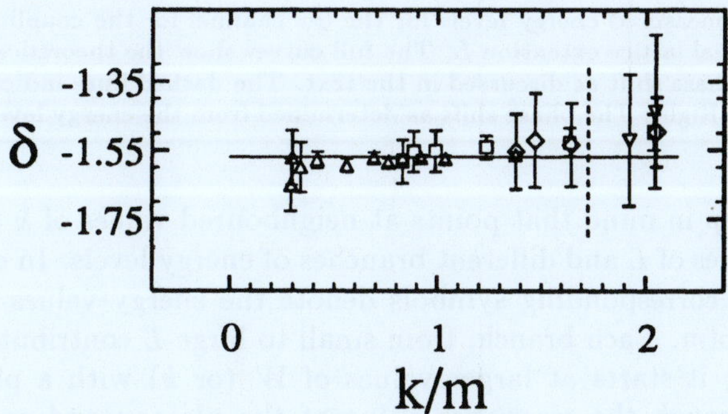

Fig. 27. **The phase shift obtained for the Ising model from the energy spectrum in fig. 26 (fig. from [46]).**

In fig. 26 we compare the results for the energy levels for the 2D Ising with those for the free, non-interacting case. We find that the lowest level moves from the threshold (in the free case) to higher values in the Ising situation. In the figure we also plot the results of the (decoupled) single η state at $m_\eta = 0.5$. The level crossing therefore is fictitious and does not really occur, since the $\varphi\varphi$ channel is decoupled from the η-channel ($g = 0$). We note that we have only a few energy eigenstates between the elastic threshold (at 0.38) and the inelastic threshold (at 0.76).

The critical 2D Ising model has an S-matrix equal to -1 and we therefore expect a phase shift $\delta_{\text{Ising}} = -\frac{\pi}{2} \pmod{\pi}$ in the scaling regime. Fig. 27 shows the results for the phase shift $\delta(k)$ determined from the data for the Ising model (fig. 26(b)) as a function of the dimensionless momentum k/m_φ.

Figs. 28 exhibit results of the 2D resonance model in the coupled situation, for $g \neq 0$. For most of the points the errors are smaller than the symbols, typically 0.5% of the energy value. The degeneracies in the energy spectrum have now disappeared, resulting in avoided level crossing with a gap that grows with g.

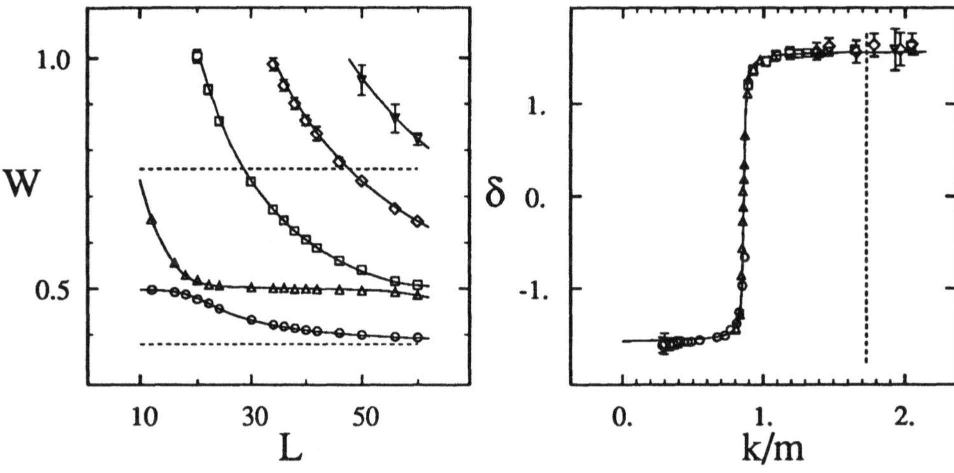

Fig. 28. Left: The measured energy levels for the $\varphi\varphi$ channel for the coupling constant values $g = 0.02$, vs. the spatial lattice extension L. The full curves show the theoretical expectations for the values from the phase shift as discussed in the text. The dashed lines indicate the 2- and the 4-particle thresholds. Right: The phase shift as determined from the energy levels (fig. from [46]).

One should keep in mind that points at neighboured values of k may come from quite different values of L and different branches of energy levels. In order to emphasize this situation corresponding symbols denote the energy values on one branch, i.e. for one value of n. Each branch, from small to large L contributes to the whole range of k values; it starts at larger values of W (or k) with a phase shift close to $\pi/2$, passes through the resonance value at the plateau and approaches $-\pi/2$ towards smaller k. The overall consistent behaviour is impressive. The high energy values, corresponding to large k may feel the lattice cutoff effects $O(a^2)$ as well as possible misrepresentation of the energy eigenstates by the considered operators. Within the accuracy of the data none of these effects seems to be a problem.

The observed overall behaviour has a simple interpretation: A resonating phase shift superimposes the background Ising phase shift. The resonance parameters may be determined from the phase shift like from experimental data, e.g. by a K-matrix fit or an effective range approximation.

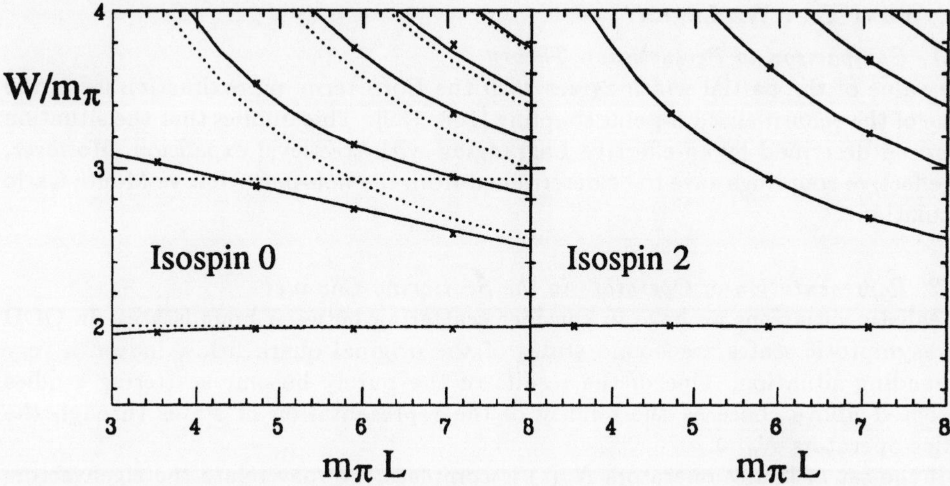

Fig. 29. The energy levels in both s-wave channels of the 4D $O(4)$ model for a particular choice of couplings ($m_\pi = 0.295, m_\sigma = 0.906$); the dashed curves denote results due to renormalized perturbation theory (from [148]).

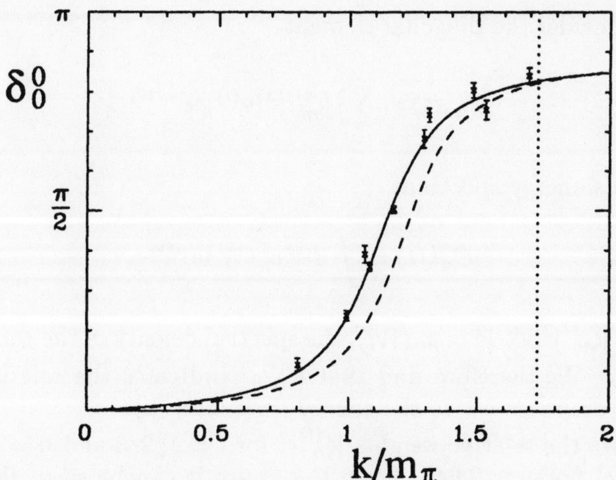

Fig. 30. The isospin 0 s-wave phase shift with the Higgs resonance determined from the energy spectrum in a non-perturbative Monte Carlo simulation (from [148]).

Similar results were obtained in the study of the 4D $O(4)$ model. In 4 dimensions the numerical task is much more demanding and therefore one could not study as broad a range of lattice sizes as in the 2D case. However, the region of the first avoided level crossing could be reached, as demonstrated in fig. 29. Fig. 30 gives the isospin 0 s-wave phase shift; this is the channel with the Higgs resonance.

4.4. What Can We Learn?

4.4.1. *Comparison to Perturbation Theory*

The value of the partial width agrees with the Born term approximation using the value of the renormalized 3-point coupling [148, 139]. This implies that the situation is indeed described by an effective Lagrangian with tree-level expansion. However, the effective couplings have to be determined from the non-perturbative Monte Carlo calculation.

4.4.2. *Representation of Operators in the Scattering Channel*

In realistic situations we have to consider scattering between boundstates: In QCD the asymptotic states are bound states of the original quark fields, indeed a very demanding situation. One of the results of the purely bosonic scattering studies presented above concerns the quality of the representation of states through the lattice operators $N_n(t)$.

If the set of lattice operators $N_n(t)$ is complete, we may relate the eigenvectors of the diagonalization (4.8) to the physical eigenvectors $\zeta^{(k)}(t) \propto \mathbf{v}^{(l)}$ [45]. Let us express the original approximation (4.5) to $C(t)$ in terms of these eigenvectors,

$$C_{nm}(t) = \sum_{l=1}^{r} \zeta_n^{(l)*} \zeta_m^{(l)} |\mathbf{v}^{(l)}|^2 e^{-tW_l} , \quad n, m = 1, 2 \ldots r . \tag{4.26}$$

In particular consider the diagonal element

$$C_{nn}(t) = \sum_{l=1}^{r} |\zeta_n^{(l)}|^2 |\mathbf{v}^{(l)}|^2 e^{-tW_l} . \tag{4.27}$$

For a continuous energy spectrum,

$$C_{nn}(t) = \int dW \rho_n(W) e^{-tW} , \tag{4.28}$$

and we identify $|\zeta_n^{(l)}|^2 |\mathbf{v}^{(l)}|^2 \propto \rho_n(W_l)$, the spectral density of the correlation function of operator N_n. We therefore find that $|\zeta_n^{(l)}|^2$ indicates the relative weight of the contribution of operator N_n to the energy eigenstate $|l\rangle$.

Fig. 31 shows the relative weights $|\zeta_n^{(l)}|^2$ for $l = 1, 2, 3$ and $n = 1, 2, 3$ for the 2D resonance model (case g=0.04). From the figure it can be seen, that the most important contributions to an eigenstate come from lattice operators that have similar eigenenergies in their eigensystem. Compare the energy spectrum of fig.s 26 and 28 with fig. 31. For small L the lowest energy state W_1 is close to the resonance energy. There the state is completely dominated by the η-operator N_1 (circles in fig. 28), which describes an η-particle at rest. For increasing L, W_1 decreases and approaches the 2-particle threshold. This manifests itself in a drastic decline of the contribution of N_1, accompanied by an increase of the amplitudes for two φ-particles at rest

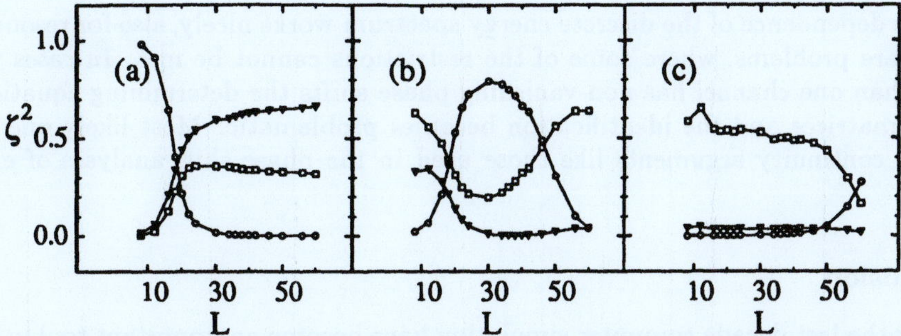

Fig. 31. Contributions of the operators N_1, N_2 and N_3 (circles, triangles, squares) to the energy eigenstates (a) $|1\rangle$, (b) $|2\rangle$ and (c) $|3\rangle$, displayed as a function of the lattice size L (fig. from [46]).

(operator N_2, triangles) and with relative unit-momentum (operator N_3, squares). For lattice size larger than 20, the energy W_1 is already too far below the resonance energy so that the energetically higher η at rest cannot contribute much to $|1\rangle$.

In this way we may discuss the other levels as well. E.g. the contribution of the η-operator N_1 (circles) to the second lowest state $|2\rangle$ assumes peak values where it crosses the resonance energy plateau.

Also, with this consideration we can identify misrepresentation of energy eigenstates in terms of the considered operators. At higher energies one observes a shift of the weight factors $|\zeta_n^{(l)}|^2$ towards higher operators N_n. If one has a situation, where the center of weight lies outside the scope of operators considered, more operators should be included for a good representation of that state [45]. The study of the representation of scattering states by the operators entering the correlation functions therefore sheds light on the problem of constructing good approximations for the energy eigenstates and it allows to control unwanted truncation effects.

4.4.3. *Lattice Artifacts*

In [44] the calculations for the Ising model were repeated for different coupling, corresponding to coarser lattice spacing. One found a stronger deviation from the expected phase shift at $am_\varphi = 0.5$ than at $am_\varphi = 0.19$, compatible with the expected corrections $O(a^2)$. However, as mentioned earlier, the d.r. (4.15) gives the total energy of the asymptotic 2-particle state which (under the assumption of localized interaction region) is just twice the energy of the outgoing particles. Now the $O(a^2)$ corrections of the single particle d.r. have been nicely described by replacing the continuum d.r. by the lattice relation (4.29). If one replaces (4.15) by the corresponding lattice expression,

$$W_n = 2\,\mathrm{arcosh}(1 - \cos k_n + \cosh m) \;, \tag{4.29}$$

the data for W_n now produce slightly different values of k_n and δ, in much better agreement with a constant value of $-\pi/2$. One concludes, that the leading $O(a^2)$ corrections can be expressed by replacing the continuum d.r. by the lattice relation. This observation was confirmed in the study of the 4D model [148].

In summary, it seems that Lüscher's method to determine phase shifts from the volume dependence of the discrete energy spectrum works nicely, also for resonances. There are problems, where some of the restrictions cannot be met. In cases where more than one channel has non-vanishing phase shifts the determining equation involves matrices and the identification becomes problematic. Most likely one needs further continuity arguments like those used in the phase shift analyses of experiments.

Conclusion

During the last decade computer simulation have become an important tool in QFT. Within the realm of purely bosonic field theory efficient stochastic algorithms have been developed, that allow high precision results. Almost always, however, it is insufficient to just let the programs run and wait long enough until the answer stabilizes. Finite size and relaxation problems are persistent. In certain limits we may have an idea of the structure of an underlying (possibly: effective) field theory that will have unknown parameters and a limited domain of validity. In this way different effective models may describe different limiting cases of the microscopic theory. The nonperturbative Monte Carlo calculations can provide the unknown values of the parameters in *ab initio* calculations. This has made the results based on a finite volume theory (like finite size scaling from renormalization group, chiral perturbation theory, or the finite volume dependence in the situation of phase shifts, to mention the examples discussed here) the most reliable ones.

The good agreement of Monte Carlo results with analytical expectations has opened the way for more sophisticated applications like using the Monte Carlo approach as an "experimental" test laboratory for quantum field theory models. Obviously one cannot hope to prove something (maybe one can disprove some ideas) but one can hope to identify relatively quickly promising or most-likely-bad theories. An example for this approach is the study of Yukawa-models and chiral fermions (cf. the review in [27]). The more conservative reason for Monte Carlo studies, and still the main motivation, is the possibility to determine numbers in this direct, nonperturbative quantization scheme, that are up to now not at all obtainable with other methods.

References

[1] M. Aizenman, Phys. Rev. Lett. **47** (1981) 1.

[2] M. Aizenman, Commun. Math. Phys. **86** (1982) 1.

[3] M. Aizenman and R. Graham, Nucl. Phys. B **225** **[FS9]** (1983) 261.

[4] M. N. Barber, in *Phase Transitions and Critical Phenomena*, edited by C. Domb and J. Lebowitz, volume VIII, Academic Press, New York, 1983.

[5] B. Berg and T. Neuhaus, Phys. Lett. B **267** (1991) 249.

[6] B. Berg and T. Neuhaus, Phys. Rev. Lett. **68** (1992) 9.

[7] G. Bhanot and K. Bitar, Phys. Rev. Lett. **61** (1988) 798.

[8] G. Bhanot, K. Bitar, U. M. Heller, and H. Neuberger, Nucl. Phys. B **343** (1990) 467.

[9] G. Bhanot, K. Bitar, U. M. Heller, and H. Neuberger, Nucl. Phys. B **353** (1991) 551.

[10] K. Binder, Finite size effects at phase transitions, in *Computational Methods in Field Theory Lecture Notes in Physics 409*, edited by H. Gausterer and C. B. Lang, page 59, Springer-Verlag, Berlin, Heidelberg, 1992.

[11] H. W. J. Bloete and R. H. Swendsen, Phys. Rev. B **22** (1980) 4481.

[12] K. C. Bowler et al., Phys. Lett. B **179** (1986) 375.

[13] E. Brézin, J. Physique **43** (1982) 15.

[14] E. Brézin, J. C. Le Guillou, and J. Zinn-Justin, Field theoretical approach to critical phenomena, in *Phase Transitions and Critical Phenomena*, edited by C. Domb and M. S. Green, volume VI, page 127, Academic Press: New York, 1976), 1976.

[15] D. Brydges, J. Fröhlich, and A. D. Sokal, Commun. Math. Phys. **91** (1983) 141.

[16] D. C. Brydges, J. Fröhlich, and T. Spencer, Commun. Math. Phys. **83** (1982) 123.

[17] T. W. Burkhardt and J. M. J. van Leeuwen, Progress and problems in real-space renormalization, in *Real-space Renormalization*, edited by T. W. Burkhardt and J. M. J. van Leeuwen, page 1, Springer-Verlag Heidelberg, 1982.

[18] T. W. Burkhardt and J. M. J. van Leeuwen, editors, *Real-space Renormalization*, Springer-Verlag Heidelberg, 1982.

[19] N. Cabibbo, L. Maiani, G. Parisi, and R. Petronzio, Nucl. Phys. B **158** (1979) 295.

[20] D. J. E. Callaway, Phys. Rep. **167** (1988) 241.

[21] D. J. E. Callaway and R. Petronzio, Nucl. Phys. B **240 [FS12]** (1984) 577.

[22] J. L. Cardy, in *Finite-Size Scaling*, edited by J. L. Cardy, page 1, North-Holland, Amsterdam, 1988.

[23] A. Coniglio and W. Klein, J. Phys. A **13** (1980) 2775.

[24] M. Creutz, *Quarks, Gluons and Lattices*, Cambridge Univ. Press, 1983.

[25] M. Creutz, *Quantum Fields on the Computer*, World Scientific, Singapore, 1992.

[26] R. Dashen and H. Neuberger, Phys. Rev. Lett. **50** (1983) 1897.

[27] A. K. De and J. Jersák, Yukawa models on the lattice, in *Heavy Flavours*, edited by A. J. Buras and M. Lindner, page 732, World Scientific, Singapore, 1993.

[28] C. A. de Carvalho, S. Caracciolo, and J. Fröhlich, Nucl. Phys. B **215 [FS7]** (1983) 209.

[29] T. A. DeGrand, The present and future of lattice qcd, in *Computational Methods in Field Theory Lecture Notes in Physics 409*, edited by H. Gausterer and C. B. Lang, page 159, Springer-Verlag, Berlin, Heidelberg, 1992.

[30] M. Falcioni, E. Marinari, M. L. Paciello, G. Parisi, and B. Taglienti, Phys. Lett. B **108** (1982) 331.

[31] A. M. Ferrenberg and R. H. Swendsen, Phys. Rev. Lett. **61** (1988) 2635.

[32] A. M. Ferrenberg and R. H. Swendsen, Computers in Physics **Sept./Oct.** (1989) 101.

[33] A. M. Ferrenberg and R. H. Swendsen, Phys. Rev. Lett. **63** (1989) 1195.

[34] H. R. Fiebig and R. M. Woloshyn, Nucl. Phys. B (Proc. Suppl.) **30** (1993) 883.

[35] H. R. Fiebig, R. M. Woloshyn, and A. Dominguez, Meson-meson scattering phase shifts in 2+1 dimensional lattice qed, preprint FIU-PHY-92-10-23, 1992.

[36] M. E. Fisher, in *Lectures in Theoretical Physics*, edited by W. E. Brittin, volume VIIC, page 1, Gordon and Breach, New York, 1968.

[37] M. E. Fisher, in *Critical Phenomena, Proc. of the 51th Enrico Fermi Summer School, Varena*, edited by M. S. Green, Academic Press, New York, 1972.

[38] M. E. Fisher and M. Randeria, Phys. Rev. Lett. **56** (1986) 2332.

[39] C. Frick et al., Nucl. Phys. B **331** (1990) 515.

[40] J. Fröhlich, Nucl. Phys. B **200 [FS5]** (1982) 281.

[41] J. Gasser and H. Leutwyler, Ann. Phys. **158** (1984) 142.

[42] J. Gasser and H. Leutwyler, Phys. Lett. B **184** (1987) 83.

[43] J. Gasser and H. Leutwyler, Nucl. Phys. B **307** (1988) 763.

[44] C. R. Gattringer, I. Hip, and C. B. Lang, Nucl. Phys. B (Proc. Suppl.) **20** (1993) 875.

[45] C. R. Gattringer and C. B. Lang, Phys. Lett. B **274** (1992) 95.

[46] C. R. Gattringer and C. B. Lang, Nucl. Phys. **B 391** (1993) 463.

[47] H. Gausterer and C. B. Lang, Phys. Lett. B **186** (1987) 103.

[48] K. Gawedzki and A. Kupiainen, Phys. Rev. Lett. **54** (1985) 92.

[49] J. Glimm and A. Jaffe, Ann. Inst. H. Poincaré **A22** (1975) 13.

[50] J. Glimm and A. Jaffe, Fortschr. Phys. **21** (1976) 327.

[51] J. Glimm and A. Jaffe, *Quantum Physics. A Functional Integral Point of View. 2nd Edition*, Springer-Verlag, New York, 1987.

[52] M. Göckeler, Nucl. Phys. B (Proc. Suppl.) **17** (1990) 347.

[53] M. Göckeler, H. Kastrup, T. Neuhaus, and F. Zimmermann, Nucl. Phys. B (Proc. Suppl.) **26** (1992) 516.

[54] M. Göckeler, H. A. Kastrup, T. Neuhaus, and F. Zimmermann, Nucl. Phys. B **404** (1993) 517.

[55] M. Göckeler and H. Leutwyler, Nucl. Phys. B **350** (1991) 228.

[56] R. B. Griffiths and P. A. Pearce, Phys. Rev. Lett. **41** (1978) 917.

[57] R. B. Griffiths and P. A. Pearce, J. Stat. Phys. **20** (1979) 499.

[58] R. Gupta and A. Patel, Phys. Lett. B **183** (1987) 193.

[59] T. Hara, J. Stat. Phys. **47** (1987) 57.

[60] T. Hara and H. Tasaki, J. Stat. Phys. **47** (1987) 99.

[61] A. Hasenfratz and P. Hasenfratz, Phys. Lett. B **93** (1980) 165.

[62] A. Hasenfratz and P. Hasenfratz, Nucl. Phys. B **270 [FS16]** (1986) 687.

[63] A. Hasenfratz, P. Hasenfratz, U. Heller, and F. Karsch, Phys. Lett. B **140** (1984) 76.

[64] A. Hasenfratz et al., Nucl. Phys. B **356** (1991) 332.

[65] A. Hasenfratz, K. Jansen, C. B. Lang, T. Neuhaus, and H. Yoneyama, Phys. Lett. B **199** (1987) 531.

[66] P. Hasenfratz, Nucl. Phys. B (Proc. Suppl.) **9** (1989) 3.

[67] P. Hasenfratz and H. Leutwyler, Nucl. Phys. B **343** (1990) 241.

[68] P. Hasenfratz and J. Nager, Z. Physik C **37** (1988) 477.

[69] P. Hasenfratz and F. Niedermayer, Perfect lattice action for asymptotically free theories, preprint BUTP-93/17, 1993.

[70] U. M. Heller, Nucl. Phys. B (Proc. Suppl.) **17** (1990) 649.

[71] U. M. Heller, Status of the higgs mass bound, preprint FSU-SCRI-93-144, talk presented at LATTICE 93 in Dallas, 1993.

[72] U. M. Heller, M. Klomfass, H. Neuberger, and P. Vranas, Nucl. Phys. B (Proc. Suppl.) **26** (1992) 522.

[73] U. M. Heller, M. Klomfass, H. Neuberger, and P. Vranas, Nucl. Phys. B **405** (1993) 555.

[74] U. M. Heller, H. Neuberger, and P. Vranas, Phys. Lett. B **283** (1992) 335.

[75] J. E. Hirsch and S. H. Shenker, Phys. Rev. B **27** (1983) 1736.

[76] B. Hu, Phys. Rep. **91C** (1982) 233.

[77] C. Itzykson and J.-M. Drouffe, *Statistical Field Theory*, Cambridge University Press, Cambridge, 1989.

[78] K. Jansen, J. Kuti, and C. Liu, Phys. Lett. B **309** (1993) 119.

[79] K. Jansen, J. Kuti, and C. Liu, Nucl. Phys. B (Proc. Suppl.) **30** (1993) 681.

[80] K. Jansen and C. B. Lang, Phys. Rev. Lett. **66** (1991) 3008.

[81] P. W. Kasteleyn and C. M. Fortuin, J. Phys. Soc. Jpn. (Suppl.) **26** (1969) 11.

[82] P. W. Kasteleyn and C. M. Fortuin, Physica (Utrecht) **57** (1972) 536.

[83] R. Kenna, Finite-size scaling in the $O(N)$ ϕ_4^4 model, hep-lat/9307003, 1993.

[84] R. Kenna and C. B. Lang, Phys. Lett. B **264** (1991) 396.

[85] R. Kenna and C. B. Lang, Nucl. Phys. B (Proc. Suppl.) **30** (1993) 697.

[86] R. Kenna and C. B. Lang, Nucl. Phys. B **393** (1993) 461.

[87] J. Kuti, L. Lin, and Y. Shen, Nucl. Phys. B (Proc. Suppl.) **4** (1988) 397.

[88] J. Kuti, L. Lin, and Y. Shen, Phys. Rev. Lett. **61** (1988) 678.

[89] C. B. Lang, Phys. Lett. B **155** (1985) 399.

[90] C. B. Lang, Nucl. Phys. B **265** **[FS15]** (1986) 630.

[91] C. B. Lang, Phys. Lett. B **229** (1989) 97.

[92] C. B. Lang and M. Salmhofer, Phys. Lett. B **205** (1988) 329.

[93] C. B. Lang and U. Winkler, Phys. Rev. D **47** (1993) 4705.

[94] I.-H. Lee, Nucl. Phys. B (Proc. Suppl.) **4** (1988) 373.

[95] H. Leutwyler, Phys. Lett. B **189** (1987) 197.

[96] M. Lüscher, Commun. Math. Phys. **105** (1986) 153.

[97] M. Lüscher, in *Fields, Strings and Critical Phenomena, Les Houches, Session XLIX, 1988*, edited by E. Brézin and J. Zinn-Justin, Elsevier, Amsterdam:, 1989.

[98] M. Lüscher, Nucl. Phys. B **364** (1991) 237.

[99] M. Lüscher, Nucl. Phys. B **354** (1991) 531.

[100] M. Lüscher and P. Weisz, Nucl. Phys. B **290 [FS20]** (1987) 25.

[101] M. Lüscher and P. Weisz, Nucl. Phys. B **295 [FS21]** (1988) 65.

[102] M. Lüscher and P. Weisz, Nucl. Phys. B **318** (1989) 705.

[103] M. Lüscher and U. Wolff, Nucl. Pbys. B **339** (1990) 222.

[104] S.-K. Ma, *Modern Theory of Critical Phenomena*, Benjamin/Cummings Publ. Co. Inc., Reading, MA, 1976.

[105] S.-K. Ma, Phys. Rev. Lett. **37** (1976) 461.

[106] F. Martinelli and E. Olivieri, Some remarks on pathologies of renormalization-group transformations for the ising-model, Univ. Roma preprint, 1993.

[107] N. Metropolis, A. W. Rosenbluth, A. H. Teller, and E. Teller, J. Chem. Phys. **21** (1953) 1087.

[108] I. Montvay and P. Weisz, Nucl. Phys. B **290 [FS20]** (1987) 327.

[109] G. Murthy and R. Shankar, Phys. Rev. B **32** (1985) 5851.

[110] H. Neuberger, Phys. Lett. B **199** (1987) 536.

[111] H. Neuberger, Phys. Rev. Lett. **60** (1988) 889.

[112] H. Neuberger, Nucl. Phys. B (Proc. Suppl.) **17** (1990) 17.

[113] T. Neuhaus, Nucl. Phys. B (Proc. Suppl.) **9** (1989) 21.

[114] J. Nishimura, Phys. Lett. B **294** (1992) 375.

[115] K. Osterwalder and E. Seiler, Ann. Phys. **110** (1978) 440.

[116] C. Rebbi, *Lattice Gauge Theories and Monte Carlo Simulations*, World Scientific, Singapore, 1983.

[117] G. Roepstorff, *Pfadintegrale in der Quantenphysik*, Vieweg: Braunschweig, 1991.

[118] H. J. Rothe, *Lattice Gauge Theories - An Introduction*, World Scientific, Singapore, 1992.

[119] M. Sato, T. Miwa, and M. Jimbo, Proc. Japan Acad. , Ser. A **53** (1977) 6.

[120] E. Seiler, *Gauge Theories as a Problem of Constructive Quantum Field Theory and Statistical Mechanics, Springer Lecture Notes in Physics 159*, Springer, Berlin, 1982.

[121] R. Shankar, R. Gupta, and G. Murthy, Phys. Rev. Lett. **55** (1985) 1812.

[122] S. H. Shenker and J. Tobochnik, Phys. Rev. B **22** (1980) 4462.

[123] A. Sokal, Monte carlo methods in statistical physics: Foundations and new algorithms, Lecture Notes Cours de Troisieme Cycle de la Physique en Suisse Romande, Lausanne, 1989.

[124] A. Sokal, Nucl. Phys. B (Proc. Suppl.) **20** (1991) 55.

[125] H. Stanley, *Introduction to Phase Transitions and Critical Phenomena*, Oxford University Press, New York, 1971.

[126] D. Stauffer, Phys. Rep. **54** (1979) 1.

[127] M. Sweeny, Phys. Rev. B **27** (1983) 4445.

[128] R. H. Swendsen, Phys. Rev. Lett. **42** (1979) 859.

[129] R. H. Swendsen, Phys. Rev. Lett. **52** (1984) 1165.

[130] R. H. Swendsen, Phys. Rev. Lett. **52** (1984) 2321.

[131] R. H. Swendsen and J.-S. Wang, Phys. Rev. Lett. **58** (1987) 86.

[132] K. Symanzik, J. Math. Phys. **7** (1966) 510.

[133] K. Symanzik, in *Proc. Int. School of Physics "Enrico Fermi", Varenna Course XLV*, edited by R. Jost, Academic Press, 1969.

[134] K. Symanzik, Nucl. Phys. B **226** (1983) 187.

[135] M. M. Tsypin, The effective potential of the lattice ϕ^4 theory and the upper bound on the higgs mass, Lebedev Phys. Inst. Moscow preprint 280, 1985.

[136] A. C. D. van Enter, R. Fernández, and A. D. Sokal, Phys. Rev. Let. **66** (1991) 3253.

[137] A. C. D. van Enter, R. Fernández, and A. D. Sokal, to appear in: J. Stat. Phys (1993).

[138] J.-S. Wang and D. Stauffer, Z. Phys. B **78** (1990) 145.

[139] J. Westphalen, F. Zimmermann, M. Göckeler, and H. A. Kastrup, Scattering phases in the broken phase of the 4-d o(4) non-linear σ-model, Contrib. to LATTICE 93 in Dallas, 1993.

[140] C. Whitner, PhD Thesis, Princeton University, 1984.

[141] K. G. Wilson, Phys. Rev. B **4** (1971) 3184.

[142] K. G. Wilson, Phys. Rev. D **10** (1974) 2445.

[143] K. G. Wilson and J. Kogut, Phys. Rep. **12C** (1974) 76.

[144] U. Wolff, Phys. Rev. Lett. **62** (1989) 361.

[145] U. Wolff, High precision simulations with fast algorithms, in *Computational Methods in Field Theory, Lecture Notes in Physics 409*, edited by H. Gausterer and C. B. Lang, page 127, Springer-Verlag, Berlin, Heidelberg, 1992.

[146] C. N. Yang and T. D. Lee, Phys. Rev. **87** (1952) 404.

[147] C. N. Yang and T. D. Lee, Phys. Rev. **87** (1952) 410.

[148] F. Zimmermann, Unstable particles in finite volume: the broken phase of the 4-d o(4) non-linear σ-model, RHTH Aachen PhD Thesis, 1993.

[149] F. Zimmermann, J. Westphalen, M. Göckeler, and H. A. Kastrup, Nucl. Phys. B (Proc. Suppl.) **30** (1993) 879.

BROWNIAN MOTION OVER A KAHLER MANIFOLD AND ELLIPTIC GENERA OF LEVEL N

R. Léandre

INTRODUCTION

The geometry of the free loop space of a manifold M, that is the space of smooth applications from the circle S_1 into M is closely related to the topology of the manifold by its natural circle action over it. Namely we can recover the manifold from the free loop space as the fixed point set of this circle action on the loop space. In finite dimensions, when there is a circle action, there are two types of relation between the fixed point set and the whole manifold:

-)The Berline-Vergne localisation formulas ([B.V]).

Integrals of equivariant closed forms are written as integrals over the fixed point set. The relation between the equivariant cohomology of the manifold and the cohomology of the fixed point set has should be seen from this point of view [J.P].

-)The Lefschetz formulas.

The equivariant index of an invariant operator is a finite Laurent serie in the parameter q of the circle; each term can be written as an integral over the fixed point set. It was conjectured by Witten [Wi$_2$], based on observations on the Dirac operator on loop space, that only the constant term is not zero for a large class of classical operators . Taubes has given an infinitesimal version of this result by studying infinitesimally small loop [T]. The main idea of the finite dimensional Lefschetz formulas is to localise the operator on the normal bundle of the fixed point set (See [T]). The construction of the Dirac operator by Taubes for infinitesimally small loops is an infinite dimensional copy of these finite dimensional constructions.

In the case of the free loop space, these relations appear in two version of the index theory.

-)Finite dimensional index theory.

The index formula is formaly a localisation formula over the loop space, as it was first shown by Atiyah-Witten [A]. Bismut considered the case of a twisted Dirac operator, and he pioneered the relation between the equivariant cohomology of the loop space and the index theorem ([B$_6$], [B$_7$]) by introducing the equivariant Chern character over loop space. [G.J.P] gave an algebraic interpretation , introducing the notion of the Witten current over loop space and using in this context the relation ([Ch]) between the cyclic cohomology of forms on the manifold and the equivariant cohomology of forms of the loop space by means of Chen iterated integrals.

-)Infinite dimensional index theory.

In the physicist litterature, this theory is related to conformal field theory because the

A. I. Cardoso et al. (eds.), Stochastic Analysis and Applications in Physics, 193–217.

considered indices are involved with the supersymmetric non-linear sigma model. We will now indicate four different index theorems over the free loop space.

Let us recall some background material (See[H.B.G] [Be] and [Se] for a more extensive survey). A genus ϕ in real oriented cobordism theory is determined by its logarithm $Log_\phi(z) = \sum \frac{\phi(\mathbf{P}^{2n}(\mathbf{C}))z^{2n+1}}{2n+1}$ since the cobordism ring tensorised by $\mathbf{Q}$ is spanned by the even projective complex spaces. [O] has introduced genera with a stronger multplicative property than the ordinary genera and has showed that their logarithms are elliptic integrals $\int_0^z \frac{dt}{(1-2\epsilon t^2+\eta t^4)^{\frac{1}{2}}}$. Such genera play an important role in the so called elliptic cohomology. If we consider the inverse function of this elliptic integral, we get a doubly periodic function called an elliptic function.If we normalise its period lattice to be $2\pi i\mathbf{Z} + 2\pi i\tau\mathbf{Z}$, the genus $\phi_q(M)$ is a modular form where we have set $q = exp[2i\pi\tau]$. Witten ([Wi$_2$]) has given an interpretation of this modular form in terms of the equivariant signature of the loop space. There is an obstruction in constructing the signature operator which is the existence of an orientation of the loop space. A simpler operator is given by the Dirac operator whose equivariant index gives the Witten genus : the formal obstruction to construct this is the existence of a spin structure over the loop space and vanishes if half of the first Pontryaguin class of M is equal to 0. Witten's computations ([Wi$_2$]) are linked to a polarisation over the loop space which is given by Fourier expansion : the square root of the fundamental bundle can be treated only by using a zeta function regularisation ([Wi$_2$]) . Witten noticed that all these problems disappear when we consider the Euler-Poincaré number of the loop space, which should be equals by localisation to the Euler-Poincaré number of the manifold.

Before this, [D.H.V.W] considered the string theory of a quotient of a manifold M by a finite group of symmetries G. In such cases M/G is in general not a manifold, and the Euler number of M/G can be computed by an averaged Lefschetz formula ([H.Z]). This formula is not stable by minimal desingularisation of the orbifold M/G. [D.H.V.W] introduced a formula which should give the Euler-Poincaré number of the loop space of an orbifold, and which should be stable under desingularisation (See [R] for examples).

The almost complex structure related to the Fourier serie expansion has a kernel,namely the constant terms, at least in the normal bundle of the constant loop : it is better to speak of a polarisation. In the case of a complex manifold, the loop space inherits a complex structure which is independant of the Fourier expansion. In particular, there is no topological obstruction to construct it, except for the traditional obstruction to construct a Kähler structure over the manifold. In this case, [Hi] has introduced the corresponding $\overline{\partial}_\infty$ operator and considered the corresponding equivariant Riemann-Roch theorem which is related to the theory of elliptic genera of level N.

Taubes ([T]) has given a first approximation to the Dirac operator for infinitesimally small loops: his choice is motivated by the string theory as we will see. He considers the normal bundle of the constant loop (the fixed point set of the loop space under the circle action), that is the space of loops in T_xM with non constant terms in the Fourier serie (For this he considers the metric $\int_0^1 < X_s, Y_s > ds$ over the loop space which is used in [B$_5$],[B$_6$] for others reasons). He takes the measure $dx \bigotimes dQ_{1,x}$, where $Q_{1,x}$ is the Gaussian measure of covariance $(-\Delta)^{1/2}$ over the loop space of T_xM, Δ being the laplacian over the tangent space. Since over the loop space of T_xM, there is a natural polarisation which comes from the expansion in Fourier series, it is possible to define the operator $\overline{\partial}_{\infty,x}$ over the loop space of T_xM by considering a complex refinment of classical supersymmetric Fock spaces in probability (See [Ni],[F]). The adjoint of $\overline{\partial}_{\infty,x}$ is computed for the Gaussian measure , which comes ultimately from its relation with the $1+1$ non linear sigma model ([Wi$_2$],[A.K.M.W$_1$], [A.K.M.W$_2$]).The main idea of Taubes in order to construct his operator is to take derivative in the parameter space for this family of Gaussian measures. The main problem is that this Gaussian measure lives over distribution and cannot be extended over curved loop, not even small loops.

The purpose of our work is to replace this Gaussian measure by the Brownian bridge measure, which can be curved by means of the theory of stochastic differential equation. The idea is to use the B.H.K. measure: this was initially introduced by [H.K] in the flat case, and [B$_5$] showed its importance in explaining the relation between the structure of loop spaces and index theory (See [A.H.K], [Fr],[Fre] for the case of loop groups). The beginning of an analytical interpretation of the algebraïc theory of [G.J.P] is given in [J.L$_1$]: a theory of forms L^p is given including convergences criteria of the type of the entire cyclic cohomology. In particular the Bismut Chern character belongs to all L^p. No differential operation is given in [J.L$_1$].

Such differential operations are known since a long time in the different versions of infinite dimensional analysis (See [Da.F], [H.K.P.S],[Kr],[Ma$_2$] for surveys): these are the h-derivatives ([Ca],[Gr$_1$]) and the Ornstein-Uhlenbeck operator ([Me]). [Sh] and [Ar.M] study differential forms over the Wiener space and different versions of the exterior derivative. [Sm] studied forms of finite codimension. [Ar$_1$] and [Ar$_2$] study extensively the index of the Dirac operator over the flat loop space in the case of the free field or with interacting terms: the underlying Hilbert space is $H_{1/2,2}$. They give a path integral representation of the index of such operators over the torus, strongly inspired by the work in the scalar case of Hoegh-Krohn ([H.K]) (See [J.L.W$_1$] and [J.L.W$_2$] for physics references). [Do$_1$] and [Do$_2$] do such analysis by using the white noise analysis.

For the analysis in infinite dimensional curved spaces, there exist different types of theory (We refer to [Ma$_2$] for a survey).

-)[Ku] construct a diffusion process over infinite dimensional manifolds.

-)The theory of Wiener manifolds. The reader can see Ramer's thesis ([Ra]) for the non scalar case.

-)The analysis over infinite dimensional manifold, which was used by the Russian school ([B.S],[D.F]).

-)The quasi-sure analysis ([Ge$_2$],[A.M$_1$]) which works over finite codimensional manifolds of the Wiener space. (See[K] for forms).

-)The analysis over loop groups ([A.M$_2$], [Gr$_2$], [Gr$_3$] is closer to the purpose of this paper, but uses a different tangent space, which depends deeply on the structure of the group and the quasi-invariance formulas of [A.H.K].

The present paper is more related to [L$_4$] and [L$_5$], where the case of the free loop space of a Riemannian manifold is considered : let us remark that the tangent space which allows to do an analysis in such case was introduced by Malliavin in the last paper of

[A.M_2]. The Bismut's integration by part formula over the free loop space was carried out, by using a mixture of small time asymptotics formulas (See [L_1],[L_2],[Wa_2]) and the geometric integration by parts formula for the open brownian motion of [B_3]. Connections invariant by rotation are introduced and also an equivariant Ornstein-Uhlenbeck operator is constructed which allow us to do a Sobolev-Calculus invariant by rotation. A relation between the exterior stochastic derivative and the Hochschild boundary of forms of the manifold is given, which extends a fact well known in the case of smooth loops to the case of stochastic loops. Moreover, the circle action is diagonalised as this will be needed in [J.L_2] and the Killing vector field is seen as an antisymmetric first order operator densely defined. (The reader can see [F.M] for an analogous first order calculus in the case of the based path space).

[A.L.R] construct a diffusion process, invariant by rotation over the free loop space by using technics of Fukushima [Fu].

[J.L_2] define a non equivariant regularised derivative over the full space of forms of the loop space and compute its adjoint. A rigorous conjecture for the index of the regularised de Rham operator is given. The name regularised arises from the fact that there is no complex an no cohomology group. By localisation, its index should be the Euler-Poincaré number of the manifold. Since there is no circle action needed in that framework, the first part of [J.L_2] ignored the circle symmetry in order to simplify the computations. This cannot be done for the Dirac operator, which only has an equivariant index: [J.L_2] meet the problem of constructing a bundle of spinors equivariantly and globaly. There is in [J.L_2] only an extension of the Taubes construction only over a small neighborhood of the constant loops: it is achieved by extending the Fourier serie structure. [J.L_2] used the stochastic Calculus. In order to define the "restriction" to these non scalar operators to infinitesimally small loop (that is on the family of Brownian bridges on the family of tangent spaces), [J.L_2] introduced the B.H.K measure in small time and the corresponding Bismut's dilatations, which permit a small time analysis of the operators. The computations are similar to the computations in [B_4],[L_2],[I.W_2],[Az],[Hs],[El_2], [Sh.U]. Let us remark that only the procedure of [B_3] works in such limits in law, because we consider continuous applications. [J.L_2] find limit operators at the manner of Taubes, and uses a countable set of index theorems after decomposing the supersymmetric Fock bundle into a countable set of finite dimensional bundles.

[L.R] applies methods which are very similar to [J.L_2] in the case of the Euler characteristic of the loop space of an orbifold. [L.R] used the measure intoduced by Bismut for twisted loops for studying finite dimensional Lefschetz formulas. The limit procedures are related to the Bismut proof of the Lefschetz formulas ([B_5]) rather than to the proof of the Atiyah-Singer theorem. [L.R] obtained in the limit the Dixon-Harvey-Vafa -Witten formula for the Euler-Poincaré number of the loop space of an orbifold. There is an hidden circular symmetry in the twisted loop space. This symmetry is broken in the computations of [L.R] in order to simplify the exposition , because there is not an equivariant Euler number of the twisted loop space.

In the present paper, we conjecture an equivariant Riemann-Roch formula over the free loop space. The metric over the free loop space is the metric of [J.L_1] which is invariant under rotation. We define an equivariant regularised $\overline{\partial}_\infty$ operator over the free loop space of a Kähler manifold and we compute its adjoint $\overline{\partial}_\infty^*$: both are invariant by rotation as well as their domain. We give a rigorous conjecture for the equivariant index of $\overline{\partial}_\infty + \overline{\partial}_\infty^*$, which is related to the theory of elliptic genera of level N. However we have to remark that there are some conventions in giving a meaning to the formal formula which gives the Riemann-Roch theorem over the free loop space: we do not follow the same computations as in [Hi], but in the cases considered by [Hi], we get the same result. We motivate this conjecture by considering the B.H.K measure in small time, and we perform a rescaling invariant by rotation over the tangent space for small loops . By a homotopy argument, we can consider the Hilbert structure of [J.L$_2$], which simplifies the computations. Bismut's dilatation are given over a dense linear set invariant by rotation of sections of $T^{0,1}$ forms. We get limit operators in the manner of Taubes : the polarisation which arises from the Fourier expansion is replaced by the complex structure which arises from the complex stucture of the manifold, and the Gaussian measure of covariance $(-\Delta)^{1/2}$ by the brownian bridge law.

REGULARISED DOLBEAULT OPERATOR OVER THE FREE LOOP SPACE

Let M be a compact complex manifold of dimension $2d$ endowed with a Kähler structure. If $<.,.>$ is the Riemannian structure, and if J is the complex structure

$$< ., J. >= \omega(.,.) \tag{1.1}$$

where ω is a closed two form. The complexified tangent space $TM \bigotimes \mathbb{C}$ can be divided into the space of $T^{1,0}$ vectors $\frac{\partial}{\partial z}$ which correspond to the case where $JX = iX$ and into $T^{0,1}$ vectors $\frac{\partial}{\partial \bar{z}}$ which correspond to the case where $JX = -iX$.

Let Δ be the Laplace-Beltrami operator associated to the Kähler metric and $p_t(x,y)$ the associated heat kernel. d is the riemannian distance. Let γ_t be the associated brownian motion:

$$\int f(y)p_t(x,y)dy = E[f(\gamma_t)] \tag{1.2}$$

Let τ_t be the holonomy between γ_0 and γ_t: it preserves the complex structure because the manifold is Kählerian. Let $L^\infty(M)$ be the loop space of M: it is the space of the continuous applications from the circle S_1 into the manifold M. We put over $L^\infty(M)$ the B.H.K. measure ([B$_4$],[H.K],[E.L]):

$$\frac{p_1(x,x)dx \bigotimes dP_{1,x,x}}{\int p_1(x,x)dx} = d\mu_1 \tag{1.3}$$

$P_{1,x,x}$ is the law of the Brownian brige starting from x and coming back to x in time 1. μ_1 is invariant under the circle action over the free loop space: $\gamma_. \to \gamma_{t+.}$. We will call q an element of the circle, when we will speak of the circle action.

There is a tangent space over a loop γ:

$$X_t = \tau_t H_t \tag{1.4}$$

where H_t has finite energy (See [A.M$_2$] last reference, [B$_3$],[J.L$_1$],[J.L$_2$],[L$_4$],[L$_5$]). Moreover, there is the boundary condition $X_1 = X_0$, because we are working over the free loop space. There is an almost complex structure J_∞ over the free loop space by putting $(J_\infty X)_t = \tau_t J_{\gamma_0} H_t$. This complex structure is invariant under rotation because the holonomy preserves the complex structure of the manifold. We get an Hilbert structure compatible with the almost complex structure and which is invariant under rotation ([J.L$_1$]) :

$$< X, X > = \int_0^1 < H_t, H_t > dt + \int_0^1 < H_t', H_t' > dt \tag{1.5}$$

The covariant derivative over a loop of a tangent vector X_s is $DX_s = \tau_s H_s'$.We can get an orthonormal basis of this Hilbert space by using the Gram- Schmidt procedure as in [L$_5$] for the tangent space of the based loop space.

If $n < 0$, we put $X_n^l(X_0)_s = \tau_s \frac{\sin 2\pi ns}{\sqrt{Cn^2+1}} X_0$. If $n > 0$, we put $X_n'(X_0) = 2^{1/2}\tau_s \frac{\cos 2\pi ns - 1}{n}$ X_0. By using the Gram-Schmidt procedure, we get as in [L$_5$] an orthonormal basis $X_n^l(X_0)$ of the tangent space of the based loop space by starting from an orthonormal basis X_i from the tangent space in γ_0. In [L$_5$], the Gram-Schmidt procedure starts frome $n_0 > 0$. It is in fact possible to start from 1 by considering the B.H.K measure in time ϵ^2. In this case, we take the Hilbert structure :

$$< X, X >_\epsilon = \int_0^1 < X_t, X_t > dt + \epsilon^{-2} \int_0^1 < DX_t, DX_t > dt \tag{1.6}$$

In such case, we remove X_n' by $\epsilon X_n'$: in the formula (3.32) of [L$_5$], a term in ϵ^2 appears before $\frac{1}{mn}$ and $\frac{1}{m^2}$. For $n > 0$, we get $X_n^l(X_0)_s = \tau_s H_n^l(X_0)_s$. After using some evident notations, we have:

$$H_n^l(X_0)_s = \sum_{j<n} \frac{C(j,n)}{jn} H_j'(X_0)_s + C(n,n) H_n'(X_0)_s \tag{1.7}$$

with bounded $C(j,n)$ and $C(n,n)$. $C(n,n)$ is moreover bounded below by $C > 0$. Let $\frac{\partial}{\partial z}$ a $T^{1,0}$ vector over the tangent space in γ_0 : we call by $X_n^l(\frac{\partial}{\partial z})$ the corresponding $T^{1,0}$ vector over the loop space. Analoguous notations are done for $T^{0,1}$ vectors and for forms.

Let us study the orthonormal complement of the based loop space tangent space. It has the shape :

$$\tau_t(e^{-t}\lambda + e^t\mu) = X_0'(X_0) \tag{1.8}$$

The periodicity conditions are:

$$\lambda + \mu = X_0 \qquad (1.9)$$
$$e^{-1} + e\mu = \tau_1^{-1} X_0$$

Therefore:

$$\lambda = \frac{e - \tau_1^{-1}}{e - e^{-1}} X_0 \qquad (1.10)$$
$$\mu = \frac{e^{-1} - \tau_1^{-1}}{e^{-1} - e} X_0$$

The metric is

$$< X_0^l(X_0), X_0^l(Y_0) >= G_\gamma(X_0, Y_0) \qquad (1.11)$$

where G_γ is a bounded matrix with bounded derivatives, larger than $C < .,. >$ for the initial Hilbert structure. If X_i is an orthogonal basis in T_{γ_0}, we deduce by orthogonalisation an orthonormal basis of the orthogonal of the based loop space tangent space $X_0^l(X_i)$. The component are bounded with bounded derivatives in the holonomy over the loop. We deduce a basis $X_0^l(\frac{\partial}{\partial z})$ and $X_0^l(\frac{\partial}{\partial \bar{z}})$ with respect to the almost complex structure.

The corresponding $T^{1,0}$ forms are denoted by $X_n^l(dz)$: analoguous notations are given for the $T^{0,1}$ forms. In finite dimensional geometry, $\overline{\partial}_r$ is defined over the bundle of $T^{0,1}$ forms tensorised by the bundle of $T^{1,0}$ forms of degree r. In order to define our regularised $\overline{\partial}_{\infty,r}$ operator, we take any finite sum of $X_J^l(d\bar{z}) \wedge X_{J'}^l(dz)$ ($J = (n_1, ..., n_k)$; $J' = (n_1', ..., n_r')$) in a local system of $d\bar{z}$ and of dz (The dz and the $d\bar{z}$ depend only of γ_0). Moreover $X_J^l(d\bar{z}) = X_{n_1}^l(d\bar{z}) \wedge X_{n_2}^l(d\bar{z}) \wedge ... \wedge X_{n_k}^l(d\bar{z})$ and $X_{J'}^l(dz) = X_{n_1'}^l(dz) \wedge ... \wedge X_{n_r'}^l(dz)$. The coordinates of our local sections are cylindrical functions $f(\gamma_{t_1}, ..., \gamma_{t_2})$. The set Ω_0 of sections σ where our operator is primarily defined satisfies:

$$\sum_{finite, J, J'} f_{J,J'} X_J^l(d\bar{z}) \wedge X_{J'}^l(dz) = \sigma \qquad (1.12)$$

in a local system of dz and of $d\bar{z}$. The pointwise Hilbert norm of such section is

$$\|\sigma\|^2 = \sum | f_{J,J'} |^2 \qquad (1.13)$$

where we take complex modulus because the coordinates $f_{J,J'}$ are complex valued functionals.

In order to define operators, we define connections over the tangent space. Since there is only an equivariant Riemann-Roch theorem over the free loop space, we have to consider connections invariant by rotations as in [L$_4$] [L$_5$].

Over the tangent space of the based loop space, we choose as connection :

$$\nabla^0 = \tau_t \nabla H_t \qquad (1.14)$$

∇ is the pull-back of the Levi-Civita connection over M by the evaluation map $e_0 : \gamma_. \to \gamma_0$. ∇^0 preserves the almost complex structure over the tangent space of the based loop space. Over the orthogonal complement of vectors of the based loop space, we take the projection of the previous connection. Let Π be the projection operator from the the tangent space of the path space (there is no boundary condition) over the tangent space of the loop space. It is a bounded operator with bounded derivatives in the derivatives of the holonomy. We get:

$$\nabla^0 X_0'(X_0)_s = \tau_s \Pi(e^{-s}\nabla \frac{e-\tau_1^{-1}}{e-e^{-1}} X_0 + e^s \nabla \frac{e^{-1}-\tau_1^{-1}}{e^{-1}-e} X_0) \tag{1.15}$$

In order to understand this formula, we have identified a tangent vector with its bounded energy path in T_{γ_0}. Moreover, let us recall ([B3],[L5]) that :

$$\nabla_X \tau_t = \tau_t \int_0^t \tau_s^{-1} R(d\gamma_s, X_s)\tau_s \tag{1.16}$$

The connection ∇^0 is not invariant by rotation. As in [L4],[L5], we average ∇^0 in order to get a connection invariant by rotation:

$$\nabla^{inv} = \int_0^t \nabla^t \, dt \tag{1.17}$$

where ∇^t is the rotated connection. Since τ_t preserves the complex structure, ∇^t preserves the almost complex structure over the loop space as well as ∇^{inv}. ∇^{inv} is moreover an Hilbertian connection.

DEFINITION I 1: In a system of local charts:

$$\overline{\partial}_{\infty,r} = \sum_{n,d\overline{z}_i} X_n^l(d\overline{z}_i) \wedge \nabla^{inv}_{X_n^l(\frac{\partial}{\partial \overline{z}_i})} \tag{1.18}$$

over Ω_0.

Ω_0 is not invariant by rotation. But we get:

THEOREM I 2: $\overline{\partial}_{\infty,r}$ is closable. Its closure is invariant under rotation as well as the closure of $\overline{\partial}^*_{\infty,r} + \overline{\partial}_{\infty,r}$.

Proof:We have to show that $\overline{\partial}_{\infty,r}$ is defined over Ω_0. Since in Ω_0 we consider only finite sum, we have only to show that the sum:

$$\sum_{n,dz} X_n^l(d\overline{z}) \wedge \nabla_{X_n^l(\frac{\partial}{\partial \overline{z}})} X_m^l(X_0) \tag{1.19}$$

converges. Namely the supremum norm of $\tau_t^{-1} X_n^l(dz)_t$ is smaller than $\frac{C}{|n|+1}$ and so there is no problem for the derivatives of the cylindrical functions which appear.

Moreover in a system of local charts in γ_0 and in γ_t

(1.20) $$X_m^l(d\overline{z}) = \sum_{j,d\overline{z}'_t} a_m^{j,t} X_j^{l,t}(d\overline{z}_t)$$

where $X_j^{l,t}$ is construct as X_j^l , but in time t. Each $a_m^{j,t}$ is a smooth application in τ_t, τ_1, in γ_t and in γ_0. Moreover, a simple computation shows that:

(1.21) $$| a_m^{j,t} |< \frac{C}{| mj | +1} + C\delta_{m,j} + C\delta_{m,-j}$$

as well as its derivatives in τ_t,τ_1,γ_t, γ_0. We deduce from this and from (1.15) that:

(1.22) $$\sum_j < da_m^{j,t}, X_n^l >^2 < \frac{C(t)}{n^2+1}$$

where the random variable $C(t)$ is bounded in all th L^p with norms independent of t. From (1.15), we get:

(1.23) $$\nabla^t_{X_n^l} X_j^{l,t}(d\overline{z}_t) = X_j^{l,t}(\nabla_{X^l_{n,t}} d\overline{z}_t)$$

except for $j = 0$. For $j = 0$, we have to see the derivatives of the orthogonalisation procedure in γ_t which is a smooth application in γ_t and in the holonomy between γ_t and γ_t, the circle beeing runned in the direct sense and of τ_t. From (1.15), we deduce that there is no problem of convergences.

The fact that $\overline{\partial}_{\infty,r}$ is closable comes from the fact that we can compute $\overline{\partial}^*_{\infty,r}$. Let us recall namely the Bismut's integration by parts formulas:

(1.24) $$\mu[< dF, X_n^l(\frac{\partial}{\partial \overline{z}} > \overline{G}) >] + \mu[F < dG, \overline{X_n^l(\frac{\partial}{\partial z})} >] = \mu[FG \overline{div X_n^l(\frac{\partial}{\partial z})}]$$

Therefore, the adjoint of the covariant derivative along $X_n^l(\frac{\partial}{\partial \overline{z}})$ is $-\nabla_{X_n^l(\frac{\partial}{\partial \overline{z}})} + div X_n^l(\frac{\partial}{\partial \overline{z}})$. Moreover, we have because the connection preserves the metric and the almost complex structure:

(1.25) $$< d < \sigma, \sigma' >, X > = < \nabla_X^{inv} \sigma, \sigma' > + < \sigma, \nabla_{\overline{X}}^{inv} \sigma' >$$

In consequence, we get :

(1.26) $$\nabla^*_{X_n^l(\frac{\partial}{\partial \overline{z}})} \sigma = -\nabla_{X_n^l(\frac{\partial}{\partial z})} \sigma + \sigma \, div X_n^l(\partial z)$$

Let us use the fact:

(1.27) $$< d\overline{z}, \frac{\partial}{\partial \overline{z}} > = 1$$

We deduce:

$$(X^l_n(d\overline{z})\wedge)^* = i_{X^l_n(\frac{\partial}{\partial \overline{z}})} \tag{1.28}$$

i being the interior product. We deduce that over Ω_0,

$$\overline{\partial}^*_{\infty,r} = \sum_{n,dz}(-\nabla_{X^l_n(\frac{\partial}{\partial \overline{z}})} + divX^l_n(\frac{\partial}{\partial z}))\, i_{X^l_n(\frac{\partial}{\partial \overline{z}})} \tag{1.29}$$

There is no problem of convergence, because each σ in Ω_0 is a finite sum and therefore the sum in (1.29) is finite.

It remains to show the invariance by rotation of the closure of these two operators : it is enough to show after using some clear notations that Ω_t belongs to the domain of the closure. Let us study the case of $\overline{\partial}^*_{\infty,r}$ which is the more difficult. We writte $X^{l,t}_m(d\overline{z}_t)$ in the system of $X^l_j(d\overline{z})$: we found component $b^{j,t}_m$ which are a fixed bilinear expression in a fixed function $\phi(\tau_t, \gamma_t, \gamma_0, \tau_1)$ and in H^l_j. The main difficulty is to prove the convergence of $A = \sum_{j,dz} b^{j,t}_m \, divX^l_j(\frac{\partial}{\partial z})$. The system of $\frac{\partial}{\partial z}$ is taken in γ_0. $b^{j,t}_m$ checks still (1.21). We use the integration by parts formulas for $j \neq 0$:

$$divX^l_j(X_0) = \int_0^1 < DX^l_j(X_0)_s, \delta\gamma_s > +1/2 \int_0^1 < S_{X^l_j(X_0)_s}, \delta\gamma_s > \tag{1.30}$$

where S is the Ricci tensor and $\delta\gamma_s$ defines the Ito integral over the manifold. Since the finite variation part of $X^l_j(X_0)$ is deterministic bounded by $\mid 1/j \mid$,γ_0 being fixed, there is no problem of convergences for the term in A which arises from the Ricci tensor. Let $H(t)_s$ be the process $\sum_j b^{j,t}_m H^l_{j,s}$The anticipative part in each piece of each sum is always the same. Moreover this process has finite energy. Let us consider the corresponding vector field $X(t)_s$ over the based loop space. We recognise in the first term the Ito integral $\int_0^1 < DX(t)_s, \delta\gamma_s >$ It remains to consider the expression:

$$\sum_{j,d\overline{z}} < db^{j,t}_m, X^l_j(d\overline{z}) > \tag{1.31}$$

Only the derivatives of τ_t and of γ_t play a role. (1.16) and (1.21) allow to conclude.◇

If D is an operator invariant by rotation, we denote by $Ind_q D$ its equivariant index. We can now state the following conjecture:

CONJECTURE (Hirzebruch-Witten):

*)$\overline{\partial}_{\infty,r} + \overline{\partial}^*_{\infty,r}$ has a self-adjoint extension which is invariant by rotation.

*)It is possible to regularise when y is a root of unit the expression $\chi_y(q, L^\infty(M) =$ "$\sum_r y^r Ind_q(\overline{\partial}_{\infty,r} + \overline{\partial}^*_{\infty,r})$" although each term of the sum does not exist. Moreover:

$$\chi_y(q, L^\infty(M)) = < \prod_{i=1}^d x_i \frac{1+ye^{-x_i}}{1-e^{-x_i}} \prod_{n=1}^\infty \frac{1+yq^n e^{-x_i}}{1-q^n e^{-x_i}} \frac{1+y^{-1}q^n e^{x_i}}{1-q^n e^{x_i}}, [M] > \tag{1.32}$$

[M] is the fundamental integration class over the manifold. In order to define the characteristic class which appears in (1.32), we writte an antisymmetric matrix R as a diagonal matrix of matrices $\binom{0\ -x_i}{x_i\ \ 0}$ in a direct orthonormal basis compatible with the complexe structure.

REMARK: Let us writte $(-y)^N = 1$: when $d \equiv 0\,(N)$ for N odd an when $d \equiv 0\,(2N)$ for N even, this formula is related to the elliptic genera of level N. The regularisation presented in [H.B.G] is a zeta regularisation: we group product in the formal product which gives the right formula for $\chi_y(q, L^\infty(M))$. We get the same results than [H.B.G] in the considered cases.

AN HEURISTIC PROOF OF THE HIRZEBRUCH-WITTEN CONJECTURE

We consider the B.H.K measure in small time:

$$d\mu_\epsilon = \frac{p_{\epsilon^2}(x,x)dP_{\epsilon^2,x,x}dx}{\int p_{\epsilon^2}(x,x)\,dx} \tag{2.1}$$

We take as Hilbert structure the following:

$$\int_0^1 < X_s, Y_s > ds + \epsilon^{-2}\int_0^1 < DX_s, DY_s > ds =< X, Y >_\epsilon \tag{2.2}$$

Let us put:

$$X^l_{n,\epsilon}(X_0) = \frac{\tau_s \sin(2\pi ns)}{\sqrt{Cn^2\epsilon^{-2}+1}}(X_0) \tag{2.3}$$

It is an orthonormal family of T_γ for the Hilbert structure $< . >_\epsilon$ as soon as X_0 describe an orthonormal basis of T_{γ_0}.

We apply the Gram-Schmidt procedure to the family $H'_{n,\epsilon}(X_0)_s = C\epsilon(\cos(2\pi ns) - 1)X_0$. Let us put $X'_{n,\epsilon}(X_0)_s = \tau_s H'_{n,\epsilon}(X_0)_s$.

$$< X'_{n,\epsilon}(X_0), X'_{n,\epsilon}(Y_0) >_\epsilon =< X_0, Y_0 > (\delta_{m,n} + \frac{\epsilon^2\delta_{m,n}}{Cm^2} + \frac{C\epsilon^2}{mn}) \tag{2.4}$$

We deduce that for $n > 0$:

$$H^l_{\epsilon,n}(X_0) = \sum_{0<j<n} \frac{\epsilon^2 C_\epsilon(j,n)}{jn} H'_{\epsilon,j}(X_0) + C_\epsilon(n,n)H'_{\epsilon,n}(X_0) \tag{2.5}$$

for $C_\epsilon(j,n)$ bounded and $C_\epsilon(n,n)$ tending uniforly to 1. It remains to study the corresponding distinguished vectors fields which are acting over the starting points. They are of the type $\tau_s(e^{-s\epsilon^2}\lambda + e^{s\epsilon^2}\mu)$. Moreover, we have :

$$\begin{aligned} \lambda &= \frac{e^{\epsilon^2} - \tau_1^{-1}}{e^{\epsilon^2} - e^{-\epsilon^2}} X_0 \\ \mu &= \frac{e^{-\epsilon^2} - \tau_1^{-1}}{e^{-\epsilon^2} - e^{\epsilon^2}} X_0 \end{aligned} \tag{2.6}$$

In law (this notion will be precised later), we have :

$$\lambda = -\int_0^1 R(d\gamma_{s,flat}, \gamma_{s,flat})X_0 + 1/2X_0 \tag{2.7}$$

and

$$\mu = \int_0^1 R(d\gamma_{s,flat}, \gamma_{s,flat})X_0 + 1/2X_0 \tag{2.8}$$

R is the curvature tensor and $\gamma_{s,flat}$ a flat brownian motion over $T_{\gamma_0}M$. Therefore, $H_0(X_0)_s$ is equivalent to X_0 in law when ϵ goes to 0.

But the identification of a vector with forms put in this formalism some difficulties, especialy because the choice of the orthonormal basis is not invariant by rotation. In order to overcome this problem, we will follow the lines of [J.L$_2$]. We can diagonalise the covariant derivative D over a loop γ if we complexify the tangent space in an explicit form, if we work over a small neighborhood of the constant loop. It is equivariant under the circle action and is defined by $\| \tau_1 - 1 \|^2 < \delta$: $\| . \|$ is the Hilbert norm. The equivariance arises from the following fact : $\tau_{1+s,s}$ is the holonomy starting from γ_s, the circle beeing runned in the direct sense. We have :

$$\tau_{1+s,s} = \tau_s \tau_1 \tau_s^{-1} \tag{2.9}$$

The equivariance follows from the fact that the Hilbert-Schmidt norm is invariant under the orthogonal transformation. The interest to choose such transformations is that there is a natural version of $Log\tau_1 = \sum \frac{(\tau_1 - I)^n}{n}$ over that neighborhood if δ is small enough. $Log\tau_1$ is an antisymmetric matrix. Let us choose an orthogonal basis of T_{γ_0}. We get almost an orthogonal basis of T_γ complexified by taking for $n \geq 0$

$$X_n(X_i)_s = \tau_s exp[-sLog\tau_1] \frac{\cos 2\pi ns}{\sqrt{Cn^2+1}} X_i \tag{2.10}$$

and for $n < 0$

$$X_n(X_i)_s = \tau_s exp[-sLog\tau_1] \frac{\sin 2\pi ns}{\sqrt{Cn^2+1}} X_i \tag{2.11}$$

The word almost comes from the fact if we choose the Hilbert structure (1.5), we have :

$$\begin{aligned} < X_n(X_i), X_m(X_j) > = \frac{\delta_{n,m}}{Cn^2+1} (&< Log\tau_1 X_i, Log\tau_1 X_j >_{\gamma_0} \\ &+ < X_i, X_j >_{\gamma_0} + n^2 < X_i, X_j >_{\gamma_0}) \end{aligned} \tag{2.12}$$

In order to get an orthonormal basis, we splitt τ_1 in block of two dimensional rotations, and we apply the Gram-Schmidt procedure. But the index is stable under homotopy. In order to simplify the computations, we take a simpler orthogonal structure for small loop :

$$< X_n(X_i), X_m(X_j) >= \delta_{n,m} < X_i, X_j >_{\gamma_0} \tag{2.13}$$

Moreover, $exp[-sLog\tau_1]$ keeps the complex structure, and since D is an operator invariant by rotation, the Hilbert structure is invariant by rotation. We have therefore over the previous small neighborhood of the constant loop a $\overline{\partial}'_{\infty,r})$ operator

$$\sum_{n,dz} X_n(d\overline{z}) \wedge \nabla^{inv}_{X_n(\frac{\partial}{\partial \overline{z}})} \tag{2.14}$$

We choose as connection the same than in [J.L$_2$]:

$$\nabla^0 X_n(X_i) = X_n(\nabla X_i) \tag{2.15}$$

and we average under the circle action:

$$\nabla^{inv} = \int_0^1 \nabla^t \, dt \tag{2.16}$$

For big loops, we choose the some operator as before, by considering a smooth cutoff.

Over Ω_0, $\overline{\partial}_{\infty,r}$ and $\overline{\partial}'_{\infty,r}$ are homotopic : namely, there is an homotopy between the two Hilbert structures for small loop and an homotopy between the connections invariant by rotations, since the connection space is an affine space. So we will perform by using stability of the index under homotopy our computations for $\overline{\partial}'_{\infty,r}$, and we decide to writte the two operators in the same way.

We put $\widetilde{X}_n = X_n + iX_{-n}$ if $n > 0$ and if $n < 0$, $\widetilde{X}_n = X_n - iX_{-n}$.

Let us compute ∇^s. Let $\tau_{t,s}$ be the parrallel transport from s to t, the circle being runned in the direct sense. We get modulo a normalisation constant :

$$\begin{aligned} \nabla^s_. \widetilde{X}_n(X_i)_t &= \tau_{t,s} exp[2i\pi nt] exp[-(\widetilde{t-s}) log\tau_{1+s,s}] \nabla_.(\tau_s exp[-slog\tau_1] X_i) \\ &= \tau_{t,s}\tau_s exp[-(\widetilde{t-s}) log\tau_1]\tau_s^{-1} \nabla_.(\tau_s exp[-sLog\tau_1]X_i)\, exp[2\pi int] = \\ &= exp[2\pi int]\tau_t exp[-tlog\tau_1] exp[slog\tau_1]\tau_s^{-1}\nabla_.(\tau_s exp[-slog\tau_1]X_i) \end{aligned} \tag{2.17}$$

In the previous formula, $\widetilde{t-s}$ is the distance between s and t, the circle being runned in the direct sense. We get:

(2.18)

$$\nabla^{inv}_. \widetilde{X}_n(X_i)_t = exp[2\pi int]\tau_t exp[-tlog\tau_1] \int_0^1 exp[slog\tau_1]\tau_s^{-1}\nabla_.(\tau_s exp[-slog\tau_1]X_i)\, ds$$

Modulo a smooth cutoff, we get :

$$\overline{\partial}^*_{\infty,r} = \sum_{n,dz}(-\nabla_{X_n(\frac{\partial}{\partial \overline{z}}} + div X_n(\frac{\partial}{\partial z}))i_{X_n(\frac{\partial}{\partial \overline{z}})} \tag{2.19}$$

We can perform the same analysis as before in order to study $\overline{\partial}_{\infty,r} + \overline{\partial}^*_{\infty,r}$. Since we are involved with the Bismut's dilatation which work over a smaller set, we won't do this. We will follow the line of [J.L$_2$] for this.

The vectors fields $\widetilde{X}_n$ behave badly under the circle action. Following the lines of [J.L$_2$], we introduce the vectors fields :

$$\widetilde{X}_n^{av}(X_i)_t = \int_{S_1} exp[2\pi ins]\widetilde{X}_n^s(X_i)_t ds \tag{2.20}$$

where $\widetilde{X}_n^s$ is the vector field computed from $\widetilde{X}_n$ but in time s. For this, we need a vector field X_i over M. $\widetilde{X}_n^s$ is the vector field:

$$\begin{aligned}&\widetilde{X}_n(\tau_{1,s} exp[-(1-s) log \tau_{1+s,s}] exp[2\pi in(1-s)] X_i(\gamma_s)) = \\ &\widetilde{X}_n(\tau_1 exp[-(1-s) log \tau_1]\tau_s^{-1} exp[2\pi i(1-s)n] X_i(\gamma_s))\end{aligned} \tag{2.21}$$

We deduce from this that:

$$\widetilde{X}_n^{av}(X_i)_t = \tau_1 \frac{exp[-t log \tau_1] exp[2\pi int]}{\sqrt{Cn^2+1}} \int_{S_1} \tau_1 exp[-(1-s) log \tau_1]\tau_s^{-1} X_i(\gamma_s)\, ds \tag{2.22}$$

In general, $\widetilde{X}_n^{av}(X_i)$ is not a basis of the tangent space of a loop if X_i is a basis of the tangent space in the starting point. Let us introduce a finite set of small open balls $B(x_j, r_j)$ which constitute a cover of the compact manifold M such that $B(x_j, r_j - 2\alpha) \cap B(x'_j, r'_j + \alpha) = \emptyset$ for α small enough if j is not equal to j'. Over each $B(x_j, r_j)$, let us introduce an orthonormal basis $X_i^j(x)$ of $T_x(M)$ over $B(x_j, r_j + \alpha)$ which is equals to zero outside $B(x_j, r_j + \alpha)$. We can perform the same for $T^{1,0}$ and $T^{0,1}$ vectors.

Let us choose the equivariant open set $\sup_{s<1} d(\gamma_s, x_j) < r_j + \alpha$ .The random variable $\gamma \to \sup_s d(\gamma_s, x_j)$ is not smooth. We regularise such random variable as in [J.L$_2$] : let g_j be a smooth function from $[0, r_j + \alpha/2]$ into $[1, \infty]$, which is infinite if $z > r_j + \alpha/2$, and which behaves as $\frac{1}{(r_j+\alpha/2-z)^{+,n}}$ when $z \to r_j + \alpha/2 - z$ with n big. Let be the functional $\int_0^1 g_j(d(\gamma_s, x_j))\, ds$. Let us suppose that g is equal to 1 if and only $z \le r_j$. The previous formula is equal to 1 if and only $\sup d(\gamma_s, x_j) \le r_j$ and is infinite if $\sup d(\gamma_s, x_j) \ge r_j + \alpha'$ for $\alpha' > \alpha/2$. Let G be a function from $[-\infty, \infty]$ into $[0, 1]$ which is equal to zero outside a neighborhood of 1. Let us put :

$$F_j(\gamma) = G(\int_0^1 g_j(d(\gamma_s, x_j))\, ds) \tag{2.23}$$

Let us recall (See [J.L$_2$]) that F_j is smooth if n is big enough. We choose as U_j^∞ the equivariant neighborhood $F_j(\gamma) > C$ for some $C < 1$. Let us remark that $\cup U_j^\infty = U^\infty$ is an equivariant neighborhood of the constant loop with probability larger than $1 - exp[-C\epsilon^{-2}]$ for the B.H.K. measure μ_ϵ when ϵ goes to zero.

Let be δ small. Let be $\Psi_{s,t} = \sum_{i,j} \| \tau_{s,t} X_i^j(\gamma_t) - X_i^j(\gamma_s) \|^2$ which is a smooth functional over the loop space. We can find by integrating in $dsdt$ a smooth functional G over the full loop space constructed in the same way of F_j:

$$G = H(\int_{0<s<t<1} g(\Psi_{s,t})dsdt) \tag{2.24}$$

G takes its values in [0,1], is equal to 1 if $\sup_{s,t} \Psi_{s,t} \leq \delta$ and equal to 0 if $\sup_{s,t} \Psi_{s,t} > 2\delta$. $G > C > 0$ is an open neighborhood V^∞ of the fixed point set. We can fix g such that all $\widetilde{X}_n^{av}(X_i^j)$ are for a convenient j a basis of T_γ. The same is true for a basis of $T_\gamma^{1,0,*}$ and of $T_\gamma^{0,1,*}$: $\widetilde{X}_n^{av}(dz_i)$ is a basis of $T_\gamma^{1,0,*}$ and $\widetilde{X}_n^{av}(d\overline{z}_i)$ a basis of $T_\gamma^{0,1,*}$, if dz_i is a basis of $T_{\gamma_0}^{1,0,*}$ over $B(x_j, r_j)$, which vanishes outside $B(x_j, r_j + \alpha)$.

Let us choose following [J.L$_2$] smooth real functions f_l^j equals to zero outside $B(x_j, r_j + \alpha)$ and such that any smooth functions over $B(x_j, r_j)$ can be approached uniformly by linear combinations of the f_l^j in the C^k-topology. In a system of local charts for $B(x_j, r_j)$, we choose the set of monomials in the coordinates.

Let $F_j(A)$ be a product of $\int_0^1 f_j^l(\gamma_s) exp[2i\pi l' s]ds$. The Bismut's operation is defined as follow: we divide each integral where a $l' \neq 0$ appears by ϵ.

Let $\xi_j(A, B, C)$ be the product of some $F_j(A)$ by any wedge product B in the $\widetilde{X}_n^{av}(d\overline{z}_i)$ and by a wedge product of length r C in the $\widetilde{X}_n^{av}(dz_i)$: $d\overline{z}_i$ and dz_i are convenient local basis of $T^{0,1}$ and of $T^{1,0}$ over $B(x_j, r_j + \alpha)$

Let Ω be the linear space of :

$$\sigma = \sum_j g(F_j + h_j)\lambda_j(A, B, C)\xi_j(A, B, C) + \sum_j (1 - g)(F_j + H_j)\sigma_j \tag{2.25}$$

The $\lambda_j(A, B, C)$ are constant. σ_j is a finite combination of any wedge product in $X_{n,m}(d\overline{z}_i)$ and of r wedge product in $X_{n,m}(dz_i)$. We have written:

$$X_{n,m}(X_0) = \int_{S_1} exp[2\pi ims] X_n^{l,s}(X_0)ds \tag{2.26}$$

By Fourier expansion, the set of $X_{n,m}(X_0)$ is dense in the tangent space of a loop. The component of each wedge product in σ_j are cylindrical functionals. $B_\epsilon \xi_j(A, B, C)$ is the product of $B_\epsilon F(A)$ by the same wedge products in $\widetilde{X}_n^{av}(X_0)(d\overline{z}_i)$ and of $\widetilde{X}_n^{av}(X_0)(dz_i)$ which are in $\xi_j(A, B, C)$.

Let us recall that X^s is constructed as follow: we get a vector over $\gamma_{s+.}$, we deduce a vector $X^{-^t s}$ over $\gamma_.$ and we put $X^s(\gamma_.) = X(\gamma_{s+.})^{-^t s}$. We deduce that:

$$X_{n,m}(X_0)^s = q^{-m} X_{n,m}(X_0) \tag{2.27}$$

and that :

$$X_{n,m}(\gamma.)^{t_s} = q^m X_{n,m}(\gamma_{s+}.) \tag{2.28}$$

Therefore $X_{n,m}(\gamma)$ is of multiplicity -m under the circle action. This shows us that Ω is stable under the circle action, because each F_j is of multiplicity finite under the circle action.

DEFINITION II 1: The Bismut dilatation is defined as follow. If σ is as in (2.25):

$$B_\epsilon\sigma = \sum g(F_j + H_j)B_\epsilon\xi_j(A,B,C)\lambda_j(A,B,C) + \sum(1-g)(F_j+H_j)\sigma_j \tag{2.29}$$

PROPOSITION II 2 ([J.L$_2$]): B_ϵ is well defined.

Proof: If $\sigma = 0$, $\lambda_j(A,B,,C) = 0$ by using the weak Hörmanders theorem ([Hö]).◇

In order to define a limit model, we follow the lines of Taubes ([T]). As probability space, we take the family of flat brownian bridges γ_{flat} in $T_x(M)$ starting from 0 endowed with the measure $dx \bigotimes dP_{1,x,flat}$. The tangent space is the space $X_{l,t}$ of sections in $T_{\gamma_0}M$ with finite energy. As Hilbert norm, we take:

$$\| X_{l,0} \|^2 + \int_0^1 \| X'_{l,t} \|^2 \, dt = \| X_l \|_1^2 \tag{2.30}$$

In order to define forms, we take the Hilbert norm :

$$\int_0^1 \| X_{l,t} \|^2 \, dt + \int_0^1 \| X'_{l,t} \|^2 \, dt = \| X_l \|_2^2 \tag{2.31}$$

An orthonormal basis is given by [J.L$_1$]: $\overline{X}_{n,l}(X_0)_s = \frac{\cos(2\pi ns)}{\sqrt{Cn^2+1}}X_0$ if $n \geq 0$ and $\overline{X}_{n,l}(X_0)_s = \frac{\sin(2\pi ns)}{\sqrt{Cn^2+1}}X_0$. $n < 0$. The fiber is given by combination of wedge products in $\overline{X}_{n,l}(d\overline{z})$ and of r wedge products in $\overline{X}_{n,l}(dz)$ (because we consider $\overline{\partial}_{r,l}$). An holomorphic boson is given by a combination of Wick products in $\int_0^1 < \overline{X}_{n,l}(dz)_s, \delta\gamma_{s,flat} >$, and an antiholomorphic boson is given by a Wick product of $\int_0^1 < \overline{X}_{n,l}(d\overline{z})_s, \delta\gamma_{s,flat} >$ $(\gamma_{s,flat})$ is a flat brownian motion). As core, we define sections of finite combination of wedge products in $\overline{X}_{n,l}(d\overline{z})$ and of r wedge products in $\overline{X}_{n,l}(dz)$. The components are finite combination of Wick products in $\int_0^1 < \overline{X}_{n,l)(d\overline{z})_s}, \delta\gamma_{s,flat} >$ an in $\int_0^1 < \overline{X}_{n,l}(dz)_s, \delta\gamma_{s,flat} >$ and of smooth functions $f(\gamma_0)$. $\overline{X}_{r,l}$ has two components in this case:

-)A finite dimensional component which can be written in normal coordinate because the metric is Kähler. In normal coordinate, it is:

$$\overline{\partial}_{f,r,l} = \sum \frac{\partial}{\partial \overline{z}_i} \wedge \overline{X}_{0,l}(d\overline{z}_i) \tag{2.32}$$

We take derivative over the component of the Wick products and there is no derivatives of the wedge products, because we work in normal coordinates ([T]).

-)An infinite dimensional part which is already more and less studied by [F],[Ni] and [Rö] when there is no parameter. It is:

$$\overline{\partial}_{\infty,r,l} = \sum_{n\neq 0,dz} c(n)\nabla_{X_{n,l}(\frac{\partial}{\partial \overline{z}})} \wedge \overline{X}_{n,l}(d\overline{z}) \tag{2.33}$$

where ∇ is the h-drivative.

Since in $\overline{\partial}_{\infty,r,l}$, we dont take component in the dz , $d\overline{z}$ parts of the Wick product or of $f(x)$, these two parts of the limit operator are bounded. The $c(n)$ are bounded auxiliary numbers, which are even.

We can compute easily $\overline{\partial}^*_{r,l}$. In normal local coordinates, it is :

$$\begin{aligned}\overline{\partial}^*_{r,l} &= \sum_{dz} c(0) i_{\overline{X}_{0,l}(\frac{\partial}{\partial \overline{z}})}(-\nabla_{X_{0,l}(\frac{\partial}{\partial \overline{z}})} + div X_{0,l}(\frac{\partial}{\partial z})) \\ &+ \sum_{n\neq 0,dz} c(n) i_{\overline{X}_{n,l}(\frac{\partial}{\partial \overline{z}})}(-\nabla_{X_{n,l}(\frac{\partial}{\partial \overline{z}})} + div X_{n,l}(\frac{\partial}{\partial z}))\end{aligned} \tag{2.34}$$

Therefore, $\overline{\partial}_{r,l} + \overline{\partial}^*_{r,l}$ is symmetric, therefore closable. The interior product has to be seen for the second Hilbert norm.

The limit model inherit a circle action: it is defined as followed, by beginning by studying the bosonic case. The smooth complex functionals are invariant under this action. If we perform a multiplication by a q belonging to S_1, if $n > 0$, $\int_0^1 < X'_{n,l,s} + iX'_{-n,l,s}, \delta\gamma_{s,flat} >$ is transformed by q^{-n} and $\int_0^1 < X'_{n,l,s} - iX'_{-n,l,s}, \delta\gamma_{s,flat} >$ is transformed by q^n (See [J.L$_2$]). If we take a Wick product of such functionals, the exponent of the circle action of such Wick products is the sum of the exponent of the circle action over each element of the Wick product. Such circle action can be understood as the natural limit circle action over the loop space endowed with the B.H.K measure μ_ϵ : namely in such case $\int_0^1 exp[2i\pi ns] < X(\gamma_s), \delta\gamma_s >$ is transformed by q^{-n} under the circle action ([L$_5$]). It remains to perform a Bismut dilatation to get the result at the limit, and therefore by orthogonalisation for Wick product. The limit circle action is an isometry of the scalar Hilbert space which arises from the bosonic limit model . If $n > 0$,$\overline{X}_{n,l}(X_0) + i\overline{X}_{-n,l}(X_0)$ has to be understood as the limit of $\widetilde{X}^{av}_n(X_0)$, which behaves as q^{-n} under the circle action. The same is true for $n < 0$: $\overline{X}_{n,l}(X_0) - i\overline{X}_{-n,l}(X_0)$ which behaves as q^n under the circle action. Moreover, $\overline{\partial}_{r,l}$ commute with the circle action. Let us check this assumption, after choosing some clear notations:

$$\overline{\partial}_{\infty,r,l} = -i/2 \sum_{n\neq 0,dz} c'(n)\nabla_{exp[-2i\pi ns](\frac{\partial}{\partial \overline{z}})} \wedge \overline{exp[2i\pi ns]}(d\overline{z}) \tag{2.35}$$

$c'(n)$ is equal to $-c(n)$ if $n > 0$ and to $c(n)$ if $n < 0$. In order to understand this formula, we take the h-derivative in the traditional sense: we consider the derivative of $X_{n,l}(\frac{\partial}{\partial \overline{z}})$ in time t. The wedge product acts as q^n with respect to the circle action and the h-drivative acts as q^{-n} or as 0. The limit operator commute with the circle action. We can speak of the equivariant index of $\overline{\partial}_{r,l} + \overline{\partial}_{r,l}$. We get:

THEOREM II 3 (Hirzebruch-Witten). When y is a N root of unit, we have:

$$(2.36) \qquad \chi_y(q, L^\infty(M)) =< \prod_{i=1}^{\infty} x_i \frac{1 + ye^{-x_i}}{1 - e^{-x_i}} \prod_{n=1}^{\infty} \frac{1 + yq^n e^{-x_i}}{1 - q^n e^{-x_i}} \prod_{n=1}^{\infty} \frac{y + q^n e^{x_i}}{q^n e^{x_i} - 1}.[M] >$$

Moreover, in (2.36) we group product in order to get (1.32).

Proof: Let us compute first $(\overline{\partial}_{r,l} + \overline{\partial}^*_{r,l})^2 = \Delta_{r,l}$. We have:

$$(2.37) \qquad \Delta_{r,l} = (\overline{\partial}_{f,r,l} + \overline{\partial}^*_{f,r,l})^2 + (\overline{\partial}_{\infty,r,l} + \overline{\partial}^*_{\infty,r,l})^2 = \Delta_{f,r,l} + \Delta_{\infty,r,l}$$

because the anticommutation property. The operation which associates $X_{n,l}$ to $\overline{X}_{n,l}$ is an isometry. Let $N_{B,c}(antiho)$ be the second quantised operator of $c(n)^2$ which counts the number of antiholomorphic bosons and let $N_{F,c}$ be the second quantised operator of $c(n)^2$ which counts the number of $T^{0,1}$ forms in the infinite dimensional part. We have ([F],[Ni]):

$$(2.38) \qquad \Delta_{\infty,r,l} = N_{B,c}(antiho) + N_{F,c}$$

Therefore the infinite dimensional laplacian has a set of eigenvectors which can be seen following Taubes as a countable set of finite dimensional bundles: moreover $\overline{\partial}_{b,r,l} + \overline{\partial}^*_{b,r,l}$ transforms a section of each bundle into a section of the same finite dimensional subbundle. $\overline{\partial}_{r,l} + \overline{\partial}^*_{r,l}$ has therefore a self-adjoint extension, which is compatible with the self adjoint extension od $\Delta_{r,l}$. Following Taubes, the kernel of $\Delta_{r,l}$ can be computed as followed: we take first the kernel of $\Delta_{\infty,r,l}$: this gives a countable set of finite dimensional bundles, each pieces being kept by $\Delta_{f,r,l}$. It is the set of holomorphic bosons tensorised by r $T^{1,0}$ forms. We can study the S_1 action over such set. Since a Wick algebra is a symmetric tensor algebra, we come back to a countable set of Riemann-Roch formulas. The characteristic serie of $\chi_y(q, L^\infty(M))$ should be $\prod_{i=1}^d x_i \prod_{n=-\infty}^{n=\infty} \frac{1+yq^n e^{-x_i}}{1-q^n e^{-x_i}}$. It is clear that this infinite product has no meaning. We follow [Hi] in order to handle with that : we separate the terms with positive n and negative n; we multiply the denominator and the numerator of the last term by $q^n e^{x_i}$. We get the characteristic serie given in (2.36): we group in product in order to get (1.32). It is possible because y is a N root of unit.◊

REMARK : There is in fact no equivariant index. Only formal manipulations allow to give (1.32).

Let us now motivate the introduction of this limit operator by using a limit theorem, which is analoguous to the limit theorem of [J.L$_2$] and [L.R]. The tangent space of the based loop space is canonicaly isomorphic to the tangent space of the based loop space in the constant loop with the metric $\int_0^1 \| X'_{l,s} \|^2 \, ds + \int_0^1 \| X_{l,s} \|^2 \, ds$. The normal component is isomorphic to the normal component in a constant loop by a smooth isomorphism which is non deterministic, but with derivatives bounded in the holonomy. In small time ϵ^2, we dont perform any rescaling over the forms. We perform for small loop a rescaling over $X_n(X_0)$ by multiplying it by ϵ as in [J.L$_2$]. For big loop, we don't do any transformation. We put in normal coordinates:

$$\overline{\partial}_{\epsilon,\infty,r} = \sum_{n,dz} c(n)X_n(d\overline{z}) \wedge \nabla^{inv}_{\epsilon X_n(\frac{\partial}{\partial \overline{z}})}$$

(2.39)

$$\overline{\partial}^*_{\epsilon,\infty,r} = \sum_{n,dz} c(n)(-\nabla^{inv}_{\epsilon X_n(\frac{\partial}{\partial \overline{z}})} + div\epsilon X_n(\frac{\partial}{\partial z}))i_{X_n(\frac{\partial}{\partial \overline{z}})}$$

Of course this definition works for small loop. For big loop, we keep the original operators. We perform a smooth cutoff between the both. We have:

THEOREM II 4 : Let σ be a fixed element of Ω. $B_\epsilon\sigma$ tends in law to σ_l. $(\overline{\partial}_{\epsilon,\infty,r} + \overline{\partial}^*_{\epsilon,\infty,r})B_\epsilon\sigma$ tends in law to $(\overline{\partial}_{r,l} + \overline{\partial}^*_{r,l})\sigma_l$.

Proof : -)First step. $B_\epsilon\sigma$ tends in law to σ_l .

If σ belongs to Ω, it is clear that the contribution for big loops cancels when $\epsilon \to 0$. $\frac{\int_0^1 exp[2i\pi ns]<X(\gamma_s),d\gamma_s>}{\epsilon}$ tends in law to $\int_0^1 exp[2i\pi ns] < X(\gamma_0), d\gamma_{s,flat} >$. The functional $\int_0^1 f(\gamma_s)exp[2i\pi ns]ds$ can be written in the same type after integration by parts. For an infinite combination the Bismut's procedure is necessary ([B$_3$]). The problem does not come from this but from the fact that in $\widetilde{X}^{av}_n(X_i)$, $\int_0^1 \tau_1 exp[-(1-s)log\tau_1]\tau_s^{-1}X_i(\gamma_s)ds$ appears, which tends in law to $X_i(\gamma_0)$. The first step is proved.

-)Second step: $(\overline{\partial}_{\epsilon,r} + \overline{\partial}^*_{\epsilon,r})B_\epsilon\sigma \to (\overline{\partial}_{r,l} + \overline{\partial}^*_{r,l})\sigma_l$ in law.

Let us begin to study the first contribution: $\overline{\partial}_{\epsilon,r}B_\epsilon\sigma$ .We have:

$$(2.40) \qquad < X_n(X_i), \widetilde{X}^{av}_n(X_j) >= \int_{S_1} < \tau_1 exp[-(1-s)log\tau_1]\tau_s^{-1}X_i(\gamma_s), X_j(\gamma_0) > ds$$

In normal coordinate, the first derivative of such expression tends to zero when ϵ goes to zero. This shows us that in the limit behaviour of $\overline{\partial}_{\epsilon,r}B_\epsilon\sigma$, we dont have to take care with the derivative of the component of $\widetilde{X}^{av}_n(X_i)$. Similar considerations hold for the derivative for ∇^{inv} of the canonical basis X_n because we work in normal coordinates. Since $g(F_j + H_j)$ is constant in a little neighborhood of the constant loop, we have to replace the cutoff function which appear in (2.25) by 1 , and their derivatives by 0. It remains to study the behaviour of $< dB_\epsilon F_j(A), \epsilon\widetilde{X}_n(d\overline{z}) >$ for $n \neq 0$. We proceed as in [J.L$_2$],[L.R]: if $m \neq 0$, by integration by parts, we get:

$$(2.41) \qquad \frac{-i}{2\pi m}\int_0^1 < df(\gamma_s), d\gamma_s > exp[2i\pi ms] = \int_0^1 exp[2i\pi ms]f(\gamma_s)ds$$

Therefore, in law, if $n \neq 0$

$$(2.42)\quad \frac{-i}{m\epsilon} < d\int_0^1 < df(\gamma_s), d\gamma_s > exp[2i\pi ms], \epsilon\widetilde{X}_n(X_i) > \rightarrow \delta_{n,-m}\frac{n}{m\sqrt{Cn^2+1}}\int_0^1 < df(\gamma_0), X_i(\gamma_0) >$$

and

$$(2.43)\quad B_\epsilon \int_0^1 exp[2i\pi ms] f(\gamma_s)ds \rightarrow \frac{-i}{2\pi m}\int_0^1 exp[2i\pi ms] < df(\gamma_0), d\gamma_{s,flat} >$$

Let us study the case $n = 0$. Since there is no rescaling of $X_0(X_i)$ by ϵ, we can see that in law:

$$(2.44)\quad \frac{-i}{m\epsilon} < d\int_0^1 < df(\gamma_s), d\gamma_s > exp[2i\pi ms], X_0(X_i) > \rightarrow \frac{-i}{m}\int_0^1 < \nabla_{X_i(\gamma_0)} df(\gamma_0), d\gamma_{s,flat} > exp[2i\pi ms]$$

Let us study the case $m = 0$.It is clear that if $n \neq 0$ in law:

$$(2.45)\quad < d\int_0^1 f(\gamma_s)ds, \epsilon\widetilde{X}_n(X_0) > \rightarrow 0$$

It is clear too that if $n = 0$

$$(2.46)\quad < d\int_0^1 f(\gamma_s)ds, X_0(X_0) > \rightarrow < df(\gamma_0), X_0 >$$

Therefore the limit in law holds for $\overline{\partial}_{\epsilon,r} B_\epsilon \sigma$, by using the Bismut's procedure ([B$_3$]) in order to handle with bounded continuous functionals in the component.

In order to study the limit behaviour of $\overline{\partial}^*_{\epsilon,r}$, we have only a finite sum, and the main problem is to compute $\epsilon div\widetilde{X}_n(X_0)$ if $n \neq 0$. We use the formula (1.3) of [L$_4$]. But since we work in small time, we have to divide the flat Ito term by ϵ^2, while the Ricci term is kept. We deduce that in law :

$$(2.47)\quad \epsilon div\widetilde{X}_n(X_0) \rightarrow \frac{in}{\sqrt{Cn^2+1}}\int_0^1 < exp[2i\pi ms]X_0, d\gamma_{s,flat} >$$

For $n = 0$

$$divX̃_n(X_0) \to divX_0 \tag{2.48}$$

because in normal cordinates $\tau_1 - I$ has a behaviour in ϵ^2.

Therefore the result, with $c(n) = sign \frac{n}{\sqrt{Cn^2+1}}$, by using the Bismut's procedure in order to handle with limit in law. ◊

REMARK: We have worked with forms $\widetilde{X}_n(d\bar{z})$ and vectors $\widetilde{X}_n(\frac{\partial}{\partial \bar{z}})$. Namely, we can writte:

$$\begin{aligned}\bar{\partial}_{\epsilon,r} &= \sum_{dz} X_0(d\bar{z}) \wedge \nabla_{X_0(\frac{\partial}{\partial \bar{z}})} \\ &+ i/4 \sum_{n>0,dz} \widetilde{X}_{-n}(d\bar{z}) \wedge \nabla^{inv}_{\epsilon \widetilde{X}_n(\frac{\partial}{\partial \bar{z}})} - i/4 \sum_{n<0,dz} \widetilde{X}_n(d\bar{z}) \wedge \nabla^{inv}_{\epsilon \widetilde{X}_{-n}(\frac{\partial}{\partial \bar{z}})}\end{aligned} \tag{2.49}$$

BIBLIOGRAPHY

[A.M$_1$] Airault H. Malliavin P : Quasi sure analysis. Publication Université Paris VI.

[A.M$_2$] Airault H. Malliavin P : Integration on loop groups. Publication Paris VI.

[A.H.K] Albeverio S. Hoegh-Krohn R. : The energy representation of a Sobolev Lie group. Compositio Math. 36. (1978), 37-52.

[A.L.R] Albeverio S. Léandre R. Röckner M. : Construction of a rotational diffusion on the free loop space. C.R.A.S. t.316. Série I. (1993), 287-292.

[A.K.M.W$_1$] Alvarez O. Killingback T.P. Mangano M. Windey P. : String theory and loop space index theorem. C.M.P. 111 (1987), 1-10.

[A.K.M.W$_2$]Alvarez O. Killingback T.P. Mangano M. Windey P. : The Dirac-Ramond operator in string theory and loop space index theorems. pp 189-216. In H. Hamber edit. Proc. Conf. On nonperturbative methods in field theory Jan 87. Proc. Suppl. Nuclear. Phys. B. 1. A. (1987)

[Ar$_1$]Arai A. : A path integral representation of the index of Kähler-Dirac operators on an infinite dimensional space. J.F.A. 82. (1989), 330-369.

[Ar$_1$]Arai A. : A general class of infinite dimensional operators and path representation of their index. J.F.A. 105 (1992), 342-408.

[Ar.M]Arai A. Mitoma I. : De Rham-Hodge-Kodaira decomposition in infinite dimension. Math.Ann.291 (1991), 51-73.

[At]Atiyah M. : Circular symmetry and stationary phase approximation. pp 43-59. Colloque en l'honneur de L.Schwartz. Asterisque 131 (1985).

[Az]Azencott R. : Une approche probabiliste du théorème de l'indice (d'après J.M.Bismut). Séminaire Bourbaki. Exposé 633.

[Be]Bennequin.D. : En marge de l'apologie de Witten. Gazette des mathématiciens 50 (1991), 73-93.

[B.V]Berline N. Vergne M: Zéros d'un champ de vecteurs et classes caractéristiques équiva-

riantes. Duke Math. Jour. 50 (1983), 539-548.
[B_1]Bismut J.M. : Martingales, the Malliavin Calculus and hypoellipticity under general Hörmander conditions. Z.W. 56 (1981), 469-505.
[B_2]Bismut J.M. : Mécanique aléatoire. Lecture Notes In Math. 866. Springer. (1981).
[B_3]Bismut J.M. : Large deviations and the Malliavin Calculus. Progress in Math. 45. Birkhaüser. (1984).
[B_4]Bismut J.M. : The Atiyah-Singer theorem: a probabilistic approach. J.F.A. 57 (1984), 56-99.
[B_5]Bismut J.M. : The Lefschetz fixed point formulas. J.F.A. 57 (1984), 329-348.
[B_6]Bismut J.M. : Index theorem and equivariant cohomology on the loop space. C.M.P. 98 (1985), 213-237.
[B_7]Bismut J.M. : Localisation formulas, superconnections and the index theorem for families. C.M.P. 103 (1986), 127-166.
[B.S]Bogachev V.I. Smolyanov O.G. : Analytic properties of infinite dimensional distributions. Uspekhi.Mat. Nauk. 45 (1990), 3-83.
[Ca]Cameron R.H.: The first variation of an indefinite Wiener integral. Proc. Amer. Soc. 2 (1951), 914-924.
[Ch]Chen K.T. : Iterated path integrals of differential forms and loop space homology. Ann.Math. 97 (1983), 213-237.
[Co]Connes A. : Entire cyclic cohomology of Banach algebras and characters of Θ-summable Fredholm modules. K-theory 1 (1988), 519-548.
[D.F]Dalecky.Y. Fomin V.I. : Measures and differential equations in infinite dimensional spaces. Math. and Applications. Nauka.
[D.H.V.W]Dixon L. Harvey J. Vafa C. Witten Ed. : String on orbifold. Nucl. Phys. B. 261 (1985), 678-686.
[Do_1]Doku I. : Laplace operators associated with Hida derivatives in white noise analysis. Preprint.
[Do_2]Doku I. : Index of a Dirac operators in quantum field theory which is associated with Hida derivatives. Preprint.
[Dr]Driver B. : A Cameron-Martin type quasi-invariance theorem for pinned brownian motion on a compact Riemannian manifold. Preprint.
[El_1]Elworthy K.D. : Stochastic differential equations on manifold. L.M.S. Lectures Notes Serie 20. Cambridge University Press (1982).
[El_2]Elworthy K.D.: Geometric Aspects of diffusions on manifolds. pp 277-427. In Ecole d'été de Saint-Flour 1987. P.L. Hennequin édit. Lecture Notes in Math. 1362. Springer. (1989).
[Em.L]Emery M. Léandre R. : Sur une formule de Bismut. pp 448-452. Séminaire de Probabilités XXIV. Lecture Notes in Maths. 1426. Springer (1990).
[F]Fang S. : Equations $\overline{\partial} u = \omega$ sur un espace de Wiener complexe. Preprint.
[F.M]Fang S. Malliavin P.: Stochastic Calculus on Riemannian manifold. Preprint.
[Fr]Freed D.: Thesis. M.I.T. (1984).
[Fre]Frenkel I.B. : Orbital theory for affine Lie Algebras. Invent. Math. 77 (1984), 301-352.
[Fu]Fukushima M.: Dirichlet forms and Markov process. North-Holland. (1980).
[Ge_1]Getzler E. : Cyclic homology and the path integral of the Dirac operator.Preprint.

[Ge_2]Getzler E. : Dirichlet form on a loop space. Bull. Sciences. Math. 2. 113 (1989), 157-174.

[G.J.P]Getzler E. Jones J.D.S. Petrack S. : Differential forms on a loop space and the cyclic bar complex. Topology 30 (1991), 339-373.

[Gi]Gilkey P.B. : Invariance theory, the heat-equation and the Atiyah-Singer theorem. Math.Lect.Series. 11. Publish and Perish.

[Gr.P]Grorud A. Pontier M. : Calcul anticipatif d'ordre deux. Stochastics. 42. (1993),1-23.

[Gr_1]Gross L. : Potential theory on Hilbert space. J.F.A. 1 (1967), 123-181.

[Gr_2]Gross L. : Logarithmic Sobolev inequalities on a loop group. J.F.A. 102 (1991), 268-312.

[Gr_3]Gross L. : Uniqueness of ground states for Schrödinger operators over loop groups. Preprint.

[H.K.P.S]Hida T. Kuo H.H. Potthoff J. Streit L. : White noise. An infinite dimensional Calculus. Kluwer (1993).

[Hi]Hirzebruch F. : Elliptic genera of level N for complex manifold. pp 37-63. In differential Geometrical Methods in Theoretical Physics. K.Bleuler. M.Werner edit. Kluwer. (1988)

[H.B.G]Hirzebruch F. Berger T. Jung R. : Manifolds and modular forms. Aspect of Mathematics 20. Vieweg (1993)

[H.H]Hirzebruch F. Höfer T. : On the Euler number of an orbifold. Math. Ann. 286 (1990), 255-260.

[H.K]Hoegh-Krohn R. : Relativistic quantum statistical mechanics in 2 dimensional space time. C.M.P. 38 (1974), 195-224.

[Hö]Hörmander L.: Hypoelliptic second order differential equations. Acta Mathé. 119 (1969), 147-171.

[Hs]Hsu P.: Brownian motion and the Atiyah-Singer theorem. Preprint.

[$I.W_1$]Ikeda N. Watanabe S. : Stochastic differential equations and diffusion processes. North-Holland (1981).

[$I.W_2$]Ikeda N. Watanabe S. : Malliavin Calculus for Wiener's functional and its apllication. In "From local time to global geometry". K.Elworthy edit. Pitman 1986.

[$J.L.W_1$]Jaffe A. Lesniewsi A. Weitsman J. : Index of a family of Dirac operators on loop space. C.M.P. 112 (1987), 75-88.

[$J.L.W_2$]Jaffe A. Lesniewski A. Weitsman J. : The loop space $S^1 \to \mathbb{R}$ and supersymmetric quantum fields. Ann. Phys. 183 (1988), 337-351.

[$J.L_1$]Jones J. Léandre R. : L^p Chen forms over loop spaces. pp 104-162. In Stochastic Analysis. M.Barlow. N.Bingham edit. Cambridge University Press (1991).

[$J.L_2$]Jones J. Léandre R.: A stochastic approach to the Dirac operator over the free loop space.In preparation.

[J.P]Jones J. Petrack S. : The fixed point theorem in equivariant cohomology. Preprint.

[Kr]Krée P. : La théorie des distributions en dimension quelconque et l'intégration stochastique. pp 170-233. In stochastic Analysis and related topics. Korezlioglu H. Ustunel A.S. Lectures Notes in Math. 1316. Springer. (1988).

[Ku]Kuo H.H. : Diffusion and Brownian motion on infinite dimensional manifolds. Trans. Amer. Math. Soc. 159. (1972), 439-451.

[K]Kusuoka S. : De Rham cohomology of Wiener-Riemannian manifolds. Preprint.

[L_1]Léandre R.: Applications quantitatives et qualitatives du calcul de Malliavin. pp 109-133. Proc. Col. Franco-Japonais. M. Métivier. S. Watanabe édit. Lectures Note Math. 1322. Springer. (1988). English translation. pp 173-197. Geometry of Random motion. R. Durrett. M. Pinsky edit. Contemporary Math. 73. (1988).

[L_2]Léandre R. : Sur le théorème d'Atiyah-Singer. P.T.R.F. 80 (1988), 119-137.

[L_3]Léandre R. : Sur le théorème de l'indice des familles. pp 348-414. Séminaire de Probabilités XXII. Lectures Notes in Math. 1321 (1988).

[L_4]Léandre R. : Integration by parts formulas and rotationally invariant Sobolev Calculus on free loop spaces.In infinite dimensional problem in Physic. XXVIII winter school of theoretical physics. Gielerak R. Borowiec A. edit. Journal of Geometry and Physics. 11 (1993), 517-528.

[L_5]Léandre R. : Invariant Sobolev Calculus on the free loop space. Preprint.

[L.R]Léandre R. Roan S.S. : A stochastic approach to the Euler-Poincaré number of the loop space of a developable orbifold. To be publish In Journal of Geometry and Physics.

[M.L]Mac-Laughlin D.: Orientation and string structures on loop space. Pacific. J. of Math. 155. 1. (1992), 143-156.

[Ma_1]Malliavin P. : Stochastic calculus of variations and hypoelliptic operators. pp 195-263. Proc. Inter. Con. Stoch. Differ. Equa. K. Ito edit. Wiley (1978).

[Me]Meyer P.A. : Notes sur les processus d'Ornstein-Uhlenbeck. pp 95-133. Séminaires de Probabilités XVI. Lecture Notes in Math. 920 (1982).

[Ni]Nishimura T. : Exterior bundle over a complex abstract Wiener space. Preprint.

[O]Ochanine S. : Sur les genres multiplicatifs définis par des intégrales elliptiques. Topology 26 (1987), 143-151.

[Ra]Ramer R. : On the de Rham complex of finite codimensional forms on infinite dimensional manifolds. Thesis. Warwick University. 1974.

[R]Roan S.S. : On the generalisation of Kummer surfaces. J. Dif. Geometry. (1989), 523-537.

[Rö]Röckle H. : Thesis. Universitaet Bochum (1993).

[Se]Segal G. : Elliptic cohomology (after Landweber-Stong-Ochanine- Witten and others pp 187-201. Séminaire Bourbaki Exposé 695. Astérisque 161-162. (1988).

[Sh]Shigekawa I. : De Rham-Hodge-Kodairas decomposition on an abstract Wiener space. J.Math.Kyoto.Uni. 26 (1986), 191-202.

[Sh.U]Shigekawa I. Ueki N.: A stochastic approach to the Riemann-Roch theorem. Osaka Journal of Math. 25 (1988), 759-784.

[Sm]Smolyanov O.G. : De Rham currents and Stoke's formula in a Hilbert space. Soviet Math. Dok. 33. 3 (1986), 140-144.

[T]Taubes C. : S^1 actions and elliptic genera. C.M.P. 122 (1989), 455-526.

[Wa_1]Watanabe S. : Stochastic differential equations and Malliavin Calculus. Tata Institute of fundamental Research. Springer (1984).

[Wa_2]Watanabe S. : Analysis of Wiener functionals and its applications to heat kernels. Ann. Probab. 15. 1 (1987), 1-37.

[Wi_1]Witten Ed. : Supersymmetry and Morse theory. J. Diff. Geometry 17 (1982), 661-692.

[Wi$_2$]Witten Ed. : The index of the Dirac operator in loop space. pp161-181. In elliptic curves and modular forms in algebraic topology. Landweber edi. Lecture Notes in Math. 1326. Springer (1988).

R.LEANDRE
département de Mathématiques. I.R.M.A
Université Louis Pasteur
Rue Descartes 67084 Strasbourg
FRANCE

On Lyapounov Stability Theorems for Stochastic (Deterministic) Evolution Equations[1]

V. Mandrekar
Michigan State University

0 Introduction

The purpose of this work is to present a review of the results related to some extensions of Lyapounov methods in the stability theory for the solutions of infinite-dimensional (stochastic and deterministic) evolution equations. As a consequence, one can derive results on the stability of solutions of stochastic partial and delay differential equations.

Let $\mathcal{X}$ be a Banach space and consider the Cauchy problem $\dot{u}(t) = Au(t)$, $u(0) = x \in \mathcal{X}$. In case $\mathcal{X}$ is finite-dimensional, Lyapounov proved that a solution $\{u^x(t),\ t \geq 0\}$ is stable i.e. $\|u^x(t)\| \leq \|x\|\, e^{-\gamma t}$, iff

$$\max\{Re\lambda : \det(\lambda I - A) = 0\} < 0. \tag{0.1}$$

The above condition is equivalent to the existence of a positive-definite matrix R satisfying (see [B] p. 240-243)

$$\text{(0.2)}\quad \text{(a) } A^*R + RA = -I \quad \text{and} \quad \text{(b) } (Rx, x) = \int_0^\infty \|u_t^x\|^2\, dt, \text{ in case of stability.}$$

The crucial property of R (in addition to (0.2 (a)) is R is positive-definite i.e. $(Rx, x) \geq c\,\|x\|^2$. This property is used heavily in attacking stability of non-linear systems ([K]).

For $\mathcal{X}$ infinite-dimensional Banach space, the situation is quite different. For example, the condition (0.1) does not imply the stability of the Cauchy problem unless A is bounded ([Pa], p. 117) where, of course, (0.1) is replaced by

$$\sup\{Re\lambda : \lambda \in \sigma(A)\} < 0 \tag{0.3}$$

$\sigma(A)$ the spectrum of A.

In case A generates an analytic semigroup then (0.3) implies the stability ([Pa], p. 118). Another condition for stability is the finiteness of $\int_0^\infty \|T_t x\|^p\, dt$ $(p > 1)$ ([Pa], p. 116). It is clear that for $p = 2$ and $\mathcal{X} = H$, a Hilbert space, one can define a non-negative definite

[1]Supported in part by ONR N00014-91-J-1087

A. I. Cardoso et al. (eds.), Stochastic Analysis and Applications in Physics, 219–237.

form $(Rx, x)_H = \int_0^\infty \|T_t x\|_H^2\, dt$. In [D], Datko showed that (0.2 (a)) is equivalent to stability and in case of stability (Rx, x) is as above. Unfortunately, $(Rx, x)_H \not\geq c\,\|x\|_H^2$ and hence the use of Lyapounov function (Rx, x) to study the stability of non-linear system remained elusive. Under various conditions [Z], [Cu], [I$_1$] and [H] extended and used Datko type result to study the exponential mean square stability of stochastic linear system in the case $\mathcal{X} = H$. It should be noted that in this case $(Rx, x) = E\int_0^\infty \|X_t^x\|^2\, dt$ where $\{X_t^x,\ t \geq 0\}$ is the solution of the linear equation

$$(0.4) \qquad dX_t = AX_t\, dt + B(X_t)\, dW_t \qquad X_0 = x.$$

In addition, R satisfies the equation analogues to (0.2 (a)). In [K], $(R\,x, x)$ was used to study the stability of some non-linear stochastic differential equations in the finite-dimensional case. This technique requires $(R\,x, x) \geq c\,\|x\|^2$. In case A is bounded the finite-dimensional technique goes through easily ([D-K]). However, for A unbounded and coercive an example was constructed in ([K-M], Example 3) to show that $\int_0^\infty E\,\|X_t^x\|_H^2\ dt$ is not bounded below i.e. $\geq c\,\|x\|_H^2$ and, in fact, $\Lambda(x) = \int_0^\infty E\,\|X_t^x\|_V^2\ dt$ is the appropriate Lyapounov function which allows one to study the stability of solutions of non-linear problems. Now a word about organization.

In the next section, we present the theorem of Datko mentioned earlier and in the coercive case the analogue of the results in [K-M]. This is "new" material. In Section 3, results of [D-Z], [H] and [I$_2$] are derived by a combination of techniques of [D-Z] and [K-M]. In fact the necessity in Theorem 3.1 avoids difficulties encountered in the proof in [D-Z]. Sections 4 and 5 contain material from the work in [K-M] and Section 6 contains examples from [K-M], [H] and [Z]. Finally Section 2 presents the prerequisites on stochastic integrals from [M-P] and Ito Lemma ([P]).

1 Cauchy Problem and Stability of Linear Deterministic Evolution Equation

The purpose of this section is to motivate the semi-group techniques used in the stochastic situation. We use the ideas from [D-Z]. Linear partial differential equations can be formulated as evolution equation in a Banach space $\mathcal{X}$.

$$(1.1) \qquad \frac{d}{dt}u(t) = Au(t) \qquad t \geq 0, \text{ with } u(0) = x \in \mathcal{X}.$$

Here A is a linear (not necessarily) bounded operator on $\mathcal{X}$ with dense domain. The above problem is called Cauchy Problem. It is said to be well-posed if for $x \in \mathcal{D}(A)$, there exists exactly one (strongly) differentiable function $\{u(t, x),\ t \geq 0\}$ satisfying (1.1) for $t \geq 0$ and $u(t, x_n) \to 0$ if $\{x_n\} \subset \mathcal{D}(A)$ converges to zero. If this convergence is uniform on compact subsets of $[0, \infty)$, the Cauchy problem is uniformly well-posed.

If the problem is well-posed, it is easy to see that $S(t)x = u(t, x)$ for $x \in \mathcal{D}(A)$ is a linear operator bounded on $\mathcal{D}(A)$ and has an extension (denoted by $S(t)$ again) to E. It is easy to see that $S(0) = I$, (by uniqueness) $S(t+s) = S(t)S(s)$ and $S(\cdot)x$ is continuous on $[0,\infty)$ i.e. $\{S(t),\ t \geq 0\}$ is a C_o-semigroup of bounded linear operators.

Conversely, given a C_o-semigroup with infinitesimal generator A ([Y]), we get that $u(t) = S(t)x$ is the solution of the well-posed Cauchy problem. Hence in order to study the stability of the zero solution of the Cauchy problem one needs to study the structure of the semigroup $\{S(t),\ t \geq 0\}$. In case $E = C^d$, the zero solution of Cauchy problem is exponentially stable if the operator norm $\|S(t)\| \leq c_1 e^{-\gamma t}\,(\gamma > 0)$. A necessary and sufficient condition on A in order to generate exponentially stable solution is, of course, that A has all its characteristic roots in the half-plane $Re\, z < 0$. This fact, although true for bounded operators A (see [D-K]), is not valid in general for C_0-semigroups ([Pa], p. 117) in an infinite-dimensional case. For so-called analytic semigroup the sufficient condition for stability holds [SL]. In the finite-dimensional case, Lyapounov showed ([B]) that this is equivalent to the existence of a unique positive definite hermitian matrix B satisfying the equation $A^*B + BA = -I$ (the identity matrix). This allowed him to construct the so-called Lyapounov function (Bx, x), which has played a fundamental role in the study of stability of the solutions of differential equations. The following analogue of the theorem of Lyapounov was proved by Datko ([D]) for $E = H$ a Hilbert space. The proof is from ([D], [Pa]).

Theorem 1.1 *A necessary and sufficient condition that a C_o-semigroup on a real Hilbert space H satisfies the condition $\|S(t)\| \leq c_o e^{-\gamma t}\,(r > 0)$ where $1 \leq c_o < \infty$ is the existence of a Hermitian endomorphism C on H with $C \geq 0$ such that for all $x \in \mathcal{D}(A)$, the relation $2(CAx, x)_H = -\|x\|^2$ holds.*

Proof. The necessity follows by showing $Cx = \int_0^\infty S^*(t)S(t)\,x\,dt$. For the sufficiency, one observes that the existence of C implies $\int_0^\infty \|S(t)x\|^2\,dt < \infty$. From this and the Hille-Yosida theorem, one has $\|S(t)\| \leq Me^{\alpha t}$ ($M \geq 1$ and α real) and one gets $\lim\limits_{t\to\infty} \|S(t)x\| \to 0$ for $x \in H$. Hence $\|S(t)\| \leq M$ $(M \geq 1)$. The map $x \to S(t)x$ is defined on H into $L^2([0,\infty); H)$ is linear and closed. Hence by closed graph theorem

$$\left(\int_0^\infty \|S(t)x\|^2\,dt\right)^{1/2} \leq M_1\,\|x\|.$$

Let $0 < \rho < M^{-1}$ and define

$$t_x(\rho) = \sup\{t : \|S(t)x\| \geq \rho\,\|x\| \text{ for all } s \in [0,t]\}.$$

As $\|S(t)x\| \to 0$ we get $t_x(\rho)$ is finite for all $x \neq 0$ and

$$\rho^2 t_x(\rho)\,\|x\|^2 \leq \int_0^{t_x(\rho)} \|S(t)x\|^2\,dt \leq M_1^2\,\|x\|^2$$

So $t_x(\rho) \leq (M_1/\rho)^2$. For $t > (M_1/\rho)^2$

$$\|S(t)x\| \leq \|S(t - t_x(\rho))\|\ \|S(t_x(\rho))x\| \leq M\rho\,\|x\|.$$

But $(M\rho) < 1$. Hence for $t > (M_1/\rho)^2$ we get $\|S(t)\| \leq Me^{-\gamma t}$ by the semigroup property.

Although $(C\,x, x)_H$ is a positive definite quadratic form, it does not play the role of Lyapounov function as pointed out in [Pa$_1$]. In fact, one needs $\|S(t_0)x\| \geq c_1\,\|x\|$

for $x \in H$ to get $(Cx, x) \geq c_2 \|x\|^2$ and without the lower bound it is not possible to carry through Lyapounov's method to handle non-linearities except in "hyperbolic" case ([Pa$_1$]). However, for an important subclass of analytic C_0-semigroup we can construct an appropriate Lyapounov function.

We say that a linear operator $A : \mathcal{D}(A) \subseteq H$ to H is coercive [T] if
(i) there exists a Hilbert space V densely embedded in H and a continuous bilinear form $a : V \times V \to \mathbf{R}^1$ such that

$$-2a(\nu,\nu) \geq \alpha \|\nu\|_V^2 - \lambda_0 \|\nu\|_H^2 \quad \forall \ \nu \in V \tag{1.2}$$

for $\alpha > 0$ and $\lambda \geq 0$.
(ii) $\mathcal{D}(A) = \{u \in V : a(u, \cdot)$ is continuous in the topology of $H\}$.
(iii) $a(u,v) = \langle Au, \nu\rangle \ \forall \ u \in \mathcal{D}(A)$ and $\nu \in V$.
It is known ([T]) that the coercive operator A generates an analytic C_0-semigroup $S(t)$ such that $\|S(t)\| \leq e^{\lambda_0 t}$. If a is symmetric then A is self-adjoint. One also gets the following identity for a coercive operator A

$$\int_0^t a(S(s)x, S(s)x)ds = \frac{1}{2}\left[\|S(t)x\|_H^2 - \|x\|_H^2\right]. \tag{1.3}$$

We now have the following theorem.

Theorem 1.2 *The solution of the well-posed Cauchy problem (1.1) with A coercive and symmetric is exponentially stable if there exists a continuous bilinear form Λ on H satisfying $c_2 \|x\|_H^2 \leq \Lambda(x) \leq \tilde{c}_2 \|x\|_H^2$ and $(\Lambda'(x), Ax)_H \leq -c_3\Lambda(x)$ where $c_2, \tilde{c}_2, c_3$ are positive finite constants. In particular, $2(\Lambda'(x), Ax)_H = -\|x\|_V^2$ then we get the above condition.*

Proof. $\Lambda(S(t)x) - \Lambda(S(t')\varphi) = \int_{t'}^t \frac{d}{ds}\Lambda(S(s)x)ds$. But

$$\begin{aligned} \frac{d}{ds}\Lambda(S(s)x) &= (\Lambda'(S(s)x),\ AS(s)x)_H \\ &\leq -c_3\Lambda(S(s)x). \end{aligned}$$

With $\Phi(t) = \Lambda(S(t)x)$ and the fact that $\Lambda(S(t)x)$ is continuous in t we get

$$\Phi'(t) \leq -c_3\Phi(t).$$

Hence $(\Phi(t)e^{c_3t})' \leq 0$ giving

$$\Phi(t)e^{c_3t} \leq \Phi(0).$$

Hence we get

$$\begin{aligned} c_2 \|S(t)x\|_H^2 \leq \Lambda(S(t)x) &\leq \Lambda(x)e^{-c_3t} \\ &\leq \tilde{c}_2 \|x\|_H^2 \, e^{-c_3t}. \end{aligned}$$

Our purpose now is to construct the bilinear form Λ under the condition $\|S(t)x\|_H^2 \leq c_1e^{-\gamma t}\|x\|^2$. By (1.3) and (1.2)

$$\begin{aligned}\|S(t)x\|_H^2 &= \|x\|_H^2 + \int_0^t a(S(s)x, S(s)x)ds \\ &\le \|x\|_H^2 + |\lambda| \int_0^t \|S(s)x\|^2 ds - \alpha \int_0^t \|S(s)x\|_V^2 ds.\end{aligned}$$

Hence

$$\|S(t)x\|^2 + \alpha \int_0^t \|S(s)x\|_V^2 \ ds \le \|x\|_H^2 + |\lambda| \int_0^t \|S(s)x\|_H^2 \ ds.$$

Use exponential stability condition and let $t \to \infty$ to get

(1.4)
$$\int_0^\infty \|S(s)x\|_V^2 ds \le \frac{1}{\alpha}(1 + |\lambda| c_1/\gamma) \|\varphi\|_H^2 .$$

Define

$$\Lambda(x) = \int_0^\infty \|S(s)x\|_V^2 ds \text{ and } \tilde{c}_2 = \frac{1}{\alpha}\left(1 + \frac{|x| c_1}{\gamma}\right).$$

From (1.4), we get that $\forall\ x \in H,\ S(\cdot)x \in L^2([0,\infty); V)$ and by Schwarz inequality we get that $\int_0^\infty (S(t)x,\ S(t)y)$ is a continuous bilinear form on H and hence Λ' exists and is continuous on H. Now by (1.3)

$$\|S(t)x\|_H^2 - \|x\|_H^2 = 2 \int_0^t a(S(s)x, S(s)x)ds.$$

As $S(s)x \in L^2([0,\infty) : V)$ and a is a continuous bilinear form on $V \times V$ we get

$$|a(S(s)x, S(s)x)| \le c_2' \|S(s)x\|_V^2$$

Hence we have

$$\|S(t)x\|_H^2 - \|x\|_H^2 \ge -c_2' \int_0^t \|S(s)x\|_V^2 ds$$

Letting $t \to \infty$ we get $\Lambda(x) \ge c_2 \|x\|_H^2$ i.e.

(1.5)
$$c_2 \|x\|_H^2 \le \Lambda(x) \le \tilde{c}_2 \|x\|_H^2 .$$

We note that in general $\|S(t)x\|_V$ may not be continuous.

Theorem 1.3 *Consider the Cauchy problem (1.1) with A, a coercive operator. Assume that the zero solution of the equation is exponentially stable. Then*

$$\Lambda(x) = \int_0^\infty \|S(t)x\|_V^2 dt$$

satisfies

$$c_2 \|x\|_H^2 \le \Lambda(x) \le \tilde{c}_2 \|x\|_H^2 \, and (\Lambda'(x), Ax)_H \le -c_3 \Lambda(x)$$

and if, in addition, $t \to \|S(t)x\|_V^2$ is continuous at zero for $x \in V$ then $(\Lambda'(x), Ax)_H = -\|x\|_V^2$.

Proof. Note that $S(t+s)x = S(t)\,S(s)x$. We shall prove the last property

$$\begin{aligned} \Lambda(S(t)x) &= \int_0^\infty \|S(s)\,S(t)x\|_V^2\,ds \\ &= \int_0^\infty \|S(s+t)x\|_V^2\,ds \\ &= \int_t^\infty \|S(s)x\|_V^2\,ds \\ &= \int_0^\infty \|S(s)x\|_V^2\,ds - \int_0^t \|S(s)x\|_V^2\,ds. \end{aligned}$$

Now under the second assumption, we get

$$\frac{d}{dt}\Lambda(S(t)x) = 2(S(t)x,\, AS(t)x) = -\,\|S(t)x\|_V^2\,.$$

Put $t = 0$, to get $(\Lambda'(x),\, Ax) = -\,\|x\|_V^2$. Otherwise, observe that for k constant

$$\frac{d}{dt}\Lambda(S(t)x) \leq -k\,\|S(t)x\|_H^2$$

by the continuity of the injection $V \to H$.

Using $\Lambda(x) \geq c_2\,\|x\|^2$ we get

$$\frac{d}{dt}\Lambda(S(t)x) \leq -c_3\,\|S(t)x\|_H^2\,.$$

Putting $t = 0$, we get the result.

Remark: The above proof uses ideas in [K-M]. Note that $\Lambda(x) = (Cx, x)_H$ and hence $2(Cx, Ax) = -\,\|x\|_V^2$ i.e. C is the solution of Lyapounov equation.

Note that if A is bounded then it is coercive with $V = H$ and hence the Lyapounov function is $\int_0^\infty \|S(t)x\|_H^2\,dt$ as in the finite-dimensional case. Thus one can extend the theory in [K] to this case giving analogues of results in [D-K] for the stochastic case. In section 3, we study the stability of the mild solution of the stochastic Cauchy problem using analogue of Theorem 1.1 and in section 4, we study for the coercive case, the stability of the strong solution using analogue of Theorems 1.2 and 1.3. For this we need some background in stochastic differential equation and Ito's formula. We develop them in the next section.

2 Brownian Motion and Ito Formula

Let K be a separable Hilbert space with inner product $(\ ,\ )$ and $(\Omega, \mathcal{F}, P)$ be a probability space. Let $\{W(t),\, t \geq 0\}$ be a Gaussian stochastic process on $(\Omega, \mathcal{F}, P)$, with values in K with $E\,W(t) = 0$ for all $t \geq 0$ and covariance $E(W(t), x)(W(s), y) = (t \wedge s)(Qx, y)$ for $t, s \in [0, \infty)$ and Q a non-negative trace class operator ([D-S]) on K into K. It is easy to check that $W(0) = 0$, $(W(t) - W(s))$ is independent of $\mathcal{F}_s^W = \sigma\{(W(t), x),\ t \leq s,\ x \in K\}$. Using Kolmogorov's theorem ([Lo]) one can show that the process $\{X_t, t \geq 0\}$ can be realized in the space of continuous functions. We shall henceforth assume that W has continuous sample paths. Let us denote by $L_Q(K, H)$ the set of linear operators $C : K \to H$ (possibly unbounded) such that $\mathcal{D}(C)$ (the domain of C)$\supseteq Q^{1/2}K$ and $\mathrm{tr}(CQ^{1/2})(CQ^{1/2})^*$

is finite. Such a space is complete under $\|C\|_Q$ given by the above trace and contains $L(K,H)$, the bounded linear operators on K into H. Let $0 < T < \infty$,

$$\Phi : [0,T] \times \Omega \to L_Q(K,H)$$

such that

$$\text{(2.1)} \qquad \begin{array}{l} \text{(i)} \quad \Phi(t,\cdot)Q^{1/2} \text{ is } \mathcal{F}_t^W\text{-measurable for each } t: \\ \text{and} \\ \text{(ii)} \quad \int_0^T E\,\|\Phi(t,\cdot)\|_Q^2\, dt \text{ is finite.} \end{array}$$

The above space was first introduced by Metivier (see [M-P]). Essentially, following the procedure as in one-dimensional case, one can introduce the stochastic integral

$$M_t = \int_0^t \Phi(s,w)dW(s) \text{ for each } t \le T,$$

for a function Φ satisfying (2.1) (See [M-P]). It is easy to check that $\{M_t, \mathcal{F}_t^W,\ t \ge 0\}$ is an H-valued a.s. continuous martingale with associated operator-valued "increasing" process given by $\ll M \gg_t = \int_0^t \Phi(s)Q\Phi^*(s)ds$ ([M-P], p. 172). Note that $\ll M \gg_t$ is an $L(H,H)$-valued process of finite trace and

$$(M_t,h)_H\,(M_t,k)_H - (\ll M \gg_t h,k)_H$$

is an $\mathcal{F}_t^W$-martingale.

In order to study the coercive problem we need some more concepts.

Let $V \subseteq H \subseteq V'$ be real separable Hilbert spaces with $\|\ \|_V$, $\|\ \|_H$ and $\|\ \|_{V'}$ respectively. We assume that injections $V \hookrightarrow H$ and $H \hookrightarrow V'$ are continuous. If, in addition, $\langle \nu,\nu^*\rangle = (\nu,\nu^*)_H$ for $\nu^* \in H$ and $\nu \in V$, we call (V,H,V') a Gelfand triplet. Here $\langle\ ,\ \rangle$ denotes the duality on $V \times V'$. We denote by $M^2([0,t],V)$ the class of measurable functions $u : [0,T] \times \Omega \to V$ satisfying

$$\text{(2.2)} \qquad \left\{ \begin{array}{ll} \text{(i)} & u(t,\cdot) \text{ is } \mathcal{F}_t^W\text{-measurable; and,} \\ \text{(ii)} & E\int_0^T \|u(t,w)\|_V^2\, dt \text{ is finite.} \end{array} \right.$$

$M^2([0,T],V')$ is defined similarly. The following Ito Formula is given in ([P], p. 147 Theorem 3.2).

Proposition 2.1 *Let $\varphi \in L^2(\Omega,\mathcal{F}_0^W,P;H)$, $u \in M^2([0,T],V)$, $\nu \in M^2([0,T],V')$ and M_t be a continuous square integrable H-valued martingale such that*

$$\text{(2.3)} \qquad u(t) = \varphi + \int_0^t \nu(s)ds + M_t \quad t \in [0,T].$$

If $u(t) \in H$ for all t, and ψ is a function on H, twice differentiable and satisfying:

(i) ψ, ψ', ψ'' are locally bounded on H,
(ii) ψ and ψ' are continous on H,
(iii) for all $Q \in L(H,H)$ of trace class, $\mathrm{tr}[Q_0\psi'']$ is a continuous functional on H.
(iv) If $u \in V$, $\psi'(u) \in V$; $u \to \psi'(u)$ is continuous (with strong topology) into V endowed with weak topology.
(v) $\exists k$ s.t. $\|\psi'(u)\|_V \le k(1 + \|u\|_V)\ \ \forall\, u \in V$

Then

$$\psi(u(t)) = \psi(u_0) + \int_0^t (\nu, \psi'(u(s)))ds + \int_0^t (\psi'(u(s)), dM_s)_H + \tfrac{1}{2}\mathrm{tr} \int_0^t \psi''(u(s))d \ll M \gg_s .$$

We consider now the following equation for $u \in M^2([0,T], V)$, $\varphi \in H$,

$$u(t) = \varphi + \int_0^t A(u(s))ds + \int_0^t B(u(s))dW(s) \tag{2.4}$$

where $A : V \to V'$ is an operator with $\|A(u)\|_{V'} \leq \beta \|u\|_V$, $B(u) \in L(K,H)$ for $u \in V$, and W is a B.M. with Q trace class. Then $\nu(s) = A(u(s))$ satisfies $\|\nu(s)\|_{V'} \leq \beta \|u(s)\|_V$ and $u \in M^2([0,T], V)$ gives $\nu \in M^2([0,T], V')$. Also $B(u(t)) \in L(K,H)$ implies $\int_0^t B(u(s))dW(s)$ exists and is a martingale with $\ll M \gg_t = \int_0^t (Bu(s))Q(B(u(s)))^* ds$.

Now if ψ satisfies assumptions of Proposition 2.1 and $u(t) \in H$ then

$$\begin{aligned}\psi(u(t)) = \psi(\varphi) + \int_0^t \langle \psi'(u(s)),\, A(u(s))\rangle\, ds \\ + \frac{1}{2}\mathrm{tr} \int_0^t \psi''(u(s))B(u(s))QB^*(u(s))ds \\ + \int_0^t (\psi'(u(s)),\, B(u(s))dW_s)_H .\end{aligned}$$

Thus we have the following Ito Lemma.

Theorem 2.1 *Let ψ satisfy assumptions of Proposition 2.1 and u be solution of the equation (2.4) then*

$$\psi(u(t)) = \psi(\varphi) + \int_0^t \mathcal{L}\psi(u(s))ds + \int_0^t (\psi'(u(s)),\, B(u(s)dW(s))$$

where

$$\mathcal{L}\psi(u) = \langle \psi'(u),\, A(u)\rangle + \frac{1}{2} tr(\psi''(u)B(u)QB^*(u)). \tag{2.5}$$

3 Stability of the mild solution of stochastic linear evolution equation

In this section we consider the stochastic differential equation

$$dX = AX_t + B(X_t)dW, \quad X(0) = x, \tag{3.1}$$

where A is a linear (non necessarily) bounded operator which is an infinitesimal generator of the C_0-semigroup $\{S(t),\, t \geq 0\}$, $B \in L(H, L(K,H))$ and $\{W(t),\, t \geq 0\}$ is a Brownian motion with covariance $(t \wedge s)\,(Qx, y)$, $x, y \in K$.

A mild solution of the equation (3.1) is a solution of the integral equation

$$X_t = S(t)x + \int_0^t S(t-s)B(X_s)dW(s) \tag{3.2}$$

We note that following arguments of ([D-Z]) we get that (3.2) has $\mathcal{F}_t^W$-adapted solution with values in $C([0,T]; L_2(\Omega, H))$. Hence one gets a unique solution of (3.2) in $[0,\infty)$ satisfying for $t < \infty$,

$$E\,\|X_t^x\|_H^2 < \infty.$$

We now want to present some results on exponential mean square stability of the mild solution X_t^x of (3.1). We say that a solution is exponentially mean square stable (m.s.s.) if

$$E\,\|X_t^x\|_H^2 \le c\, e^{-\gamma t}\, \|x\|_H^2\, (\gamma > 0) \text{ for all } t \ge 0.$$

We now present the results of ([H], [I$_1$], [Z]) for the above system. (See [D-Z]) Analogous to Theorem 1.1, let us assume that there exists a solution of Lyapounov type equation i.e. there exists a non-negative definite linear operator $R \in L(H,H)$ (i.e. $R \in L^+(H,H)$) so that

$$\mathcal{L}(R\,x, y)_H = -(x,y)_H \quad \forall\ x, y \in H \tag{3.3}$$

where $\mathcal{L}$ is as in Theorem 2.1 with A, B as in (3.1).

Theorem 3.1 *A solution $\{X_t^x,\ t \ge 0\}$ of (3.2) is exponentially m.s.s. iff there exists a solution $R \in L^+(H,H)$ satisfying (3.3).*

Proof. Sufficiency Assume that the solution of (3.3) exists then

$$2(R\,x, A\,y) = -((I + \Delta(R))x, y)$$

with $(\Delta(R)x, y) = \text{tr}[R\,B(x)\,Q B^*(y)]$. So $2(R\,x, A\,x)_H = -((I + \Delta(R))x, x)_H$ and $(I + \Delta(R))$ is invertible and onto gives by Theorem 1.1, the semigroup $\{S(t)\}$ satisfies

$$\|S(t)\| \le M\, e^{-\lambda t}\ (\lambda > 0)\ \forall\ t \ge 0. \tag{3.4}$$

Let $A_n = n\,A\,R(n,A) = n\,A \int_0^\infty e^{-nu} S(u) du$, so A_n is a bounded operator. Let $\{X_t^x(n)\}$ be the solution of the equation

$$dX_t^x(n) = A_n X_t^x(n) + B(X_t^x(n)) dW(t) \tag{3.5}$$

with $X_0^x(n) = x$. This has a strong solution. Applying Ito formula to

$$(RX_t^x(n),\ X_t^x(n))_H$$

we get

$$\begin{aligned} E(RX_t^x(n),\ X_t^x(n))_H &= (R\,x, x) + \int_0^t 2(RX_s^x(n),\ A_n X_s^x(n))_H ds \\ &+ \int_0^t (\Delta(R) X_s^x(n),\ X_s^x(n))\, ds. \end{aligned}$$

Since R is solution of (3.3), with $J_n = n\,R(n,A)$ we have,

$$\begin{aligned} &2(R\,X_s^x(n),\ A_n X_s^x(n))_H \\ &= -(\Delta(R) X_s^x(n),\ J_n X_s^x(n))_H - (X_s^x(n),\ J_n X_s^x(n))_H. \end{aligned}$$

So

$$\begin{aligned} E(R\,X_t^x(n),\ X_t^x(n))_H = (R\,x, x)_H - \int_0^t E(X_s^x(n),\ J_n X_s^x(n))_H \\ + \int_0^t E\{(\Delta(R) X_s^x(n),\ X_s^x(n)_H - J_n X_s^x(n))_H\} ds. \end{aligned}$$

Let $n \to \infty$, $e^{tA_n}x \to S(t)x$ for all x and the solution satisfies for $T < \infty$, $\sup_{t \le T} E\|X_t^x\|_H^2 < \infty$ and is a mild solution. Hence we get

$$E(RX_t^x, X_t^x)_H = (Rx, x)_H - E\int_0^t \|X_s^x\|_H^2\, ds. \tag{3.6}$$

Let $\psi(t) = E(RX_t^x, X_t^x)_H \le -\frac{1}{\|R\|} E\|X_t^x\|_H^2$, $\psi'(t) = -E\|X_t^x\|_H^2$ and $\psi(0) = (Rx, x)_H$. We get

$$\psi(t) \le (Rx, x)_H e^{-\frac{1}{\|R\|}t}$$

$$\begin{aligned} E\|X_t^x\|_H^2 &\le 2\|S(t)x\|_H^2 + 2E\left\|\int_0^t S(t-s)\,B(X_s^x)dW(s)\right\|_H^2 \\ &\le 2M^2e^{-2\lambda t} + 2\mathrm{tr}(W)\int_0^t M^2\|B\|^2 e^{-2\lambda(t-s)}E\|X_s^x\|_H^2\, ds \\ &\le 2M^2e^{-2\lambda t} + 2\mathrm{tr}(W)M^2\|B\|^2\int_0^t e^{-2\lambda(t-s)}\psi'(s)ds \end{aligned}$$

using integration by parts, $(Rx, x) \le \|R\|\,\|x\|^2$ and

$$\psi(t) \le (Rx, x)_H\, e^{-\frac{1}{\|R\|}t} \quad \forall\ t \ge 0$$

we get the sufficiency. For the necessity we give only the sketch. Under the condition of stability and Schwarz inequality, $\int_0^\infty E(X_t^x, X_t^y)_H\, dt$ is a bounded bilinear form on $H \times H$ and hence gives that there is a bounded operator $R : H \to H$ so that

$$(Rx, y)_H = \int_0^\infty E(X_t^x, X_t^y)_H\, dt.$$

Clearly R is non-negative. To show that R satisfies Lyapounov equation, observe that $\{X_t^x(n), t \ge 0\}$ in (3.5) is unique Markov process solution since A_n is bounded. Let $\Lambda(x) = (Rx, x)_H$

$$\begin{aligned} \Lambda(X_t^x(n)) &= \int_0^\infty E\|X_s^x(n)\|_H^2\, dt \\ &= E\int_0^\infty (E(\left\|X_s^{X_t^x(n)}(n)\right\|_H^2 \mid X_t^x(n))ds. \end{aligned}$$

Let $\mathcal{F}_t^x(n) = \sigma\{X_s^x(n),\, s \le t\}$ and use Markov property

$$E(\left\|X_s^{X_t^x(n)}(n)\right\|_H^2 \mid X_t^x(n)) = E(\left\|X_s^{X_t^x(n)}(n)\right\|_H^2 \mid \mathcal{F}_t^X(n)).$$

But by uniqueness of the solution of (3.5),

$$E(\left\|X_s^{X_t^x(n)}(n)\right\|_H^2 \mid \mathcal{F}_t^x(n)) = E(\|X_{t+s}^x(n)\|_H^2 \mid \mathcal{F}_t^x(n)).$$

Hence

$$E(\Lambda(X_t^x(n)) = \int_0^\infty E\|X_{s+t}^x(n)\|_H^2\, ds = \int_t^\infty E\|X_s^x(n)\|_H^2\, ds.$$

By Ito Formula

$$E(\Lambda(X_t^x(n))) = \Lambda(x) + \int_0^t E(\mathcal{L}_n\Lambda(X_s^x(n))\, ds$$

where $\mathcal{L}_n\Lambda(x) = (Rx, A_nx)_H + \frac{1}{2}(\Delta(R)x, x)$. Hence taking derivative on both sides we get $(Rx, x) = \Lambda(x) = (Rx, A_nx) + \frac{1}{2}(\Delta(R)x, x)$. Letting $n \to \infty$ we get for $x \in \mathcal{D}(A)$ the result.

4 Mean square stability of linear systems

We study now the analogue of the stability problem in Section 2 for the case of stochastic linear system. Let us assume that $V \subseteq H \subseteq V'$ is a Gelfand triplet of Hilbert spaces and $\{W(t),\, t \geq 0\}$ is a Brownian motion taking values in a separable Hilbert space K. We consider the linear system given by

$$dX_t = A\,X_s\,ds + B(X_s)\,dW(s) \quad \text{with} \quad X_0 = x \tag{4.1}$$

where $A \in L(V, V')$, $B \in L(V, L(K, H))$ satisfying coercivity condition: There exists $\alpha > 0$ and λ real

$$2\,\langle \nu, A\nu\rangle + \operatorname{tr}[(B\nu)Q(B\nu)^*] \leq \lambda\,\|\nu\|_H^2 - \alpha\,\|\nu\|_V^2 \tag{4.2}$$

for $\nu \in V$.

It is known that (4.1) has a unique strong solution $X_t^x \in L^2(\Omega;\, C([0,T], H)) \cap M^2([0,T], V)$ by [P] and [R].

Theorem 4.1 *Consider the equation (4.1) satisfying (4.2) and $\{X_t^x,\ t \geq 0\}$ be its unique solution. Then it is exponentially m.s.s. if there exists a function $\Lambda : H \to \mathbf{R}_+$ satisfying conditions of Ito Lemma, with*

$$c_2\,\|x\|_H^2 \leq \Lambda(x) \leq \tilde{c}_2\,\|x\|_H^2 \text{ and } \mathcal{L}\Lambda(x) \leq -c_3\Lambda(x) \tag{$*$}$$

for $c_2, \tilde{c}_2, c_3$ are positive finite real numbers.

<u>Proof</u>. Apply Ito Lemma to $\Lambda(X_t^x)$ and take expectation. Then

$$\begin{aligned} E(\Lambda(X_t^x) - \Lambda(X_{t'}^x) &= E\int_{t'}^t \mathcal{L}\Lambda(X_s^x)ds \\ &\leq -c_3\int_{t'}^t E\Lambda(X_s^x)ds. \end{aligned}$$

Setting $\Phi(t) = E\Lambda(X_t^x)$ and the fact that $\Lambda(X_t^x)$ is continuous in t, we get from the above inequality

$$\Phi(t) \leq e^{-c_3 t}\Phi(0). \tag{4.3}$$

Using $c_2\,\|x\|_H^2 \leq \Lambda(x) \leq \tilde{c}_2\,\|x\|_H^2$ and (4.3),

$$c_2\,\|x\|_H^2 \leq E\Lambda(X_t^x) \leq \Lambda(x)\,e^{-c_3 t} \leq \tilde{c}_2\,\|x\|^2\ e^{-c_3 t}.$$

The above result as stated in the introduction, was proved in [K-M]. However, (4.3) was proved in [I$_2$] earlier for mild solution. We note that to obtain (4.3), we only used the property $\mathcal{L}\Lambda(x) \leq -c_3\Lambda(x)$. Using $\{X_t^x(n), t \geq 0\}$ introduced in Section 3 and the above argument (i.e. unique strong solution and Ito Lemma) we observe by (4.3) that

$$e^{c_3 t}E\Lambda(X_t^x(n)) \leq \Lambda(x).$$

As $\{X_t^x(n),\ t \geq 0\}$ converges to $\{X_t^x,\ t \geq 0\}$ the mild solution in $L_2(\Omega, H)$ we get by dominated convergence theorem the result of [I$_1$] (Theorem 3.1) in the linear case. It should be noted that however that this does not give exponential m.s.s.

Following [K] we call Λ satisfying conditions of Theorem 4.1 the Lyapounov function.

The major difficulty is to construct such a function. As we show in our example (Example 3 [K-M]) that $(R\,x, x) = \int_0^\infty E\,\|X_t^x\|_H^2\ dt$ is not a Lyapounov function. As far as I know the first successful attempt to construct such a function was in [K-M]. In fact the results of Section 2 are motivated from this work. We now give the construction of Lyapounov function. For this we need the following energy equality for the solution of (4.1) under (4.2).

$$
\begin{aligned}
E\,\|X_t^x\|_H^2 = \|x\|_H^2 \ &+\ 2E\int_0^t \langle A\,X_s^x,\ X_s^x\rangle\, ds \\
&+\ E\int_0^t \mathrm{tr}[(B\,X_s^x)Q(B\,X_s^x)^*]ds
\end{aligned}
\tag{4.4}
$$

which can be derived by applying Ito Lemma to $\psi(x) = \|x\|^2$.

From (4.4) and coercivity condition (4.2), we have

$$
E\,\|X_t^x\|_H^2 \leq \|x\|_H^2 + |\lambda| \int_0^t E\,\|X_s^x\|_H^2\ ds - \alpha\int_0^t E\,\|X_s^x\|_V^2\ ds.
$$

Hence by exponential m.s.s.,

$$
E\,\|X_t^x\|_H^2 + \alpha\int_0^t E\,\|X_s^x\|_V^2\ ds \leq \|x\|_H^2\,(1 + |\lambda|\,c_1\int_0^t\ e^{-\gamma s}\,ds).
$$

Which, letting $t \to \infty$, gives

$$
\int_0^\infty E\,\|X_s^x\|_V^2\ ds \leq \frac{1}{\alpha}(1 + |\lambda|\,c_1/\gamma)\,\|x\|_H^2\,.
$$

We now show that

$$
\Lambda(x) = \int_0^\infty E\,\|X_s^x\|_V^2\ ds
$$

satisfies conditions of Ito Lemma.

Using linearity in x of X_s^x and Schwarz inequality it is easy to check that

$$
T(x,y) = \int_0^\infty E(X_s^x,\ X_s^y)_V\,ds \qquad x, y \in H
$$

is a continuous bilinear form on $H \times H$ and hence equals $(\tilde{R}\,x, y)_H$ for a bounded linear operator $\tilde{R}$ and $\Lambda(x) = (\tilde{R}\,x, x)$. Hence properties (i), (ii), (iii) in Ito Lemma are satisfied by Λ. To prove property (iv) one has for $x,\ y \in V$, using continuity of embedding $V \hookrightarrow H$ that $T(x,y)$ is a continuous bilinear form on $V \times V$. As before this gives $\Lambda'(x) \in V$ and Λ' is continuous on V and $\Lambda'(x) \leq 2\,\|T\|\,(1 + \|x\|_V)$. It remains to prove (*) of Theorem 4.1.

$$
\begin{aligned}
\mathcal{L}\,\|\nu\|_H^2 \ &=\ 2\,\langle \nu, A\nu\rangle + \mathrm{tr}((B\nu)Q(B\nu)^*) \\
\left|\mathcal{L}\,\|\nu\|_H^2\right| \ &\leq\ c_0\,\|\nu\|_V^2 + \|B\|_{L(K,H)}^2\,(\mathrm{tr}Q)\,\|\nu\|_V^2 \\
&\leq\ c_2'\,\|\nu\|_V^2
\end{aligned}
\tag{4.5}
$$

Apply Ito Formula to $\psi(x) = \|x\|_H^2$ and take expectation and use (4.6), then

$$E\|X_t^x\|_H^2 - \|x\|_H^2 = \int_0^t E\mathcal{L}\|X_s^x\|_H^2\, ds \geq -c_2'\int_0^t \|X_s^x\|_V^2\, ds.$$

By exponential m.s.s., letting $t \to \infty$, one has

(4.6) $$\Lambda(x) \geq c_2\|x\|_H^2.$$

It remains to prove the last condition in (*), to get

Theorem 4.2 *Consider the equation (4.1) satisfying (4.2) and let the solution $\{X_t^x,\ t \geq 0\}$ of it be exponentially m.s.s. Then $\Lambda(x) = \int_0^\infty E\|X_t^x\|_V^2\ dt$ satisfies (*) of Theorem 4.1. If, in addition, $t \to E\|X_t^x\|_V^2$ is continuous at zero for $x \in V$, then $\mathcal{L}\Lambda(x) = -\|x\|_V^2$.*

Remark: The last condition is the exact analogue of Lyapounov equation if H and hence V is finite dimensional. In this case of course A is bounded and hence coercive as all norms are equivalent.

Proof. Observe

$$E\Lambda(X_t^x) = E\int_0^\infty E(\left\|X_s^{X_t^x}\right\|_V^2 \mid X_t^x)ds.$$

But by Markov property of the solution of (4.1) this equals with $\mathcal{F}_t^x = \sigma\{X_u^x,\ 0 \leq u \leq t\}$

$$\int_0^\infty E[E(\left\|X_x^{X^x}\right\|_V^2 \mid \mathcal{F}_t^X)]ds.$$

The uniqueness of the solution implies

$$E(\left\|X_x^{X_t^x}\right\|_V^2 \mid \mathcal{F}_t^x) = E(\|X_{s+t}^x\|_V^2 \mid \mathcal{F}_t^x).$$

Hence by the continuity of the map $V \hookrightarrow H$,

$$E\Lambda(X_t^x) = \int_t^\infty E\|X_s^x\|_V^2\, ds \leq \int_0^\infty E\|X_s^x\|_V^2\, ds - k\int_0^t E\|X_s^x\|_H^2\, ds.$$

As $\{X_t^x, t \geq 0\}$ is a Markov process with infinitesimal generator $\mathcal{L}$ we have

$$\mathcal{L}\Lambda(x) \leq -k\|x\|_H^2 \leq -\frac{k}{\tilde{c}_2}\Lambda(x).$$

The other part follows from the above equation and the continuity of the map $t \to E\|X_t^x\|_V^2$ at zero.

5 Stability in Probability of Non-linear Equations

Let us consider the non-linear stochastic equation for $X \in M^2([0,T];V)$

$$X(t) = x + \int_0^t \tilde{A}(X(s))\,ds + \int_0^t \tilde{B}(X(s))\,dW(s), \tag{5.1}$$

where $x \in H$, $\tilde{A} : V \to V'$ is (not necessarily linear) operator with $\left\|\tilde{A}(x)\right\|_{V'} \le \beta \|x\|_V$, $\tilde{B} : V \to L(K,H)$ is an operator and W is Brownian motion as before. Here $V \subset H \subset V'$ is Gelfand triplet.

Let $\{\tilde{X}_t^x,\ t \ge 0\}$ be a solution of (5.1) for $x \in H$. If for each $\epsilon > 0$

$$\lim_{\|x\|_H \to 0} P\left\{\sup_t \left\|\tilde{X}_t^x\right\|_H > \epsilon\right\} = 0, \tag{5.2}$$

then we say that zero solution is stable in probability following [K]. We note that (4.1) is a special case of (5.1).

Theorem 5.1 *Let $\Lambda(x)$ be a functional on $\{\|x\| < \delta\}$ having the following properties*

(i) Λ *satisfies conditions of Ito Lemma.*

(ii) $\Lambda(x) \to 0$ *as* $\|x\|_H \to 0$.

(iii) $\inf_{\|x\|_H > \epsilon} \Lambda(x) = \lambda_\epsilon > 0$.

(iv) $\mathcal{L}\Lambda(x) \le 0$ *for* $x \in V$ *with* $\|x\|_H < \delta$.

where $\mathcal{L}\Lambda(x) = \left\langle \Lambda'(x), \tilde{A}(x)\right\rangle + \frac{1}{2}\, tr(\Lambda''(x)\,\tilde{B}(x)\,QB^*(x))$.

Then (5.2) holds.

Proof. We first obtain the inequality

$$P\left(\sup_{t>0} \|X_t^x\|_H > \epsilon\right) \le \frac{\Lambda(x)}{\lambda_\epsilon}, \qquad \text{for } x \in V. \tag{5.3}$$

Let $O_\epsilon = \{x \in H : \|x\|_H < \epsilon\}$ and

$$\tau_\epsilon = \inf\{t : \|X_t^x\|_H > \epsilon\}.$$

The conditions (i), (iv) and Ito Lemma imply for $x \in O_\epsilon$, $\Lambda(X_{t\wedge\tau_\epsilon}^x)$ is a supermartingale. Hence

$$\Lambda(x) \ge E\,\Lambda(X_{t\wedge\tau_\epsilon}^x) \ge \lambda_\epsilon\, P(\tau_\epsilon < t).$$

For $x \in V$, we get the above inequality. For $x_o \in H$, let $\{x_n\} \subset V$ such that $\|x_n - x_o\|_H \to 0$ as $n \to \infty$. By Theorem 2, p. 111 of [R]

$$\lim_n P\left\{\sup_t \|X_t^{x_n}\|_H > \epsilon\right\} = P\left\{\sup_t \|X_t^x\| > \epsilon\right\}. \tag{5.4}$$

From (5.4), (5.3) and continuity of Λ on H (in $\|\ \ \|_H$),

$$P\left(\sup_t \|X_t^{x_o}\| > \epsilon\right) \leq \frac{\Lambda(x_o)}{\lambda_\epsilon} \text{ for } x_o \in H.$$

Let $\|x_o\|_H \to 0$ and use condition (ii) to get the proof.

For the linear system (4.1) satisfying (4.2), the Lyapounov function $\Lambda(x)$ constructed in Section 4 satisfies conditions of theorem 5.1. Hence we get the following theorem.

Theorem 5.2 *Assume that A, B in (4.1) satisfy (4.2) and the solution $\{X_t^x,\ t \geq 0\}$ of (4.1) is exponentially m.s.s. Then it satisfies (5.2).*

Now we state theorems on stability through first order approximation.

Theorem 5.3 *Assume that A, B in (4.1) satisfy (4.2) and the solution $\{X_t^x,\ t \geq 0\}$ of (4.1) is exponentially m.s.s. and $t \to E\|X_t^x\|^2$ is continuous. Let $\{\tilde{X}_t^x,\ t \geq 0\}$ be a solution of (5.1) with $x \in V$. If*

$$\left\|\tilde{A}(\nu) - A\nu\right\|_{V'}^2 + |\ tr[\tilde{B}(\nu)\,Q\tilde{B}^*(\nu) - (B\nu)\,Q(B\nu)^*] \leq c_4\,\|\nu\|_V^2$$

for c_4 small enough in a sufficiently small neighbourhood of $\nu = 0$ in $\|\ \ \|_V$. Then $\{\tilde{X}_t^x,\ t \geq 0\}$ satisfies (5.2).

Proof. Use the Lyapounov function constructed in Theorem 4.2. Then using Theorem 4.2

$$\tilde{\mathcal{L}}\,\Lambda(x) \leq -\|x\|_V^2 + 4c_4\,\|T\|\ \|x\|_V^2$$

where $T(x,y) = E\int_0^\infty (X_s^x,\ X_s^y)\,ds,\ x,y \in V$, $\|T\|$ is the norm of this bilinear form and c_4 as in the assumption above. Use Theorem 5.1 to complete the proof.

Similarly we also get

Theorem 5.4 *Assume that A, B in (4.1) satisfy (4.2) and the solution $\{X_t^x,\ t \geq 0\}$ of (4.1 is exponentially m.s.s. If $\tilde{A}$, $\tilde{B}$ in (5.1) satisfy*

$$\left\|\tilde{A}(\nu) - A(\nu)\right\|_{V'}^2 + \ |\ tr[\tilde{B}(\nu)\,Q\,\tilde{B}(\nu)^* - (B\nu)\,Q(B\nu)^*] \leq c_4\,\|\nu\|_H^2$$

for c_4 small enough in sufficiently small neighbourhood of $\nu = 0$ in $\|\ \ \|_H$. Then a solution of $\{\tilde{X}_t^x,\ t \geq 0\}$ (5.1) satisfies (5.2).

6 Examples

1. Heat Equation. Let D be a bounded open domain with smooth boundary $\subseteq \mathbf{R}^n$. Let $H = L^2(D)$ and V be the sobolev space $H_0^1(D)$ with $u/\partial D = 0$ for $u \in H_0^1(D)$. Consider the equation

$$\frac{\partial u(t,x)}{\partial t} = Au(t,x) \qquad u(0,x) = \varphi(x).$$

Assume that A is strictly elliptic linear differential operator of second order on D. Then by Garding inequality coercivity condition is satisfied ([Pa], 7.2). Hence A is a generator of an analytic semigroup and if

$$\sup_{\substack{\|x\|=1 \\ x\in H_0^1(D)}} \frac{((A+A^*x,\, x)_H}{\|x\|_H^2} = \lambda_0 < 0$$

then the semigroup is stable. Because of coercivity we can construct a Lyapounov function $\Lambda(x) = \|x\|_H^2$. Let us consider stochastic case ([K-M]), let

$$\{W(t,x),\ t \geq 0,\ x \in D\}$$

be H-valued Wiener process with associated Q given by a positive definite kernel

$$q(x,y) \in L^2(D \times D),\ q(x,x) \in L^2(D).$$

Consider the stochastic equation

$$dX_t = AX_t\ dt + B(X_t)\ dW(t).$$

Where $B(\varphi)\ f(\cdot) = \varphi(\cdot)\, f(\cdot)$ for $f \in L^2(D)$.

In this case $\|x\|_H^2 = \Lambda(x)$ and we get

$$\begin{aligned}
\mathcal{L}\,\|x\|_H^2 &= 2(Ax,\, x)_H + \operatorname{tr}[(Bx)\,Q\,(Bx)^*] \\
&= 2(Ax\, x)_H + \int_D q(t,\, t)\, x^2(t)\, dt \\
&= ((A + A^* + q)x,\, x)_H
\end{aligned}$$

(A^* conjugate of A). Hence we get, if

(6.1) $$\lambda_0 = \max_{\substack{\|x\|=1 \\ x\in H_0^1(D)}} \frac{((A+A^*+q)x,\, x)_H}{\|x\|^2} < 0$$

then we get stability.

We can also obtain the stability in probability results from Section 5. Consider non-linear equation in D

(6.2) $$d\,u(t,x) = \tilde{A}(x, u(t,x)) + \tilde{B}(u(t,x))\, dW_Q(t,x)$$

with $u(t,x)\mid_{\partial D} = 0$, $\quad u(0,x) = \varphi(x)$.

Assume that

$$\tilde{A}(x,u) = A(x)u + \alpha_1(x,u) \qquad \tilde{B}(x,u) = B(u) + \alpha_2(x,u)$$

where $\alpha_i(x,u)$ are functions satisfying the conditions

$$\sup_{x\in D} |\alpha_i(x,u_2) - \alpha_i(x,u_1)| < c\,|u_2 - u_1|$$

$$\alpha_i(x,0) = 0$$

$$\sup_{x\in D} |\alpha_i(x,u)| = 0(|u|) \qquad u \to 0.$$

Suppose A is strictly elliptic satisfying (6.1), then we get that coercivity and monotonicity are satisfied for (6.2) ([K-R], [P]) and hence we get that the solution of (6.2) satisfies (5.2).

We note that this example includes earlier ones given in ([H], [C]).

2. Delay Systems: ([H], [Z]). Consider delay system of the form

$$dx_t = \int_{-r}^{0} dN(s)\, x_{t+s} + B(x_t)\, dW(t) \qquad x_t = 0 \quad t < 0$$

where $x_t \in \mathbf{R}^n$ and $N(\cdot)$ is a left continuous function of bounded variation on $[-r,0]$ into $n \times n$ matrices.

Let $H = \mathbf{R}^n \times L_2([-r,0],\mathbf{R}^n)$ ($r > 0$ and is called maximum delay). Let $\mathcal{D}(A) = H^2([-r,0],\mathbf{R}^n)$. Then $\mathcal{D}(A)$ can be embedded in H, $(f(0), f(\cdot)) \in H$ for $f \in \mathcal{D}(A)$. Clearly $\mathcal{D}(A)$ is dense in H

$$Af = \left(\int_{-r}^{0} dN(s)\, f(s), -\dot{f}(\cdot) \right).$$

Then it is known that A generates a C_0-semigroup on H. If $X_t = (x(t),\ x(t+\cdot)) \in H$ and $B(X_t)u = (B(x_t)u, 0)$ then

$$dX_t = AX_t\, dt + B(X_t)\, dW_t.$$

The semigroup S_t is stable if

$$\sup\left\{ Re\lambda : \det\left(\int_{-r}^{0} e^{\lambda s}\, dN(s) - \lambda I \right) = 0 \right\} < 0$$

and existence of R follows from ([I$_2$]) under the condition (for $K = \mathbf{R}^d$)

$$\sup_{\|\varphi\|=1} \int_0^\infty \sum_{j=1}^{d} |B_j u^\varphi(t)|^2\, dt < 1$$

where $u^\varphi(t)$ is solution at time t of

$$\dot{u}_t + \int_{-r}^{0} dN_s u_{s+t} = 0 \qquad t \geq 0$$

with $u(t,\varphi) = \varphi(t)$ for $t \leq 0$, $\quad \varphi \in \mathcal{D}(A)$.

References

[B] Bellman R., (1960), Introduction to matrix analysis, McGraw Hill, New York.

[C] Chow P.L., (1982), Stability of non-linear stochastic evolution equations, *J. Math. Anal. and Applns.*, **89**, pp. 400-409.

[Cu] Curtain R.F., (1981), Stability of stochastic partial differential equation, *J. Math. Anal. and Applns.*, **79**, pp. 352-369.

[D-K] Daletskii Ju.L., Krein M.G., (1974), Stability of solutions of differential equations in Banach space, *Transl. of Amer. Math. Soc.* **43**, Providence, R.I.

[D-Z] Daprato G. and Zabezyk J., (1992), Stochastic equations in infinite-dimensions, Cambridge University Press, New York.

[D] Datko, R., (1970), Extending a theorem of A.M. Lyapounov to Hilbert space, *J. Math. Anal. and Applns*, **39**, pp. 610-616.

[D-S] Dunford N. and Schwartz J.T., (1963), Linear operators II Interscience, New York.

[H] Hausman U.G., (1978), Asymptotic stability of the linear Ito equation in infinite dimensions, *J. Math. Anal. Applns.*, **65**, pp. 219-235.

[I_1] Ichikawa A., (1979), Dynamic programing approach to stochastic evolution equations, *SIAM J. Control and Optim.*, **17**, pp. 152-174.

[I_2] Ichikawa A., (1982), Stability of semilinear stochastic evolution equations, *J. Math. Anal. Applns.*, **90**, pp. 12-44.

[I_3] Ichikawa A., (1984), Semilinear stochastic evolution equations: Boundedness, stability and invariant measures, *Stochastics*, **12**, pp. 1-39.

[K] Khasminskii R.Z., (1980), Stochastic stability of differential equations, Sijthoff and Noordoff, Netherlands.

[K-M] Khasminskii R.Z. and Mandrekar V., On stability of solutions of stochastic evolution equations. (To appear).

[K-R] Krylov N.V. and Rozovskii B.L., (1981), Stochastic evolution equations, *J. Soviet Math.*, **16**, pp. 1233-1277.

[L] Loéve, M., (1959), Probability Theory. Van Nostrand, Princeton..

[M-P] Metivier M. and Pellaumail J., (1980), Stochastic Integration, Academic Press, New York.

[P] Pardoux E., (1979), Stochastic partial differential equations and filtering of diffusion processes, *Stochastics*, **3**, pp. 127-167.

[Pa] Pazy A., (1983), Semigroups of linear operators and applications to partial differential equations, Springer-Verlag, New York.

[Pa_1] Pazy A., (1972), On the applicability of Lyapounov's theorem in Hilbert space, *SIAM J. Math. and Anal.*, **3**, pp. 291-294.

[R] Rozovskii B.L., (1990), Stochastic evolution systems, Kuwer Academic Publishers, Boston.

[S] Slemrod, M., (1976), Asymptotic behaviour of C_0-semigroups as determined by the spectrum of the generator, Indiana University, *Math. J.*, 783-792.

[T] Tanabe H., (1979), Equations of evolution, Pitman.

[Y] Yosida K., (1965), Functional analysis, Springer-Verlag, New York.

[Z] Zabezyk J., (1979), On the stability of infinite-dimensional linear stochastic systems, Banach Center Publ., **5**, PWN-Polish Scientific Publishers.

Department of Statistics and Probability
Michigan State University
East Lansing, MI 48824

Stochastic Techniques in Condensed Matter Physics

R. Vilela Mendes

Centro de Física da Matéria Condensada

Av. Gama Pinto 2, 1699 Lisboa Codex, Portugal

1. Introduction

The purpose of these lectures is to describe a few applications of stochastic techniques to selected problems in condensed matter physics. These applications are described in the sections 3,4 and 6. However, to make the lectures reasonably self-contained, I have included in the sections 2 and 5 two short reviews of the sign problem in Quantum Monte Carlo and of the deviations from scaling in fully developed turbulence. These reviews should not in any way be considered as exhaustive or even well-balanced and fair to the many authors that have contributed to these fields. They are simply a kind of minimum knowledge required to understand the applications that are developed in the other sections.

Finally sections 7 and 8 are of a slightly different nature in the sense that, rather than applying standard stochastic techniques to condensed matter problems, I have attempted to show how the notions of quantum group and q-deformation, that first appeared in the context of exactly integrable systems in condensed

A. I. Cardoso et al. (eds.), Stochastic Analysis and Applications in Physics, 239–281.

matter physics, may now be carried over to the theory of stochastic processes. Section 7 is a naive introduction to the notions of q-deformations and q-algebras which, for the sake of brevity, is not based on the symmetries of exactly integrable models but, instead, on the calculus of finite differences.

2. Quantum Monte-Carlo and the sign problem

Numerical simulation is an important technique to explore many of the systems of physical importance in Condensed Matter, which cannot either be solved exactly or dealt with by approximation techniques. Among the numerical simulation methods, Quantum Monte-Carlo plays a predominant role. It is used both to compute expectation values of operators and to estimate the ground state wave function. For a recent extensive review see Ref.[1].

Expectation values. The worldline method

The aim is to compute

$$< A > = \frac{1}{Z} \mathrm{Tr}\, A\, e^{-\beta H} \tag{2.1}$$

where

$$Z = \mathrm{Tr}\, e^{-\beta H} = \sum_{i_1} < i_1 |\, e^{-\beta H} \,| i_1 > \qquad \text{and} \qquad \beta = \frac{1}{KT} \tag{2.2}$$

To compute the partition function Z we divide the interval $(0,\beta)$ into L subintervals of length $\Delta\tau = \frac{\beta}{L}$

$$Z = \sum_{i_1,\ldots,i_L} < i_1 |\, e^{-\Delta\tau H} \,| i_L > < i_L |\, e^{-\Delta\tau H} \,| i_{L-1} > \ldots < i_2 |\, e^{-\Delta\tau H} \,| i_1 > \tag{2.3}$$

Now split the Hamiltonian into diagonalizable pieces

$$H = H_1 + H_2 \tag{2.4}$$

and use Trotter's formula[2]

$$e^{-\Delta\tau H} = e^{-\Delta\tau H_2}\, e^{-\Delta\tau H_1} \left\{1 + O(\Delta\tau^2)\right\} \tag{2.5}$$

to obtain

$$Z \simeq \sum_{i_1,\ldots,i_{2L}} < i_1 |\, U_1 \,| i_{2L} > < i_{2L} |\, U_2 \,| i_{2L-1} > \ldots < i_2 |\, U_2 \,| i_1 > \tag{2.6}$$

where $U_i = e^{-\Delta\tau H_i}$. The sum (2.6) is now computed by importance sampling, that is, one generates a sequence of distributions $\{i_1(k) \ldots i_{2L}(k)\}$, $k = 1,2,3, \ldots$, with probabilities

$$p(k) = \frac{<i_1(k)|\, U_1 \,|\, i_{2L}(k)> \ \ldots \ <i_2(k)|\, U_2 \,|\, i_1(k)>}{\sum\limits_{i_1,\ldots,i_{2L}} <i_1|\, U_1 \,|\, i_{2L}> \ \ldots \ <i_2|\, U_2 \,|\, i_1>} \tag{2.7}$$

The distributions are, for example, selected by the heat bath algorithm. After an allowed move in the worldlines one computes the ratio R of the numerators in Eq.(2.7). The new configuration is accepted with probability $P = R/(1+R)$.

We have described the computation of the partition function. To compute the expectation values of the operators we have to consider three different cases:
(i) Operators that are diagonal in the occupation number representation: We simply average over the intermediate states after statistical equilibrium is attained.
(ii) Non-diagonal operators A conserving particle number within a two-site block: Decompose $A = A_1 + A_2$ and compute $<i_k|\, A_i\, U_i|\, i_{k+1}>$.
(iii) Operators that do not conserve the particle number: The calculation now involves disconnected world lines and is more delicate[3].

Consider as an example a one-dimensional chain of spin-polarised fermions with nearest-neighbour interactions

$$H = -t \sum_i \left(c_i^\dagger c_{i+1} + c_{i+1}^\dagger c_i \right) + V \sum_i \left(n_i - \tfrac{1}{2}\right)\left(n_{i+1} - \tfrac{1}{2}\right) = \sum_{i=1}^{N} H_{i,i+1} \tag{2.8}$$

with $n_i = c_i^\dagger c_i$. Choose

$$H_1 = \sum_{i \text{ odd}} H_{i,i+1} \qquad\qquad H_2 = \sum_{i \text{ even}} H_{i,i+1} \tag{2.9}$$

The terms in H_1 and H_2 commute. Then

$$U_1 = e^{-\Delta\tau\, H_1} = \prod_{i \text{ odd}} e^{-\Delta\tau\, H_{i,i+1}} \tag{2.10}$$

and the computation of the matrix elements is a simple two-site problem. Using a basis $|\, n_i\, n_{i+1}>$, where n_i=0,1, one obtains

$$e^{-\Delta\tau\, H_{i,i+1}} \,|\, 00> \ = \ |\, 00>\, e^{-\Delta\tau \frac{V}{4}}$$
$$e^{-\Delta\tau\, H_{i,i+1}} \,|\, 11> \ = \ |\, 11>\, e^{-\Delta\tau \frac{V}{4}}$$

$$e^{-\Delta\tau H_{i,i+1}} \mid 10> \; = \left\{\cosh(t\Delta\tau) \mid 10> + \sinh(t\Delta\tau) \mid 01>\right\} e^{\Delta\tau\frac{V}{4}}$$
$$e^{-\Delta\tau H_{i,i+1}} \mid 01> \; = \left\{\cosh(t\Delta\tau) \mid 01> + \sinh(t\Delta\tau) \mid 10>\right\} e^{\Delta\tau\frac{V}{4}}$$

It follows that, with this choice for H_1 and H_2, all terms in Eq.(2.6) are positive. The interpretation of of the quantities p(k) as probabilities is consistent. Unfortunately for higher dimensional cases the situation is not so favourable.

Projector Monte-Carlo

For the ground state ($T\to 0$, $\beta\to\infty$) the computation of the partition function is unreliable because $\Delta = \frac{\beta}{L}$ is large. Instead one may use the fact that the quantity

$$e^{-\tau H} \mid \psi> \; = e^{-\tau E_0}\left\{ \mid 0> <0 \mid \psi> + \sum_{n>0} e^{-\tau(E_n - E_0)} \right\} \tag{2.11}$$

tends to a constant times the ground state when $\tau\to\infty$, provided $<0 \mid \psi> \neq 0$. Therefore using Trotter's formula to compute $e^{-\tau H} \mid \psi>$, we obtain a numerical computation of the ground state. As initial state $|\psi>$ it is convenient to start already from a reasonable approximation of the ground state, for example a variational wave function.

Hubbard-Stratonovich transformation

For the practical applications of the Quantum Monte-Carlo method, the Hubbard-Stratonovich transformation is an useful tool. It replaces a two-body propagator by a coherent superposition of one-body terms containing auxiliary classical fields. For the Hubbard model, for example

$$H = K + V \tag{2.12a}$$

$$K = -t \sum_{<ij>,\sigma} \left(c^{\dagger}_{i\sigma} c_{j\sigma} + c^{\dagger}_{j\sigma} c_{i\sigma} \right) \tag{2.12b}$$

$$V = U \sum_i \left(n_{i\Uparrow} - \tfrac{1}{2}\right)\left(n_{i\Downarrow} - \tfrac{1}{2}\right) - \sum_i \left(n_{i\Uparrow} + n_{i\Downarrow}\right) \tag{2.12c}$$

the transformation is

$$e^{-\Delta\tau U n_{i\Uparrow} n_{i\Downarrow}} = \mathrm{Tr}_{\sigma}\, e^{\left\{\lambda\sigma\,(n_{i\Uparrow} - n_{i\Downarrow}) - \Delta\tau \frac{U}{2}(n_{i\Uparrow} + n_{i\Downarrow})\right\}} \tag{2.13}$$

with $\lambda = 2\,\arctan\sqrt{\tanh(\Delta\tau U/4)}$. Therefore a quartic form in the electron operators is traded for a quadratic form plus a summation over the auxiliary spin variables σ that take values ± 1.

The (minus) sign problem

If not all the terms in Eq.(2.6) are positive, their interpretation as probabilities for the importance sampling calculation is not straightforward. We may still write

$$< A > = \frac{1}{Z} \operatorname{Tr}(AP) = \frac{\operatorname{Tr}(A\,|P|\,\operatorname{sign}P)}{\operatorname{Tr}(|P|\,\operatorname{sign}P)} = \frac{< A\,\operatorname{sign}P >_{|P|}}{< A\,\operatorname{sign}P >_{|P|}} \qquad (2.14)$$

and use |P| as a probability for the Monte-Carlo simulation. However at low temperature the weight of the positive and negative terms is such that both the numerator and the denominator are small quantities. Therefore the statistical errors make the evaluation of $< A >$ quite unreliable.

The physical origin of the sign problem[4]

In the projector Monte-Carlo method, for example, the quantity

$$\psi_\tau = e^{-\tau H} |\phi_{\tau=0}>$$

tends to the ground state when $\tau \to \infty$. The weights that are used for the statistical sampling are proportional to the wave function ψ_τ. However we know that, for fermions, ψ_τ cannot be positive for all configurations. Decompose the wave function into positive and negative parts

$$\psi_\tau = \psi_\tau^+ - \psi_\tau^-$$

At low energies both ψ_τ^+ and ψ_τ^- have non-vanishing projections on the "hard-core" bosonic ground state. Hence at low temperatures the simulation spends most of the time below the actual energy of the fermionic ground state. These configurations have to be cancelled in the sum, hence the minus signs. A similar effect is unavoidable in frustrated spin systems.

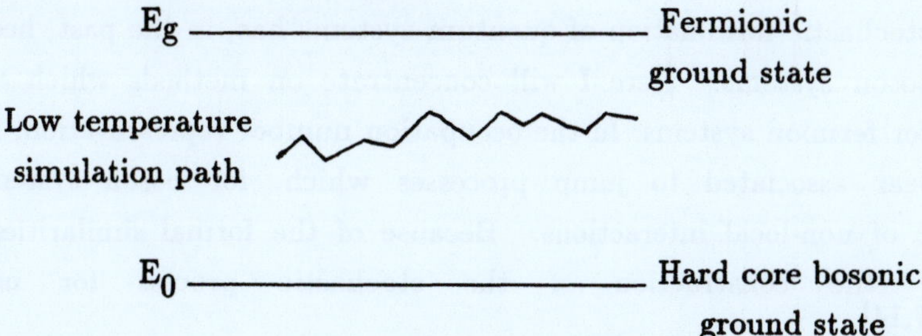

One therefore sees that at low temperatures the Quantum Monte-Carlo method may be quite unreliable and a need is felt for alternative simulation methods.

3. Ground state processes and numerical simulations[5]

The ground state and the Euclidian dynamics of a quantum system are uniquely associated to a stochastic process. Let ρ and G be the unique invariant measure density and the generator of stochastic process such that

$$- \hbar \, \rho^{\frac{1}{2}} \, G \, \rho^{-\frac{1}{2}} = H \tag{3.1}$$

is the Hamiltonian operator of a physical system. Then $\rho^{1/2}$ is the (zero energy) ground state , i. e. $H\rho^{1/2}=0$[6,7] $(\rho^{1/2} \doteq \phi)$. From the representation $T=\exp\{-tG\}$ for the operator semigroup

$$\left(T_t \, f\right)(x) = \int P(t,x,dy) \, f(y) \tag{3.2}$$

associated to the transition functions of the Markov process, it follows that the Euclidian correlations of the theory (Schwinger functions) are obtained from the stochastic correlations of the process. The stochastic process may therefore be used to study the dynamical properties of the quantum system.

Existence of a process with unique invariant measure insures the existence of a Schrödinger picture for the quantum system and stochastic techniques like the theory of small random perturbations of Wentzell and Freidlin[8,9], may be used to obtain rigorous non-perturbative results, in general not accessible through functional analytic methods[10,11]. The stochastic process involved in this construction is a classical stochastic process which happens to have $\rho = |\phi|^2$ as its invariant measure and generates a set of Euclidian correlations. It should not however be confused with the *quantum probability*[12,13] process associated to real-time quantum evolutions.

This stochastic formulation of quantum systems has, in the past, been used mostly for boson systems. Here I will concentrate on methods which are also appropriate for fermion systems. In the occupation number representation, fermion systems appear associated to jump processes which, for boson systems, are characteristic of non-local interactions. Because of the formal similarities I will first review the construction of the stochastic process for non-local interactions[7,14].

Let H be an Hamiltonian operator and $\phi(x)$ its (real) lowest energy eigenstate (which is adjusted to zero energy by the addition of a constant to H).

$$\left(H\,\phi\right)(x) = -\frac{\hbar^2}{2m}\,\Delta\phi(x) + U(x)\,\phi(x) + \int d^n y\; V(x,y)\,\phi(y) = 0 \tag{3.3}$$

with V(x,y) real finite and V(x,y) = V(y,x).

From (3.1) one obtains for the generator of the ground state process with invariant measure $\rho=\phi^2$,

$$\left(G\,f\right)(x) = \left(\frac{\hbar}{m}\frac{\nabla\phi}{\phi}\,\nabla\right) f + \frac{\hbar}{2m}\,\Delta f - \frac{1}{\hbar}\int d^n y\; \phi^{-1}(x)\,V(x,y)\,\phi(y)\,\{f(y)-f(x)\} \tag{3.4}$$

Now we compare with the general differential Chapman-Kolmogorov equation

$$\partial_t\, p(xt|zt') = -\nabla\cdot\{b(x,t)\,p(xt|zt')\} + \frac{1}{2}\sum_{i,j}\frac{\partial^2}{\partial x_i\partial x_j}\{a_{ij}(x,t)\,p(xt|zt')\}$$
$$+ \int d^n y\,\{W(x|y)\,p(yt|zt') - W(y|x)\,p(xt|zt')\} \tag{3.5}$$

for which the generator $Gf = \lim_{\Delta t\downarrow 0}\dfrac{(T_{\Delta t}\,f - f)}{\Delta t}$ is

$$\left(Gf\right)(x) = b(x,t)\,\nabla_x f(x) + \frac{1}{2}\,a_{ij}(x,t)\,\frac{\partial^2}{\partial x_i\partial x_j}\,f(x) + \int d^n y\; W(y|x)\,\{f(y)-f(x)\} \tag{3.6}$$

Comparing Eqs. (3.4) and (3.6) one concludes that the ground state process associated to the non-local Hamiltonian has drift, diffusion coefficient and jumping kernel, respectively

$$b = \frac{\hbar}{m}\frac{\nabla\phi}{\phi} \tag{3.7}$$

$$a_{ij} = \frac{\hbar}{m}\,\delta_{ij} \tag{3.8}$$

$$W(y|x) = -\frac{1}{\hbar}\,\phi^{-1}(x)\,V(x,y)\,\phi(y) \tag{3.9a}$$

Notice the association, in processes ruled by non-local interactions, of diffusion and jumping[1]. This will also be typical of processes involving bosons and fermions (see below).

Likewise, if the state space of an Hamiltonian is spanned by a discrete set $\{|f\rangle\}$ of basis states one sees, by analogy with (3.9a), that the jumping process with kernel

$$W(f|f') = -\frac{1}{\hbar}\,\mathrm{Re}\left\{\frac{1}{\langle f'|\phi\rangle}\,\langle f'|\,H\,|f\rangle\,\langle f|\phi\rangle\right\} \tag{3.9b}$$

has invariant density $\rho(f)=|<f|\phi>|^2$ as may easily be checked from the stochastic equation

$$\partial_t \rho(f) = \sum_{f'} \{ W(f|f') \rho(f') - W(f'|f) \rho(f) \} \tag{3.10}$$

Notice that in this equation the kernel is defined only up to terms const. $\times \delta_{f,f'}$.

In the Eqs. (3.7) and (3.9) the drift and the jumping kernel are both functions of the ground state wave function ϕ which in general is not known for a given Hamiltonian. It is however possible to formulate the dynamical problem in such a way that the drift and the jumping kernel are obtained directly without having to solve the equation $H \phi(x) = \lambda_0 \phi(x)$ for the lowest eigenvalue λ_0. Let ϕ_0 be some initial state non-orthogonal to the ground state. Then, in the $t \to \infty$ limit, the solution $\phi_t = \exp\{-\frac{1}{\hbar}Ht\} \phi_0$ of the Euclidian equation

$$\hbar \partial_t \phi_t(x) = -H \phi_t(x) \tag{3.11}$$

converges to the ground state. Namely $\exp\{\frac{1}{\hbar}\lambda_0 t\}\phi_t \to \phi$. Defining

$$b_t = \frac{\hbar}{m} \frac{\nabla \phi_t}{\phi_t} \quad ; \quad W_t(y|x) = -\frac{1}{\hbar} \phi_t^{-1}(x) V(x,y) \phi_t(y) \tag{3.12}$$

where ϕ_t is a solution of the Euclidian equation (3.11), one obtains evolution equations for b_t and $W_t(y|x)$

$$\partial_t b_t(x) = \frac{1}{2} \nabla b_t^2(x) + \frac{\hbar}{2m} \nabla\left(\nabla \cdot b_t(x)\right) - \frac{1}{m} \nabla U(x) + \frac{\hbar}{m} \nabla_x \int d^n y \, W_t(y|x) \tag{3.13a}$$

---

[1]In the modelization of quantum systems, there are potentials that supply the simplest examples for each type of physical behaviour. For example the harmonic oscillator is typical of a pure point spectrum, the Coulomb potential typical of systems with both point and continuous spectrum and accumulation of levels at the ionisation threshold, etc. For roton excitations the non-local convolution potentials[14], of which the simplest one is the symmetric translation potential $(V\psi)(x) = \lambda\psi(x+a) + \lambda^*\psi(x-a)$, provide perhaps the simplest model. In the stochastic interpretation of the Euclidian process the roton is the situation where slow forward diffusion and backward jumps exactly compensate each other. This may be thought of as the rigorous formulation of an old image of Feynman according to which one type of roton excitation may be viewed as a group of children going down a slide, then going around and coming back to the slide again.

$$\partial_t W_t(y|x) = W_t(y|x) \int d^n z \left\{W_t(z|y) - W_t(z|x)\right\} + \tfrac{1}{2} W_t(y|x) \Big\{2\, U(x) - \tfrac{m}{\hbar}\, b_t^2(x)$$

$$- \nabla \cdot b_t(x) - 2\, U(y) + \tfrac{m}{\hbar}\, b_t^2(y) + \nabla \cdot b_t(y)\Big\} \qquad (3.13b)$$

As ϕ_t converges to the ground state, the solutions of Eqs.(3.13) converge to the drift and jumping kernels associated to the ground state process. Therefore these quantities may be obtained directly by iteration of Eqs.(3.13), without having to solve the Schrödinger equation. In the same way, for the pure jump case associated to the discrete basis $\{|f>\}$ representation, one obtains the (Euclidian) evolution equation

$$\partial_t K_t(f|f') = \sum_g K_t(f|f') \left\{K_t(g|f) - K_t(g|f')\right\} \qquad (3.14a)$$

for the quantity

$$K_t(f|f') = -\frac{1}{\hbar} \frac{1}{<f'|\phi_t>} <f'|\, H\, |f> <f|\phi_t> \qquad (3.14b)$$

the jumping kernel being

$$W_t(f|f') = \mathrm{Re}\, K_t(f|f')$$

Starting from an initial W_0, consistent with (3.14b), this equation converges in the $t \to \infty$ limit to the jumping kernel of the ground state process. We may start for example from

$$W_0(f|f') = -\frac{1}{\hbar} \mathrm{Re} < f|\, H\, |\, f' >$$

which corresponds to consider the initial ϕ_t to have the same projection on all basis states. Although Eq.(3.14a) by itself contains no explicit dynamical information on the interaction potential, it is able to generate the ground state kernel because it preserves under time evolution all the information introduced by the initial condition, namely $\partial_t \{W_t(f'|f)\, W_t(f|f')\} = 0$. That is, what Eq.(3.14a) does is simply to balance the transition probabilities to be consistent with an invariant measure while, at the same time, preserving the information about the Hamiltonian that is supplied by the initial condition.

For a many-body quantum system, because the number of possible states is very large, the Euclidian equations (3.14a) become a very large system. If the interactions are short-range, a possible solution is to make the iteration on a small cluster simulation of the system to obtain an approximation of the ground state stochastic kernels. These kernels may then be used to obtain the correlation functions by simulating the process in a much larger lattice. The simulation of the process uses Eqs. (3.5) and (3.10) or the corresponding stochastic differential equations.

Expectation values of observables for the process X_t defined by Eqs. (3.7-3.9) are the quantum expectation values in the ground state

$$E\{A(X_t)\} = \int A(x)\,\rho(x)\,dx = \int \phi(x)\,A(x)\,\phi(x)\,dx = (\phi, A\phi) \quad (3.15)$$

The multi-time correlations of the stochastic process are the Euclidian correlations of the quantum system. For example

$$E\{A(X_t)A(X_0)\} = \int \left(T_t A\right)(x) A(x) \rho(x)\,dx = \int \phi^2(x)\,A(x)\left(e^{-\frac{t}{\hbar}(\phi^{-1}H\phi)}A\right)(x)\,dx$$
$$= \left(\phi,\, A\, e^{-\frac{t}{\hbar}H}\, A\,\phi\right) \quad (3.16)$$

and similarly for the general n-point correlation function.

Assume that one chooses to compute the stochastic kernels in a basis $\{|f>\}$ that diagonalizes the observable A for which one wishes to obtain the correlation functions. Then, the calculation above also shows that, in such an "adapted occupation number basis", there is no essential difference in the procedures to be used for boson or for fermion systems. In particular the "sign problem" will be avoided.

Examples

Here I will consider two examples where the ground state may be explicitly represented as a functional of the operators of the theory. The study of the examples clarifies the nature of the stochastic processes associated to the ground state. The first example is the BCS model. The BCS ground state

$$|\phi> = \prod_{k,\sigma} \left(u_k + v_k\, b^\dagger_{k\sigma}\, b^\dagger_{-k,-\sigma} \right) |0> \qquad u_k^2 + v_k^2 = 1 \quad (3.17)$$

is a zero energy eigenstate of the Hamiltonian

$$H = \sum_{k,\sigma} E_k u_k^2 \left(b^\dagger_{k\sigma} - \alpha_k\, b_{-k,-\sigma}\right)\left(b_{k\sigma} - \alpha_k\, b^\dagger_{-k,-\sigma}\right) \quad (3.18)$$

with

$$\alpha_k = -\alpha_{-k} = \frac{v_k}{u_k} \tag{3.19}$$

Let $\{ |c_{k\sigma}\rangle = b^\dagger_{k\sigma}\, b^\dagger_{-k,-\sigma}\, |0\rangle \ ;\ k_i>0 \}$ be a basis of Cooper pairs. Then from (3.9b) the ground state process is a jumping process with kernel

$$W(\{c_{k\sigma}\}|\{c_{k'\sigma'}\}) = \frac{\prod_{q\in\{c_{k\sigma}\}} v_q}{\prod_{q\in\{c_{k'\sigma'}\}} v_q} \Big\{ -\sum_{q\in\{c_{k\sigma}\}} 2\sqrt{E_q^2-\Delta_q^2}\ \delta_{\{c_{k\sigma}\},\{c_{k'\sigma'}\}} +$$

$$\Delta_p\, \delta_{\{c_{k'\sigma'}\},\{c_{k\sigma}\}+c_p} + \Delta_p\, \delta_{\{c_{k'\sigma'}\},\{c_{k\sigma}\}-c_p} \Big\} \frac{1}{\hbar} \tag{3.20}$$

with $\Delta_p = 2\, E_p\, u_p\, v_p$. $\{c_{k\sigma}\}$ denotes a set of occupied Cooper pairs and $\{c_{k\sigma}\} \pm c_p$ is the same set with one more or one less occupied pair. Eq.(3.20) is obtained from (3.9b) neglecting a term of the type const. $\times\, \delta_{\{c_{k\sigma}\},\{c_{k'\sigma'}\}}$

Another formulation of the ground state process which has, at least, formal interest is obtained using a representation of the creation and annihilation operators in Grassman variables η^*_k . The representation is

$$b^\dagger_{k\sigma} \to \eta^*_{k\sigma} \qquad\qquad b_{k,\sigma} \to \left(\frac{d}{d\eta^*_{k\sigma}}\right)_L \tag{3.21}$$

where

$$\eta_{k\sigma}\, \eta^*_{q\sigma'} + \eta^*_{q\sigma'}\, \eta_{k\sigma} = \eta^*_{k\sigma}\, \eta^*_{q\sigma'} + \eta^*_{q\sigma'}\, \eta^*_{k\sigma} = 0$$

and the subscript L in the differentiation symbol means that the variable $\eta^*_{k\sigma}$ in a monomial must be pushed to the extreme left before the differentiation is applied.

The representation (3.21) acts on functions of $\{\eta^*_{k\sigma}\}$ alone. In this representation the BCS ground state is

$$\phi(\eta^*) = \prod_{\substack{k,\sigma \\ (k_i>0)}} \left(u_k + v_k\, \eta^*_{k\sigma}\, \eta^*_{-k-\sigma} \right) \tag{3.22}$$

and the Hamiltonian

$$H = \sum_{k,\sigma} E_k u_k^2 \left(\eta^*_{k\sigma} - \alpha_k \frac{d}{d\eta^*_{-k-\sigma}} \right) \left(\frac{d}{d\eta^*_{k\sigma}} - \alpha_k\, \eta^*_{-k-\sigma} \right) \tag{3.23}$$

Performing on H the transformation induced by $\phi(\eta^*)$ one obtains

$$-\phi^{-1}\, H\, \phi = \sum_{k,\sigma} E_k\, u_k \frac{d}{d\eta^*_{-k-\sigma}} \frac{d}{d\eta^*_{k\sigma}} - \sum_{k,\sigma} E_k\, \eta^*_{k\sigma} \frac{d}{d\eta^*_{k\sigma}} \tag{3.24}$$

Comparing with the standard form of the elliptic operator,

$$\frac{1}{2} \sum_{i,j} \alpha_{ij}(x,t) \frac{\partial^2}{\partial x_i \partial x_j} + \sum_i \beta_i \frac{\partial}{\partial x_i}$$

generator of a diffusion process, one concludes that formally the BCS ground state process may be interpreted as a diffusion process in Grassman variables, with drift

$$\beta_{k\sigma} = - E_k \, \eta^*_{k\sigma} \tag{3.25a}$$

and diffusion coefficient

$$\alpha_{k\sigma,p\sigma'} = 2 \, E_k \, u_k \, \delta_{k,-p} \, \delta_{\sigma,\sigma'} \tag{3.25b}$$

For practical purposes however the formulation of the ground state as a jump process in an occupation number basis is probably more useful.

For the second example consider a supersymmetric hopping model defined on a 2-dimensional square lattice by

$$H = \frac{1}{2}\sum_{(ij)}\left(-\hbar^2 \frac{\partial^2}{\partial\phi_{(ij)}\partial\phi_{(ij)}} + \frac{\partial W}{\partial\phi_{(ij)}} \frac{\partial W}{\partial\phi_{(ij)}}\right) + \frac{1}{2}\hbar \sum_{(ij)(kl)} \frac{\partial^2 W}{\partial\phi_{(ij)}\partial\phi_{(kl)}} [c^\dagger_{(ij)}, c_{(kl)}] \tag{3.26}$$

where $\phi_{(ij)}$ and $c_{(ij)}$ are bosonic and fermionic fields at the lattice point (ij) satisfying

$$[\phi_{(ij)}, \frac{\hbar}{i} \frac{\partial}{\partial\phi_{(ij)}}] = i \, \hbar \, \delta_{(ij),(kl)} \qquad \{c^\dagger_{(kl)}, c_{(ij)}\} = \delta_{(ij),(kl)} \tag{3.27}$$

For the superpotential $W(\phi)$

$$W(\phi) = -\frac{t}{\hbar} \sum_{[(ij),(kl)]} \phi_{(ij)} \, \phi_{(kl)} \tag{3.28}$$

with [(ij),(kl)] denoting a sum over nearest neighbours, the Hamiltonian becomes

$$H = \frac{1}{2}\sum_{(ij)}\left(-\hbar^2 \frac{\partial^2}{\partial\phi_{(ij)}\partial\phi_{(ij)}} + \left(\sum_{[(kl),(ij)]} \phi_{(kl)}\right)^2\right) - \frac{t}{2} \sum_{[(ij)(kl)]} [c^\dagger_{(ij)}, c_{(kl)}] \tag{3.29}$$

Denote by |0> and |−> the empty and the filled sectors in the Fermion Fock space. Then the state

$$|\psi> \; = \; e^{\frac{W(\phi)}{\hbar}} \sum_{(ij)} (-1)^{i+j} \, c_{(ij)} \, |-> \tag{3.30}$$

is the ground state in the one-hole sector and is a zero-energy eigenstate of the operator $\bar{H} = H + 4t$. Computing the drift and the jumping kernel from Eqs. (3.7) and (3.9) one obtains

$$b_{(ij)} \; = \; - \frac{t}{\hbar \, m} \sum_{[(kl),(ij)]} \phi_{(kl)} \tag{3.31a}$$

$$W(\phi \; i \; j \mid \phi' \; k \; l) \; = \; - \frac{t}{\hbar} \, \delta_{[(kl),(ij)]} \, \delta_{\phi,\phi'} \tag{3.31b}$$

The sum in $b_{(ij)}$ is over all lattice points (kl) that are nearest neighbours to (ij) and $\delta_{[(kl),(ij)]}$ vanishes unless (ij) and (kl) are nearest neighbours. The jumping kernel is computed in the basis

$$|\phi \; i \; j> \; = \; |\phi> \; c_{(ij)}|->$$

One sees that the boson-fermion nature of the problem implies that the corresponding process involves both diffusion and jumping.

Numerical simulation of the ground state process. The Hubbard model

As explained above the essential step is the calculation of the jumping kernels $W(f'|f)$ from the asymptotic solution of Eq.(3.14). Notice that $|f)$ runs over all states of the system and therefore, for a system with a large number of degrees of freedom, Eq.(3.14) becomes a differential equation of very high dimension. However in a system like the Hubbard model, where the elementary interactions are short-range, the kernels may, with good approximation, be obtained by solving the equation for a small cluster, and then storing the solution in a look-up table that is used in a much larger lattice to extract the global properties of the ground state. Notice that the small-cluster does not even have to be very small because, for each set of parameters, Eq.(3.14) has to be solved only once.

Consider the Hubbard model. The Hamiltonian is

$$H = - t \sum_{\{i,j\}} c^{\dagger}(i) \; c(j) \; + \; U \sum_{i} n_{\uparrow}(i) \; n_{\downarrow}(i)$$

With $W_0(f'|f) = -\text{Re} < f'| H | f >$ the initial values for the Eq.(3.14) are:

$W_0(f'\|f) = U N_p$	$f'=f$
$W_0(f'\|f) = -t$	if f' differ from f by an electron jump to a neighbouring site
$W_0(f'\|f) = 0$	for all other pairs f,f'

and from the (Euclidian) evolution equation (3.14) the only non-zero W's are those which are non-zero for the initial condition. Therefore, in this case, a 3×3 lattice with periodic boundary conditions already provides an unbiased environment for the elementary jumps. However a 3×3 lattice already carries 4^9 possible different configurations, considering at each site no-electron, one-electron(↑ or ↓) or a pair. Using the symmetry properties these configurations may be grouped into classes, a class being defined as the configurations that can be transformed into one another by translations, rotations or mirror reflections. This finally leads to 4480 different classes which are classified in the table below according to the number of up and down electrons.

Number of electrons ↑

		0	1	2	3	4	5	6	7	8	9
	0	1	1	2	4	5	5	4	2	1	1
	1	1	3	8	16	23	23	16	8	3	1
Number	2	2	8	26	54	78	78	54	26	8	2
of	3	4	16	54	118	170	170	118	54	16	4
elect.	4	5	23	78	170	250	250	170	78	23	5
↓	5	5	23	78	170	250	250	170	78	23	5
	6	4	16	54	118	170	170	118	54	16	4
	7	2	8	26	54	78	78	54	26	8	2
	8	1	3	8	16	23	23	16	8	3	1
	9	1	1	2	4	5	5	4	2	1	1

We may therefore limit ourselves to the study of a representative element in each class and their neighbouring configurations. The system of equations corresponding to the possible, and topologically distinct, transitions between the different classes is now solved numerically and the W(f′|f)'s corresponding to its (rapidly) convergent large time solution are stored in a look-up table. Look-up tables were generated for several U/t values in the range 0 to 10.

Two remarks should be made concerning the calculation of the jumping kernels. The ground state stochastic equations were derived under the assumption that the energy of the ground state is normalised to zero. Because the ground state energy is in general not known it means that the diagonal kernels W(f′|f) are defined only up to an additive constant. This has no effect in the evolution of Eq.(3.14) because the additive constant cancels in sum of the right-hand side, but it means that the diagonal element values stored in the look-up table are not the actual W(f|f)'s. The way to obtain its actual values for the stochastic evolution in the ground state process is through

$$W(f|f) = 1 - \sum_{g \neq f} W(g|f)\, \Delta t$$

provided Δt is sufficiently small compared with the W(g|f)'s.

To obtain now the properties of the ground state the simulation starts from a lattice with a certain number of up and down electrons and the evolution of the stochastic ground state process is carried out by each time picking up a site at random and having its 3×3 neighbourhood evolve according to the jumping kernels of the look-up table. In practice all compatible W's are piled-up according to their sizes as intervals in the line and the decision taken by a random number uniformly distributed in the interval $[0, \sum W]$.

I now present some results of a study of pairing and short range antiferromagnetic correlations in the ground state of the Hubbard model. The simulations were carried out in a 20×20 lattice with 198 up electrons and 198 down electrons. The boundary conditions are periodic and the parameter values explored were $U/t = 0,1,4$ and 10. The initial electron distribution is random (constrained by the Pauli principle). For each case the results we present correspond to 10^6 productive evolutions (i.e. evolutions where there was an actual change in the lattice configuration). Measures of average values are taken only after the number of double occupancies is stabilised. The following quantities were measured:

- number of double occupied sites
- number of hole pairs
- short range antiferromagnetic correlation
- medium range antiferromagnetic correlation

The antiferromagnetic correlation is measured by looking at a neighbourhood of each point and counting +1 for the antiferromagnetically correlated neighbours, −1 for the ferromagnetic ones and zero for those that are either empty or have double occupancy. For the "short range" one considers a 3×3 neighbourhood and for the "medium range" a 5×5 neighbourhood. The results are the following:

	U/t			
	0	1	4	10
double occupancy	99	93	74	50
hole pairs	54	46	28	14
short rang antiferr.	53	84	92	117
medium rang antif.	33	75	76	32

A critical issue in models of high-temperature superconductivity is whether or not the effect is due to a type of Bose condensation of pre-existing local pairs of carriers. In particular the observed short coherence lengths seem to favour such a local pair picture. It is therefore important to search for any symptoms of dynamical hole pairing in the Hubbard model. One sees from the table that the number of hole pairs decreases with U/t, but that is to expected from the statistical effect of the decreasing double occupancy which also decreases the number of holes. To disentangle the statistics from dynamical effects compare, in the following table, the average distribution of holes in the Hubbard model ground state (at $U/t = 4$) with a purely random distribution for the same total number of holes.

Number of holes in the cluster	Hubbard model	Random distribution
1	37.61	32.97
2	9.19	8.11
3	3.51	3.44
4	1.75	1.54
5	0.67	0.72
6	0.46	0.35
7	0.32	0.17
8	0.13	0.09
9	0.12	0.04
10	0.03	0.02
11	0.08	0.01
12	0.03	0.00

The first column defines the size of the hole clusters and the second and the third contain the average number of such clusters in the Hubbard model and in a random distribution. The conclusion is that there no essential distinction between the random unbiased distribution of holes and the Hubbard model. Therefore we find no evidence for hole pair formation in the Hubbard model. Incidentally this agrees with the analytical estimates of Ref.[15] where it was concluded that the Hubbard model by itself does not lead to pair formation and that an additional pairing interaction seems to be needed in models of high-T_C superconductivity.

What the results show however is evidence for short range antiferromagnetic correlation, although already at a modest medium range the correlation seems to be lost.

4. Probabilistic diagonalization

For most non-trivial many-body Hamiltonians it is extremely difficult to obtain the exact eigenstates or even reliable analytical approximations. It may however be possible to make probabilistic estimates of the coefficients of the exact eigenstates in some Hilbert space basis. The difficulty of obtaining exact analytical or numerical results increases with the number of degrees of freedom. However the accuracy of probabilistic estimates also increases with the number of degrees of freedom. This idea was developed in Ref.[15] for models of strongly correlated electron systems. Consider the Hamiltonian

$$H_0 = -t \sum_{\substack{<i,j> \\ \sigma}} (1 - n_{i-\sigma}) c^\dagger_{i\sigma} c_{j\sigma} (1 - n_{j-\sigma}) \tag{4.1}$$

with $n_{i\sigma} = c^\dagger_{i\sigma} c_{i\sigma}$. H_0 is the part of the Hubbard Hamiltonian that transfers electrons from single occupied sites to empty sites without changing the eigenvalue of $D = \sum_i n_{i\uparrow} n_{i\downarrow}$. In the dynamics of the ground state for a less than half-filled band, in the D=0 subspace, H_0 controls the hole dynamics.

Let t be > 0 and consider the diagonalization problem for H_0 in the D=0 sector. Take a less than half filled band with N sites and n_h holes. The situations we are mostly concerned with are situations where there is a macroscopic number of holes, i. e. n_h/N is of order at least a few per cent. States with n_h holes will be denoted by $|i,\beta>$, i and β being the labels respectively for the hole configuration and the electron spin configuration. H_0 has matrix elements between $|i,\beta>$ and $|j(i),\beta>$, j(i) being the hole configurations that are connected to i by the motion of an hole to a nearest neighbour site. Let $|\psi>$ be an eigenstate of H_0

$$H_0|\psi> = \lambda|\psi> = \lambda \sum_{i,\beta} c_{i\beta}|i,\beta> = \sum_{i,\beta} \sum_{j(i)} \frac{c_{j(i)\beta}}{c_{i\beta}} c_{i\beta} \, |i,\beta>$$

Then

$$\sum_{j(i)} \frac{c_{j(i)\beta}}{c_{i\beta}} = \text{constant} = \lambda \tag{4.2}$$

The energy of the ground state will be minimal if $\text{sign}(c_{i\beta}) = \text{sign}(c_{j(i)\beta})$, i. e. configurations differing by the motion of an hole should have the same sign.

Successive applications of H_0 will connect states with all hole configurations and all electron spin distributions (preserving only the number of holes and the number of up and down spins). The eigenstates of H_0 are thus coherent superpositions of all hole configurations. The number of states to which a given $|i,\beta>$ is connected by H_0 is called its "connectivity number" ν(i) and varies from $4n_h$ for a state without hole clusters to about $4\sqrt{n_h}$ for a state with a global cluster.

The connectivity number ν(i) is the number of terms in the sum of Eq.(4.2).

Because for an eigenstate the sum in (4.2) is a constant for all $|i,\beta>$ and the connectivity varies with the formation of clusters, this provides a way to estimate the coefficients, hence the probability of cluster formation. An exact analytic calculation, i. e. an exact diagonalization of H_0 is not practical for a large number N of sites. However the very fact that N is a large number may be turned to our profit to obtain "exact" results in the probabilistic sense, i. e. to obtain ratios for the coefficients $c_{i\beta}$ which hold in the average.

In the expansion of the ground state $|\psi_o>$ of H_0 consider the coefficients of the states with p hole pairs, which we denote by c_{2p}. When, starting from a given hole configuration, one moves one hole, the probability that a hole pair is formed by this motion is $3(n_h-1)/(N-1) \simeq 3n_h/N$. Noticing that a hole pair contributes 6 to the connectivity as opposed to 8 for two isolated holes, one obtains from (4.2) the following, probabilistically rigorous, equation

$$4(n_h-2p)\left\{(1-\frac{3n_h}{N})+\frac{3n_h}{N}\frac{c_{2p+2}}{c_{2p}}\right\}+6p\left\{(1-\frac{3n_h}{N})\frac{c_{2p-2}}{c_{2p}}+\frac{3n_h}{N}\right\}=$$

$$=4\,n_h\left\{(1-\frac{3n_h}{N})+\frac{3n_h}{N}\frac{c_2}{c_0}\right\} \tag{4.3}$$

Considering the addition of a hole pair to have an uniform effect on the ratio of the coefficients

$$\frac{c_{2p+2}}{c_{2p}}=\frac{c_{2p-2}}{c_{2p}}=\frac{c_2}{c_0}=z \tag{4.4}$$

and defining $x=\frac{3n_h}{N}$, one obtains

$$-4(1-x)-4xz+3(1-x)\frac{1}{z}+3x=0 \tag{4.5}$$

The solution is $z=\frac{3}{4}$ (the other solution of (4.5) being negative for $x<1$). This implies

$$c_{2p}=\left(\frac{3}{4}\right)^p c_0 \tag{4.6}$$

which means that the probability to find a macroscopic number of hole pairs $(\frac{2p}{n_h}\sim O(1))$ vanishes in the thermodynamic limit ($N\to\infty$). A similar reasoning would lead to a similar conclusion for the probability of occurrence of other hole clusters in macroscopic numbers. The ground state $|\psi_o>$ of H_0 is therefore a coherent superposition dominated by sparse hole configurations with coefficients of similar magnitude and the same sign. For each fixed number of up and down spins one finds in $|\psi_0>$ all possible spin configurations β with coefficients of similar magnitude, determined only by the cluster structure of the hole distribution. One therefore concludes that the introduction of a macroscopic number of holes destroys any antiferromagnetic long range order that might exist in the half-filled band. The ground state of H_0 is degenerate in the sense that we will have a set of states of equal energy $|\psi_0 n_+ n_- >$ labelled by the number of up and down spins.

The other eigenstates of H_0 are also degenerate coherent superpositions of all hole and spin configurations. The highest eigenvalue will correspond to coefficients of magnitude identical to the ones in $|\psi_0 n_+ n_- >$ and such that configurations differing by the motion of a hole have opposite signs. The basis of H_0 – eigenstates is called the "resonant hole configuration"(RHC) basis and the diagonalization of the full effective Hamiltonian of the Hubbard model may be carried out in this basis. For more details and the application to other interaction terms I refer to [15].

A different probabilistic idea to carry out the diagonalization of the Hamiltonians in many-body problems has been proposed by De Raedt, von der Linden and Frick[16,17] and called "Monte-Carlo diagonalization" by these authors.

5. Scaling laws and deviations from scaling in fully developed turbulence

The modern understanding of fully developed turbulence is very much related to the old idea of the Richardson cascade[18]. In this cascade an hierarchy of eddies transfers the energy from the macroscopic scale to smaller and smaller scales until the fragmentation process is stopped by dissipation. From the Navier-Stokes equation

$$\partial_t v + (v \cdot \nabla) v = - \frac{1}{\rho} \nabla p + \nu \Delta v + f \tag{5.1}$$

one sees that dissipation is only important when Δv is large, that is, at small scales. We may therefore distinguish the following regimens or ranges:

The macroscopic (external) range

The inertial range (Δv small)

The dissipation range (Δv large)

In the inertial range the viscosity term $\nu \Delta v$ may be neglected and the simplified

Navier-Stokes equation is scale invariant, that is, invariant for the transformations

$$\vec{r} \to \lambda \vec{r} \qquad \vec{v} \to \lambda^{h} \vec{v} \qquad t \to \lambda^{1-h} t \tag{5.2}$$

This means that, in the inertial range of fully developed turbulence, there is scale invariance in a statistical sense. Let

$$\delta v(r,\ell) = v(r+\ell) - v(r)$$

be the increment in the velocity field along ℓ. From homogeneity, isotropy and the assumption of finite mean rate of dissipation per unit mass (ϵ), Kolmogorov in 1941[19] derived the following relation for the third momentum of the velocity increment parallel to ℓ (*third-order longitudinal structure function*)

$$S_3(\ell) = \langle (\delta v_{||}(r,\ell))^3 \rangle = -\frac{4}{5} \epsilon \ell \tag{5.3}$$

Comparing with (5.2) we conclude, from scale invariance, that $h = 1/3$. Using scale invariance once more we may now derive the form of the higher order structure functions, namely

$$S_p(\ell) = \langle (\delta v_{||}(r,\ell))^p \rangle = C_p \, \epsilon^{\frac{1}{3}p} \, \ell^{\frac{1}{3}p} \tag{5.4}$$

Notice that C_p depends on the geometry of the turbulence producing mechanism. That is, there is scaling but not universality[20]. Defining

$$\langle |\delta v_{||}(r,\ell)|^p \rangle \sim \ell^{\zeta(p)} \tag{5.5}$$

Kolmogorov's (41) theory therefore predicts

$$\zeta(p) = \frac{1}{3} p \tag{5.6}$$

It was only in the eighties that this prediction could be experimentally checked in an accurate way. The experimental data has shown deviations from Kolmogorov scaling. Instead of (5.6), the data[21] shows:

(i) For small p: $\quad \zeta(p) \simeq \frac{0.83}{3} p + 0.17$

(ii) For large p: $\quad$ is not linear in p

Fractal hypothesis

The deviation at small p is explained by the fractal hypothesis[22,23]. This states that the energy transfer does not occur in the whole volume but only in a set of dimension $d_f < 3$. Numerical simulations of the Navier-Stokes equation[24] seem also to confirm this hypothesis. Let S be the set that contains the active eddies at scale ℓ

$$\delta v(r,\ell) \sim \begin{cases} \ell^{h} & r \in S \\ 0 & r \notin S \end{cases}$$

For the time, energy and rate of dissipation per unit mass, at scale ℓ, we have the following estimates

$$t(\ell) \sim \frac{\ell}{\delta v(\ell)} \qquad E(\ell) \sim \delta v^2(\ell) \qquad \epsilon(\ell) \sim \frac{E(\ell)}{t(\ell)} \sim \frac{\delta v^3(\ell)}{\ell} \tag{5.7}$$

Then

$$\int \frac{\delta v^3(\ell)}{\ell}\, d^3y \sim \ell^{3h+2}$$

implies $h = (d_f - 2)/3$ and

$$\zeta(p) = \frac{d_f - 2}{3}\, p + 3 - d_f \tag{5.8}$$

Eq.(5.8) fits the data at small p for $d_f \simeq 2.83$. The (single) fractal model however does not explain the non-linearity at large p.

Multifractal models[25,26]

Instead of a single scaling exponent ℓ^h assume that there is a range of scaling exponents $I = (h_{min}, h_{max})$. For each $h \in I$ there is a set $S_h \in \mathbf{R}^3$ of Hausdorff dimension D(h). Then

$$< |\delta v(\ell)|^p > \sim \int_I d\mu(h)\, \ell^{ph+3-D(h)} \tag{5.9}$$

For $\ell \to 0$ a steepest descent argument implies that the smallest exponent dominates

$$< |\delta v(\ell)|^p > \sim \ell^{\zeta(p)} \qquad \zeta(p) = \min_h (ph + 3 - D(h)) \tag{5.10}$$

Therefore $\zeta(p)$ is the Legendre transform of the dimension function D(h).

The random β-model

A concrete implementation of the multifractal idea is the random β-model. Let, for example, the length scales be related by powers of two.

$$\ell_n = 2^{-n}\, \ell_0$$

where ℓ_0 is the macroscopic injection scale. Let $\delta v_n = \delta v(\ell_n)$ and call β the ratio of the volume occupied by the active eddies at scale ℓ_{n+1} to the volume of the parent eddy at scale ℓ_n. Because the energy transfer rate between scales must be constant in the inertial range, but the total mass of the active eddies is multiplied by β, we have for the transfer rate per unit mass

$$\epsilon_n = \beta\, \epsilon_{n+1}$$

Now, to implement the multifractal idea, one assumes that there are eddies of different types with different β's for each type of transfer. Then, recalling the estimate (5.7) for ϵ, one obtains

$$\frac{\delta v_n^3(k)}{\ell_n} = \beta_{n+1}(k)\, \frac{\delta v_{n+1}^3(k)}{\ell_{n+1}} \tag{5.11}$$

where each k denotes a particular type of energy transfer. In this way, at step n, the velocity fluctuation inside an eddy depends on the precise history of its fragmentation. It is determined by the sequence $[\beta_1, \dots, \beta_n]$. The volume that it occupies is $\prod_{i=1}^{n} \beta_i \ell_0^3$. Then

$$\delta v_n \sim \ell_n^{1/3} \left(\prod_{i=1}^{n} \beta_i \right)^{-1/3} \tag{5.12}$$

and

$$< |\delta v_n|^p > \sim \ell_n^{p/3} \int \prod_{i=1}^{n} d\beta_i\, \beta_i^{1-p/3}\; P(\beta_1, \dots, \beta_n) \tag{5.13}$$

where $P(\beta_1, \dots, \beta_n)$ is the probability of the fragmentation $[\beta_1, \dots, \beta_n]$.

If one assumes independent fragmentations

$$P(\beta_1, \dots, \beta_n) = \prod_{i=1}^{n} P(\beta_i) \tag{5.14}$$

one obtains

$$\zeta(p) = \frac{p}{3} - \ln_2 < \beta^{1-p/3} > \tag{5.15}$$

where $< >$ denotes average in the probability $P(\beta_1, \dots, \beta_n)$. The simplest model for P is a binomial model

$$P(\beta) = \gamma\, \delta_1 + (1-\gamma)\, \delta_{1/2} \tag{5.16}$$

The two values of β, $\beta_0=1$ and $\beta_1=1/2$, correspond respectively to dimensions $d_f=3$ and $d_f=2$. That is, the binomial model considers two types of eddy energy transfers, namely volume – to – volume and volume – to – surface. With (5.16) one now obtains

$$\zeta(p) = \frac{p}{3} - \ln_2 \left\{ \gamma + \frac{1-\gamma}{2^{1-p/3}} \right\} \tag{5.17}$$

which actually fits the existing data, for $\gamma = 0.875$.

6. Asymmetric stochastic processes on a tree and the turbulent cascade[27]

In Section 5 we have seen that, at least at the level of precision of the existing experiments, the statistics of the velocity fluctuations in the turbulent cascade are well described by multifractal models like the random β-model. However the models and the quantities that are computed do not involve time-correlations and therefore are not a detailed test of the dynamics of the turbulent cascade. Indeed assumptions like (5.14) and (5.16) only define the probability distribution ρ_i at each level of the cascade tree (Fig.1). They make no statement concerning the time evolution and the time scales of the eddies in the cascade. In fact it is known that only a few restrictions are imposed on the form of the velocity fields by the predictions of the statistical models described in Section 5 and many distinct velocity fields may be compatible with the spectra and scaling laws. One way to further the research on the dynamical properties of the velocity field is to study time-correlations for the observables. To make a connection of the models with the time dependence of each observable one should first consider the dynamical aspects of the energy cascade itself. To do this a time hierarchy in the development of the turbulent cascade must be defined. In this section one deals with this issue.

For a binary cascade tree (Fig.1) we may use a dyadic labelling for the possible states at each level. The *state space* V_n at level ℓ_n is the set of all products $\beta^{(1)} \cdot \ldots \cdot \beta^{(n)}$ with $\beta^{(i)} \in \beta_0, \beta_1$ (In Eq.(5.16) $\beta_0 = 1$ and $\beta_1 = 0.5$). There are 2^n elements in V_n and the probability of the state i is

$$\rho_i = P(\beta^{(1)} \cdot \ldots \cdot \beta^{(n)}) = \gamma^{n_0(i)} (1-\gamma)^{n_1(i)} \tag{6.1}$$

where $n_0(i)$ and $n_1(i)$ are the number of zeros and ones in the dyadic labelling of the state i.

From the random β-model all one obtains is a statement about these probabilities. This suffices to interpret most of the current experimental results which concern mostly velocity increments over a distance r and turbulent energy dissipation over a box of size r. These quantities do not involve time-correlations and therefore do not make a detailed test of the dynamics of the cascade, they only test its invariant probability measure. To identify the physical mechanisms behind the structure of fully developed turbulence more information is needed. If one

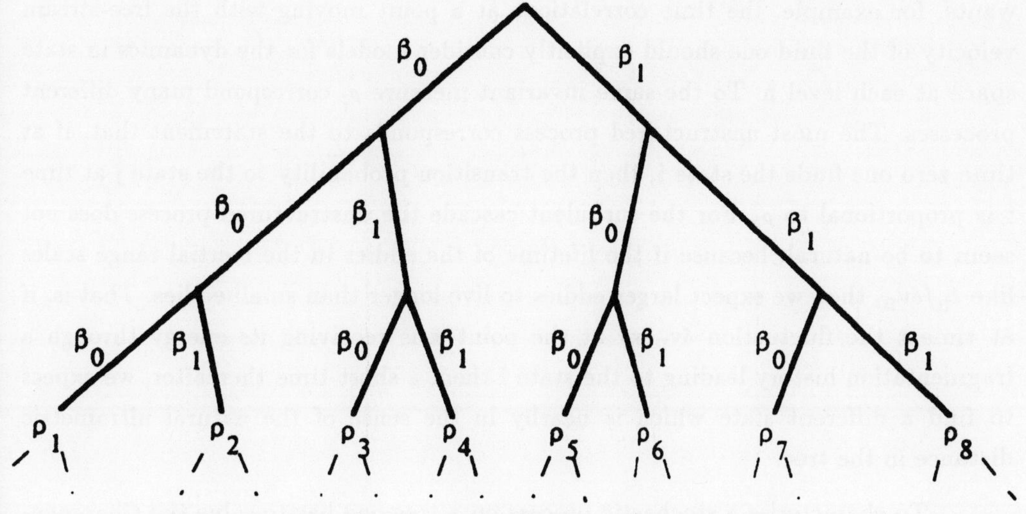
β_0
β_1
β_0
β_1
β_0
β_1
β_0
β_1
β_0
β_1
β_0
β_1
β_0
β_1
ρ_1
ρ_2
ρ_3
ρ_4
ρ_5
ρ_6
ρ_7
ρ_8

FIG. 1

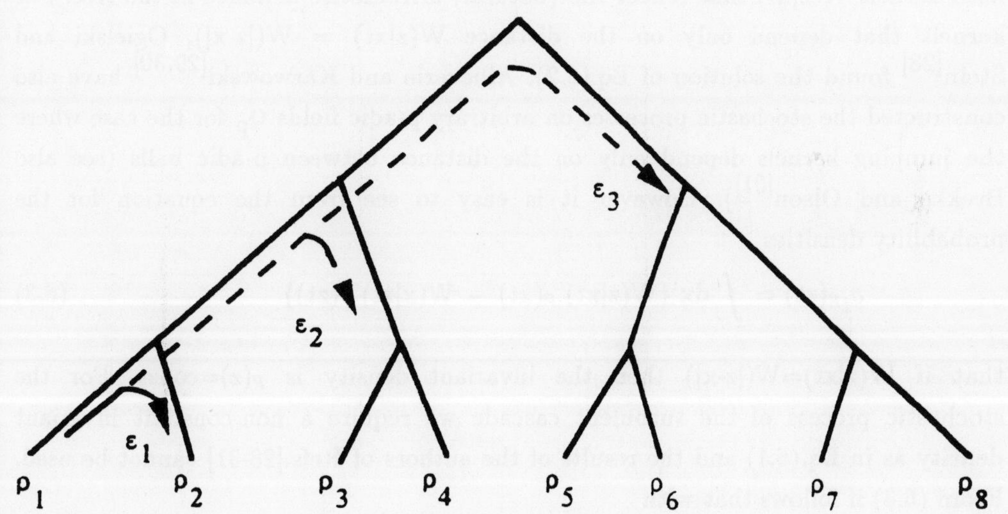
ε_3
ε_2
ε_1
ρ_1
ρ_2
ρ_3
ρ_4
ρ_5
ρ_6
ρ_7
ρ_8

FIG. 2

wants, for example, the time correlations at a point moving with the free-stream velocity of the fluid one should explicitly consider models for the dynamics in state space at each level n. To the same invariant measure ρ_i correspond many different processes. The most unstructured process corresponds to the statement that, if at time zero one finds the state i, then the transition probability to the state j at time t is proportional to ρ_j. For the turbulent cascade the unstructured process does not seem to be natural, because if the lifetime of the eddies in the inertial range scales like $\ell_n/\delta v_n$, then we expect larger eddies to live longer than small eddies. That is, if at time t the fluctuation $\delta v_n(x)$ at the point x is receiving its energy through a fragmentation history leading to the state i then, a short time thereafter, we expect to find a different state which is nearby in the sense of the natural ultrametric distance in the tree.

To characterise a stochastic process on a tree one has to solve the Chapman-Kolmogorov equation for the transition probabilities

$$\partial_t p(zt|y0) = \int dx \, \{W(z|xt)\, p(xt|y0) - W(x|zt)\, p(zt|y0)\} \tag{6.2}$$

with kernels W(z|xt) that reflect the (natural) ultrametric distance in the tree. For kernels that depend only on the distance $W(z|xt) = W(|z-x|)$, Ogielski and Stein[28] found the solution of Eq.(6.2). Albeverio and Karwowski[29,30] have also constructed the stochastic processes on arbitrary p-adic fields $\mathbf{Q}_p$ for the case where the jumping kernels depend only on the distance between p-adic balls (see also Brekke and Olson[31]). However it is easy to see from the equation for the probability densities

$$\partial_t \rho(zt) = \int dx \, \{W(z|xt)\, \rho(xt) - W(x|zt)\, \rho(zt)\} \tag{6.3}$$

that if $W(z|xt)=W(|z-x|)$ then the invariant density is $\rho(z)$=const. For the stochastic process of the turbulent cascade we require a non-constant invariant density as in Eq.(6.1) and the results of the authors of Refs.[28-31] cannot be used. From (6.3) if follows that with

$$W(z|xt) = \rho(z)\, f(|z-x|) \tag{6.4}$$

the invariant density is $\rho(z)$ and, at the same time, full account is taken of the dependence of the transition probability on the distance between the points z and x in state space. First we characterise the solutions of the Chapman-Kolmogorov equation for kernels of the form (6.4) and then show how to use these solutions to compute (or parametrise) the time correlations of the turbulent cascade.

Random walk on a tree with asymmetric distance-dependent jumping kernels

Eq.(6.3) in matrix form is

$$\frac{\partial}{\partial t}\rho(t) = W\ \rho(t) \tag{6.5}$$

where W is the matrix

$$\begin{bmatrix}
W_{11} & \rho_1\epsilon_1 & \rho_1\epsilon_2 & \rho_1\epsilon_2 & \rho_1\epsilon_3 & \rho_1\epsilon_3 & \rho_1\epsilon_3 & \rho_1\epsilon_3 & \cdot & \cdot \\
\rho_2\epsilon_1 & W_{22} & \rho_2\epsilon_2 & \rho_2\epsilon_2 & \rho_2\epsilon_3 & \rho_2\epsilon_3 & \rho_2\epsilon_3 & \rho_2\epsilon_3 & \cdot & \cdot \\
\rho_3\epsilon_2 & \rho_3\epsilon_2 & W_{33} & \rho_3\epsilon_1 & \rho_3\epsilon_3 & \rho_3\epsilon_3 & \rho_3\epsilon_3 & \rho_3\epsilon_3 & \cdot & \cdot \\
\rho_4\epsilon_2 & \rho_4\epsilon_2 & \rho_4\epsilon_1 & W_{44} & \rho_4\epsilon_3 & \rho_4\epsilon_3 & \rho_4\epsilon_3 & \rho_4\epsilon_3 & \cdot & \cdot \\
\rho_5\epsilon_3 & \rho_5\epsilon_3 & \rho_5\epsilon_3 & \rho_5\epsilon_3 & W_{55} & \rho_5\epsilon_1 & \rho_5\epsilon_2 & \rho_5\epsilon_2 & \cdot & \cdot \\
\rho_6\epsilon_3 & \rho_6\epsilon_3 & \rho_6\epsilon_3 & \rho_6\epsilon_3 & \rho_6\epsilon_1 & W_{66} & \rho_6\epsilon_2 & \rho_6\epsilon_2 & \cdot & \cdot \\
\rho_7\epsilon_3 & \rho_7\epsilon_3 & \rho_7\epsilon_3 & \rho_7\epsilon_3 & \rho_7\epsilon_2 & \rho_7\epsilon_2 & W_{77} & \rho_7\epsilon_1 & \cdot & \cdot \\
\rho_8\epsilon_3 & \rho_8\epsilon_3 & \rho_8\epsilon_3 & \rho_8\epsilon_3 & \rho_8\epsilon_2 & \rho_8\epsilon_2 & \rho_8\epsilon_1 & W_{88} & \cdot & \cdot \\
\cdot & \cdot & \cdot & \cdot & \cdot & \cdot & \cdot & \cdot & \cdot & \cdot \\
\cdot & \cdot & \cdot & \cdot & \cdot & \cdot & \cdot & \cdot & \cdot & \cdot
\end{bmatrix}$$

Notice that the matrix has increasingly larger non-diagonal blocks of size $2^i \times 2^i$ which have the common factor ϵ_i. These blocks correspond to jumps to a ultrametric distance i. A matrix element in a block of size $2^i \times 2^i$ at line k equals $\rho_k\epsilon_i$, ϵ_i being the value of f(|z-x|) in Eq.(6.4) for a jump to a distance i. The elements W_{ii} in the diagonal are such that the columns add to zero.

As in the symmetric case studied by Ogielski and Stein[28] one finds the complete set of eigenvectors of the matrix W. For a matrix of dimension 2^n, which describes the stochastic process at the nth level of the tree, the eigenvectors are:

(i) The eigenvector

$$\begin{bmatrix} \rho_1 \\ \rho_2 \\ \cdot \\ \cdot \\ \cdot \\ \rho_{2^n} \end{bmatrix}$$

with eigenvalue $\lambda_o = 0$;

(ii) n classes with 2^{n-k} (k=1 ... n) eigenvectors each, where each eigenvector has only 2^k non-zero elements. The first 2^{k-1} elements are positive and the others are negative. The 2^k non-zero elements of an eigenvector have a common ancestor, in the tree, at the level n-k. The non-zero elements of an (unnormalized) eigenvector are formed by multiplying the corresponding ρ_i by the sum of the ρ_j's of the complementary group in the non-zero set of elements. The formation rule is easier to understand from an example. Let n=3 (Fig. 2). There are then three classes of eigenvectors in the group (ii), typical examples of which are:

a)

$$\begin{bmatrix} \rho_1\rho_2 \\ -\rho_2\rho_1 \\ 0 \\ 0 \\ 0 \\ 0 \\ 0 \\ 0 \end{bmatrix}$$

$\lambda_1 = -(\epsilon_1(\rho_1+\rho_2)+\epsilon_2(\rho_3+\rho_4)+\epsilon_3(\rho_5+\rho_6+\rho_7+\rho_8))$

Four eigenvectors of this type corresponding to the independent groups of two elements with a common ancestor at level 2.

b)

$$\begin{bmatrix} \rho_1(\rho_3+\rho_4) \\ \rho_2(\rho_3+\rho_4) \\ -\rho_3(\rho_1+\rho_2) \\ -\rho_4(\rho_1+\rho_2) \\ 0 \\ 0 \\ 0 \\ 0 \end{bmatrix}$$

$\lambda_5 = -(\epsilon_2(\rho_1+\rho_2+\rho_3+\rho_4)+\epsilon_3(\rho_5+\rho_6+\rho_7+\rho_8))$

Two eigenvectors of this type.

c)

$$\begin{bmatrix} \rho_1(\rho_5+\rho_6+\rho_7+\rho_8) \\ \rho_2(\rho_5+\rho_6+\rho_7+\rho_8) \\ \rho_3(\rho_5+\rho_6+\rho_7+\rho_8) \\ \rho_4(\rho_5+\rho_6+\rho_7+\rho_8) \\ -\rho_5(\rho_1+\rho_2+\rho_3+\rho_4) \\ -\rho_6(\rho_1+\rho_2+\rho_3+\rho_4) \\ -\rho_7(\rho_1+\rho_2+\rho_3+\rho_4) \\ -\rho_8(\rho_1+\rho_2+\rho_3+\rho_4) \end{bmatrix}$$

$\lambda_7 = \ -\epsilon_3(\rho_1+\rho_2+\rho_3+\rho_4+\rho_5+\rho_6+\rho_7+\rho_8) \ = \ -\epsilon_3$

One eigenvector of this type.

The rule of formation for the eigenvalues is clear from the example above. Each eigenvalue is a sum of terms

$$\lambda_i = -\sum_{j=l(i)}^{n} \epsilon_j \sum_k \rho_k \tag{6.6}$$

where $\epsilon_{l(i)}$ is the first ϵ which covers all the non-zero elements of the vector and the sum $\sum_k \rho_k$ contains all the probability densities of the states reached by an ϵ_j – jump. A solution of Eq.(6.5) is an arbitrary superposition

$$\rho(t) = \sum c_i \, e^{-\lambda_i t} \, v_i \tag{6.7}$$

of the eigenvectors above. From (6.7) it is now easy to construct the general solution of the Chapman-Kolmogorov equation. To obtain the transition probability p(jt|i0) from the state i at time zero to the state j at time t, one chooses the coefficients c_i in (6.7) in such a way that, at time zero, only the i-th component is non-zero and then read the value of the j-th component at time t. Before stating the general result we illustrate it by writing the transition probabilities for transitions between typical states in each group for the case n=3.

$$p(1t|10)=\frac{\rho_1}{\rho_1+\dots+\rho_8}+\frac{\rho_1\rho_2}{\rho_1(\rho_1+\rho_2)}\,e^{-t\lambda_1}+\frac{\rho_1(\rho_3+\rho_4)}{(\rho_1+\rho_2)(\rho_1+\rho_2+\rho_3+\rho_4)}\,e^{-t\lambda_5}+$$

$$\frac{\rho_1(\rho_5+\dots+\rho_8)}{(\rho_1+\dots+\rho_4)(\rho_1+\dots+\rho_8)}\,e^{-t\lambda_7}$$

$$p(2t|10)=\frac{\rho_2}{\rho_1+\dots+\rho_8}-\frac{\rho_2\rho_1}{\rho_1(\rho_1+\rho_2)}\,e^{-t\lambda_1}+\frac{\rho_2(\rho_3+\rho_4)}{(\rho_1+\rho_2)(\rho_1+\rho_2+\rho_3+\rho_4)}\,e^{-t\lambda_5}+$$

$$\frac{\rho_2(\rho_5+...+\rho_8)}{(\rho_1+...+\rho_4)(\rho_1+...+\rho_8)} e^{-t\lambda_7}$$

$$p(3t|10)=\frac{\rho_3}{\rho_1+...+\rho_8} - \frac{\rho_3(\rho_1+\rho_2)}{(\rho_1+\rho_2)(\rho_1+\rho_2+\rho_3+\rho_4)} e^{-t\lambda_5} +$$

$$\frac{\rho_3(\rho_5+...+\rho_8)}{(\rho_1+...+\rho_4)(\rho_1+...+\rho_8)} e^{-t\lambda_7}$$

$$p(5t|10)=\frac{\rho_5}{\rho_1+...+\rho_8} - \frac{\rho_5(\rho_1+...+\rho_4)}{(\rho_1+...+\rho_4)(\rho_1+...+\rho_8)} e^{-t\lambda_7} \tag{6.8}$$

Of course in this case $\rho_1+...+\rho_8=1$ but we have kept this term to emphasise the rule of formation of the coefficients. The general rule for the transition probability p(jt|i0) between two states i and j at the level n in the tree is the following:

(i) p(jt|i0) is a sum of terms, the first of which is ρ_j (the target probability), and has as many terms as the number of eigenvectors that have non-zero elements in both the i and the j positions.

$$p(j\ t|i\ 0) = \rho_j + \sum_k c_k e^{-t\lambda_k} \tag{6.9}$$

(ii) The exponential factor in each term contains the eigenvalue of the associated eigenvector.

(iii) The coefficients all contain in the numerator the target probability ρ_j multiplied by the sum of the probabilities of non-zero elements of the corresponding eigenvectors in the half that does not contain j. The denominator is the sum of the half that contains i multiplied by the sum of all probabilities associated to the non-zero elements of the eigenvector.

(iv) The sign of the coefficient is the product of the signs of the i and j entries in the eigenvector.

Here I have dealt with an asymmetric stochastic process on a finite tree. If however the regular tree is infinite the appropriate labelling set for the states is a field $\mathbf{Q}_p$ of p-adic numbers. The stochastic processes on the infinite tree are stochastic processes in $\mathbf{Q}_p$[32]. Processes on finite trees may actually be obtained from the processes on $\mathbf{Q}_p$ by considering p-adic balls of appropriate radius.

Time correlations in the turbulent cascade

Once the transition probabilities $p(zt|y0)$, solutions of the Chapman-Kolmogorov equation are known, the time correlations of the process are obtained from

$$< x(t)\, x(0) > \; = \int dy\, dx\;\; y\;\; p(yt|x0)\;\; x\, \rho(x) \tag{6.10}$$

or

$$< x(t)x(0) > \; = \sum_{i,j} x_j\; p(jt|i0)\; x_i\; \rho_i \tag{6.11}$$

for a discrete state space. Using the results derived above (Eq.(6.9)) one sees that at large times the time correlation at level n will be dominated by the largest non-zero eigenvalue $\lambda_{2^n-1} = -\epsilon_n$. Assuming that the dynamics of the turbulent cascade is controlled by the decay of the eddies, the largest non-zero eigenvalue will always be the same, associated to the mean lifetime of large eddies. However the asymptotic long-time correlation will be difficult to measure because of the small values of $< x(t)x(0) >$ at large t. If, as proposed, the time correlations in the turbulent cascade are described by a stochastic process with kernels that depend on (ultrametric) tree distances, a first qualitative prediction is the occurrence of several exponential slopes as the time increases. Of special interest is the slope of the short-time correlation which is controlled by the smallest eigenvalue. Using the dyadic expansion to label the points x_i in state space

$$x_i = \beta_0^{n_0(i)}\; \beta_1^{n_1(i)} \tag{6.12}$$

$$\rho_i = \gamma^{n_0(i)}\; (1-\gamma)^{n_1(i)} \tag{6.13}$$

where $n_0(i)$ and $n_1(i)$ are the number of zeros and ones in the dyadic expansion of i. Assuming $\gamma > (1-\gamma)$, the smallest eigenvalue for the dynamics at level n is

$$\lambda_1^{(n)} = -\{\epsilon_1^{(n)}(\rho_1^{(n)}+\rho_2^{(n)})+\epsilon_2^{(n)}(\rho_3^{(n)}+\rho_4^{(n)})+\epsilon_3^{(n)}(\rho_5^{(n)}+\ \dots\ +\rho_8^{(n)})+$$

$$\epsilon_4^{(n)}(\rho_9^{(n)}+\ \dots\ +\rho_{16}^{(n)}\,)+\ \dots\} \tag{6.14}$$

If the dynamics of the turbulent cascade is associated to the decay of the eddies of different sizes, it is reasonable to assume that

$$\epsilon_i^{(n-1)} = \epsilon_{i+1}^{(n)} \tag{6.15}$$

Using this relation and the relations between the probability densities at the levels n and n-1 one obtains

$$\lambda_1^{(n)} - \lambda_1^{(n-1)} = -\left(\epsilon_1^{(n)} - \epsilon_1^{(n-1)}\right)\rho_1^{(n-1)} \tag{6.16}$$

One concludes that the ratio of short-time correlations measures the difference between the lifetimes of the structures at different length scales. If the dynamics of the cascade is controlled by the decay of the eddies and these have different lifetimes at different scales, the ultrametric stochastic model is an appropriate way to parametrise the dynamics and to characterise it in quantitative terms. Other models yield different correlation structures. Notice that here we are concerned with the time fluctuations of the turbulent cascade itself, not with the changes induced by the overall motion of the fluid. This means that for a fluid in motion with free-stream velocity $\vec{U}$ the correlations to measure, for an observable Δ, are $< \Delta(x+\vec{U}t,t)\ \Delta(x,0) >$.

7. A short introduction to q-bialgebras

The concept of symmetry plays an essential role in the description of physical phenomena. In most cases this symmetry is related to covariance under the transformations induced by a Lie algebra. A generalisation of this mathematical structure, the q – deformed (or quantum) algebras, has recently emerged[33 – 39]. q – deformed algebras, first discovered in the context of integrable lattice models, were later identified as an underlying mathematical structure in topological field theories[40] and rational conformal field theories[41]. Other attempts to apply the notion of q – deformed algebras cover a wide range of different domains, from space – time symmetries[42 – 44] to gauge fields[45], to quantum chemistry[46].

The usual way to introduce q-algebras is through the discussion of the symmetries of exactly integrable models. Here, for the sake of those less familiar with exactly integrable models, I have decided to use the q-calculus approach.

q-calculus ≡ calculus of finite differences on an exponential net

Consider the following finite-difference operators

$$D_+ f(x) = \frac{f(x_{n+1}) - f(x_n)}{x_{n+1} - x_n} \tag{7.1}$$

$$D_- f(x) = \frac{f(x_n) - f(x_{n-1})}{x_n - x_{n-1}} \tag{7.2}$$

$$D_0 f(x) = \frac{f(x_{n+1/2}) - f(x_{n-1/2})}{x_{n+1/2} - x_{n-1/2}} \tag{7.3}$$

For numerical integration, for example, one in general considers a uniform net $x_n = a+nh$. However to explore very large or very small scales it is conceivable that an exponential net might be more convenient, that is, $x_n = a\, q^{2n}$. Then the finite-difference operators take the form:

$$D_+ f(x) = \frac{f(q^2x) - f(x)}{(q^2-1)x} \tag{7.4}$$

$$D_- f(x) = \frac{f(x) - f(xq^{-2})}{(1-q^{-2})x} \tag{7.5}$$

$$D_0 f(x) = \frac{f(qx) - f(xq^{-1})}{(q-q^{-1})x} \tag{7.6}$$

Applying the symmetric difference operator to a power of x

$$D_0\, x^n = \frac{q^n - q^{-n}}{q-q^{-1}}\, x^{n-1} = [n]_q\, x^{n-1} \tag{7.7}$$

The rule of derivation of powers is similar to the usual one if one defines *q-numbers* $[n]_q = (q^n - q^{-n})/(q-q^{-1})$. Notice that $[n]_q \to 1$ when $q \to 1$.

Some properties of q – numbers

$$[-n]_q = -[n]_q \qquad [n]_{q^{-1}} = [n]_q$$

$$[a+b]_q = [a]_q\, q^b + q^{-a}\, [b]_q$$

$$[a+1]_q\, [b+1]_q - [a]_q [b]_q = [a+b+1]_q$$

$$[a]_q\, [b+c]_q = [a+c]_q\, [b]_q + [a-b]_q\, [c]_q$$

$$[a]_q^2 - [b]_q^2 = [a-b]_q\, [a+b]_q$$

From now on, whenever no confusion arises, I will drop the subscript q in $[n]_q$, writing only [n]. The Leibnitz rule for q – calculus is

$$\begin{aligned} D_0 \{f_1(x) f_2(x)\} &= f_1(qx)\, D_0 f_2(x) + f_2(q^{-1}x)\, D_0 f_1(x) \\ &= f_1(q^{-1}x)\, D_0 f_2(x) + f_2(qx)\, D_0 f_1(x) \end{aligned}$$

We now use the q – calculus derivative operator to define the standard algebras.

Define creation and annihilation operators in the usual way

$$a^\dagger = x \qquad\qquad a = D_0$$

They act on the basis

$$|\,n> = \frac{x^n}{\sqrt{[n]\,!}} \qquad\qquad [n]\,! = [n][n-1]\,...\,[1] \tag{7.8}$$

as follows

$$a^\dagger\,|\,n> \;=\; \sqrt{[n+1]}\;\;|\,n+1> \qquad\qquad a\,|\,n> \;=\; \sqrt{[n]}\;|\,n-1> \tag{7.9}$$

Together with the number operator, $N\,|\,n> = n\,|\,n>$, they form the *q – deformed Heisenberg algebra*

$$a\,a^\dagger - q^{-1}\,a^\dagger a = q^N \tag{7.10a}$$

$$[N,a^\dagger] = a^\dagger \tag{7.10b}$$

$$[N,a] = -a \tag{7.10c}$$

q – calculus is symmetric for the replacement $q \to q^{-1}$. Hence

$$a\,a^\dagger - q\,a^\dagger a = q^{-N} \tag{7.10d}$$

In the undeformed case, the algebra of SU(2) may be obtained from a pair of mutually commuting boson operators (Jordan-Schwinger map)

$$J_+ = a_1\,a_2^\dagger \qquad\qquad J_- = a_2\,a_1^\dagger \qquad\qquad J_z = \tfrac{1}{2}\,(N_1 - N_2) \tag{7.11}$$

Using the same definition with q – deformed creation and annihilation operators one obtains the $SU_q(2)$ algebra

$$[J_3^{(q)},J_\pm^{(q)}] = \pm\,J_\pm^{(q)} \tag{7.12a}$$

$$[J_+^{(q)},J_-^{(q)}] = [2\,J_3^{(q)}]_q = \frac{q^{2J_3^{(q)}} - q^{-2J_3^{(q)}}}{q - q^{-1}} \tag{7.12b}$$

The $SU_q(2)$ irreducible representations have a structure similar to the SU(2) case

$$J_3^{(q)}\,|\,j\;m> \;=\; m\,|\,j\;m> \tag{7.13a}$$

$$J_+^{(q)}\,|\,j\;m> \;=\; \sqrt{[j-m][j+m+1]}\;|\,j\;m+1> \tag{7.13b}$$

$$J_-^{(q)}\,|\,j\;m> \;=\; \sqrt{[j+m][j-m+1]}\;|\,j\;m-1> \tag{7.13c}$$

with Casimir operator

$$C_2 = J_-^{(q)}J_+^{(q)} + [J_3^{(q)} + \tfrac{1}{2}]_q^2$$

$$C_2 \mid j\, m> \;=\; [j+\tfrac{1}{2}]_q^2 \mid j\, m>$$

The construction leading to $SU_q(2)$ may be generalised to all classical simple algebras. Let $\{X_i^{\pm}, H_i\}$ be the generators in the Chevalley-Weyl basis and (a_{ij}) the Cartan matrix. Then the q-deformed algebra is

$$[H_i, H_j] = 0 \qquad\qquad [H_j, X_i^{\pm}] = a_{ij}\, X_i^{\pm}$$

$$[X_i^{+}, X_j^{-}] = \delta_{ij}\, [H_i]_q \quad \text{or} \quad [X_i^{+}, X_i^{-}] = 0 \quad \text{if } a_{ij} = 0$$

The q – deformation of the Lie algebras is not, in a strict sense, a deformation of the algebra because the commutation relations involve elements that are not in the algebra but in the universal enveloping algebra. Besides the commutation relations what else changes in the algebraic structure, under q – deformation? Suppose that we want to couple two of the q – deformed angular momentum representation defined in (7.13). One finds that for the commutation relations to be maintained for the total angular momentum, they have to be coupled in the following way:

$$J_3^{(q)}(1,2) = J_3^{(q)}(1) \otimes 1 + 1 \otimes J_3^{(q)}(2)$$

$$J_{\pm}^{(q)}(1,2) = J_{\pm}^{(q)}(1) \otimes q^{J_3^{(q)}(2)} + q^{-J_3^{(q)}(1)} \otimes J_{\pm}^{(q)}(2)$$

Therefore, besides the modification in the algebra product $\mathcal{A} \otimes \mathcal{A} \to \mathcal{A}$, there is also a modification in the coupling operation $\mathcal{A} \to \mathcal{A} \otimes \mathcal{A}$. The coupling of angular momenta is an example of a general operation called *comultiplication.* One sees that q – deformation involves in fact an algebraic structure with two operations, that is, a *bialgebra.* More precisely a q – deformed algebra (or q – bialgebra) is a q – deformation of a Hopf algebra.

<u>Hopf algebras</u>

A Hopf algebra is a bialgebra $(\mathcal{A}, m, \eta, \Delta, \epsilon)$ with an antipode S. m denotes the multiplication, Δ the comultiplication, η the unit and ϵ the counit. These operations are defined by the following commutative diagrams

$m : \mathcal{A} \otimes \mathcal{A} \to \mathcal{A}$

$\eta : \mathbf{C} \to \mathcal{A}$

m and η are linear

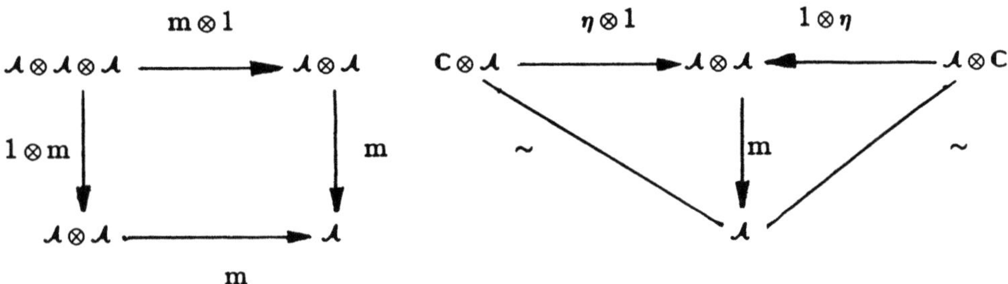

$$\Delta : \mathcal{A} \to \mathcal{A} \otimes \mathcal{A}$$

$$\epsilon : \mathcal{A} \to \mathbf{C}$$

Δ and ϵ are linear and $\Delta(ab) = \Delta(a)\,\Delta(b)$

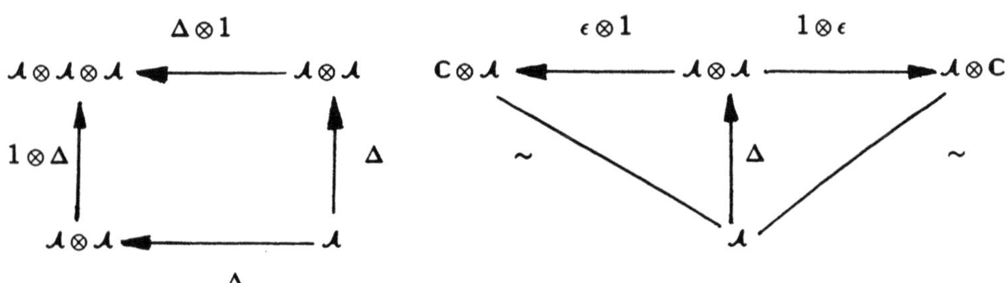

$$S : \mathcal{A} \to \mathcal{A}$$

$$m(S \otimes 1)\Delta \; = \; m(1 \otimes S)\Delta \; = \; \eta \circ \epsilon$$

Examples of Hopf algebras

(i) Continuous functions on a topological group G

$$\big(f \cdot h\big)(g) \; = \; f(g)\, h(g)$$

$$\big(\Delta(f)\big)(g_1 \otimes g_2) \; = \; f(g_1 g_2)$$

$$\eta(x) \; = \; x\, 1 \qquad x \in \mathbf{C}$$

$$\epsilon(f) = f(e)$$

$$S(f)(g) = f(g^{-1})$$

is a commutative Hopf algebra.

(ii) Universal enveloping algebra $\mathcal{U}(\mathcal{L})$ of a Lie algebra $\mathcal{L}$

multiplication in $\mathcal{U}(\mathcal{L})$

$$\Delta(x) = x \otimes 1 + 1 \otimes x$$

$$\eta(\alpha) = \alpha\, 1$$

$$\epsilon(1) = 1 \quad \text{and zero otherwise}$$

$$S(x) = -x$$

is a co-commutative Hopf algebra.

8. q – deformed Brownian motion[47]

The q-deformation idea leads to interesting structures in the context of Lie algebras and superalgebras. It is therefore natural to ask whether the notion of q-deformation can also be extended to other (non-algebraic) mathematical structures. Here I will extend it to stochastic processes. The starting point is the well-known embedding of Brownian motion in the Fock space of bosonic free fields in one dimension[48,49]. Extending this correspondence to a one-parameter family of creation and annihilation operators satisfying a q – deformed algebra one establishes a q – deformation of Brownian motion.

Because q – deformation of the boson operators is invariant under the replacement $q \to q^{-1}$, the algebra may be written in an explicitly symmetric form

$$aa^{\dagger} - \tfrac{1}{2}(q+q^{-1})\, a^{\dagger}a = \tfrac{1}{2}(q^{N} + q^{-N}) \tag{8.1}$$

Consider now a family $\{a_{\tau}, a_{\tau}^{\dagger}\}$ of q – deformed operators labelled by a continuous time parameter and a scalar field

$$\phi_{\tau} = a_{\tau} + a_{\tau}^{\dagger} \tag{8.2}$$

For the $\{a_{\tau}, a_{\tau}^{\dagger}\}$ family generalise the relations (8.1) and (7.10b,c) to

$$a_{\tau_1} a_{\tau_2}^{\dagger} - \tfrac{1}{2}(q+q^{-1})\; a_{\tau_2}^{\dagger} a_{\tau_1} = \tfrac{1}{2}(q^{N}+q^{-N})\; \delta(\tau_1 - \tau_2) \tag{8.3}$$

$$[N, a^\dagger_\tau] = a^\dagger_\tau \qquad [N, a_\tau] = -a_\tau \tag{8.4}$$

This is the simplest extension of the relations to a family of q – deformed operators labelled by a continuous parameter. Other generalisations are possible, involving for example braid relations at different times. Notice also that, for the purpose of construction of a stochastic process, no assumptions are needed concerning the commutation properties of $a_{\tau_1}\ a_{\tau_2}$ and $a^\dagger_{\tau_1}\ a^\dagger_{\tau_2}$ at different times.

Smearing the fields with characteristic functions $\chi_{[0,t]}$ of the interval [0,t]

$$a_q(t) = a_q(\chi_{[0,t]}) = \int_0^t d\tau\ a_\tau \tag{8.5a}$$

$$a^\dagger_q(t) = a^\dagger_q(\chi_{[0,t]}) = \int_0^t d\tau\ a^\dagger_\tau \tag{8.5b}$$

$$\phi_q(t) = \phi_q(\chi_{[0,t]}) = \int_0^t d\tau\ \phi_\tau \tag{8.6}$$

the algebraic relations become

$$a_q(t_1)\ a^\dagger_q(t_2) - \tfrac{1}{2}(q+q^{-1})\ a^\dagger_q(t_2)\ a_q(t_1) = \tfrac{1}{2}(q^N+q^{-N})\ <\tau_1|\tau_2> \tag{8.7}$$

where $<\tau_1|\tau_2> = \min(t_1,t_2)$.

One now uses (8.7) to construct a q – deformation of Brownian motion. Let (Ω,F_t,μ,B_t) be the usual Brownian motion. Ω is the set of continuous functions vanishing at $t=0$, μ is the Wiener measure and F_t is the σ-ring generated by $\{B_s: 0\le s\le t\}$. On the other hand let $(\mathcal{H},A_t,\psi_0,\phi_1(t))$ be the free quantum field over $K=L^2([0,\infty),\mathbf{R})$. $\phi_1(t)=\phi_1(\chi_{[0,t]})$ (for $q=1$), $\mathcal{H}$ is the symmetric Fock space over K, ψ_0 is the Fock vacuum and A_t is the W*-algebra generated by $\{\phi_1(s):0\le s\le t\}$. Then[49], interpreting B_t as a multiplication operator in $L^2(\Omega,F_t,\mu)$, there is a unitary operator $V : L^2(\Omega,F_t,\mu)\to\mathcal{H}$ such that $VB_tV^{-1}=\phi(t)$. I.e. $\phi(t)$ as a stochastic process with expectation

$$E(f(\phi(t))) = <\psi_0,f(\phi(t))\ \psi_0> \tag{8.8}$$

coincides with Brownian motion. For this identification of the free scalar field with Brownian motion it is useful to characterise the filtration A_t by the conditional expectation of Wick products[50]

$$E(\ :\phi_1(u_1)\ \dots\ \phi_1(u_n):\ |\ A_t) = \ :\phi_1(\chi_{[0,t]}u_1)\ \cdots\ \phi_1(\chi_{[0,t]}u_n): \tag{8.9}$$

Recall that the Wick products span the algebra generated by $\phi_1(u)$. Hence, by linearity, definition on Wick products suffices to define conditional expectations on the complete algebra.

One now uses a minimal version of this correspondence to define q – deformed Brownian motion.

Definition : q – deformed Brownian motion is the process $(\Omega, F_t, \mu_{\psi_0}, \phi_q(t))$ where

(i) $\phi_q(t)$ is the operator defined by (8.5 – 8.7)

(ii) Expectations of field functionals $f(\phi_q)$ are obtained by

$$E(f(\phi_q)) = \langle \psi_0, f(\phi_q)\, \psi_0 \rangle, \; \psi_0 \text{ being defined by } a_q\, \psi_0 = 0 \tag{8.10}$$

(iii) The filtration F_t is characterised by the conditional expectations of Wick products

$$E(:\phi_q(t_1) \ldots \phi_q(t_n): \mid F_s) = :\phi_q(\chi_{[0,s]}\chi_{[0,t_1]}) \cdots \phi_q(\chi_{[0,s]}\chi_{[0,t_n]}): \tag{8.11}$$

Notice that the algebraic relations (8.7) allow all elements of the algebra generated by $\phi_q(t)$ to be reduced to Wick products, hence all conditional expectations may be computed. Notice also that in this minimal definition the family F_t of measurable events is fixed in advance and we avoid an explicit realisation of the probability space Ω.

Theorem 8.1 : q – deformed Brownian motion has :

(i) zero mean, $E(\phi_q(t)) = 0$

(ii) variance $E(\phi_q(t)\, \phi_q(s)) = \min(s,t)$

(iii) independent increments in the sense

$$E(\{\phi_q(t_1) - \phi_q(t_2)\}\{\phi_q(t_3) - \phi_q(t_4)\}) = E(\phi_q(t_1) - \phi_q(t_2))\; E(\phi_q(t_3) - \phi_q(t_4)) = 0$$

if there is no overlap between the intervals $[t_1,t_2]$ and $[t_3,t_4]$

(iv) is a martingale

Properties (i) to (iii) follow by a simple computation using (8.6) (8.7) and (8.10). The martingale property is a consequence of (8.11). If $s < t$

$$E(\phi_q(t) \mid F_s) = \phi_q(\chi_{[0,s]}\chi_{[0,t]}) = \phi_q(s)$$

Theorem 8.1 summarises the similarities of q – deformed Brownian motion to the usual Brownian motion. The next result displays their main differences as well as an explicit characterisation in terms of cumulants.

Theorem 8.2 :

(i) q – deformed Brownian motion is not a Gaussian process

(ii) The cumulants are

$$E_T(\phi_q(t_1) \dots \phi_q(t_n)) = \sum c_{(i_1 i_2)\dots(i_{n-1} i_n)} < t_{i_1} | t_{i_2} > \dots < t_{i_{n-1}} | t_{i_n} > \quad (8.12)$$

where $< t_i | t_j > = \min(t_i , t_j)$, the sum is over all $\binom{n}{2}$ different partitions of the set $(t_1 \dots t_n)$ into pairs and the coefficients $c_{(i_1 i_2)\dots(i_{n-1} i_n)}$ are obtained by the following graphical rules

(ii.a) For each term in (8.12) one draws the $< t_{i_k} | t_{i_{k+1}} >$ contractions as follows

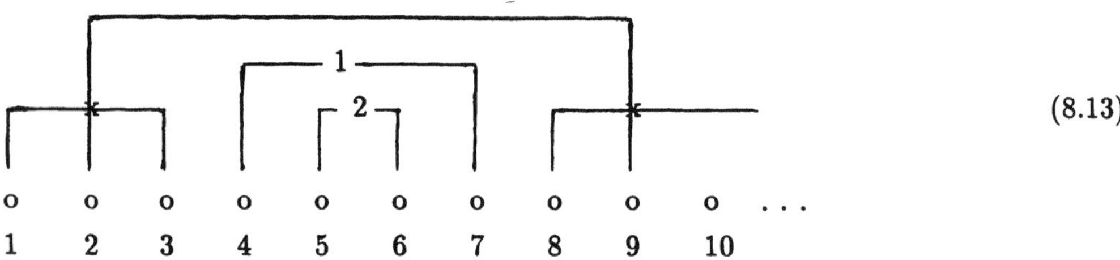

(8.13)

(ii.b) For each crossing of lines there is a factor

$$\{\tfrac{1}{2}(q+q^{-1}) - 1\} \text{ in } c_{(i_1 i_2)\dots(i_{n-1} i_n)}$$

(ii.c) For each contraction contained at depth α inside other contractions there is a factor

$$\{\tfrac{1}{2}(q^\alpha + q^{-\alpha}) - 1\} \text{ in } c_{(i_1 i_2)\dots(i_{n-1} i_n)}$$

(ii.d) If there are no crossings nor inner contractions the coefficient $c_{(i_1 i_2)\dots(i_{n-1} i_n)}$ vanishes.

In the example (8.13) the contribution of the diagram to the coefficient is

$$\{\tfrac{1}{2}(q+q^{-1}) - 1\}^2 \ \{\tfrac{1}{2}(q^2+q^{-2}) - 1\} \ \{\tfrac{1}{2}(q+q^{-1}) - 1\}$$

the first factor coming from the crossings and the last two from the inner contractions at level 2 and 1.

A necessary and sufficient condition for a process to be Gaussian is that it possesses cumulants of all orders and that they vanish for order higher than 2. Computing the 4-time correlation one obtains using (8.7) and (8.10)

$$E(t_1t_2t_3t_4) = E(\phi_q(t_1)\,\phi_q(t_2)\,\phi_q(t_3)\,\phi_q(t_4)) =$$

$$= \langle t_1|t_2\rangle \langle t_3|t_4\rangle + \tfrac{1}{2}(q+q^{-1})\langle t_1|t_3\rangle \langle t_2|t_4\rangle + \tfrac{1}{2}(q+q^{-1})\langle t_1|\, t_4\rangle \langle t_2|t_3\rangle$$

implying that the cumulant

$$E_T(t_1t_2t_3t_4) = E(t_1t_2t_3t_4) - E(t_1t_2)E(t_3t_4) - E(t_1t_3)E(t_2t_4) - E(t_1t_4)E(t_2t_3)$$

does not vanish. Hence the process is not Gaussian.

The explicit expression for the cumulants of arbitrary orders is obtained by systematic reduction of the expectation values using the algebraic relations (8.7). The "crossing lines" factor comes from the coefficient of the second term in the l.h.s. of (8.7) and the "inner contractions" factor from the r.h.s. together with (8.4).

Bozejko and Speicher[51] have also proposed a generalisation of Brownian motion using a modified algebra of boson operators. The algebra they use, however, is not the $q-$deformed Heisenberg algebra but the quon commutation relations of Greenberg[52] which interpolate between bosons and fermions.

As a final remark I point out that using $q-$fermions b and $b^\dagger$ with

$$bb^\dagger + q\, b^\dagger b = q^M \quad ; \quad [M,b^\dagger] = b^\dagger \quad ; \quad [M,b] = -b \tag{8.14}$$

and a generalisation along the lines of (8.2 – 8.7) one may construct a $q-$deformation of the non-commutative Clifford process[50].

References

[1] W. von der Linden; Phys. Reports 220 (1992) 53.

[2] M. Suzuki; Prog. Theor. Phys. 56 (1976) 1454.

[3] J. E. Hirsch, R. L. Sugar, D. J. Scalapino and R. Blankenbecler; Phys. Rev. B26 (1982) 5033.

[4] I. Morgenstern; Z. Phys. B 77 (1989) 267.

[5] S. Eleutério and R. Vilela Mendes; *Stochastic ground state processes*, Lisbon preprint IFM/4/93, to appear in Phys. Rev. B.

[6] S. Albeverio, R. Høegh-Krohn and L. Streit; J. Math Phys. 18 (1977) 907; 21 (1980) 1636.

[7] R. Vilela Mendes; J. Math. Phys. 27 (1986) 178.

[8] A. D. Wentzell and M. I. Freidlin; Russ. Math. Surv. 25 (1970) 1; *Random Perturbations of Dynamical Systems* (Springer, Berlin, 1984).

[9] A. Friedman; *Stochastic Differential Equations and Applications* (Academic

Press, New York, 1976), Vol. 2, Chap. 14.

[10] G. Jona-Lasinio, F. Martinelli and E. Scoppola; Commun. Math. Phys. 80 (1981) 223; Phys. Rep. 77 (1981) 313.

[11] R. Vilela Mendes; Phys. Rev. D37 (1988) 1608; Phys. Lett. B202 (1988) 369 ; Phys. Lett. B223 (1989) 83.

[12] R. L. Hudson and K. R. Parthasarathy; Commun. Math. Phys. 93 (1984) 301.

[13] P. A. Meyer; "*Élements de Probabilités Quantiques*" in "*Séminaire de Probabilités XX - XXIII*", Lecture Notes in Mathematics 1204, 1247, 1321, 1372, Springer 1986, 1987, 1988, 1989.

[14] R. Vilela Mendes; Phys. Lett. A113 (1985) 187.

[15] R. Vilela Mendes; Journ. of Phys.: Condensed Matter 3 (1991) 6757.

[16] H. De Raedt and W. von der Linden; Phys. Rev. B45 (1992) 8787.

[17] H. De Raedt and M. Frick; Phys. Reports 231 (1993) 107.

[18] L. F. Richardson; Proc. R. Soc. London A110 (1926) 709.

[19] A. N. Kolmogorov; Dokl. Akad. Nauk SSSR 30,301; 31,338; 31,99; 32,16 (1941)

[20] U. Frisch; Proc. R. Soc. London A434 (1991) 89.

[21] F. Anselmet, Y. Cagne, E. J. Hopfinger and R. A. Antonia; J. Fluid Mech. 140 (1984) 63.

[22] B. B. Mandelbrot; J. Fluid Mech. 62 (1974) 331.

[23] U. Frisch, P. Sulem and M. Nelkin; J. Fluid Mech. 87 (1978) 719.

[24] E. D. Siggia; J. Fluid Mech. 107 (1981) 375.

[25] U. Frisch and G. Parisi; in "*Turbulence and Predictability in Geophysical Fluid Dynamics*" (ed. M. Gil, R. Benzi and G. Parisi) pp. 84-88, North Holland, 1985

[26] R. Benzi, G. Paladin, G. Parisi and A. Vulpiani; J. Phys. A: Math. Gen. 17 (1984) 3521.

[27] R. Lima and R. Vilela Mendes; "*A stochastic process for the dynamics of the turbulent cascade*", Marseille preprint CPT – 93/P.2965.

[28] A. T. Ogielski and D. L. Stein; Phys. Rev. Lett. 55 (1985) 1634.

[29] S. Albeverio and W. Karwowski; "Diffusion on p-adic numbers", Proceedings of Nagoya Conference, ed. T. Hida, 1991

[30] S. Albeverio and W. Karwowski; "A random walk on p-adics - The generator and its spectrum", Bochum preprint SFB 237/no.135 (1991)

[31] L. Brekke and M. Olson; "p-adic diffusion in glasses", Chicago preprint EFI-89-23 (1989)

[32] W. Karwowski and R. Vilela Mendes; "*Hierarchical structures and asymmetric stochastic processes on p-adics and adeles*", preprint IHES/M/93/65, to appear in J. Math. Phys.

[33] E. K. Sklyanin; Funct. Anal. Appl. 16 (1982) 263.

[34] P. P. Kulish and N. Yu. Reshetikhin; J. Sov. Math. 23 (1983) 2435.

[35] V. G. Drinfel'd; Sov. Math. Dokl. 32 (1985) 254; 36 (1988) 212; J. Sov. Math 41 (1988) 898.

[36] M. Jimbo; Lett. Math Phys. 10 (1985) 63; 11 (1986) 247.

[37] S. L. Woronowicz; Comm. Math. Phys. 111 (1987) 613.

[38] L. D. Faddeev, N. Yu. Reshetikhin and L. A. Takhtajan; Alg. Anal. 1 (1988) 129.

[39] Yu. I. Manin; Ann Inst. Fourier 37 (1987) 191; Comm. Math. Phys. 123 (1989) 163.

[40] E. Witten; Comm. Math. Phys. 121 (1989) 351.

[41] L. D. Faddeev; Comm. Math. Phys. 132 (1990) 131 and references therein.

[42] P. Podles and S. L. Woronowicz; Commun. Math. Phys. 130 (1990) 381.

[43] C. Gomez and G. Sierra; Phys. Lett. B255 (1991) 51.

[44] J. Lukierski, A. Nowicki, H. Ruegg and V. N. Tolstoy; Phys. Lett. B264 (1991) 331.

[45] I. Ya. Aref'eva and I. V. Volovich; Mod. Phys. Lett A6 (1991) 893; Phys. Lett. 264B (1991) 62.

[46] M. Kibler and T. Négadi; Lyon preprint LYCEN 9121 (1991).

[47] V. I. Man'ko and R. Vilela Mendes; Phys. Lett. A180 (1993) 39.

[48] T. Hida, "Brownian motion", Springer, Berlin 1980.

[49] R. F. Streater, Acta Physica Austriaca, Suppl. XXVI (1984) 53.

[50] R. F. Streater, "The fermion stochastic calculus I" in "Stochastic Processes – Mathematics and Physics", S. Albeverio, Ph. Blanchard and L. Streit (Eds.), Springer Lecture Notes in Mathematics 1158, Berlin, 1986.

[51] M. Bozejko and R. Speicher; Commun. Math. Phys. 137 (1991) 519.

[52] O. W. Greenberg; Bull. Am. Phys. Soc. 35 (1990) 981; Phys. Rev. D43 (1991) 4111.

STOCHASTIC PARTIAL DIFFERENTIAL EQUATIONS AND APPLICATIONS TO HYDRODYNAMICS

BERNT ØKSENDAL
Dept. of Mathematics
University of Oslo
Box 1053, Blindern
N-0316 Oslo, NORWAY

CONTENTS

A. I. Cardoso et al. (eds.), Stochastic Analysis and Applications in Physics, 283–305.

CHAPTER 0. INTRODUCTION

The purpose of these notes is to explain how white noise and related methods can be used to model and describe important random dynamical phenomena. Because of personal interest and activity so far emphasis will be put on stochastic partial differential equations arizing in hydrodynamics, but the methods we give are general and not at all restricted to such equations. In this survey we will concentrate on *multidimensional* white noise; its construction, methods and applications.

Since the purpose is to describe nature mathematically it is important to focus on modelling, an aspect which has been neglected by many authors. Important questions are:

1. What features are we trying to describe?
2. What kind of observations do we have as a basis for our model?
3. When the mathematical model is established, how do proceed mathematically to solve the equations?
4. What information can be deduced from the solutions? Does this information agree with the observations?

This presentation is based on joint works with several authors: J. Gjerde, H. Gjessing, H. Holden, T. Lindstrøm, O. Martio, J. Ubøe and T.-S. Zhang. The starting point for our work was a project supported by VISTA, a research cooperation between the Norwegian Academy of Science and Letters and Den Norske Stats Oljeselskap A.S. (Statoil). The purpose of this project is to describe mathematically the flow of oil, gas and water in porous rocks. Such multi-phase flow can in general be described by a system of nonlinear partial differential equations. These equations represent an enormous challenge even if all the parameters of the system were known. However, in almost all practical applications the permeability properties of the rock are essentially unknown, and this makes the description several levels harder. The lack of information makes it natural to introduce stochastic models, but in what way and in what sense stochastics should enter the equations is in no way canonical. This is where the modelling questions arize.

Having settled with a stochastic partial differential equation (SPDE) as our model, the next step is to perform rigorous matematical calculus on it to produce the solution. Here we would like not just a proof of the existence (and uniqueness), but also some probabilistic properties of the solution, so that we can use the solution to get more information and more insight about the physical situation we are trying to describe.

In the following we will first (in Chapter 1) discuss the basic equations and some observed features of fluid flow in porous media, together with some results based on deterministic methods from the theory of weighted Sobolev spaces and quasiconformal mappings.

Then in Chapter 2 we will review some of the terminology and theory necessary for the stochastic model and its calculus. In particular, we construct multidimensional white noise and establish some of its properties.

In Chapter 3 we give a general scheme for solving multidimensional white noise equations with Wick products and we illustrate the method with an example.

Then in Chapter 4 we return to some of the specific SPDE's related to fluid flow where some of the parameters are modelled by functionals of multidimensional white noise.

CHAPTER 1. FRACTAL BOUNDARIES IN MOVING BOUNDARY PROBLEMS

1a) The basic equations for multi-phase fluid flow in porous media. The moving boundary problem.

One of the basic equations describing one phase fluid flow in a porous medium is the *Darcy law*, which states that

$$\vec{q}_t(x) = -\frac{1}{\mu} K(x) \bigtriangledown p_t(x) \tag{1.1}$$

where $\vec{q}_t(x)$ is the (seepage) velocity of the fluid at the time t and at the point $x \in \mathbf{R}^d$, $p_t(x)$ is the pressure of the fluid at t and x (the gradient is taken w.r.t. x), μ is the viscosity of the fluid and $K(x) = [K_{ij}(x)]_{i,j=1}^d$ is the *permeability matrix* of the medium at x. For a general heterogenous medium the value of $K(x)$ will vary with x, often very irregularly. Moreover, in general the medium may be *anisotropic*, in the sense that the macroscopic flow properties depend on the direction of the flow. Element number i, j of $K(x)$, $K_{ij}(x)$, can be interpreted as the fluid velocity induced in direction i by a unit pressure gradient in direction j (setting $\mu = 1$). It is generally assumed (and it can also be deduced from more basic assumptions) that this permeability matrix $K(x)$ is *symmetric* and *positive* (= non-negative) *definit*. In the *isotropic* case we have $K(x) = k(x)I$ where I is the $d \times d$ identity matrix and $k(x) \geq 0$.

In the case of m-phase fluid flow it is customary to assume that a similar kind of Darcy law holds for the fluid of each phase of the fluid. Combined with the continuity equation this leads to equations of the form

$$\frac{\partial \theta_i}{\partial t} + div(M_i(x,\theta) \bigtriangledown p_t^{(i)}(x)) = \xi_t^{(i)}(x)\,; \quad i = 1, 2, \cdots, m \tag{1.2}$$

where $\theta_i = \theta_i(t,x), p_t^{(i)}(x)$ is the saturation and pressure, respectively, of phase $i, \theta = (\theta_1, \cdots, \theta_m), \xi_t^{(i)}(x)$ is the source rate of phase i and the divergence is taken w.r.t. x.

Summing over i and putting $\theta = \sum_{i=1}^m \theta_i, \xi_t = \sum_{i=1}^m \xi_t^{(i)}$ and assuming that the pressure p is the same in all phases we get an equation of the form

$$\frac{\partial \theta}{\partial t} + div(N(x,\theta) \bigtriangledown p_t) = \xi_t(x) \tag{1.3}$$

If we - as a first approximation - ignore the dependence of θ in N, i.e. we put $N(x,\theta) = N(x)$, we get

$$\frac{\partial}{\partial t}\theta(t,x) + div(N(x) \bigtriangledown p_t(x)) = \xi_t(x) \tag{1.4}$$

If we assume that the fluid occupies the whole volume, then $\theta(t,x) \equiv 1$ and we get

$$div(N(x) \bigtriangledown p_t(x)) = \xi_t(x) \tag{1.5}$$

Combined with either specified values of $p_t(\cdot)$ at the boundary of the domain considered or with specified values of the flux $\bigtriangledown p_t(\cdot)$ at the boundary, we see that (1.5) represents a *linear, second order semi-elliptic boundary value problem of Dirichlet or Neumann type*, respectively.

Substituting the solution p of (1.5) into the original equation (1.2) we get equations of the form

$$\frac{\partial \theta_i}{\partial t} + div(P_i(x,\theta)) = \xi_t^{(i)}\,; \quad i = 1,2,\cdot,m \tag{1.6}$$

which is a system of *nonlinear, first order partial differential equations in the unknowns* $\theta_1, \cdots, \theta_m$.

We see that this approximation procedure splits the original equation (1.2) into two separate systems of equations: The pressure equation (1.5) and the saturation equations (1.6).

We will return to (a special case of) (1.6) in Chapter 3. In the following let us take a closer look at the original equation (1.2) in the case when $m = 1$, i.e. there is only one phase of the fluid. Then, if we assume that $\mu = 1$ as before and also that the density of the fluid is set equal to 1, we get

$$\frac{\partial}{\partial t}\theta(t,x) - div(K(x) \bigtriangledown p_t(x)) = \xi_t(x) \tag{1.7}$$

where $K(x)$ is the permeability matrix.

As stated above the permeability matrix $K(x)$ often varies irregularly with x and therefore (and for other reasons) it is necessary to consider the weak (variational) form of (1.7), namely

$$\begin{aligned} &-\int\int \theta(t,x)\psi(x)\phi'(t)dxdt + \int\int [\bigtriangledown\psi(x)]^T K(x) \bigtriangledown p_t(x)\phi(t)dxdt \\ &= \int\int \xi_t(x)\psi(x)\phi(t)dxdt \end{aligned} \tag{1.8}$$

for all $\phi \in C_0^\infty(\mathbf{R}), \psi \in C_0^\infty(\mathbf{R}^d)$ (where dx, dt denotes Lebesgue measure on $\mathbf{R}^d$, $\mathbf{R}$, respectively). Now let us assume that $\theta(t,x)$ can only assume two values:

$$\theta(t,x) = \theta_0(x) > 0 \quad \text{(complete saturation at } x\text{)}$$

or

$$\theta(t,x) = 0 \quad \text{(the medium is dry at } x\text{)}$$

Put

$$D_t = \{x; \theta(t,x) = \theta_0(x)\} \quad \text{(the wet region)}$$

Then

$$-\int\int \theta(t,x)\psi(x)\phi'(t)dxdt = \int \frac{\partial}{\partial t}(\int_{D_t} \theta_0(x)\psi(x)dx)\phi(t)dt \tag{1.9}$$

Letting $\phi(\cdot)$ approach $\chi_{[0,t]}(\cdot)$ and substituting (1.9) in (1.8), we get

$$\int_{D_t \setminus D_0} \theta_0(x)\psi(x)dx + \int [\nabla\psi(x)]^T K(x) \nabla u_t(x)dx \\ = \int \psi(x)(\int_0^t \xi_s(x)ds)dx, \tag{1.10}$$

where

$$u_t(x) = \int_0^t p_s(x)ds \quad \text{(the Baiocchi transformation)} \tag{1.11}$$

Let us now assume that the (measurable), symmetric, positive definite matrix function $K(x)$ satisfies the requirement that there exist a *p-admissible weight* $m(x) \geq 0$ such that

$$c_1 \cdot m(x)|y|^2 \leq y^T K(x) y \leq c_2 \cdot m(x)|y|^2 \tag{1.12}$$

for all $y \in \mathbf{R}^d$, where $c_1, c_2 > 0$ are constants (see [HKM] for definition and properties of p-admissible weights). If $U \subset \mathbf{R}^d$ is open we let $H_0^1(U,m)$ denote the weighted Sobolev space consisting of the closure of $C_0^\infty(U)$ with respect to the norm

$$\|\psi\|_{H_0^1} = [\int_U (\psi^2(x) + |\nabla\psi(x)|^2)dx]^{\frac{1}{2}}$$

From now on we assume that there exists a bounded open set $U \subset \mathbf{R}^d$ such that for all times t considered, say $t \in [0,T]$, we have $D_t \subset U$.

Equations (1.10), (1.11) together with the assumptions

$$u_t \geq 0 \text{ and } u_t(\cdot) \in H_0^1(D_t, m) \quad \forall t \geq 0 \tag{1.13}$$

is called *(the weak formulation of) the moving boundary problem* for D_t and $p_t(x)$ (or $u_t(x)$).

This name is better understood if we consider the following *classical* (strong) formulation of the problem: Under sufficient smoothness conditions one can show that (1.10), (1.11), (1.13) lead to the 3 equations

$$div(K(x) \nabla p_t(x)) = -\xi_t(x) \quad ; x \in D_t \tag{1.14}$$

$$p_t(x) = 0 \qquad x \in \partial D_t \tag{1.15}$$

$$\theta_0(x) \cdot \frac{d}{dt}(\partial D_t) = -N^T(x)K(x) \bigtriangledown p_t(x) \; ; \; x \in \partial D_t \tag{1.16}$$

where $N^T(x) = (N_1(x), \cdots, N_d(x))$ is the outer unit normal of D_t at $x \in \partial D_t$.

For more information we refer to [R], [MØ]. In contrast to the classical formulation, the weak version always has a unique solution (if (1.12) holds):

THEOREM 1 [R, Sec.2] Suppose $\xi_t(x) \geq 0$. Then there is a unique solution $u_t(x), D_t$ of the weak formulation (1.10), (1.11), (1.13) of the moving boundary problem.

1b) Physical experiments. The existence of porous media whose moving wet boundaries are fractal.

Since the permeability function $K(x)$ is so irregular, it may be tempting to replace it by its (constant) average $\overline{K}$ in the equations above. However, if K is constant, then it is well known ([Gu]), [BG]) that the boundary ∂D_t of the wet region will be smooth for all $t > 0$ if $\xi_t(x) > 0$. And this is far from what is being observed in physical experiments of fluid flow in porous rocks (see e.g. [LØU 3]). In fact, it has been conjectured that ∂D_t is a fractal for all $t > 0$ and that the Hausdorff dimension of ∂D_t, $dim_{\mathcal{H}}(\partial D_t)$, has the approximate values

$$dim_{\mathcal{H}}(\partial D_t) \approx 1.7 \quad \text{if} \quad d = 2 \tag{1.23}$$

$$dim_{\mathcal{H}}(\partial D_t) \approx 2.5 \quad \text{if} \quad d = 3 \tag{1.24}$$

See e.g. [MFJ].

Therefore, to be able to describe fluid flow in porous rocks realistically, it is necessary to be able to handle irregular permeability functions $K(x)$. A natural question is: If we allow $K(x)$ to be sufficienty irregular, can we deduce from the mathematical model above that ∂D_t is fractal?

This is a question about the connection between smoothness/irregularity properties of $K(x)$ and the corresponding moving boundary ∂D_t and this connection is so far not well understood. In fact, from private communications we have got the impression that several applied mathematicians doubt that *any* $K(x)$ - not matter how irregular - can produce a fractal moving boundary. In view of this the following result may be surprising to some:

THEOREM 2 [MØ]. For all dimensions d and all $\epsilon > 0$ there exists a porous medium (i.e. a permeability matrix $K(x)$ and a maximal saturation function $\theta_0(x) > 0$) such that if D_t solves the corresponding moving boundary problem (with D_0 equal to the unit ball centered at the origin and $\xi_t(x)$ being the unit point mass at the origin) then

$$dim_{\mathcal{H}}(\partial D_t) > d - \epsilon \tag{1.25}$$

for at least one $t > 0$.

This medium is constructed as a distortion of a homogeneous medium by a suitable quasiconformal mapping. For details we refer to [MØ].

So ∂D_t is not only fractal, but its Hausdorff dimension can be made arbitrary close to d, for at least *one* $t > 0$. It is not clear for how many values of t this holds.

Note however, that in this theorem the medium (or $K(x)$) is *constructed* for the purpose of obtaining (1.25). The big question is: If we are given a physical medium, and hence an irregular $K(x)$, can we prove that the moving boundary is fractal (in some generic sense)? One way of making this question precise would be to introduce some kind of *stochastic* model for $K(x)$ and investigate the probability that the corresponding ∂D_t is fractal.

Such an approach is also useful from the point of view of other applications, because in many practical situations (for example in oil reservoirs) the values of $K(x)$ are virtually unknown and difficult to measure. This lack of information makes it natural to model $K(x)$ by some kind of *noise*. More precisely, we will represent $K(x)$ as a certain *functional of white noise*. By choosing the functional properly we can achieve that $K(x)$ has the right probabilistic properties. Moreover, we can benefit from the established theory and calculus of white noise analysis to solve the corresponding stochastic partial differential equations.

We will make this more precise in the next chapters. First we review briefly the necessary terminology and results from white noise theory.

CHAPTER 2. MULTIDIMENSIONAL WHITE NOISE, WICK PRODUCTS AND STOCHASTIC PERMEABILITY

The general reference for 1-dimensional white noise theory is [HKPS]. See also several of the contributions in this volume. Of particular interest to us here is the multidimensional white noise, which has been thoroughly discussed in [Gj]. We review some of the basic notations, terminology and results:

In the following we let $\mathcal{S}(\mathbf{R}^d)$ denote the Schwartz space of rapidly decreasing smooth functions on $\mathbf{R}^d$ with the usual topology and we let $\mathcal{S}'(\mathbf{R}^d)$ denote the dual space (the space of tempered distributions). Let $\mathcal{B}(\mathcal{S}'(\mathbf{R}^d))$ be the Borel subsets of $\mathcal{S}'(\mathbf{R}^d)$ in the weak star topology. If m is a natural number we put

$$\mathcal{S} := \prod_{i=1}^{m} \mathcal{S}(\mathbf{R}^d),\ \mathcal{S}' := \prod_{i=1}^{m} \mathcal{S}'(\mathbf{R}^d),\ \mathcal{B} := \prod_{i=1}^{m} \mathcal{B}(\mathcal{S}'(\mathbf{R}^d))$$

Since $\mathcal{S}$ is a countably Hilbert space [Gj] there exists by the Minlos theorem [HKPS] a probability measure μ on $\mathcal{B}$ such that

$$\int_{\mathcal{S}'} e^{i<\omega,\phi>} d\mu(\omega) = e^{-\frac{1}{2}\|\phi\|^2} \quad \text{for all } \phi \in \mathcal{S} \tag{2.1}$$

Here $<\omega,\phi>=<\omega_1,\phi_1> + \cdots + <\omega_m,\phi_m>$ is the action of $\omega = (\omega_1,\cdots,\omega_m) \in \mathcal{S}'$ on $\phi = (\phi_1,\cdots,\phi_m) \in \mathcal{S}$ ($<\omega_k,\phi_k>$ being the action of $\omega_k \in \mathcal{S}'(\mathbf{R}^d)$ on $\phi_k \in \mathcal{S}(\mathbf{R}^d)$) and

$$\|\phi\| = \|\phi\|_{\mathcal{K}} = [\sum_{k=1}^{m} \|\phi_k\|^2_{L^2(\mathbf{R}^d)}]^{\frac{1}{2}} = [\sum_{k=1}^{m} \int_{\mathbf{R}^d} \phi_k^2(x)dx]^{\frac{1}{2}}$$

is the norm of ϕ in the Hilbert space $\mathcal{K} = \bigoplus_{k=1}^{m} L^2(\mathbf{R}^d)$ (the orthogonal sum).

One can show that (see e.g. [Gj])

$$\mu = \mu_1 \times \mu_2 \times \cdots \times \mu_m,$$

where μ_k is the corresponding Minlos measure on $\mathcal{S}'(\mathbf{R}^d)$ (the 1-dimensional white noise probability measure).

We call $(\mathcal{S}', \mathcal{B}, \mu)$ *the multi-dimensional white noise probability space.*

The *m-dimensional white noise* $\vec{W} : \mathcal{S} \times \mathcal{S}' \to \mathbf{R}^m$ is defined by

$$\vec{W}_\phi(\omega) = (< \omega_1, \phi_1 >, \cdots, < \omega_m, \phi_m >) \in \mathbf{R}^m \tag{2.2}$$

if $\omega = (\omega_1, \cdots, \omega_m) \in \mathcal{S}, \phi = (\phi_1, \cdots, \phi_m) \in \mathcal{S}'$.

The connection to m-dimensional (d-parameter) Brownian motion $(B_x^{(1)}, \cdots, B_x^{(m)})$ is that

$$\vec{W}_\phi(\omega) = \left(\int \phi_1(x) dB_x^{(1)}, \cdots, \int \phi_m(x) dB_x^{(m)}\right); \ \phi \in \mathcal{S} \tag{2.3}$$

The *Hermite polynomials* $h_n(x)$ are defined by

$$h_n(x) = (-1)^n e^{\frac{1}{2}x^2} \cdot \frac{d^n}{dx^n}(e^{-\frac{1}{2}x^2}); \ n = 0, 1, 2, \cdots \tag{2.4}$$

and the *Hermite functions* $\xi_n(x)$ are defined by

$$\xi_n(x) = \pi^{-\frac{1}{4}}((n-1)!)^{-\frac{1}{2}} e^{-\frac{1}{2}x^2} h_{n-1}(\sqrt{2}x); \ n = 1, 2, \cdots \tag{2.5}$$

Let $\beta^{(j)} = (\beta_1^{(j)}, \beta_2^{(j)}, \cdots, \beta_d^{(j)})$ be multi-index nr. j in some fixed ordering of all d-dimensional multi-indices $\beta = (\beta_1, \cdots, \beta_d)$ and define

$$\eta_j = \xi_{\beta_1^{(j)}} \otimes \cdots \otimes \xi_{\beta_d^{(j)}} \quad ; j = 1, 2, \cdots \tag{2.6}$$

Then $\{\eta_j\}_{j=1}^{\infty}$ is an orthonormal basis of $L^2(\mathbf{R}^d)$. From this we can construct an orthonormal basis $\{e^{(n)}\}_{n=1}^{\infty}$ of $\mathcal{K}$ by choosing some linear ordering of the family of all m-vectors $\zeta_{ij} = (0, 0, \cdots, 0, \eta_j, 0, \cdots, 0)$ with η_j as coordinate nr. i, $1 \le i \le m, j = 1, 2, \cdots$.

To be specific let us put

$$e^{(n)} = \zeta_{ij} \tag{2.7}$$

where $i \in \{1, 2, \cdots, m\}$ and $j \in \{1, 2, \cdots\}$ are the unique integers such that

$$n = m(j-1) + i.$$

For $\alpha = (\alpha_1, \cdots, \alpha_n)$ a multi-index and $\omega \in \mathcal{S}'$ define

$$H_\alpha(\omega) := h_{\alpha_1}(< \omega, e^{(1)} >) h_{\alpha_2}(< \omega, e^{(2)} >) \cdots h_{\alpha_n}(< \omega, e^{(n)} >) \tag{2.8}$$

Then $\{H_\alpha\}_\alpha$ constitute an orthogonal basis for $L^2(\mu)$ and

(2.9) $$\|H_\alpha\|^2_{L^2(\mu)} = \alpha! := \alpha_1! \cdots \alpha_n!$$

(See [HKPS]). Therefore any $f \in L^2(\mu)$ has a (unique) representation

(2.10) $$f(\omega) = \sum_\alpha c_\alpha H_\alpha(\omega)$$

where

(2.11) $$\|f\|^2 = \sum_\alpha \alpha! c_\alpha^2$$

This is the Wiener-Ito chaos expansion of f (for multi-dimensional white noise).

Next we define the multi-dimensional analogue of the Albeverio-Kondratiev-Streit spaces (AKS spaces, for short) (see [AKS] or [HLØUZ 3]):

DEFINITION 2.1 [AKS]

Part a): (The stochastic test function spaces). For $0 \leq \rho \leq 1$ let $(\mathcal{S})^\rho$ consist of all

$$f = \sum_\alpha c_\alpha H_\alpha \in L^2(\mu)$$

such that

(2.12) $$\|f\|^2_{\rho,k} := \sum_\alpha c_\alpha^2 (\alpha!)^{1+\rho} (2\mathrm{N})^{\alpha k} < \infty \quad \text{for all } k,$$

Here

(2.13) $$(2\mathrm{N})^\alpha = \prod_{j=1}^n (2^d \beta_1^{(j)} \cdots \beta_d^{(j)})^{\alpha_j} \quad \text{if} \quad \alpha = (\alpha_1, \cdots, \alpha_n)$$

where $\beta^{(j)} = (\beta_1^{(j)}, \cdots, \beta_d^{(j)})$ is as described above.

Part b): (The stochastic distribution spaces). For $0 \leq \rho \leq 1$ let $(\mathcal{S})^{-\rho}$ consist of all formal expansions

$$F = \sum_\alpha b_\alpha H_\alpha$$

such that

(2.14) $$\sum b_\alpha^2 (\alpha!)^{1-\rho} (2\mathrm{N})^{-\alpha q} < \infty \quad \text{for some } q < \infty$$

The family of seminorms $\|f\|_{\rho,k}; k = 1, 2, \cdots$ gives rize to a topology on $(\mathcal{S})^\rho$ and we can regard $(\mathcal{S})^{-\rho}$ as the dual of $(\mathcal{S})^\rho$ by the action

(2.15) $$< F, f >= \sum_\alpha b_\alpha c_\alpha \alpha!$$

if $F = \sum_\alpha b_\alpha H_\alpha \in (\mathcal{S})^{-\rho}$, $f = \sum_\alpha c_\alpha H_\alpha \in (\mathcal{S})^{\rho}$.

REMARK For general $\rho \in [0,1]$ we have

$$(\mathcal{S})^1 \subset (\mathcal{S})^\rho \subset (\mathcal{S})^0 \subset (\mathcal{S})^{-0} \subset (\mathcal{S})^{-\rho} \subset (\mathcal{S})^{-1}$$

Observe in particular that with this notation $(\mathcal{S})^0$ and $(\mathcal{S})^{-0}$ are different spaces. Indeed, comparing Definition 2.1 with Zhang's characterization [Z] one can see that $(\mathcal{S})^0$ and $(\mathcal{S})^{-0}$ coincide with the *Hida test function space* $(\mathcal{S})$ and the *Hida space of generalized white noise functionals* $(\mathcal{S})^*$, respectively. See [HKPS] for more information on these spaces.

We can now define the *Wick product*:

DEFINITION 2.2 The Wick product $F \diamond G$ of two elements

$$F = \sum_\alpha b_\alpha H_\alpha,\ G = \sum_\alpha a_\alpha H_\alpha \in (\mathcal{S})^{-1}$$

is defined by

$$F \diamond G = \sum_{\alpha,\beta} a_\alpha b_\beta H_{\alpha+\beta}$$

The following is a consequence of Definition 2.1:

LEMMA 2.3

(i) $F, G \in (\mathcal{S})^{-1} \Rightarrow F \diamond G \in (\mathcal{S})^{-1}$
(ii) $F, G \in (\mathcal{S})^{1} \Rightarrow F \diamond G \in (\mathcal{S})^{1}$

The *Hermite transform*, introduced in [LØU 1-3], has a natural extension to $(\mathcal{S})^{-1}$:

DEFINITION 2.4 If $F = \sum_\alpha b_\alpha H_\alpha \in (\mathcal{S})^{-1}$ then the *Hermite transform* of F, $\mathcal{H}F = \tilde{F}$, is defined by

$$\tilde{F}(z) = \mathcal{H}F(z) = \sum_\alpha b_\alpha z^\alpha \quad \text{(whenever convergent)}$$

where $z = (z_1, z_2, \cdots) \in \mathbf{C}^{\mathbf{N}}$ (the set of all finite sequences of complex numbers) and

$$z^\alpha = z_1^{\alpha_1} z_2^{\alpha_2} \cdots z_n^{\alpha_n} \quad \text{if} \quad \alpha = (\alpha_1, \cdots, \alpha_n)$$

One can show that (see [HLØUZ 3, (2.32)])

$$\tilde{F}(z_1, \cdots, z_k) = \langle F, \eta \rangle \tag{2.16}$$

where $\eta \in (\mathcal{S})^1_{\mathbf{C}}$ is the (complex) stochastic test function

$$\eta(\omega) = Exp < \omega, z_1 e^{(1)} + \cdots + z_k e^{(k)} > := \sum_{n=0}^{\infty} \frac{1}{n!} < \omega, z_1 e^{(1)} + \cdots + z_k e^{(k)} >^{\diamond n}. \tag{2.17}$$

Combining Definition 2.4 with Definition 2.2 we get

LEMMA 2.5 If $F, G \in (\mathcal{S})^{-1}$ then

$$\mathcal{H}(F \diamond G)(z) = (\mathcal{H}F)(z) \cdot (\mathcal{H}G)(z) \quad \text{(when convergent)}$$

where the product on the right hand side is the usual complex product in $\mathbf{C}$.

There are several reasons why the Wick product is important:

1) The Wick product (or a related variant of it) has been used for a long time in quantum physics
2) The Wick product is a natural multiplication in $(\mathcal{S})^{-1}$ and it coincides with ordinary multiplication if one of the factors is deterministic
3) The Wick product is closely connected to stochastic integration via the formula

$$\int_0^T Y_t \delta B_t = \int_0^T Y_t \diamond W_t dt \tag{2.18}$$

where the integral on the left is the Skorohod integral (which coincides with the Ito integral if the integrand $Y_t(\omega)$ is adapted) and the integral on the right is defined as a Pettis integral in $(\mathcal{S})^*$. See e.g. [LØU 2] and see [B].

It is easily seen that if $F \in (\mathcal{S})^{-\rho}$ for some $\rho < 1$ then $\tilde{F}(z)$ converges for all finite sequences $z = (z_1, \cdots, z_n)$ of complex numbers (see e.g. [HLØUZ 3]).

For $(\mathcal{S})^{-1}$ however, the situation is different:

For $q < \infty, R < \infty$ define the "ellipsoids" $\mathbf{B}_q(R)$ by

$$\mathbf{B}_q(R) = \{(\zeta_1, \zeta_2, \cdots) \in \mathbf{C}^{\mathbf{N}}; \sum_{\alpha \neq 0} |\zeta^\alpha|^2 (2\mathbf{N})^{\alpha q} < R^2\} \tag{2.19}$$

Note that $q_1 < q_2 \Rightarrow \mathbf{B}_{q_2}(R) \subset \mathbf{B}_{q_1}(R)$.

LEMMA 2.6 [AKS] (Characterization lemma for $(\mathcal{S})^{-1}$)

a) If $F(\omega) = \sum_\alpha a_\alpha H_\alpha(\omega) \in (\mathcal{S})^{-1}$ then there exists $q < \infty, M_q < \infty$ such that

$$|\tilde{F}(z)| \leq \sum_\alpha |a_\alpha||z^\alpha| \leq M_q \cdot [\sum_\alpha (2\mathbf{N})^{\alpha q} |z^\alpha|^2]^{\frac{1}{2}} \quad \text{for all } z \in \mathbf{C}^{\mathbf{N}}$$

In particular, $\tilde{F}$ is a bounded analytic function on $\mathbf{B}_q(R)$ for all $R < \infty$.

b) Conversely, suppose $g(z) = \sum_\alpha b_\alpha z^\alpha$ is a given power series of $z \in \mathbf{C}^{\mathbf{N}}$ such that there exists $q < \infty$ and $\delta > 0$, such that ($g(z)$ is absolutely convergent when $z \in \mathbf{B}_q(\delta)$ and)

$$\sup_{z \in \mathbf{B}_q(\delta)} |g(z)| < \infty. \tag{2.20}$$

Then there exists $G \in (\mathcal{S})^{-1}$ such that $\tilde{G} = g$, namely

$$G(\omega) = \sum_\alpha b_\alpha H_\alpha(\omega).$$

Proof.

a) We have

$$|\tilde{F}(z)| \leq \sum_{\alpha} |a_\alpha| \cdot |z^\alpha| \leq [\sum_{\alpha} |a_\alpha|^2 (2\mathrm{N})^{-\alpha q}]^{\frac{1}{2}} [\sum_{\alpha} |z^\alpha|^2 (2\mathrm{N})^{\alpha q}]^{\frac{1}{2}}.$$

Since $F \in (\mathcal{S})^{-1}$, we see that $M_q := [\sum_{\alpha} |a_\alpha|^2 (2\mathrm{N})^{-\alpha q}]^{\frac{1}{2}} < \infty$ if q is large enough.

b) Conversely, assume that (2.20) holds. Then

(2.21) $$\sum_{\alpha} |b_\alpha| \cdot |z^\alpha| < \infty \text{ for all } z \in \mathbf{B}_q(\delta), \text{ for all } q \geq q_0, \text{ say}$$

For $r < \infty$ choose $z = z^{(r)} = (z_1, z_2, \cdots)$ with

$$z_j = (2^d \beta_1^{(j)} \cdots \beta_d^{(j)})^{-r} \; ; \; j = 1, 2, \cdots$$

Then

$$\sum_{\alpha} (2\mathrm{N})^{\alpha r} |z^\alpha|^2 = \sum_{\alpha} (2\mathrm{N})^{-\alpha r} < \delta^2$$

is r is large enough,say $r \geq q_1$. Hence $z \in \mathbf{B}_r(\delta)$ for $r \geq q_1$ so by (2.21) we have, for $q \geq \max\ (q_0, q_1)$,

$$\sum_{\alpha} |b_\alpha| (2\mathrm{N})^{-\alpha q} = \sum_{\alpha} |b_\alpha| z^\alpha \leq \sum_{\alpha} |b_\alpha| \cdot |z^\alpha| < \infty.$$

In particular,

$$K := \sup_{\alpha} |b_\alpha| (2\mathrm{N})^{-\alpha q} < \infty.$$

This gives

$$\sum_{\alpha} |b_\alpha|^2 (2\mathrm{N})^{-2\alpha q} \leq K \cdot \sum_{\alpha} |b_\alpha| (2\mathrm{N})^{-\alpha q} < \infty,$$

so $G := \sum_{\alpha} b_\alpha H_\alpha \in (\mathcal{S})^{-1}$ as claimed.

EXAMPLE 2.7

Choose $F = W_\phi^{(i)}$, the i'th component of m-dimensional white noise, $\phi = (\phi_1, \cdots, \phi_m) \in \mathcal{S}$. Then $F(\omega) =< \omega_i, \phi_i >$ if $\omega = (\omega_1, \cdots, \omega_m) \in \mathcal{S}$. Since $\phi_i \in L^2(\mathbf{R}^d)$ we can write

$$\phi_i = \sum_{j} \lambda_{ij} \eta_j$$

and then we get

$$\begin{aligned} W_\phi^{(i)}(\omega) &= \sum_{j} \lambda_{ij} < \omega_i, \eta_j > = \sum_{j} \lambda_{ij} < \omega, e_{m(j-1)+i} > = \sum_{j} \lambda_{ij} h_1(< \omega, e_{m(j-1)+i} >) \\ &= \sum_{j} \lambda_{ij} H_{\epsilon_{m(j-1)+i}}(\omega), \end{aligned}$$

where $\epsilon_k = (0, 0, \cdots, 0, 1, 0, \cdots)$ with 1 on k'th place.

So the Hermite transform of $W_\phi^{(i)}$ is

$$\mathcal{H}W_\phi^{(i)}(z_1, z_2, \cdots) = \sum_j \lambda_{ij} z_{m(j-1)+i} \tag{2.22}$$

We see that the Hermite transforms of the components $W_\phi^{(i)}$ of $\vec{W}_\phi$ involve disjoint families of z_k-variables.

EXAMPLE 2.8 The white noise $W_\phi^{(i)}$ presented so far may be regarded as a ϕ-smoothed version of a *singular* white noise $W_x^{(i)} \in (\mathcal{S})^{-1}$ defined by

$$W_x^{(i)}(\omega) = \sum_j \eta_j(x) H_{\epsilon_{m(j-1)+i}}(\omega) \quad ; x \in \mathbf{R}^d \tag{2.23}$$

The $\mathcal{H}$-transform of $W_x^{(i)}$ is

$$\mathcal{H}W_x^{(i)}(z) = \sum_j \eta_j(x) z_{m(j-1)+i} \quad ; \quad z = (z_1, z_2, \cdots) \tag{2.24}$$

The m-dimensional singular white noise is denoted by $\vec{W}_x(\omega) = (W_x^{(1)}(\omega), \cdots, W_x^{(m)}(\omega))$.

Note that it follows from Lemma 2.6b) that the analytic functions operate on the $\mathcal{H}$-transforms of $(\mathcal{S})^{-1}$ (see also [GHLØUZ]).

LEMMA 2.9 [AKS] Suppose $g = \mathcal{H}X$ for some $X \in (\mathcal{S})^{-1}$. Let $f : V \subset \mathbf{C} \to \mathbf{C}$ be analytic, where V is a neighbourhood of $\zeta_0 = E[X] = g(0)$ in $\mathbf{C}$. Then there exists $Y \in (\mathcal{S})^{-1}$ such that

$$\mathcal{H}Y = f \circ g$$

Finally we mention another important result based on [AKS] and Lemma 2.6:

THEOREM 2.10 [AKS]

The following are equivalent:

(i) $X_n \to X$ in $(\mathcal{S})^{-1}$
(ii) $\exists \delta > 0, q < \infty$ such that

$$\tilde{X}_n(z) \to \tilde{X}(z) \quad \text{pointwise boundedly in } \mathbf{B}_q(\delta)$$

A model for stochastic permeability

Following the approach of [Gj] we now apply the above to construct a model for the stochastic permeability matrix $K(x, \omega)$. As mentioned earlier $K(x, \omega)$ should be a symmetric, non-negative definite $d \times d$-matrix for each x and ω. Moreover, it is usually assumed that K has, at least approximately, the following probabilistic features:

(2.25) (Independence) If $x_1 \neq x_2$ then $K(x_1,\cdot)$ and $K(x_2,\cdot)$ are independent

(2.26) (Lognormality) For each x the eigenvalues $\lambda_1(x,\cdot),\cdots,\lambda_d(x,\cdot)$ of $K(x,\cdot)$ have a lognormal distribution

(2.27) (Stationarity) For all $x_1,\cdots,x_n \in \mathbf{R}^d$ and $h \in \mathbf{R}^d$ the random variable

$$Y = (K(x_1+h,\cdot),\cdots,K(x_n+h,\cdot))$$

has a distribution which is independent of h.

A natural candidate for such a $K(x,\omega)$ is constructed as follows:

Let $\mathcal{W}_x(\omega)$ be the stochastic $d \times d$ matrix

$$\mathcal{W}_x(\omega) = \begin{bmatrix} W_x^{1,1}(\omega) & \cdots & W_x^{1,d}(\omega) \\ \vdots & & \vdots \\ W_x^{d,1}(\omega) & \cdots & W_x^{d,d}(\omega) \end{bmatrix}, \tag{2.28}$$

where $W_x^{i,j}(\omega) = W_x^{j,i}(\omega)$ for all i,j,x,ω and $\vec{W}_x := (W^{i,j})_{i\leq j}$ is $\frac{1}{2}d(d+1)$-dimensional white noise. $\mathcal{W}_x(\omega)$ is called the *symmetric white noise matrix.*

Let $\mathcal{W}_x^{\diamond n}$ denote the *Wick matrix power of order n*, so that

$$\mathcal{W}^{\diamond 2} = \mathcal{W} \diamond \mathcal{W} = [\sum_{j=1}^{d} W^{i,j} \diamond W^{j,k}]_{i,k} \quad \text{and so on.} \tag{2.29}$$

Then define

$$K(x,\omega) := Exp\mathcal{W}_x(\omega) := \sum_{n=0}^{\infty} \frac{1}{n!}\mathcal{W}_x^{\diamond n}(\omega) \tag{2.30}$$

Just as for white noise itself there is a *smoothed* variant $K_\phi(\omega)$ of the singular stochastic permeability matrix $K(x,\omega)$. The smoothed variant is defined by

$$K_\phi(\omega) = Exp\mathcal{W}_\phi(\omega) \text{ where } \mathcal{W}_\phi(\omega) = \begin{bmatrix} W_\phi^{1,1}(\omega) & \cdots & W_\phi^{1,d}(\omega) \\ \vdots & & \vdots \\ W_\phi^{d,1}(\omega) & \cdots & W_\phi^{d,d}(\omega) \end{bmatrix} \tag{2.31}$$

The smoothing is more than a technical convenience: The ideal requirement of independence, (2.25), is unrealistic. However, if we put

$$K_\phi(x,\omega) := K_{\phi_x}(\omega) \tag{2.32}$$

where

$$\phi_x(y) = \phi(y-x) \tag{2.33}$$

is the x-shift of the test function (or "window") ϕ, we see that K_ϕ satisfies (2.26), (2.27) and

(2.25)' (Weak independence) If $\text{supp}\phi_{x_1} \cap \text{supp}\phi_{x_2} = \phi$, then $K_\phi(x_1, \cdot)$ and $K_\phi(x_2, \cdot)$ are independent.

Smoothing by ϕ_x represents taking the macroscopic ϕ-average shifted to x and in a physical application $\text{supp}\phi$ should be chosen large compared to the pores of the medium but small compared to the macroscopic properties of the flow. See the discussion in [HLØUZ 3], [LØU 1].

If K_ϕ is given by (2.31) then clearly $K_\phi(x,\omega)$ is symmetric for each x,ω. By Prop. 1.31 in [G] $K_\phi(x,\omega)$ is positive definite and the eigenvalues $\lambda_i(x,\omega)$ of $K_\phi(x,\omega)$ can be written on the form

$$\lambda_i(x,\omega) = (\xi^{(i)})^T K_\phi(x,\omega)\xi^{(i)} = ExpW_\psi(\omega)$$

for suitable choice of $\xi^{(i)} \in \mathbf{R}^d$, $|\xi^{(i)}| = 1$, and $\psi = \psi(x, \phi, \xi^{(i)}) \in \mathbf{R}^d$ (T denotes transposed) where $W_\psi(\omega)$ is 1-dimensional white noise.

Thus we see that $K_\phi(x,\omega)$ given by (2.31) does indeed satisfy the requirements (2.25)', (2.26) and (2.27) and from now on we will use this as a model for the stochastic permeability matrix of a porous anisotropic medium.

CHAPTER 3. A GENERAL SCHEME FOR SOLVING STOCHASTIC PARTIAL DIFFERENTIAL EQUATIONS INVOLVING FUNCTIONALS OF WHITE NOISE

The results established in Chapter 2 can be applied to a wide class of stochastic partial (or ordinary) differential equations, not just those associated to fluid flow in porous media. The general procedure is the following:

STEP 1: Assume that the "noise" in the equation can be modelled as some (Wick) functional of (possibly multidimensional) white noise and interpret all products/functions involved as Wick products/Wick functions. (This corresponds to a Skorohod interpretation of the stochastic integrals involved See e.g. [LØU 1-3], [B])

STEP 2: Apply the Hermite transform $\mathcal{H}$ to the equation. This turns the original stochastic equation involving the unknown random field $x \to X(x,\omega)$ into a *deterministic* partial differential equation (with usual products) with the unknown $x \to v(x,z) := \tilde{X}(x,z)$, where $z = (z_1, z_2, \cdots) \in \mathbf{C}^{\mathbf{N}}$ is a complex parameter.

STEP 3: Solve this deterministic partial differential equation, for all $z \in \mathbf{B}_q(0,R)$ with $q < \infty$ sufficiently large and all $R < \infty$.

STEP 4: Verify that the solution $v(x,z)$ is a bounded analytic function on $\mathbf{B}_q(0,R)$ for all $R < \infty$. Then by the Characterization Lemma (2.5) $v(x,z)$ is indeed the Hermite transform of an element $X(x,\cdot) \in (\mathcal{S})^{-1}$, which then solves the original equation.

We illustrate this procedure on some examples:

EXAMPLE 3.1

A commonly used model for exponential growth under uncertainty of m correlated quantities $X_t^{(1)}(\omega),\ldots,X_t^{(m)}$ is

$$\frac{dX_t^{(i)}(\omega)}{dt}=(\alpha_i(t)+\sum_{j=1}^{n}\beta_{ij}(t)\cdot W_t^{(j)}(\omega))\cdot X_t^{(i)}(\omega);\ 1\le i\le n \tag{3.1}$$

where $\vec{W}_t(\omega)=(W_t^{(1)}(\omega),\cdots,W_t^{(n)}(\omega))$ is n-dimensional white noise and $\alpha_i(t),\beta_{ij}(t)$ are bounded deterministic functions. In view of (2.18) we interpret the product in (3.1) as a Wick product:

$$\frac{dX_t^{(i)}}{dt}=(\alpha_i+\sum_{j=1}^{n}\beta_{ij}W_t^{(j)})\diamond X_t^{(i)}\ ;\ 1\le i\le n \tag{3.2}$$

Taking $\mathcal{H}$-transform we get the following equations for $v^{(i)}(t,z)=\tilde{X}_t^{(i)}(z)$:

$$\frac{dv^{(i)}(t,z)}{dt}=(\alpha_i+\sum_{j=1}^{n}\beta_{ij}\tilde{W}_t^{(j)})v^{(i)}(t,z)\ ;\ 1\le i\le n \tag{3.3}$$

These equations have the solution

$$v^{(i)}(t,z)=v^{(i)}(0,z)\cdot\exp(\int_0^t\alpha_i(s)ds+\sum_{j=1}^{n}\int_0^t\beta_{ij}(s)\tilde{W}_s^{(j)}(\omega)ds) \tag{3.4}$$

Taking inverse $\mathcal{H}$-transform we end up with the solution

$$X_t^{(i)}(\omega)=X_0^{(i)}(\omega)\diamond Exp[\int_0^t\alpha_i(s)ds+\sum_{j=1}^{n}\int_0^t\beta_{ij}(s)dB_j(s)];1\le i\le n \tag{3.5}$$

Note that this is valid regardless of whether the initial values $X_0^{(i)}(\omega)$ are adapted or not. If $X_0^{(i)}(\omega)=x_i$ is deterministic, the solution coincides with the solution obtained by ordinary Ito calculus, because

$$Exp(\int_0^t\beta_{ij}(s)dB_s)=\exp(\int_0^t\beta_{ij}(s)dB-\frac{1}{2}\int_0^t\beta_{ij}^2(s)ds) \tag{3.6}$$

(see e.g. [GHLØUZ]).

REMARK. If we replace the singular white noise $\vec{W}_t(\omega)$ by the smoothed variant $\vec{W}_\phi(t,\omega)$ (and keep the Wick product) in the example above, we get the solution $X_t^{(\phi,i)}(\omega)$ given by

$$(3.7) \qquad X_t^{(\phi,i)}(\omega) = X_0^{(i)}(\omega) \diamond Exp[\int_0^t \alpha_i(s)ds + \sum_{j=1}^n \int_0^t \beta_{ij}(s) W_\phi^{(j)}(s,\omega)ds] \; ; \; 1 \leq i \leq n$$

In this case it is easily seen that $X_t^{(\phi,i)}(\omega) \to X_t^{(i)}(\omega)$ in $L^2(\mu)$ as $\phi \to \delta_0$ (the Dirac measure at 0). Bur for many stochastic partial differential equations this limit will not exist, at least not in $L^p(\mu)$. Examples of such equations are given in Chapter 4. On the other hand, such equations make perfectly good sense in the ϕ-smoothed case and there is no physical reason that the limit should exist as ϕ approaches the Dirac measure.

CHAPTER 4. APPLICATIONS TO HYDRODYNAMICS

4a) The pressure equation for one phase fluid flow in a stochastic medium.

Finally we return to the equations for fluid flow in a porous medium discussed in Chapter 1, but now for the stochastic case. We represent the stochastic permeability K by the Wick exponential of the symmetric smoothed white noise matrix, as suggested by (2.25):

$$(4.1) \qquad K = K_\phi(x,\omega) = Exp\mathcal{W}_{\phi_x}(\omega) \quad ; \phi \in \mathcal{S}$$

In view of the discussion in Chapter 2 we interpret all products in the relevant equations as Wick products. We first consider the stochastic version of the moving boundary problem (1.14)-(1.16). We will here not discuss the general case but only consider the situation for one fixed time $t \geq 0$, a given bounded open set $D_t = D$ and a given source rate $\xi_t = \xi$. Then the problem is to find $p(x,\omega)$ such that

$$(4.2) \qquad div(K_\phi(x,\cdot) \diamond \nabla p(x,\cdot)) = -\xi(x) \; ; \; x \in D$$

and

$$(4.3) \qquad p(x,\cdot) = 0 \; ; \; x \in \partial D$$

The precise meaning of (4.2) is that $p(x,\cdot) \in (\mathcal{S})^{-1}$ for each x and that for each *stochastic* test function $\eta \in (\mathcal{S})^1$ we have

$$(4.4) \qquad \langle \text{div}(K_\phi(x,\cdot) \diamond \nabla p(x,\cdot)), \eta \rangle = -\xi(x) E[\eta]; \quad x \in D$$

Since the exponential test functions

$$(4.5) \qquad \eta(\omega) = \text{Exp}\langle \omega, \varphi \rangle; \quad \omega \in \mathcal{S}', \; \varphi \in \mathcal{S}$$

are dense in $(\mathcal{S})^{-1}$, we see by (2.16) and Lemma 2.5 that (4.4) is equivalent to

$$(4.6) \qquad \text{div}(\tilde{K}_\phi(x,z) \cdot \nabla \tilde{p}(x,z)) = -\xi(x); \; x \in D, \; z \in \mathbf{C}_0^{\mathbf{N}}$$

In the case when the medium is *isotropic* we have

$$K_\phi(x,\omega) = k_\phi(x,\omega)\cdot I = \mathrm{Exp}W_\phi(x,\omega)\cdot I \tag{4.7}$$

where I is the $d\times d$ identity matrix and W_ϕ is 1-dimensional white noise. In this case it is shown in [HLØUZ 3] that (4.2)-(4.3) has a unique solution $p(x,\cdot)\in(\mathcal{S})^{-1}$. This solution is expressed in terms of an auxiliary Brownian motion $b_t(\hat\omega)$; $\hat\omega\in\hat\Omega$ in $\mathbf{R}^d$ as follows:

$$p(x,\omega) = \frac{1}{2}\mathrm{Exp}(-\frac{1}{2}W_\phi(\omega))\diamond\hat E^x\Big[\int_0^\tau \xi(b_t(\hat\omega))\mathcal{G}(t,b_t(\hat\omega),\omega)dt\Big] \tag{4.8}$$

where

$$\mathcal{G}(t,x,\omega) = \mathrm{Exp}\{\frac{1}{2}W_{\phi_x}(\omega) - \frac{1}{4}\int_0^t\Big[\frac{1}{2}(\nabla W_{\phi_x})^{\diamond 2}(\omega) + \Delta W_{\phi_x}(\omega)\Big]ds\}, \tag{4.9}$$

$\tau=\tau(\hat\omega)=\inf\{t>0; b_t(\hat\omega)\notin D\}$ and $\hat E^x$ denotes expectation with the respect to the law of $b_t(\hat\omega)$ starting at x (the gradient and the Laplacian are both taken with respect to x).

4b) Two phase fluid flow in a stochastic isotropic medium

If the medium is stochastic but isotropic then the following model for two phase fluid flow has been suggested by [Gj]:

We first consider the deterministic equations, assuming for simplicity that both fluids have viscosity and density equal to 1:

(4.10) Darcy's law for two phase flow:

$$\vec q_1 = -k_{11}\nabla p_1 - k_{12}\nabla p_2$$
$$\vec q_2 = -k_{21}\nabla p_1 - k_{22}\nabla p_2$$

where $\vec q_i(t,x)$ is the seepage velocity of fluid number i, $p_i(t,x)$ is the pressure of fluid number i and $k_{ij}(x)$ is the relative permeability with respect to fluid number i and j. It is usually assumed (and it can be proved under more basic assumptions) that $k_{ij}=k_{ji}$.

(4.11) The continuity equations:

$$\frac{\partial\theta_i}{\partial t} = -div(\vec q_i) + \xi_i; \quad i=1,2$$

where $\theta_i(t,x)$ is the saturation of fluid number i and $\xi_i(t,x)$ is the source rate of fluid number i.

Combining (4.10)-(4.11) we get

$$\frac{\partial \theta_i}{\partial t} = \sum_{j=1}^{2} div(k_{ij} \nabla p_j) + \xi_i; \quad i = 1,2 \tag{4.12}$$

As in the one phase situation let us now assume that at each time the saturation $\theta_i(t,x)$ is either 0 or has a maximal value $\tilde{\theta}_i(x) > 0$.

Define

$$D_t^i = \{x; \theta_i(t,x) = \tilde{\theta}_i(x)\}; \quad i = 1,2 \quad \text{(the wet region for fluid } i\text{)} \tag{4.13}$$

If we assume that the fluids are immiscible we have

$$D_t^1 \cap D_t^2 = \emptyset \quad \text{for all } t \geq 0 \tag{4.14}$$

Now we can proceed as before in Chapter 1. By the same argument as in Chapter 2 we use the representation

$$k_{ij}(x,\omega) = \text{Exp}W_{ij}^{\phi_x}(\omega); \quad k_{12} = k_{21}$$

where $\tilde{W} = (W_{11}, W_{12}, W_{22})$ is 3-dimensional white noise. Then we end up with the following system of stochastic moving boundary problems ([Gj, §3.2])

$$\text{div}(k_{ij} \diamond \nabla p_j) = -\delta_{ij}\xi_j \mathcal{X}_{D_t^j}; \quad i,j = 1,2 \tag{4.15}$$

$$p_j(t,x,\cdot) = 0 \quad \text{for } x \in D_t^j \tag{4.16}$$

$$\tilde{\theta}_i \cdot \frac{d}{dt}(\partial D_t^i) = -\sum_{j=1}^{2} k_{ij} \nabla p_j \quad \text{for } x \in \partial D_t^i;\ i = 1,2. \tag{4.17}$$

4c) The Burgers equation with a stochastic force

Next we turn to an important special case of the first order nonlinear saturation equation (1.6), namely the *Burgers equation*

$$\frac{\partial u_k}{\partial t} + \lambda \sum_{j=1}^{n} u_j \frac{\partial u_k}{\partial x_j} = \nu \Delta u_k \tag{4.18}$$

where $(t,x) \in \mathbf{R}^{d+1}$ and λ, ν are constants, $\nu > 0$.

In addition to fluid flow this equation has several applications. For example, it appears in the study of the growth of interfaces, e.g. the way solids form through growth processes on the surface.

A stochastic variant of this equation based on Wick products has been studied in [HLØUZ 2]:

$$\frac{\partial u_k}{\partial t} + \lambda \sum_{j=1}^{d} u_j \diamond \frac{\partial u_k}{\partial x_j} = \nu \Delta u_k + w_k \tag{4.19}$$

where $w = (w_1, \ldots, w_d)$ is d-dimensional noise, interpreted as a stochastic force. It is shown that under certain conditions the equation (4.19) can be transformed into a stochastic heat equation, by a Wick version of the Cole-Hopf transformation. This stochastic heat equation can be solved using the procedure outlined in Chapter 3. We refer to [HLØUZ 2] for details. The stochastic Burgers equation (4.19) with $d = 1$ and ordinary product instead of Wick product has been studied in [BCJ-L]. See also [HR].

4d) The transport equation in a turbulent medium

Consider a substance dissolved in a moving fluid in $\mathbf{R}^d$, being exposed to both a molecular diffusion and to a drift coming from the movement of the fluid. If the fluid is turbulent, a natural model for its velocity is d-dimentional white noise $\vec{W}_{\varphi_x}$. The consentration $u(t, x, \omega)$ of the substance at (t, x) will then satisfy the stochastic partial differential equation

$$\frac{\partial u}{\partial t} = \frac{1}{2}\nu^2 \Delta u + \vec{W}_{\varphi_x} \diamond \nabla u \tag{4.20}$$

where $\nu > 0$ is the molecular viscosity of the fluid.

In [GjHØUZ] this equation is solved given the initial values

$$u(0, x, \omega) = f(x, \omega) \tag{4.21}$$

under certain smoothness conditions on the given function f (f may be anticipating).

With Wick product replaced by usual product this equation has been studied by several authors. See e.g. [Po] and also the references there.

Acknowledgements

I am grateful to J. Kondratiev for useful communication, in particular concerning the [AKS] spaces. This work is supported by VISTA, a research cooperation between The Norwegian Academy of Science and Letters and Den Norske Stats Oljeselskap A.S. (Statoil).

REFERENCES

[AKS] S. Albeverio, J. Kondratiev and L. Streit: Spaces of white noise distributions: Constructions, Descriptions, Applications II. Manuscript 1993.

[AP] J. Ash and J. Potthoff: Ito's lemma without non-anticipatory conditions. Probab.Th.Rel.Fields 88 (1991), 17-46.

[B] F.E. Benth: Integrals in the Hida distribution space $(\mathcal{S})^*$. In T. Lindstrøm, B. Øksendal and A. S. Ustunel (editors): Stochastic Analysis and Related Topics. Gordon & Breach 1993 (to appear).

[BCJ-L] L. Bertini, N. Cancrini and G. Jona-Lasinio: The stochastic Burgers equation. Manuscript 1993.

[BG] H. Begehr and R.P. Gilbert: Hele-Shaw type flows in $\mathbf{R}^n$. Nonlinear Analysis 10 (1986), 65-85.

[CP] G. Cochran and J. Potthoff: Fixed point principles for stochastic partial differential equations. To appear in Proc. "Dynamics of Complex and Irregular Systems", Bielefeld 1991.

[CY] R.A. Carmona and J.A. Yan: In preparation.

[DH] E. Dikow and U. Hornung: A random boundary value problem modelling spatial variability in porous media flow. In M.F. Wheeler (editor): Numerical Simulation in Oil Recovery. IMA-Vol. II, Springer 1988, pp. 111-117.

[Gj] J. Gjerde: Multidimensional noise. Cand.Scient Thesis, Univ. of Oslo 1993.

[Gu] B. Gustafsson: Applications of variational inequalities to a moving boundary problem for Hele-Shaw flows. SIAM J.Math.Anal. 16 (1985), 279-300.

[GHLØUZ] H. Gjessing, H. Holden, T. Lindstrøm, J. Ubøe and T.-S. Zhang: The Wick product. To appear in A. Melnikov (editor): "Frontiers in Pure and Applied Probability". TVP Publishers, Moscow.

[GjHØUZ] J. Gjerde, H. Holden, B. Øksendal, J. Ubøe and T.-S. Zhang: An equation modelling transport of a substance in a stochastic medium. Manuscript 1993.

[HKM] J. Heinonen, T. Kilpeläinen and O. Martio: Nonlinear Potential Theory of Degenerate Elliptic Equations. Oxford University Press 1993.

[HKPS] T. Hida, H.-H. Kuo, J. Potthoff and L. Streit: White Noise Analysis. Kluwer 1993.

[HLØU 1] H. Holden, T. Lindstrøm, B. Øksendal and J. Ubøe: Discrete Wick calculus and stochastic functional equations. Potential Analysis 1 (1992), 291-306.

[HLØU 2] H. Holden, T. Lindstrøm, B. Øksendal and J. Ubøe: Discrete Wick products. In T. Lindstrøm, B. Øksendal and A.S. Ustunel (editors): Stochastic Analysis and Related Topics. Gordon & Breach 1993 (to appear).

[HR] H. Holden and N.H. Risebro: Conservation laws with a random source. Manuscript 1992.

[HP] T. Hida and J. Potthoff: White noise analysis - an overview. In T. Hida, H.-H. Kuo, J. Potthoff and L. Streit (eds.): White Noise Analysis. World Scientific 1990.

[HR] H. Holden and N.H. Risebro: Stochastic properties of the scalar Buckley-Leverett equation. SIAM J. Appl. Math. 51 (1991), 1472-1488.

[HLØUZ 1] H. Holden, T. Lindstrøm, B. Øksendal, J. Ubøe and T.-S. Zhang: Stochastic boundary value problems: A white noise functional approach. Probab. Th. Rel. Fields 95 (1993), 391-419.

[HLØUZ 2] H. Holden, T. Lindstrøm, B. Øksendal, J. Ubøe and T.-S. Zhang: The Burgers equation with a noisy force. To appear in Communications PDE.

[HLØUZ 3] H. Holden, T. Lindstrøm, B. Øksendal, J. Ubøe and T.-S. Zhang: The pressure equation for fluid flow in a stochastic medium. Manuscript 1993.

[IW] N. Ikeda and S. Watanable: Stochastic Differential Equations and Diffusion Processes (2nd edition). North-Holland/Kodansha 1989.

[LØU 1] T. Lindstrøm, B. Øksendal and J. Ubøe: Stochastic differential equations involving positive noise. In M. Barlow and N. Bingham (editors): Stochastic Analysis. Cambridge Univ. Press 1991, 261-303.

[LØU 2] T. Lindstrøm, B. Øksendal and J. Ubøe: Wick multiplication and Ito-Skorohod stochastic differential equations. In S. Albeverio et al (editors): Ideas and Methods in Mathematical Analysis, Stochastics, and Applications. Cambridge Univ. Press 1992, pp. 183-206.

[LØU 3] T. Lindstrøm, B. Øksendal and J. Ubøe: Stochastic modelling of fluid flow in porous media. In S. Chen and J. Yong (editors): Control Theory, Stochastic Analysis and Applications. World Scientific 1991, pp. 156-172.

[LØUs] T. Lindstrøm, B. Øksendal and S. Ustunel (editors:) Stochastic Analysis and Related Topics. Gordon & Breach 1993 (to appear).

[MFJ] K.J. Måløy, J. Feder and T. Jøssang. Phys.Rev.Lett. 55 (1985), 2688-2691.

[MØ] O. Martio and B. Øksendal: Fluid flow in a medium distorted by a quasiconformal map can produce fractal boundaries. Manuscript 1993.

[NZ] D. Nualart and M. Zakai: Generalized stochastic integrals and the Malliavin calculus. Probab.Th.Rel.Fields 73 (1986), 255-280.

[NP] D. Nualart and E. Pardoux: Stochastic calculus with anticipating integrands. Probab.Th.Rel.Fields 78 (1988), 535-581.

[Ø] B. Øksendal: A stochastic approach to moving boundary problems. In M. Pinsky (editor): Diffusion Processes and Related Problems in Analysis. Birkhäuser 1990, pp. 201-218.

[ØZ] B. Øksendal and T.-S. Zhang: The stochastic Volterra equation. In D. Nualart and M. Sanz Solé (editors): Barcelona Seminar on Stochastic Analysis. Birkhäuser 1993, pp. 168-202.

[Po] J. Potthoff: White noise methods for stochastic partial differential equations. Manuscript 1991.

[R] Y. Reichelt: Moving boundary problems arizing from degenerate elliptic equations. Manuscript 1993.

[Wa] J.B. Walsh: An introduction to stochastic partial differential equations. In R. Carmona, H. Kesten and J.B. Walsh (editors): École d'Été de Probabilités de Saint-Flour XIV-1984. Springer LNM 1180 (1986), 265-437.

[Z] T.-S. Zhang: Characterizations of white noise test functions and Hida distributions. Stochastics 41 (1992), 71-87.

WHITE NOISE APPROACH TO PARABOLIC STOCHASTIC PARTIAL DIFFERENTIAL EQUATIONS

J. POTTHOFF
Lehrstuhl für Mathematik V
Universität Mannheim
D–68131 Mannheim
Germany

ABSTRACT. A certain class of stochastic partial differential equations of parabolic type is studied within white noise analysis.

1. Introduction

In [Ch 89], P.L. Chow proposed stochastic partial differential equations of the form

$$\frac{\partial}{\partial t}u(t,x) - Lu(t,x) = \dot{\eta}(t,x)\nabla u(t,x), \tag{1.1}$$

as a model for the transport of a substance in a turbulent medium. L is an elliptic second order differential operator acting in the space variable $x \in \mathbb{R}^d$. Thus the left hand side of (1.1) describes the diffusion (governed by L) of the density $u(t,x)$ (at x and at time $t > 0$) of the substance. The right hand side of (1.1) expresses the coupling of the velocity field of the substance to a random field $\dot{\eta}(t,x)$, which models the turbulent velocity field of the medium. This field is the noise of $\eta(t,x)$ which will be assumed to be of the form $\eta(t,x) = \int_0^t \sigma(s,x)\,dB(s)$ where σ is a deterministic function and B is a d–dimensional Brownian motion.

As Chow observed in [Ch 89], generalized random variables arise quite naturally in the discussion of (1.1), and therefore white noise analysis [HK 93] seems to be a natural framework for this type of equation. In fact, as we shall see in Section 4, the coupling of the gradient of u to $\dot{\eta}$ brings a singularity into the corresponding integral equation, which does not allow (at least not in a straightfoward manner) for a formulation in the conventional framework of the Itô calculus. On the other hand, equation (1.1) can indeed be formulated and solved within white noise analysis. In the present paper, explicit solutions for (1.1) will be found and studied for a special

A. I. Cardoso et al. (eds.), Stochastic Analysis and Applications in Physics, 307–327.

choice of L and σ (Section 3), and existence and uniqueness of weak solutions of (1.1) will be proved in the one–dimensional case (Section 4). The necessary ingredients from white noise analysis will be collected in Section 2.

I want to remark that a very large amount of work concerning stochastic partial differential equations (SPDE's) within white noise analysis has been done by the group around B. Øksendal in the recent years. The interested reader is referred to Øksendal's contribution in this volume.

In the present article, I did not try to present the most general mathematical results concerning (1.1). Rather, it was my aim to show with a minimal technical effort how classical fixed point theorems combine with some of the tools of white noise analysis (S–transform, characterization theorem, and differential operators), to give existence and uniqueness of solutions. Moreover, I wanted to point out the peculiar behaviour of the solutions of (1.1) with respect to different interpretations ("Itô versus Stratononvich") of the noise term (cf. Section3). A more general theory along the lines of this paper will be published elsewhere [DP 94].

2. Review of White Noise Analysis

In this section I give a sketch of white noise analysis which is suited for the applications which are intended in later sections. For a general account on white noise analysis, the interested reader is referred to [HK 93], the references quoted there and to the contribution of L. Streit in this volume.

Consider the white noise probability space $(\mathcal{S}'(\mathbb{R}), \mathcal{B}, \mu)$ where $\mathcal{S}'(\mathbb{R})$ is the Schwartz space of tempered distributions, $\mathcal{B}$ is its weak Borel σ–algebra and μ is the centered Gaussian measure on $\mathcal{B}$ whose covariance is given by the inner product of $L^2(\mathbb{R}, dt)$ (where dt denotes Lebesgue measure). As is well-known, this probability space is a model for the time derivative of Brownian motion. We denote by (L^2) the L^2–space over this probability space.

(L^2) contains a dense subspace $(\mathcal{S})$ of smooth random variables which carries a certain metric topology (e.g., [HK 93]). Its dual $(\mathcal{S})^*$ is called the space of *Hida distributions*. $(\mathcal{S})^*$ is sufficiently large to contain many generalized random variables which arise in applications. The canonical bilinear pairing between $(\mathcal{S})^*$ and $(\mathcal{S})$ will be denoted by $\langle\!\langle \cdot, \cdot \rangle\!\rangle$. In order to give elements in $(\mathcal{S})^*$ a "visualization", we introduce the S–transform: for $\Phi \in (\mathcal{S})^*$, $\xi \in \mathcal{S}(\mathbb{R})$, we set

$$S\Phi(\xi) := \langle\!\langle \Phi, : e^{\langle \cdot, \xi \rangle} : \rangle\!\rangle, \tag{2.1}$$

which makes sense because for $\xi \in \mathcal{S}(\mathbb{R})$, the normalized exponential $\omega \longmapsto :\exp(\langle \omega, \xi \rangle) :\equiv \exp(\langle \omega, \xi \rangle - \frac{1}{2}|\xi|_2^2)$, $\omega \in \mathcal{S}'(\mathbb{R})$, belongs to $(\mathcal{S})$. (We denote by $\langle \cdot, \cdot \rangle$ the dual pairing between $\mathcal{S}'(\mathbb{R})$ and $\mathcal{S}(\mathbb{R})$, by $| \cdot |_2$ the norm of $L^2(\mathbb{R})$.) Indeed, it turns out that the function (2.1) on $\mathcal{S}(\mathbb{R})$ has very nice properties, namely, it is a function of the following type.

Definition 2.1. Consider a mapping F from $\mathcal{S}(\mathbb{R})$ into $\mathbb{C}$. F is called a *U–functional* if it satisifes the following two properties:

C.1. F is everywhere ray entire on $\mathcal{S}(\mathbb{R})$, i.e., for all η, $\xi \in \mathcal{S}(\mathbb{R})$, the mapping $\lambda \longmapsto F(\eta + \lambda\xi)$, $\lambda \in \mathbb{R}$, has an entire extension,

C.2. there exist constants K_1, $K_2 > 0$ and a continuous norm $|\cdot|$ on $\mathcal{S}(\mathbb{R})$ so that for all $\xi \in \mathcal{S}(\mathbb{R})$, $z \in \mathbb{C}$,

$$|F(z\xi)| \leq K_1 \exp\left(K_2 |z|^2 |\xi|^2\right). \tag{2.2}$$

The space of all such mappings is denoted by $\mathcal{U}$.

Remark. It is interesting to note that $F \in \mathcal{U}$ implies that F has an entire extension to the complexified Schwartz space $\mathcal{S}_{\mathbb{C}}(\mathbb{R})$ (i.e., F extends to a function on $\mathcal{S}_{\mathbb{C}}(\mathbb{R})$ which is continuous and such that C.1 holds for complex η, ξ). This has been proved in [KL 94].

The following theorem [PS 91] (cf. also [KL 94]) shows that the elements in $(\mathcal{S})^*$ are characterized by their S–transform and U–functionals.

Theorem 2.1. The S–transform is a bijection from $(\mathcal{S})^*$ onto $\mathcal{U}$.

There are many variants, generalizations and extensions of this result. The interested reader is referred to [HK 93, KL 94] and the references quoted there. In passing, I want to mention a result in [BT 94] which says that $\mathcal{U}$ can be equipped with a topology (arising naturally from (2.2)) so that the S–transform becomes even a homeomorphism.

Next we discuss differential operators acting on functions on the white noise space.

Let φ be a complex valued function on $\mathcal{S}'(\mathbb{R})$, and $\theta \in \mathcal{S}'(\mathbb{R})$. The Gâteaux derivative $D_\theta\varphi(\omega)$ of φ in direction θ at $\omega \in \mathcal{S}'(\mathbb{R})$ is given by

$$D_\theta\varphi(\omega) := \frac{d}{d\lambda}\varphi(\omega + \lambda\theta)\bigg|_{\lambda=0},$$

whenever the derivative exists. It can be shown (e.g., [HK 93, Chap. 5]) that for all $\theta \in \mathcal{S}'(\mathbb{R})$, D_θ extends to a continuous operator on $(\mathcal{S})$. For the special choice $\theta = \delta_t$, $t \in \mathbb{R}$, this operator will be denoted also by ∂_t. ∂_t, $t \in \mathbb{R}$, can be considered as a gradient on the white noise space.

By ∂_t^*, $t \in \mathbb{R}$, we shall denote the adjoint of ∂_t which acts on $(\mathcal{S})^*$. It is remarkable (though easy to prove) that both operators have very simple intertwining relations with the S–transform:

$$S\partial_t\varphi(\xi) = \frac{\delta}{\delta\xi(t)}S\varphi(\xi), \tag{2.3}$$

$$S\partial_t^*\Phi(\xi) = \xi(t)S\Phi(\xi), \tag{2.4}$$

where $\varphi \in (\mathcal{S})$, $\Phi \in (\mathcal{S})^*$, and $\frac{\delta}{\delta\xi(t)}$ denotes the Fréchet functional derivative with respect to ξ.

It is well-known (e.g., [HK 93, Chap. 8]) that ∂_t^* implements a stochastic integral, the ***Hitsuda–Skorokhod integral***, with respect to Brownian motion which extends the Itô integral. Heuristically speaking, the reason is that for $\Phi \in (\mathcal{S})^*$, $\partial_t^*\Phi$ is equal to the Wick product [HK 93] (cf. also the contribution of B. Øksendal to this volume) $\dot{B}(t)\diamond\Phi$ of white noise $\dot{B}(t) \in (\mathcal{S})^*$ with Φ (which is again in $(\mathcal{S})^*$). If X is a mapping from $\mathbb{R}$ into $(\mathcal{S})^*$ such that $t \longmapsto \partial_t^* X(t)$ is weakly integrable (with respect to Lebesgue measure), then we denote

$$\begin{aligned}\int_{\mathbb{R}} X(t)\,dB(t) &:= \int_{\mathbb{R}} \partial_t^* X(t)\,dt,\\ &= \int_{\mathbb{R}} X(t)\diamond\dot{B}(t)\,dt\end{aligned}$$

where the last two integrals are in the sense of Pettis and belong to $(\mathcal{S})^*$ (e.g., [HK 93]). From (2.4) we obtain

$$S\Big(\int_{\mathbb{R}} X(t)\,dB(t)\Big)(\xi) = \int_{\mathbb{R}} SX(t)(\xi)\xi(t)\,dt, \quad \xi \in \mathcal{S}(\mathbb{R}). \tag{2.5}$$

Thus the S–tranform turns the $dB(t)$–integral into one with respect to $\xi(t)\,dt$. Relation (2.5) and Theorem 2.2 will be at the basis of our strategy to solve equations of the type (1.1).

3. Explicit Example

We consider the following spatially homogeneous case of the turbulent transport SPDE (1.1)

$$\frac{\partial}{\partial t}u(t,x) - \frac{1}{2}\nu(t)\Delta u(t,x) = \dot{\eta}(t) * \nabla u(t,x), \tag{3.1}$$

with $(t,x) \in \mathbb{R}_+ \times \mathbb{R}$ and $\nu \in L^1_{\text{loc}}(\mathbb{R}_+,dt), \nu > 0, \sigma \in L^2_{\text{loc}}(\mathbb{R}_+,dt)$, and the initial value $u(0,\cdot) = \delta_0$, i.e., the Dirac distribution at the origin. (Without loss of generality for the following computations, we consider only the case of one space dimensions.) The $*$ indicates that we shall use as well a Stratonovich interpretation ("$* = \circ$") as an Itô interpretation ("$* = \diamond$") of (3.1). First we consider (3.1) in Stratonovich sense. Then its Itô form is given by

$$\frac{\partial}{\partial t}u(t,x) - \frac{1}{2}(\nu(t) + \sigma(t)^2)\Delta u(t,x) = \sigma(t)\dot{B}(t) \diamond \nabla u(t,x). \tag{3.2}$$

Let the S–transform of u be denoted by $\widehat{u}$, then (3.2) becomes for $\xi \in \mathcal{S}(I\!R)$ (cf. (2.4)),

$$\frac{\partial}{\partial t}\widehat{u}(t,x)(\xi) - \frac{1}{2}(\nu(t) + \sigma(t)^2)\Delta\widehat{u}(t,x)(\xi) = \sigma(t)\xi(t)\nabla\widehat{u}(t,x)(\xi). \tag{3.3}$$

Now apply the Fourier transform $\mathcal{F}$ (in the space variable x), and denote $\mathcal{F}\widehat{u}$ by $\widetilde{u}$, then we get the following equation

$$\frac{\partial}{\partial t}\widetilde{u}(t,p)(\xi) + \frac{1}{2}(\nu(t) + \sigma(t)^2)p^2\widetilde{u}(t,p)(\xi) = ip\sigma(t)\xi(t)\widetilde{u}(t,p)(\xi), \tag{3.4}$$

$(t,p) \in I\!R_+ \times I\!R$. We have $\widetilde{u}(0) = (2\pi)^{-\frac{1}{2}}$, and the solution of (3.4) is readily computed. If we let $\tau(t) := \int_0^t (\nu(s) + \sigma(s)^2)ds, t \in I\!R_+$, then we obtain

$$\widetilde{u}(t,p)(\xi) = (2\pi)^{-\frac{1}{2}} \exp\Big(-\frac{1}{2}p^2\tau(t) + ip\int_0^t \sigma(s)\xi(s)\,ds\Big), \tag{3.5}$$

as the solution of the initial value problem for $\widetilde{u}$. It is easy to invert the Fourier transform in (3.5) with the result

$$\begin{aligned} Su(t,x)(\xi) &= p_{\tau(t)}\big(x, (\sigma_t, \xi)_{L^2(I\!R)}\big) \\ &= (2\pi\tau(t))^{-\frac{1}{2}} \exp\Big(-(2\tau(t))^{-1}\big(x - \int_0^1 \sigma(s)\xi(s)\,ds\big)^2\Big), \end{aligned} \tag{3.6}$$

where $p_t(x,y)$ is the usual heat kernel of the Laplacian and $\sigma_t = \sigma \cdot 1_{[0,t)}$. In the next step we would like to invert the S–transform in (3.6). To this end we use the following more general result. We let $X.$ denote the canonical coordinate process on $L^2(I\!R)$, i.e., the continuous extension of $X. : \mathcal{S}(I\!R) \longrightarrow (L^2)$, $X_\xi(\omega) = \langle \omega, \xi \rangle$, $\omega \in \mathcal{S}'(I\!R)$, to $L^2(I\!R)$.

Lemma 3.1. Let $\lambda_1, \lambda_2 \in \mathbb{C}$, $f, g \in L^2(\mathbb{R})$. Assume that $|\lambda_1|\,|f|_2^{-2} < 1$. Then for $\xi \in \mathcal{S}(\mathbb{R})$,

$$S\big(\exp(\tfrac{1}{2}\lambda_1 X_f^2 + \lambda_2 X_g)\big)(\xi) = C(\lambda_1, \lambda_2, f, g) \exp\big(\tfrac{1}{2}a(f,\xi)^2 + (G,\xi)\big), \qquad (3.7a)$$

where

$$C(\lambda_1, \lambda_2, f, g) = (1 - \lambda_1\,|f|_2^2)^{-\frac{1}{2}} \exp\Big(\frac{1}{2}\frac{\lambda_1\lambda_2^2}{1-\lambda_1\,|f|_2^2}(f,g)^2 + \frac{1}{2}\lambda_2^2\,|g|_2^2\Big), \qquad (3.7b)$$

$$a = \frac{\lambda_1}{1-\lambda_1\,|f|_2^2}, \qquad (3.7c)$$

$$G = \frac{\lambda_1\lambda_2}{1-\lambda_1\,|f|_2^2}(f,g)f + \lambda_2 g. \qquad (3.7d)$$

The relations (3.7) help to find the inverse S–transforms of exponentials like on the right hand side of (3.7a) via the following. For given a, f and G, one finds

$$\lambda_1 = \frac{a}{1 + a\,|f|_2^2}, \quad \lambda_2 = 1, \qquad (3.8a)$$

$$g = G - \frac{a(G,f)}{1 + a\,|f|_2^2} f. \qquad (3.8b)$$

The proof of Lemma 3.1 is a straightforward calculation (it seems to be of advantage to decompose g orthogonally in $L^2(\mathbb{R})$ along f). Relations (3.8) follow from trivial algebra.

Corollary 3.2. Let $\tau > 0, x \in \mathbb{R}, f \in L^2(\mathbb{R})$. Then for $\xi \in \mathcal{S}(\mathbb{R})$,

$$S\,p_\tau(x, X_f)(\xi) = p_{\tau + |f|_2^2}(x, (f,\xi)_{L^2(\mathbb{R})}). \qquad (3.9)$$

(3.9) follows from (3.7) by a simple computation.

Now we are in position to compute the inverse S–transform of (3.6): from (3.9) we can read off that $f = \sigma_t$, and with $\gamma(t) = \int_0^t \nu(s)ds$,

$$\begin{aligned} u(t,x) &= p_{\gamma(t)}(x, X_{\sigma_t}) \\ &= (2\pi\gamma(t))^{-\frac{1}{2}} \exp\Big(-(2\gamma(t))^{-1}\big(x - \int_0^t \sigma(s)\,dB(s)\big)^2\Big). \end{aligned} \qquad (3.10)$$

Of course, it is the same result as in [Ch 89], obtained there in a slightly different way.

Let us return to equation (3.1) and now take $* = \diamond$, i.e., we consider (3.1) in Itô sense. In the preceding calculations this amounts to replacing $(\nu(t) + \sigma(t)^2)$ by $\nu(t)$. Thus, in this case we find for the S–transform of the solution instead of (3.6) the formula

$$v(t,x)(\xi) = p_{\gamma(t)}\big(x, (\sigma_t, \xi)_{L^2(\mathbb{R})}\big), \tag{3.11}$$

and v solves the equation

$$\frac{\partial}{\partial t} v(t,x;\xi) - \frac{1}{2}\nu(t)\Delta v(t,x;\xi) = \sigma(t)\xi(t)\nabla v(t,x;\xi), \tag{3.12}$$

for every $\xi \in \mathcal{S}(\mathbb{R})$. We can still apply Corollary 3.2 to find the inverse S–transform of (3.11), but this time *only if* $\int_0^t \nu(s)ds - \int_0^t \sigma(s)^2 ds > 0$. In that case,

$$u(t,x) = p_{\kappa(t)}\big(x, \int_0^t \sigma(s)\, dB(s)\big), \tag{3.13}$$

where $\kappa(t) = \int_0^t \nu(s)ds - \int_0^t \sigma(s)^2 ds$. If $\int_0^t \nu(s)ds = \int_0^t \sigma(s)^2 ds$, we recognize (3.11) as the S–transform of Donsker's delta function:

$$u(t,x) = \delta_x\big(\int_0^t \sigma(s)\, dB(s)\big). \tag{3.14}$$

In view of the characterization theorem, Theorem 2.2, we see that also for $\kappa(t) < 0$, (3.11) has an inverse S–transform (in $(\mathcal{S})^*$). However, it is no longer given by (3.13): for $\kappa(t) < 0$ this expression is no longer real, but since (3.11) is, so must be $u(t,x) \in (\mathcal{S})^*$. I have tried a number of times to find an explicit expression for $u(t,x)$ but did not succeed (although it is not hard to write down its chaos expansion).

Note that for $t \in \mathbb{R}_+$ with $\kappa(t) > 0$ we get a very smooth solution: $u(t,x) \in L^p(\mu)$ for all $p \geq 1$, which for $\kappa(t) \downarrow 0$ becomes Donsker's delta of X_{σ_t}, and then – as $\kappa(t)$ gets negative – becomes a generalized random variable: one can prove that $u(t,x)$ does not belong to $L^p(\mu)$ for any $p \geq 1$. Of course, we can arrange the coefficients ν and σ such that $u(t,x)$ "oscillates" in any given way between being a smooth and a genuinely generalized random field, each time passing through a Donsker delta.

This quite different behaviour of the solution with respect to the interpretation of the multiplication by the noise in (3.1) might be of help concerning questions of modelling

In the remainder of this section we study in more detail the question, in which sense $u(t,x) = S^{-1}v(t,x)$ is a solution of (3.1) with $* = \diamond$ (which we assume from now on). Obviously we have that for $\zeta \in \mathcal{S}_{\mathbb{C}}(\mathbb{R})$,

$$v(t,x;\zeta) = (2\pi\gamma(t))^{-\frac{1}{2}} \exp\Big(-\frac{1}{2\gamma(t)}\big(x - \int_0^t \sigma(s)\zeta(s)\, ds\big)^2\Big),$$

is continuously differentiable in $t \in (0, +\infty)$ for all $x \in \mathbb{R}$, and for $t \in (0, +\infty)$ it is C^∞ in $x \in \mathbb{R}$, and all derivatives are continuous on $(0, +\infty) \times \mathbb{R}$. It is also obvious that $\frac{\partial}{\partial t} v(t, x; \cdot)$, $\frac{\partial}{\partial x} v(t, x; \cdot)$, $\frac{\partial^2}{\partial x^2} v(t, x; \cdot)$ belong to $G^2_{lu}((0, +\infty) \times \mathbb{R}; \mathcal{U})$, where this space is defined in Appendix 1. It follows from Lemma A.1.3, that

$$\begin{aligned} S^{-1} \frac{\partial}{\partial t} v(t, x) &= \frac{\partial}{\partial t} u(t, x) \\ S^{-1} \frac{\partial}{\partial x} v(t, x) &= \frac{\partial}{\partial x} u(t, x) \\ S^{-1} \frac{\partial^2}{\partial x^2} v(t, x) &= \frac{\partial^2}{\partial x^2} u(t, x), \end{aligned}$$

where the right hand sides denote weak derivatives in $(\mathcal{S})^*$, and they are all weakly continuous functions from $(0, +\infty) \times \mathbb{R}$ into $(\mathcal{S})^*$. The fact that $v(t, x; \zeta)$ solves for every $\xi \in \mathcal{S}(\mathbb{R})$ equation (3.12) means that $u(t, x)$ solves (3.1) weakly on the dense subspace $\mathcal{E}$ which is the (complex) linear span of $\{\exp(< \cdot, \xi >), \xi \in \mathcal{S}(\mathbb{R})\}$. But then $u(t, x)$ solves (3.1) weakly on $(\mathcal{S})$. Now we consider the weak limit of $u(t, x)$ as t falls to zero. Let $f \in \mathcal{S}(\mathbb{R})$ and consider for $\xi \in \mathcal{S}(\mathbb{R})$,

$$\int f(x) v(t, x; \xi)\, dx = (2\pi)^{-\frac{1}{2}} \int f(\gamma(t)^{\frac{1}{2}} x)\, e^{-\frac{1}{2}(x - \gamma(t)^{-\frac{1}{2}} \int_0^t \sigma(s)\xi(s) ds)^2}\, dx.$$

Obviously, the last expression converges to $f(0)$ as $t \downarrow 0$. Moreover, for $t > 0$, $z \in \mathbb{C}$, $\xi \in \mathcal{S}(\mathbb{R})$, we find

$$\left| \int f(x) v(t, x; z\xi)\, dx \right| \leq \|f\|_\infty\, e^{\frac{1}{2}|z|^2 \gamma(t)^{-1} (\int_0^t \sigma(u)\xi(u)\, du)^2},$$

as well as

$$|v(t, x; z\xi)| \leq (2\pi\gamma(t))^{-\frac{1}{2}}\, e^{\frac{1}{2}|z|^2 \gamma(t)^{-1} (\int_0^t \sigma(u)\xi(u)\, du)^2}.$$

In view of Lemma A.1.1, the last estimate shows that for $t > 0, \{u(t, x), x \in \mathbb{R}\}$ is bounded in $(\mathcal{S})_{-p}$ for some $p \in \mathbb{N}_0$. Moreover, for $t > 0$, $x \longmapsto v(t, x; \xi)$ is continuous for every $\xi \in \mathcal{S}(\mathbb{R})$. Thus, by Lemma A.1.2, $x \longmapsto u(t, x)$ is weakly continuous, and therefore

$$\int f(x) u(t, x)\, dx$$

exists as a Pettis integral (actually as a Bochner integral), and its S–transform is given by

$$\int f(x) v(t, x; \xi)\, dx, \quad \xi \in \mathcal{S}(\mathbb{R}).$$

But this expression converges to $f(0)$ as $t \downarrow 0$ for every ξ, which entails that $\int f(x) u(t, x)\, dx$ converges weakly to $f(0)$ on $\mathcal{E}$. On the other hand the estimate on $\left|\int f(x)v(t,x;z\xi)\,dx\right|$ above proves that (cf. Lemma A.1.1) $\int f(x)u(t,x)\,dx$ is bounded in some $(\mathcal{S})_{-p}, p \in \mathbb{N}_0$, as $t \downarrow 0$. Hence it converges weakly to $f(0)$. In this sense, $u(t,x)$ admits the initial condition $u(t,\cdot) \to \delta_0$ as $t \downarrow 0$.

In the case of a general equation of type (3.1) our strategy will be to solve an associated integral equation. Let us investigate the concept of a weak solution for the integral equation of (3.1), in order to prepare the next section.

The fundamental solution $q_{t,s}(x,y)$ of $\frac{\partial}{\partial t} - \frac{1}{2}\nu(t)\Delta$ is given by

$$q_{t,s}(x,y) = p_{\gamma(t)-\gamma(s)}(x,y),$$

where as before $p_t(x,y)$ is the heat kernel. Thus the initial value problem for (3.1) $(* = \diamond)$ with $u(0,\cdot) = u_0$ (deterministic again, for simplicity) is reformulated as the integral equation (cf. also [Ch 89])

$$u(t,x) = (P_{\gamma(t)}u_0)(x) + \int_0^t (P_{\gamma(t)-\gamma(s)}\nabla u(s,\cdot))(x)\sigma(s) \diamond dB(s), \tag{3.15}$$

$(P_t f(x) = \int p_t(x,y)f(y)dy)$. Taking informally the S–transform of (3.15) we arrive at the integral equation

$$v(t,x;\xi) = (P_{\gamma(t)}u_0)(x) + \int_0^t (P_{\gamma(t)-\gamma(s)}\nabla v(s,\cdot\,;\xi))(x)\sigma(s)\xi(s)\,ds. \tag{3.16}$$

Let again $u_0 = \delta_0$, so that the first term on the right hand side of (3.16) becomes $p_{\gamma(t)}(0,x)$. It is not hard (though a bit tedious and certainly no surprise) to verify that $v(t,x;\xi)$ given in (3.11) solves (3.16). $S^{-1}v(t,x) = u(t,x)$ is the weak solution of (3.15), as we shall argue now. We had already seen in our previous discussion that for $t > 0$, $u(t,x)$ is C^1 in t and C^∞ in x, and that ∇ and the S–transform commute, i.e. $S^{-1}\nabla v(t,x) = \nabla u(t,x)$. In order to show that the integral $(P_{\gamma(t)-\gamma(s)}\nabla u(s,\cdot))(x)$ exists as a Pettis integral in $(\mathcal{S})^*$, we estimate $v(s,x;z\xi)$ for $s > 0$, $x \in \mathbb{R}$, $z \in \mathbb{C}$, $\xi \in \mathcal{S}(\mathbb{R})$, as follows:

$$\begin{aligned}|\nabla v(s,x;z\xi)| &= \left|\gamma(s)^{-1}(x - z(\sigma_s,\xi))(2\pi\gamma(s))^{-\frac{1}{2}} \exp\Big(-\frac{1}{2}\gamma(s)^{-1}(x - z(\sigma_s,\xi))^2\Big)\right| \\ &\le (2\pi\gamma(s)^3)^{-\frac{1}{2}}(|x| + |z|\,|(\sigma_s,\xi)|)\cdot \\ &\quad \cdot \exp\Big(-\frac{1}{2}\gamma(s)^{-1}(x^2 - 2\,|x|\,|z|\,|(\sigma_s,\xi)| - |z|^2\,(\sigma_s,\xi)^2\Big),\end{aligned}$$

where we denoted $\sigma_s = \sigma\cdot 1_{[0,s)}$. Now apply the inequality $\pm 2ab \le \varepsilon a^2 + \varepsilon^{-1}b^2, \varepsilon > 0$, with $\varepsilon = \frac{1}{2}$:

$$2\,|x|\,|z|\,|(\sigma_s,\xi)| \le \frac{1}{2}x^2 + 2\,|z|^2\,(\sigma_s,\xi)^2$$

so that

$$|\nabla v(s,x;z\xi)| \le (2\pi\gamma(s)^3)^{-\frac{1}{2}}(|x| + |z|\,|(\sigma_s,\xi)|)\, e^{-\frac{1}{4}\gamma(s)^{-1}x^2} e^{\frac{3}{2}\gamma(s)^{-1}|z|^2(\sigma_s,\xi)^2}$$
$$\le (2\pi)^{-\frac{1}{2}}\gamma(s)^{-1}\left(1+\gamma(s)^{-\frac{1}{2}}\,|z|\,|(\sigma_s,\xi)|\right) e^{\frac{3}{2}\gamma(s)^{-1}|z|^2(\sigma_s,\xi)^2}.$$

The last estimate shows that for $s > 0$, $\nabla v(s,\cdot;\cdot) \in G^2_u(\mathbb{R};\mathcal{U})$ (cf. Appendix 1), and so by Lemmas A.1.1, A.1.2, for $s > 0$, $x \longmapsto \nabla u(s,x)$ is weakly continuous and bounded in some $(\mathcal{S})_{-p}, p \in \mathbb{N}_0$. Consequently, this mapping is Pettis integrable against a finite measure on the real axis, and hence $(P_{\gamma(t)-\gamma(s)}\nabla u(s,\cdot))(x)$ exists as a Hida distribution. Now consider the ds–integral in (3.16). For $0 \le s < t$ we have

$$\left(P_{\gamma(t)-\gamma(s)}\nabla v(s,\cdot;z\xi)\right)(x) = -\gamma(t)^{-1}(x - z(\sigma_s,\xi))\, p_{\gamma(t)}\left(x, z(\sigma_s,\xi)\right),$$

which is obviously in $G^2_u([0,t];\mathcal{U})$ in the s–variable for $x \in \mathbb{R}$, $t > 0$. Therefore, by a similar argument as above ($z\sigma(s)\xi(s)$ is trivially taken into account) we have that for $u(t,x) = S^{-1}v(t,x)$ the integral on the right hand side of (3.15) is a well-defined expression in $(\mathcal{S})^*$. (3.16) implies that (3.15) holds weakly on a total subspace of $(\mathcal{S})$, and consequently $u(t,x)$ solves (3.15) in weak sense.

4. The One–Dimensional Case

Let Ω denote $\mathbb{R}$ or a domain in $\mathbb{R}$. Consider $D := [0,1] \times \Omega$ and denote typical elements in D by $(t,x),(s,y)$, etc. where $t,s \in [0,1]$, $x,y \in \Omega$. We are interested in second order differential operators L of the following type in Ω

$$Lf(x) = a(t,x)f''(x) + b(t,x)f'(x), \quad f \in C^2(\Omega),\ (t,x) \in D.$$

We assume that L is uniformly elliptic: there exists $\varepsilon > 0$ so that for all $(t,x) \in D$, $a(t,x) > \varepsilon$. We also need a smoothness assumption on the coefficients a, b, and for convenience we suppose that they are in $C_b^\infty(D)$. Under these conditions, the heat equation $\frac{\partial}{\partial t} - L = 0$ for L has a fundamental solution $p(t,x;s,y)$ which admits the following bounds (see, e.g., [Fr 83]): for $t > s \ge 0$,

$$|p(t,x;s,y)| \le K(t-s)^{-\frac{1}{2}} e^{-\lambda\frac{(x-y)^2}{t-s}},$$
$$|\frac{\partial}{\partial x}p(t,x;s,y)| \le K(t-s)^{-1} e^{-\lambda\frac{(x-y)^2}{t-s}},$$
$$|\frac{\partial}{\partial y}p(t,x;s,y)| \le K(t-s)^{-1} e^{-\lambda\frac{(x-y)^2}{t-s}},$$

where K and λ are appropriate strictly positive constants.

We shall also write for $f \in C_b(\Omega)$ and a kernel q on $D \times D$.

$$(Q_{t,s}f)(x) := \int_\Omega q(t,x;s,y)f(y)dy,$$

and for $u \in C_b(D)$

$$(Qu)(t,x) := \int_0^t \Big(\int_\Omega q(t,x;s,y)u(s,y)dy\Big)ds.$$
$$= \int_0^t (Q_{t,s}u(s,\cdot))(x)ds,$$

and use similar notations for kernels p, r etc. and their associated integral operators $P_{z,s},\ P,\ R_{t,s},\ R$ etc.

Let $\dot\eta(t,x)$ denote the (informal) noise of

$$\eta(t,x) := \int_0^t \sigma(s,x)\,dB(s), \quad (t,x) \in D,$$

where B is a Brownian motion, $\sigma \in C_b^1(D)$.

Our main interest is to study the informal equation

$$\frac{\partial}{\partial t}u(t,x) - Lu(t,x) = \dot\eta(t,x)\frac{\partial}{\partial x}u(t,x), \quad (t,x) \in D, \tag{4.1}$$

for some random function u on D. Obviously, (4.1) needs interpretation, and the examples and computations in Section 3 suggest to allow also for generalized random fields $u(t,x)$, $(t,x) \in D$. The "multiplication" by the noise $\dot\eta$ in (4.1) will be interpreted in Hitsuda–Skorokhod sense throughout the present section: $\dot\eta(t,x)w \equiv \sigma(t,x)\dot B(t) \diamond w$ for any (generalized) random variable w. Moreover, I was not able to show that the solution u which will be constructed below is differentiable in x. Thus we reformulate this equation in a way that can be given a proper sense. For simplicity, we shall consider here only initial conditions for (4.1) which are given by a deterministic function $u_0 \in C_\infty(\Omega)$. (Here and in the following $C_\infty(X)$, X locally compact, is the space of continuous functions (with values in $\mathbb{R}$ or $\mathbb{C}$) which vanish outside every compact, i.e., if $f \in C_\infty(X)$, then for given $\varepsilon > 0$ there exists a compact C in X with $|f(x)| < \varepsilon$ for all $x \in C^c$. In particular, every such f is uniformly continuous.) As in [Ch 89], we rewrite (4.1) as an integral equation:

$$u(t,x) = \int_\Omega p(t,x;0,y)u_0(y)\,dy + \int_0^t \int_\Omega P(t,x;s,y)\dot\eta(s,y)\frac{\partial}{\partial y}u(s,y)\,dy\,ds,$$

or

$$u(t,x) = (P_{t,0}u_0)(x) + \int_0^t (P_{t,s}\sigma(s,\cdot)\nabla u(s,\cdot))(x)\,dB(s). \tag{4.2}$$

Due to the above mentioned difficulties about the differentiability of u, we do an informal integration by parts in the convolution with P, and write this as

$$u(t,x) = (P_{t,0}u_0)(x) + \int_0^t (Q^\sigma_{t,s}u(s,\cdot))(x)\,dB(s), \tag{4.3a}$$

where $Q^{\sigma}_{t,s}$ has the kernel:

$$q^{\sigma}(t,x;s,y) = -\left[p(t,x;s,y)\frac{\partial}{\partial y}\sigma(s,y) + (\frac{\partial}{\partial y}p(t,x;s,y))\sigma(s,y)\right]. \tag{4.3b}$$

Using the above given estimates on $p(t,x;s,y)$ and its y–derviative, we can bound the kernel by

$$|q^{\sigma}(t,x;s,y)| \leq K(|\nabla\sigma|_{\infty} + |\sigma|_{\infty})(t-s)^{-1}e^{-\lambda\frac{(x-y)^2}{t-s}}. \tag{4.4}$$

Here and below, $|\cdot|_{\infty}$ denotes the sup–norm (on D or on Ω, depending on the context). Note that for $s \sim t$, q^{σ} produces a singularity of the type $(t-s)^{-\frac{1}{2}}$, which is not square–integrable in s. (The other factor $(t-s)^{-\frac{1}{2}}$ is for the "normalization" of the Gaussian kernel. Also, the case $L = \Delta$ shows that this singularity does not arise from the lack of a better estimate.) Hence the stochastic integral in (4.3a) cannot be interpreted easily in the usual sense, and we shall interprete in Hitsuda–Skorokhod sense.

We shall say that a mapping $u : D \to (\mathcal{S})^*$ is a weak solution of (4.1) in $C_{\infty}(D;(\mathcal{S})^*)$, if for every $\varphi \in (\mathcal{S})$, $\langle u, \varphi\rangle \in C_{\infty}(D)$, and if (4.3a) holds. Here, we define

$$Q^{\sigma}_{t,s}u(s,\cdot)(x)$$

as the element in $(\mathcal{S})^*$ which for $\varphi \in (\mathcal{S})$ is given by $(t > s)$

$$\langle Q^{\sigma}_{t,s}u(s,\cdot)(x), \varphi\rangle = \int_{\Omega} q^{\sigma}(t,x;s,y)\langle u(s,y), \varphi\rangle\, dy,$$

i.e., by the integral operator $Q^{\sigma}_{t,s}$ with an integral in the sense of Pettis.

We study (4.3a) by taking its S–transform. Let $\xi \in \mathcal{S}(\mathbb{R})$ and $Su(t,x) \equiv \widehat{u}(t,x)$. Then the S–transform of (4.3a) at ξ is

$$\widehat{u}(t,x)(\xi) = (P_{t,0}u_0)(x) + \int_0^t \left(Q^{\sigma}_{t,s}\widehat{u}(s,\cdot)(\xi)\right)(x)\xi(s)\, ds. \tag{4.5}$$

It will be useful to consider (4.5) for fixed ξ but to generalize to $\xi \in \mathcal{S}_{\mathbb{C}}(\mathbb{R})$ and a more general "initial" value $v_0(t,x)$. I.e., we consider (the ξ–dependence of v being suppressed to simplify the notation): $v_0 \in C_{\infty}(D)$ and the equation

$$v(t,x) = v_0(t,x) + \int_0^t \left(R_{s,t}v(x,\cdot)\right)(x)\xi(s)\, ds, \tag{4.6}$$

for fixed $\xi \in \mathcal{S}_{\mathbb{C}}(\mathbb{R})$ and a kernel $r(t,x;s,y)$ of $R_{s,t}$ which is continuous on $D \times D \setminus \{\text{diagonal}\}$ (the diagonal in $D \times D$ is the subset $\{(t,x;s,y),\ t = s,\ x = y\}$), which satisfies

$$|r(t,x;s,y)| \leq K_1(t-s)^{-1}e^{-\lambda\frac{(x-y)^2}{t-s}}, \tag{4.7}$$

where K_1 is some constant. We are going to solve (4.7) by a fixed point method on the (complex) space $C_\infty(D)$. Due to the facts that the kernel r has an integrable singularity, and that it falls off rapidly at infinity, it follows by standard estimates that $R_{s,t}$ maps $C_\infty(D)$ into itself. Thus the mapping T_ξ defined by

$$(T_\xi w)(t,x) := v_o(t,x) + \int_0^t (R_{s,t}w(s,\cdot))(x)\xi(s)\,ds \tag{4.8}$$

maps $C_\infty(D)$ into itself, and the solution of (4.6) is a fixed point of T_ξ. In order to show that T_ξ has a unique fixed point in $C_\infty(D)$, we equip $C_\infty(D)$ with the following norm

$$|w|_{\xi,\infty} := \sup_{(t,x)\in D} |w(t,x)|\,\psi^{-1}_{1,\rho(\xi)}(t), \tag{4.9}$$

where $\psi_{1,\rho(\xi)} > 0$ is the unique solution of

$$\psi_{1,\rho(\xi)}(t) = 1 + \rho(\xi)\int_0^t (t-s)^{-\frac{1}{2}}\psi_{1,\rho(\xi)}(s)\,ds, \tag{4.10}$$

with $\rho(\xi) := 1 + K_2|\xi|_\infty$, K_2 some constant which we choose below. In Appendix 2 it is shown that $|\cdot|_{\xi,\infty}$ is equivalent to the usual sup–norm. In particular, $(C_\infty(D), |\cdot|_{\xi,\infty})$ is a Banach space, and we only need to prove that T_ξ is a strict contraction with respect to $|\cdot|_{\xi,\infty}$. But this follows from the following estimation for $v, w \in C_\infty(D)$: let $(t,x)\in D$, then

$$\begin{aligned}
&|T_\xi v(t,x) - T_\xi w(t,x)| = \\
&\quad = |\int_0^t R_{t,s}(v(s,\cdot) - w(s,\cdot))(x)\xi(s)\,ds| \\
&\quad \le K_1|\xi|_\infty \int_0^t\int_\Omega (t-s)^{-1}e^{-\lambda\frac{(x-y)^2}{t-s}}\,|v(s,y) - w(s,y)|\,dy\,ds \\
&\quad \le K_1|\xi|_\infty \int_0^t (t-s)^{-\frac{1}{2}}|v(s,\cdot) - w(s,\cdot)|_\infty\,ds \int_{I\!R}(t-s)^{-\frac{1}{2}}e^{-\lambda\frac{(x-y)^2}{t-s}}\,dy \\
&\quad = K_1\sqrt{\frac{\pi}{\lambda}}\,|\xi|_\infty \int_0^t (t-s)^{-\frac{1}{2}}|v(s,\cdot) - w(s,\cdot)|_\infty ds \\
&\quad = K_1\sqrt{\frac{\pi}{\lambda}}|\xi|_\infty \int_0^t (t-s)^{-\frac{1}{2}}\psi_{1,\rho(\xi)}(s)\psi^{-1}_{1,\rho(\xi)}(s)|v(s,\cdot) - w(s,\cdot)|_\infty\,ds \\
&\quad \le K_1\sqrt{\frac{\pi}{\lambda}}\,|\xi|_\infty|v-w|_{\xi,\infty}\int_0^t (t-s)^{-\frac{1}{2}}\psi_{1,\rho(\xi)}(s)\,ds \\
&\quad = K_1\sqrt{\frac{\pi}{\lambda}}\,|\xi|_\infty|v-w|_{\xi,\infty}\rho(\xi)^{-1}(\psi_{1,\rho(\xi)}(t) - 1) \\
&\quad \le K_1\sqrt{\frac{\pi}{\lambda}}\,|\xi|_\infty\rho(\xi)^{-1}|v-w|_{\xi,\infty}\psi_{1,\rho(\xi)}(t),
\end{aligned}$$

where we used $\psi_{1,\rho(\xi)} > 0$ (cf. Lemma A.2.2), and equation (4.10) in the last equality. Thus we get

$$|T_\xi v - T_\xi w|_{\xi,\infty} \leq K_1 \sqrt{\frac{\pi}{\lambda}} |\xi|_\infty \rho(\xi)^{-1} |v - w|_{\xi,\infty}$$

and it suffices to choose $K_2 = K_1\sqrt{\frac{\pi}{\lambda}}$ so that T_ξ is a strict contraction. (Recall that $K_1 = K(|\nabla\sigma|_\infty + |\sigma|_\infty)$ in our original problem.) Hence we have proved the following

Lemma 4.1. For every $\xi \in \mathcal{S}_{\mathbb{C}}(\mathbb{R})$, (4.6) has a unique solution $\widehat{u}(t,x;\xi)$, which – as a function of $(t,x) \in D$ – belongs to $C_\infty(D)$.

Next we study $\widehat{u}$ as a function of $\xi \in \mathcal{S}_{\mathbb{C}}(\mathbb{R})$ in more detail. First we derive an estimate of Gronwall type. Similarly as in the proof that T_ξ is a contraction, we can bound $\widehat{u}(t,x)(\xi)$ as follows:

$$\begin{aligned}|\widehat{u}(t,x)(\xi)| &\leq |v_0|_\infty + K_1|\xi|_\infty \int_0^t (t-s)^{-\frac{1}{2}} |\widehat{u}(s,\cdot)(\xi)|_\infty \, ds \int_{\mathbb{R}} (t-s)^{-\frac{1}{2}} e^{-\lambda\frac{(x-y)^2}{t-s}} \, dy \\ &= |v_0|_\infty + K_1\sqrt{\frac{\pi}{\lambda}} |\xi|_\infty \int_0^t (t-s)^{-\frac{1}{2}} |\widehat{u}(s,\cdot)(\xi)|_\infty \, ds,\end{aligned}$$

and hence we have

$$|\widehat{u}(t,\cdot)(\xi)|_\infty \leq |v_o|_\infty + K_1\sqrt{\frac{\pi}{\lambda}} |\xi|_\infty \int_0^t (t-s)^{-\frac{1}{2}} |\widehat{u}(s,\cdot)(\xi)|_\infty \, ds.$$

Since $(t,x) \longmapsto \widehat{u}(t,x)(\xi)$ is for every $\xi \in \mathcal{S}_{\mathbb{C}}(\mathbb{R})$ in $C_\infty(D)$, it follows that $t \longmapsto |\widehat{u}(t,\cdot)(\xi)|_\infty$ is continuous on $[0,1]$. Thus we may apply Lemma A.2.4 in Appendix 2 to conclude the following result.

Lemma 4.2. For every $t \in [0,1], \xi \in \mathcal{S}_{\mathbb{C}}(\mathbb{R})$, the unique solution of (4.6), $\widehat{u}(t,x)(\xi)$, admits the following bound

$$|\widehat{u}(t,\cdot)(\xi)|_\infty \leq |v_0|_\infty (1 + K_1\pi\, |\xi|_\infty \sqrt{\frac{t}{\lambda}}) \exp(K_1^2 \pi^2 \frac{t}{\lambda} |\xi|_\infty^2). \tag{4.11}$$

In particular, for some constants $K_3, K_4 > 0$, one has

$$|\widehat{u}(\xi)|_\infty \leq K_3 |v_0|_\infty e^{K_4|\xi|_\infty^2}. \tag{4.12}$$

Now let ξ be of the form $\xi = \eta + z\zeta, \eta, \zeta \in \mathcal{S}(\mathbb{R}), z \in \mathbb{C}$, and write

$$\widehat{u}(t,x)(\xi) \equiv \widehat{u}(t,x;z).$$

We are interested in the analytic properties of $z \longmapsto \widehat{u}(t,x;z)$. If $z = z_1 + iz_2$, $z_1, z_2 \in I\!R$, it is not hard to use the same methods as before to show that

$$\frac{\partial}{\partial z_i}\widehat{u}(t,x;z), \quad i=1,2,\ (t,x) \in D,$$

exist. But then $\frac{\partial}{\partial \bar{z}}\widehat{u}(t,x;z)$ satisfies

$$\frac{\partial}{\partial \bar{z}}\widehat{u}(t,x;z) = \int_0^t \left[Q^\sigma_{t,s}\left(\frac{\partial}{\partial \bar{z}}\widehat{u}(s,\cdot\,;z)\right)(\xi)\right](x)\xi(s)\,ds,$$

and hence Lemma 4.2 implies that $\frac{\partial}{\partial \bar{z}}\widehat{u}(t,x;z) = 0$. In other words:

Lemma 4.3. For all $(t,x) \in D, \eta, \zeta \in \mathcal{S}(I\!R)$, the mapping

$$z \longmapsto \widehat{u}(t,x)(\eta + z\zeta)$$

is entire.

By Theorem 2.2, it follows that $\widehat{u}(t,x)$ is the S–transform of an element $u(t,x) \in (\mathcal{S})^*$. Using the bound (4.12) and the continuity of $(t,x) \mapsto \widehat{u}(t,x;\xi)$, $\xi \in \mathcal{S}(I\!R)$, we can now apply the same reasoning as at the end of Section 3 to obtain the final result:

Theorem 4.4. For every initial condition $u_0 \in C_\infty(D)$, equation (4.1) has a unique weak solution in $C_\infty(D;(\mathcal{S})^*)$.

It is clear that Theorem 4.4 can be generalized in many ways. For instance, one can choose more general initial conditions, the conditions on the coefficients a, b, σ, can be improved, the method of proof generalizes to d–dimensional domains and certain non–linear terms can be added to (4.1). Moreover, the arguments leading to Theorem 4.4 can also be used to discuss other equations, such as anticipating stochastic differential and stochastic Volterra equations (cf. also [CL 93]). All this will be worked out in forthcoming papers [DP 94].

Appendix 1: Weak Differentiability

In this appendix we establish some results concerning the calculus of Hida distributions which depend on a variable in $I\!R^d$. The notions of continuity and differentiability used here are formulated in the weak topologhy of $(\mathcal{S})^*$, although most of what is stated here holds also (with a little more care in the arguments) for the strong topology.

Let O be an open set in $I\!R^d$, and consider a mapping f from O into $(\mathcal{S})^*$. Let $x_0 \in O$ and $k \in \{1,\ldots,d\}$. We shall say that *f has a weak partial derivative in direction k at x_0*, if there exists an element in $(\mathcal{S})^*$ denoted by $(D_k f)(x_0)$, so that

$$h^{-1}\big(f(x_0 + he_k) - f(x_0)\big)$$

converges weakly in $(\mathcal{S})^*$ to $(D_k f)(x_0)$ as h converges to zero. (e_k is the unit vector in $\mathbb{R}^d$ in direction k.) If f has at all points $x \in O$ weak partial derivatives in all directions $k = 1, \ldots, d$, such that for all $k = 1, \ldots, d$, $x \longmapsto D_k f(x)$ is weakly continuous from O into $(\mathcal{S})^*$, we say that f *is weakly continuously differentiable in* O. The space of all such mappings will be denoted by $C^1_w(O; (\mathcal{S})^*)$. Obviously, if $f \in C^1_w(O; (\mathcal{S})^*)$, then for all $\varphi \in (\mathcal{S})$, $x \longmapsto \langle f(x), \varphi \rangle$ belongs to $C^1(O)$.

Assume that for all $x \in O$, $k = 1, \ldots, d$, $D_k f(x) \in (\mathcal{S})^*$ exists. Then we can repeat the preceding considerations: for $n \in \mathbb{N}$ we shall say that $f : O \to (\mathcal{S})^*$ is n *times weakly continuously differentiable in* U, if for every multi-index $\alpha = (\alpha_1, \ldots, \alpha_d)$ of length $|\alpha| = \alpha_1 + \ldots + \alpha_k \leq n$, the weak partial derivative $D^\alpha_f(x) = (D_1^{a_1} \cdots D_d^{\alpha_d} f)(x)$ exists for all $x \in O$, and is a weakly continuous function in O with values in $(\mathcal{S})^*$. The class of all such mappings is denoted by $C^n_w(O; (\mathcal{S})^*)$. $C^0_w(O; (\mathcal{S})^*) \equiv C_w(O; (\mathcal{S})^*)$ is the space of all weakly continuous mappings from O into $(\mathcal{S})^*$, $C^\infty_w(O; (\mathcal{S})^*)$ the space of those $f : O \to (\mathcal{S})^*$, such that for every $n \in \mathbb{N}$, $f \in C^n_w(O; (\mathcal{S})^*)$.

In the following we shall give criteria for when a U–functional which depends on $x \in O$ has an inverse S–transform in $C^n_w(O; (\mathcal{S})^*)$. First we make the following definitions.

Let v be a mapping from O into the space $\mathcal{U}$ of U–functionals (cf. Section 2). We say that v *is locally of uniform growth less than two*, if every $x \in O$ has a neighborhood O_x in O, such that there exist $K_1, K_2 > 0$ and $p \in \mathbb{N}_0$, so that for all $x' \in O_x$, $\xi \in \mathcal{S}(\mathbb{R})$, $z \in \mathbb{C}$, we have the estimate

$$|v(x'; z\xi)| \leq K_1 e^{K_2 |z|^2 |\xi|^2_{2,p}}.$$

The space of all mappings from O into $\mathcal{U}$ which are locally of uniform growth less than two will be denoted by $G^2_{lu}(O; \mathcal{U})$. If O_x above can be chosen as O for every $x \in O$, then we simply say that v *is uniformly of growth less than two*, and denote the corresponding space by $G^2_u(O; \mathcal{U})$.

The following is a re-statement of a result in [PS 91] (cf. also [KL 94]).

Lemma A.1.1. Assume that $v \in G^2_u(O; \mathcal{U})$. Then the family $\{f(x) = S^{-1}v(x); x \in O\}$ of Hida distributions is bounded in $(\mathcal{S})_{-q}$ for some $q \in \mathbb{N}_0$.

Next we prove a criterium for weak continuity of a family of Hida distributions. Denote by $C_{lu}(O; \mathcal{U})$ the subspace of $G^2_{lu}(O; \mathcal{U})$ consisting of those v so that for all $\xi \in \mathcal{S}(\mathbb{R})$, $x \longmapsto v(x; \xi)$ is continuous from O into $\mathbb{C}$.

Lemma A.1.2. If $v \in C_{lu}(O; \mathcal{U})$, then $S^{-1}v \in C_w(O; (\mathcal{S})^*)$.

Proof. Let $x \in O$. Then $v \in G^2_u(O_x, \mathcal{U})$ for some neighborhood O_x of x. By Lemma A.1.1 we know that $\{f(x'); x' \in O_x\}$ is bounded in some $(\mathcal{S})_{-q}$, say by $M > O$.

Let $\mathcal{E}$ be the (complex) linear span of $\{\exp(\langle\cdot,\xi\rangle);\ \xi\in\mathcal{S}(\mathbb{R})\}$, which is a dense subspace of $(\mathcal{S})$. By hypothesis, for every $\psi\in\mathcal{E}$, the mapping $x\longmapsto\langle f(x),\psi\rangle$ is continuous. Now let $\varepsilon>0$ and $\varphi\in(\mathcal{S})$ be given. Choose $\psi\in\mathcal{E}$ so that $\|\varphi-\psi\|_{2,q}<\frac{\varepsilon}{4M}$. Choose $\delta>0$ small enough, so that $\{x';|x-x'|<\delta\}\subset O_x$, and such that $|x-x'|<\delta$ implies

$$|\langle f(x),\psi\rangle-\langle f(x'),\psi\rangle|<\frac{\varepsilon}{2}.$$

Then

$$\begin{aligned}|\langle f(x),\varphi\rangle-\langle f(x'),\varphi\rangle| &\le |\langle f(x),\psi\rangle-\langle f(x'),\psi\rangle|+2M\,\|\varphi-\psi\|_{2,q}\\ &<\varepsilon.\end{aligned}$$

□

Now we turn to weak differentiability. Let $n\in\mathbb{N}$ and consider the subspace $C^n_{lu}(O;\mathcal{U})$ of $G^2_{lu}(O;\mathcal{U})$ which consists of those v such that: (i) for every $\zeta\in\mathcal{S}_{\mathbb{C}}(\mathbb{R})$, $x\longmapsto v(x;\zeta)$ belongs to $C^n(O)$, and (ii) for every multiindex α, $|\alpha|\le n$, $D^\alpha v\in G^2_{lu}(O;\mathcal{U})$.

Lemma A.3. Assume that $v\in C^n_{lu}(O;\mathcal{U}), n\in\mathbb{N}$ or $n=+\infty$. Then $S^{-1}v\in C^n_w(O;(\mathcal{S})^*)$.

Proof. We only prove the case $n=1$, the general case follows by iteration. Denote $D_kf:=S^{-1}D_kv$, $k=1,\dots,d$. By assumption and Lemma A.1.2 we know that $D_kf\in C_w(O;(\mathcal{S})^*)$. It remains to show that for all $x\in O$, $\varphi\in(\mathcal{S})$, $k=1,\dots,d$, we have

$$\lim_{h\to0}\left[h^{-1}(\langle f(x+he_k),\varphi\rangle-\langle f(x,\varphi)\rangle)-\langle D_kf(x),\varphi\rangle\right]=0,$$

and our hypothesis implies that this holds for all $\varphi\in\mathcal{E}$. We can therefore conclude the argument as in the proof of Lemma A.1.2, if we can show that there exists $\delta>0$, so that

$$\{h^{-1}(f(x+he_k)-f(x));\ 0<|h|<\delta\}$$

is bounded in $(\mathcal{S})_{-q}$ for some $q\in\mathbb{N}_0$.

Let $x\in O$ and $O_{x,k}$ a neighborhood of x so that $D_kv\in G^2_u(O_{x,k};\mathcal{U})$. Choose $\delta>0$ small enough such that $0<|h|<\delta$ implies $x+he_k\in O_{x,k}$. Let $z\in\mathbb{C},\xi\in\mathcal{S}(\mathbb{R})$, then

$$\left|h^{-1}(v(x+he_k;z\xi)-v(x;z\xi))\right|=|(D_kv)(x';z\xi)|$$

for some $x'\in O_{x,k}$ by the mean–value theorem. But the last expression is of uniform growth less than two on $O_{x,k}$. Thus $\{h^{-1}(f(x+he_k)-f(x));\ 0<|h|<\delta\}$ is bounded in $(\mathcal{S})_{-q}$, for some $q\in\mathbb{N}_0$, by Lemma A.1.1. □

Remark. The criteria for weak continuity and for weak continuous differentiability are not quite "symmetric" with respect to the role of the Schwartz space variable. Perhaps one can combine the results in [KL 94] with a Vitali type argument so that in the last lemma continuous differentiability of $x\longmapsto v(x;\xi)$, for real $\xi\in\mathcal{S}(\mathbb{R})$ is sufficient.

Appendix 2: On the Comparison Function

In this apendix we study properties of the solution $\psi_{a,b}$, the *comparison function*, of the integral equation ($a, b \in \mathbb{R}$)

$$f(t) = a + b(Kf)(t), \quad t \in [0,1], \tag{A.2.1}$$

where K is the integral operator given by

$$Kf(t) = \int_0^t (t-s)^{-\frac{1}{2}} f(s)\, ds. \tag{A.2.2}$$

We will consider K as an operator on $C([0,1])$ (equipped with the sup–norm or an equivalent norm, see below).

Lemma A.2.1. Let $a, b \in \mathbb{R}$ and set

$$\psi_{a,b}(t) := a \sum_{n=0}^{\infty} \Gamma(\frac{n}{2}+1)^{-1} (b^2 \pi t)^{\frac{n}{2}}, \quad t \in [0,1]. \tag{A.2.3}$$

Then $\psi_{a,b}$ is the unique solution of (A.2.1) in $C([0,1])$.

Proof. First of all, note that the right hand side of (A.2.3) converges uniformly in $t \in [0,1]$, and hence it defines a continuous function: $\psi_{a,b} \in C([0,1])$. For $b = 0$ it is obvious that $\psi_{a,b} = a$ solves (A.1.1). Let $b \neq 0$, and for $n \in \mathbb{N}_0$ consider the integral

$$\int_0^t (t-s)^{-\frac{1}{2}} s^{\frac{n}{2}}\, ds$$

which is a beta function, and its value is equal to

$$\sqrt{\pi}\,\Gamma(\frac{n}{2}+1)\Gamma(\frac{n+1}{2}+1)^{-1} t^{\frac{n+1}{2}}.$$

Thus the series

$$a \sum_{n=0}^{\infty} \Gamma(\frac{n}{2}+1)^{-1} (b^2\pi)^{\frac{n}{2}} \int_0^t (t-s)^{-\frac{1}{2}} s^{\frac{n}{2}}\, ds$$

converges absolutely and uniformly in $t \in [0,1]$, and Fubini's theorem implies that it is equal to $(K\psi_{a,b})(t)$. Hence we have

$$\begin{aligned} K\psi_{a,b}(t) &= \frac{a}{b} \sum_{n=0}^{\infty} \Gamma(\frac{n+1}{2}+1)^{-1} (b^2\pi t)^{\frac{n+1}{2}} \\ &= \frac{1}{b}(\psi_{a,b}(t) - a), \end{aligned}$$

by insertion of the value of the above integral. Thus $\psi_{a,b}$ is a solution of (A.2.1).

If φ, ψ are solutions of (A.2.1) in $C([0,1])$, then their difference $g \in C([0,1])$ solves

$$g(t) = b(Kg)(t), \quad t \in [0,1],$$

and thus we have for all $n \in \mathbb{N}, t \in [0,1]$,

$$g(t) = b^n(K^n g)(t).$$

In particular, we get the estimates

$$|g(t)| \leq |g|_\infty |b|^n (K^n 1)(t)$$

for all $n \in \mathbb{N}, t \in [0,1]$. $(K^n 1)(t)$ can easily be computed by using the Laplace transform, and the result is

$$(K^n 1)(t) = \Gamma(\frac{n}{2}+1)^{-1}(\pi t)^{\frac{n}{2}}.$$

Therefore we have for all $n \in \mathbb{N}$,

$$|g|_\infty \leq |g|_\infty |b|^n \pi^{\frac{n}{2}} \Gamma(\frac{n}{2}+1)^{-1}.$$

Consequently $g = 0$, i.e., $\psi_{a,b}$ is the unique solution of (A.2.1). □

Lemma A.2.2. Let $a, b \in \mathbb{R}$. For all $t \in [0,1]$ the solution $\psi_{a,b}$ of (A.2.1) admits the bounds

$$|a| e^{t\pi b^2} \leq |\psi_{a,b}(t)| \leq |a|(1 + |b|\sqrt{\pi t}) e^{t\pi b^2}. \tag{A.2.4}$$

Proof. Without loss of generality let $a, b \geq 0$ so that $\psi_{a,b} \geq 0$. Then

$$\begin{aligned} \psi_{a,b}(t) &= a \sum_{n=0}^{\infty} \Gamma(\frac{n}{2}+1)^{-1}(t\pi b^2)^{\frac{n}{2}} \\ &\geq a \sum_{n=0}^{\infty} \Gamma(n+1)^{-1}(t\pi b^2)^n \\ &= a\ e^{t\pi b^2}. \end{aligned}$$

For the upper bound estimate as follows:

$$\begin{aligned} \psi_{a,b}(t) &= a\Big(e^{t\pi b^2} + \sum_{n=0}^{\infty} \Gamma(n+\frac{3}{2})^{-1}(t\pi b^2)^{n+\frac{1}{2}}\Big) \\ &\leq a(1 + (t\pi b^2)^{\frac{1}{2}}) e^{t\pi b^2}. \end{aligned}$$

□

On $C([0,1])$ we define a family of norms $\{|\cdot|_{b,\infty},\ b\geq 0\}$ by

$$|f|_{b,\infty} := \sup_{t\in[0,1]} |f(t)\psi_{1,b}^{-1}(t)|.$$

By Lemma A.2.2 it is clear that for every $b \geq 0$, $|\cdot|_{b,\infty}$ is equivalent to the sup–norm $|\cdot|_{\infty}$.

Next we derive a Gronwall type estimate.

Lemma A.2.3. Assume that $f \in C([0,1])$ satisfies for some $b > 0$ the inequality

$$f(t) \leq b(Kf)(t), \quad t \in [0,1]. \tag{A.2.5}$$

Then $f(t) \leq 0$ for all $t \in [0,1]$.

Proof. Since K has a positive kernel, (A.2.5) implies $f(t) \leq b(Kf_+)(t)$ for all $t \in [0,1]$, where $f_+ = \max(f,0) \in C([0,1])$. Therefore (A.2.5) yields for all $c > 0$,

$$\begin{aligned} f_+(t) &\leq b(Kf_+)(t) \\ &= b(K\psi_{1,c}\psi_{1,c}^{-1}f_+)(t) \\ &\leq b|f_+|_{c,\infty}(K\psi_{1,c})(t) \\ &= b|f_+|_{c,\infty}\frac{1}{c}(\psi_{1,c}(t)-1) \\ &\leq \frac{b}{c}|f_+|_{c,\infty}\psi_{1,c}(t). \end{aligned}$$

Hence we get for all $c > 0$,

$$|f_+|_{c,\infty} \leq \frac{b}{c}|f_+|_{c,\infty}.$$

We only need to choose $c = 2b$ to conclude $|f_+|_{c,\infty} = 0$, which entails $f_+ = 0$. □

Lemma A.2.4. Let $f \in C([0,1])$ be non–negative and satisfy

$$f(t) \leq a + b(Kf)(t), \quad t \in [0,1], \tag{A.2.6}$$

for $a \geq 0, b \geq 0$. Then

$$f(t) \leq \psi_{a,b}(t) \leq a(1+b\sqrt{\pi t})e^{t\pi b^2}, \quad t \in [0,1]. \tag{A.2.7}$$

Proof. We assume that $a, b > 0$, otherwise the statement is trivial. Then we have by hypothesis

$$f(t) - \psi_{a,b}(t) \leq b\ K(f-\psi_{a,b})(t).$$

By Lemma A.2.3, $f(t) \leq \psi_{a,b}(t)$ for all $t \in [0,1]$. □

Remark. Of course, estimate (A.2.7) can also be derived by iteration of (A.2.6), and an application of the Laplace transform to compute the multiple convolution integrals.

ACKNOWLEDGEMENT. It is a pleasure to thank G. Cochran and J.–S. Lee for fruitful discussions.

References.

[BT 94] Benth, F. and Timpel, M.: Topological aspects of the characterization of Hida distributions – A remark; *Preprint* (1994)

[CL 93] Cochran, G., Lee, J.–S. and Potthoff, J.: Stochastic Volterra equations with singular kernels; *Preprint* (1993)

[Ch 89] Chow, P.L.: Generalized solution of some parabolic equations with a random drift; *J. Appl. Math. Optimization* **20** (1989) 81–96

[DP 94] Deck, Th. and Potthoff, J.: Parabolic stochastic partial differential equations; in preparation

[Fr 83] Friedman, A.: *Partial Differential Equations of Parabolic Type.* Malabar: Krieger (1983)

[HK 93] Hida, T., Kuo, H.–H., Potthoff, J. and Streit, L.: *White Noise – An Infinite Dimensional Calculus.* Dordrecht: Kluwer (1993)

[KL 94] Kondratiev, Yu.G., Leukert, P., Potthoff, J., Streit, L. and Westerkamp, W.: Generalized functionals in Gaussian spaces – The characterization theorem revisited; *Preprint* (1994)

[PS 91] Potthoff, J. and Streit, L.: A characterization of Hida distributions; *J. Funct. Anal.* **101** (1991) 212–229

COLOMBEAU GENERALIZED FUNCTIONS AND STOCHASTIC ANALYSIS

FRANCESCO RUSSO[1] [2]

1 Introduction

The class of Colombeau generalized functions has been introduced in the early eightees. A good pedagogical survey about this class and its related calculus can be found in [12]. Recently [13] has presented the state of the art on the actual theory and it has developed some numerical analysis applications. For Colombeau general theory, the reader can also consult [9], [30], [34] and references therein. The purpose was to treat non–linear problems which naturally involve Schwartz distributions. Several examples come out from quantum field theory, see [16], [43] ; there, the possibility of defining a good multiplication and non–linear operation for distributions is fundamental. In [45], also some connections with non–standard analysis have been investigated.

In the past, there was a lot of heuristic computations carried out by physicists which somehow needed a mathematical justification. Among the first mathematical temptatives to define multiplication of distributions, we find [21] and [22].

However, a mathematician could ask the following questions : is there a natural sense to treat a non–linear partial differential equation with distributional initial condition? This, in spite of the Schwartz impossibility theorem which affirms the impossibility of finding an algebra containing the Schwartz distribution which extends the multiplication of ordinary function in C^1 so that the Leibnitz rule for derivatives hold, see [41]. The approach of Colombeau avoids this difficulty, see [13].

Our interest is the application to stochastic analysis of Colombeau framework. Non-linear problems and distributions appear quite naturally since white noise and other related fields are random Schwartz distributions. We will illustrate this phenomenon with two basic examples.

Let us consider one ordinary stochastic differential equation :

$$\dot{u}(t) = f(u(t))\dot{W}(t) + g(u(t)), \quad u(0) = x, \tag{1}$$

[1]Université Paris Nord, Département de Mathématiques, Institut Galilée, av. J.B. Clément, F–93430 Villetaneuse

[2]Universität Bielefeld, BIBOS, D–33501 Bielefeld 1

A. I. Cardoso et al. (eds.), Stochastic Analysis and Applications in Physics, 329–349.

where $x \in \mathbb{R}$, f, g are smooth functions and $\dot{W}$ is a white noise on $\mathbb{R}$. If $f \equiv 1$, we can interpret (1) in the sense of distribution and look for continuous processes solving (1). Obviously (1) has an equivalent integral form

$$u(t) = x + \int_0^t g(u(s))\, ds + B(t) \tag{2}$$

where B is a Brownian motion on $\mathbb{R}_+$. $B(t)$ is formally $\int_0^t \dot{W}(s)\, ds$. By Picard methods (2) has a unique solution u which is a continuous process but its paths are not in the Sobolev space H^1. The paths of $\dot{W}$ are in H^{-1} so that it is impossible to hope to define the product $f(u(t))\dot{W}(t)$ with the help of distributions theory.

The classical stochastic analysts consider the integral form

$$u(t) = x + \int_0^t f(u(s))\dot{W}(s)\, ds + \int_0^t g(u(s))\, ds. \tag{3}$$

They give a sense to $\dot{W}(s)\, ds = dB(s)$ and the integral is intended in the sense of Ito, backward or Stratonovich; the different interpretation of the stochastic integrals corresponds to a different product a $f(u(s))\dot{W}(s)$.

If the initial condition is anticipating then we can interpret (3) even the sense of Skorohod which is related with the Wick product.

Remark that the Colombeau approach furnishes a pathwise approach which can be requested in model (1) by the engineer, the physicist or the finance operator.

The second example is given by stochastic PDE's. In [2] we have studied the case of the two–space dimensional wave equation which is perturbed by a white noise. This result can be extended to the one of the stochastic heat equation. Let f be a Lipschitz function, L the wave operator or the heat operator on $\mathbb{R}_+ \times \mathbb{R}^2$ and we consider the equation

$$LX = \lambda f(X) + \dot{W} \tag{4}$$

with zero initial and boundary conditions. (4) has a unique solution in the sense of Colombeau and it is somehow "bounded". If f is a difference of positive functions, [2] have proved that the solution X is associated with the solution of the free problem ($\lambda \equiv 0$). If we operate a Wick type normalization on $f(X)$, f being of trigonometric type, each term of the expansion in λ is associated with a random Schwartz distribution.

In this paper, we will examine some aspects concerning the two quoted examples. The first one is the consequence of joint works with P. Vallois, see [36], [37], [38], [39], see section 4. The second one has been extracted from joint works with S. Albeverio and Z. Haba, see [1], [2], see section 3.

A paper which makes use of similar tools as Colombeau generalized functions, but in the framework of Wiener functionals is [18]. In this paper however, bilinear models are considered and the multiplication of noises (which are random distributions) intervene. Concerning the relation between stochastic analysis and Colombeau framework, we remind [26], which definesa space of generalized functions on Wiener space. [31] proposes one approach for studying Colombeau generalized stochastic processes which is based on the convergence in probability.

2 Generalities about Colombeau generalized functions

Let n be a non-negative integer, T an open subset of $\mathbb{R}^n$. $\mathbb{R}^0 = \{0\}$ by convention. $C^\infty(\overline{T})$ (resp. $C(\overline{T})$) will stand for the complex valued function defined on $\mathbb{R}^n$ and vanishing outside $\overline{T}$. $\mathbb{C}$ will be for many purposes identified with $\mathbb{R}^2$.

$\mathcal{D}(\mathbb{R}^n)$ (resp. $\mathcal{D}(T)$) will denote the set of smooth functions with compact support (resp. with support included in T). $\mathcal{D}(\overline{T})$ is the space of $f \in C^\infty(\overline{T})$ with compact support. $C_b^\infty(\mathbb{R}^n)$ is the space of smooth functions whose all derivatives are bounded. $\mathcal{D}'(\overline{T})$ corresponds to the spaces of Schwartz distributions with support in $\overline{T}$.

We consider the basic family of real valued mollifiers

$$\mathcal{A}_0 = \left\{ \phi \in \mathcal{D}(\mathbb{R}) | \int_{\mathbb{R}} \phi(y)\, dy = 1 \right\}.$$

Any Friedrich mollifier belongs ot $\mathcal{A}_0$. In analogy with [12], for any $q \in \mathbb{N}$, we define

$$\mathcal{A}_q = \left\{ \phi \in \mathcal{A}_0 | \int_{\mathbb{R}} y^r \phi(y)\, dy = 0, \quad r \leq q \right\}$$

If $\phi \in \mathcal{A}_q$, we denote $\phi_n \in \mathcal{D}(\mathbb{R}^n)$ by

$$\phi_n(x_1, \ldots, x_n) = \phi(x_1) \ldots \phi(x_n).$$

In particular

$$\int_{\mathbb{R}^n} \phi_n(x)\, dx = 1, \qquad \int_{\mathbb{R}^n} x^r \phi_n(x)\, dx = 0$$

for any multi-index r over $\mathbb{R}^n$ such that $|r| \leq q$.

$\mathcal{E}(\overline{T})$ denotes the set of all function $R : \mathcal{A}_0 \times \overline{T} \to \mathbb{C}$ which belong to $C^\infty(\overline{T})$ for each fixed $\phi \in \mathcal{A}_0$. $\mathcal{E}(\overline{T})$ is an algebra for the multiplication. More important is the subalgebra of moderate elements. A *moderate* element is characterized by the following property : for every compact subset $K \subset \mathbb{R}^n$ and every derivation operator D , there is $N \in \mathbb{N}$ such that, if $\phi \in \mathcal{A}_N$ then there is $\eta, c > 0$ such that

(5) $$|JDR(\phi^\varepsilon, x)| \leq c\varepsilon^{-N}, \text{ if } \varepsilon \leq \eta.$$

If $\phi \in \mathcal{D}(\mathbb{R})$, we define

$$\begin{aligned} \phi^\varepsilon(y) &= \frac{1}{\varepsilon}\phi(\frac{y}{\varepsilon}), \quad y \in \mathbb{R} \\ \phi_n^\varepsilon(x) &= \frac{1}{\varepsilon^n}\phi_n(\frac{x}{\varepsilon}), \quad x \in \mathbb{R}^n. \end{aligned}$$

We remark that here and in the sequel, D is allowed to be of order zero; in this case D is the identity .

An element in $\mathcal{N}(\overline{T})$ (*null* element) is defined as follows : for every compact subset $C \subset \overline{T}$ and any derivation operator, there are $N \in \mathbb{N}$, a sequence $(\alpha(q))$ so that

i) $\lim_{q\to\infty} \alpha(q) = \infty$

ii) There are $\eta, c > 0$ such that

(6) $$\phi \in \mathcal{A}_q,\ q \geq N \Rightarrow |DR(\phi^\varepsilon,\ x)| \leq c\varepsilon^{\alpha(q)-N},\ \forall x \in C,\ \varepsilon \leq \eta.$$

The derivative is taken with respect to x.

$\mathcal{G}(\overline{T})$ is defined as the quotient algebra between $\mathcal{E}(\overline{T})$ and $\mathcal{N}(\overline{T})$. An element of $\mathcal{G}(\overline{T})$ will be called a (Colombeau) generalized function (defined on $\overline{T}$). A Colombeau generalized function is in a sense the "equivalence class of all possible regularizations".

$\mathcal{D}'(\mathbb{R}^n)$ is naturally imbedded into $\mathcal{G}(\mathbb{R}^n)$. If ζ is a Schwartz distribution, the respective generalized function, also denoted by ζ, is represented by

$$R_\zeta(\phi, x) = (\phi_n * \zeta)(x) = \zeta(\phi_n(x - \cdot)).$$

The equality in $\mathcal{G}(\overline{T})$ is very strong. It corresponds to the equality in $C^\infty(\overline{T})$, so it is in particular stronger than the a.e. equality. There is a weaker concept which is the association. Two (Colombeau) generalized functions G_1 and G_2 are *associated* if for any $\alpha \in \mathcal{D}(\mathbb{R}^n)$,

(7) $$\lim_{\varepsilon\to 0} \int_{\mathbb{R}^n} (R_{G_1} - R_{G_2})(\phi^\varepsilon, x)\alpha(x)\, dx = 0$$

for any $\phi \in \mathcal{A}_q$ and q large enough. The association defines an equivalence relation in $\mathcal{G}(\overline{T})$. The respective quotient is a linear space ; an important subspace is constituted by the objects which are associated with a Schwartz distribution. Remark that there are non null elements which are associated with the zero distribution, see [12], p. 64. An element of $\mathcal{G}(\overline{T})$ is at most associated with one distribution.

[12] writes about generalized analysis. [12] defines $\bar{\mathbb{C}}$ as the algebra of generalized numbers. According to our simplified formalism $\bar{\mathbb{C}}$ equals $\mathcal{G}(\mathbb{R}^0)$. There is an obvious inclusion $\mathbb{C} \to \bar{\mathbb{C}}$: If $z \in \mathbb{C}$, a representative of the corresponding generalized number given by $\phi \to z$. If $G \in \mathcal{G}(\overline{T})$, C being compact in $\overline{T}$, we define $\int_C G(x)\, dx$ as the complex generalized number

$$\phi \to \int_C R_G(\phi, x)\, dx$$

If $G \in \mathcal{G}(\overline{T})$ and C is a compact subset of $\overline{T}$, we define $\sup_{x\in C} |R_G(x)|$ as the generalized number represented by $\sup_{x\in C} |R_G(\phi, x)|$. We say that G is bounded on C, if, for any partial derivation operator D, there is $z > 0$ such that $\sup_{x\in C} |DR_G(\phi, x)| \leq R_z(\phi)$, where $R_z \in \mathcal{E}_M(\mathbb{R}^0)$ represents z.

On $\mathcal{G}(\overline{T})$ we can define non linearity operations and a natural product. If $G_1, G_2 \in \mathcal{G}(\overline{T})$, $G_1 \cdot G_2$ is the class of the representatives product $R_{G_1} R_{G_2}$. If $a \in C^\infty(\mathbb{R})$, with polynomial increase at infinity then $a(G_1)$ is represented by $(\phi, x) \to a(R_{G_1}(\phi, x))$.

$\mathcal{G}(\overline{T})$ is an algebra and its zero element coincides with $\mathcal{N}(\overline{T})$ and it is absorbing for the product. Observe however that it may happen that the product of a generalized function associated with zero and another generalized function is not associated with zero.

It is straightforward to define restrictions of a generalized function to a subset which is a closure of an open set of $\mathbb{R}^n$. How do we define the restriction of $G \in \mathcal{G}(\mathbb{R}^n)$ to a submanifold? Here, for simplicity, we will just consider the case $S = \{0\}^m \times \mathbb{R}^{n-m}$. We set

$$R_{G|_S}(\phi, x) = R_G(\phi; 0, x), \quad \phi \in \mathcal{A}_0.$$

We observe that $G|_S$ belongs to $\mathcal{G}(\mathbb{R}^m)$.

Let us discuss now the concept of random generalized function. Let $(\Omega, \mathcal{F}, P)$ be a complete probability space. A map $Y : \Omega \to \mathcal{G}(\overline{T})$ is said to be a *random (Colombeau) generalized function* if there is $R : \Omega \times \mathcal{A}_0 \times \overline{T} \to \mathbb{C}$ denoted R_Y such that

(8)
i) $R(\omega, \cdot)$ represents $Y(\omega)$ for almost all $\omega \in \Omega$

ii) For every $\phi \in \mathcal{A}_0$, $(\omega, x) \to R(\omega, \phi, x)$ is a measurable process.

R will be said to be a representative of Y. The class of random generalized function on $\overline{T}$ will be denoted by $\mathcal{G}_\Omega(\overline{T})$.

The Colombeau theory of generalized functions can be generalized to the case where the set values in a Frechet vector algebra B equipped with some homogeneous metric ρ ; let us consider for example

i) $B = L^{\infty,-}(\Omega) = \bigcap_{p \geq 1} L^p(\Omega)$.

ii) $B = \bigcap_{p \geq 1} \mathcal{D}_{p,1}$

iii) $B = \mathcal{D} = \bigcap_{p \geq 1, s \in \mathbb{R}} \mathcal{D}_{p,s}$ where $\mathcal{D}_{p,s}$ are the Sobolev–Watanabe spaces, see [47], [48]. In all these example strong derivatives and Bochner integrals we involved. We recall that any function $f : \overline{T} \to B$ has a version $\tilde{f} : \Omega \times \overline{T} \to \mathbb{C}$ which is measurable, see [14].

The Colombeau theory can be adapted for generalized function with values in previous B. This using the application $d(0, \cdot)$ instead of the $\mathbb{C}$–modulus in (5) or (6). We can define the algebra of *moderate* elements $\mathcal{E}_M(\overline{T}, B)$, the ideal $\mathcal{N}(\overline{T}, B)$ the quotient $\mathcal{G}(\overline{T}, B)$.

We observe that in the case i), ii), iii), $\rho(0, \cdot)$ can be replaced by $\|\cdot\|_\ell$, where ℓ is an index running over the norms of the Banach spaces which define the projective limit.

Vector B valued distributions defined on $\mathbb{R}^n$, see [42] can be identified with elements of $\mathcal{G}(\mathbb{R}^n, B)$ through a similar relation as in the scalar case.

Similary as for $\overline{\mathbb{C}}$ we can define $\overline{B}$.

If $B = L^{\infty,-}(\Omega)$, one can be interested to know if a B–valued generalized function has a modification with paths in $\mathcal{G}(\overline{T})$. $R \in \mathcal{E}(\overline{T}, B)$ will be said to have an $\mathcal{E}(\overline{T})$ modification if there is $\tilde{R} : \Omega \to \mathcal{E}(\overline{T})$ such that

i) $\tilde{R}(\omega, \cdot) \in \mathcal{E}(\overline{T})$ for a.a.ω

(9) ii) $(\omega, x) \to \tilde{R}(\omega, \phi, x)$ is measurable.

iii) $R(\phi, x) = \tilde{R}(\cdot, \phi, x)$ a.s. $\quad \forall x \in \overline{T},\ \phi \in \mathcal{A}_0$

A B–valued generalized function G will be said to have a samplewise version if there is a random generalized function X such that R_X is a modification R_X and R_G being suitable representatives.

An example of random generalized field is the *white noise*, see for instance [46], [17]. Formally, it is a Gaussian random field $(\dot{W}(x),\, x \in \mathbb{R}^n)$ with covariance $\delta(x-y)$. Rigorously, it can be realized as a Hilbert valued orthogonally scattered measure $W : \mathcal{B}(\mathbb{R}^n) \to L^2(\Omega)$ where $\mathcal{B}(\mathbb{R}^n)$ is the δ–ring of Borel subsets in $\mathbb{R}^n$ whose Lebesgue measure is finite. Finally $W(A) = \int_A \dot{W}(y)\, dy$. W is a Gaussian field such that $E(W(A)W(B))$ is the Lebesgue measure of $A \cap B$. If $g \in L^2(\mathbb{R}^n)$, $\int g\, dW$ is the integral of g with respect to the identified with an element of $\mathcal{G}(\overline{T}, B)$, $B = L^{\infty,-}(\Omega)$ by

$$R_W(\phi, x) = \int dW(y)\phi(x-y),$$

$\overline{W}$ can be observed as a random generalized function through

$$R_{\overline{W}}(\phi, x) = \overline{W}(\phi(x - \cdot)).$$

Of course R_W represents an element of $\mathcal{G}(\overline{T}, B)$ whose modification is given by $R_{\overline{W}}$. $\overline{W}$ is a samplewise version of W.

Remark 2.1 *Let $X \in \mathcal{G}(\overline{T}, B)$ having a samplewise version $\overline{X} \in \mathcal{G}_\Omega(\overline{T})$. If C is a compact of $\overline{T}$, D a partial derivative, then $\sup_{X \in C} |D\overline{X}(x)| \in \overline{B}$.* □

A suitable topology on $\mathcal{G}_\Omega(\overline{T})$ can be given by the uniform convergence in probability on each compact (ucp).

In particular if $G_1, G_2, \in \mathcal{G}_\Omega(\overline{T})$ will be said to be ucp associated if for any bounded subset S of $\mathcal{D}(\overline{T})$, when $\varepsilon \to 0$ we have

$$\sup_{\alpha \in S} |\int (R_{G_1} - R_{G_2})(\phi^\varepsilon, t)\alpha(t)\, dt \longrightarrow 0$$

in probability for all $\phi \in \mathcal{A}_q$ and q large enough.

If $G_1 \in \mathcal{G}_\Omega(\overline{T})$ and G_2 is a continuous process, G_1 and G_2 are uc.p associated if for any compact subset C of $\overline{T}$, when $\varepsilon \to 0$ we have

$$\sup_{t \in C} |R_{G_1}(\phi^\varepsilon, t) - G_2(t)| \longrightarrow 0 \tag{10}$$

in probability.

We observe that the space elements of $\mathcal{G}_\Omega(\overline{T})$, which are ucp associated with a random distribution, is a linear space. Any random generalized function is at most ucp associated with a random distribution.

Another product which plays a significant role in stochastic analysis is the Wick product (:) see for instance [10], [36] or [25].

In particular for $n \in \mathbb{N}$, $g \in L^2(T^n)$, let $I_n(g)$ be the multiple Wiener–Ito integral. The Wick product among multiple Wiener integrals is very natural.

$$I_n(f) \, : \, I_n(g) = I_{n+m}(f \otimes g). \tag{11}$$

We recall one basic property. Let Y be a Wiener functional, δ be the Skorohod integral, see [28], then

$$Y\delta(g) = \delta(Yg), \tag{12}$$

see for instance [36], (7). Through representatives it is possible to define Wick products of B-valued generalized functions, where B is for instance $\mathcal{D}$. Now $\mathcal{D}$ is also an algebra for the Wick product; in other cases, the Wick product is not always defined, but it does not matter.

3 Non–linear stochastic partial differential equations

The details of this section are described in [2]. All along this section B will be $L^{\infty-} = \bigcap_{p\geq 1} L^p(\Omega)$.

We will focuse our attention on a wave equation with non linear forcing which is perturbed by a white noise :

$$\left\{ \begin{array}{rcl} Lu & = & f(u) + \dot{W} \text{ on } \mathbb{R}_+ \times \mathbb{R}^d \\ u(0,\cdot) & \equiv & \dfrac{\partial u}{\partial t}(0,\cdot) \equiv 0, \end{array} \right. \tag{13}$$

where $\dot{W}$ is a Gaussian white noise, $L = \Box + m^2$, $\Box$ is the D'Alembert operator, $n \geq 0$. For the case $d = 1$, (13) has the following integral form.

$$u(t,x) = \int_{D(t,x)} dsdy \, f(u(s,y)) + \tilde{W}(t,x), \tag{14}$$

where $D(t,x)$ is the triangle whose vertices are $(0, x+t)$, $(0, x-t)$ and

$$\tilde{W}(t,x) = \int_{\mathbb{R}_+ \times \mathbb{R}} 1_{D(t,x)} dW.$$

$\tilde{W}$ is a modified Brownian sheet and thus a continuous ordinary process. Also the solution u of (14) can be realised as a continuous ordinary process. This is so because the Green function associated to the D'Alembert operator is square integrable ; it is given by the indicator function of $D(t,x)$.

The case $d = 1$ has been studied by [46], [11], [24] ; these authors have also allowed a multiplicative perturbation of the white noise.

If $d \geq 2$, the solution is not well–defined. Since the Green function is not square integrable, the free problem ($f \equiv 0$) has for solution a (Gaussian) random Schwartz distribution. In order to study (13), we need to give a proper sense to a non–linear function applied to a distribution. For this we operate with Colombeau calculus.

In this section, T will be $]0,\infty[\times\mathbb{R}^2$. The Green function associated with L with zero initial condition is

$$\begin{aligned} K(t,\mathbf{x};\,(s,\mathbf{y})) &= \frac{1}{2\pi}1_{C(,\mathbf{x})}(s,\mathbf{y})\frac{\cos(m|||x-y|||)}{|||x-y|||} \\ &= \frac{1}{2\pi}1_{C^*(s,\mathbf{y})}(t,\mathbf{x})\frac{\cos(m|||x-y|||)}{|||x-y|||} \end{aligned}$$

where we make use of the hyperbolic distance $|||x-y||| = \sqrt{(t-s)^2-|\mathbf{x}-\mathbf{y}|^2}$ and $x = (t,\mathbf{x}) \in T\times\mathbb{R}^2$, $y=(s,\mathbf{y})\in T\times\mathbb{R}^2$. $C(t,\mathbf{x})$ is the following backward light cone

$$\{(s,\mathbf{y})\in\mathbb{R}^3 | s\leq t,\ |\mathbf{y}-\mathbf{x}|\leq t-s\}$$

and $C^*(s,\mathbf{y})$ the forward light cone

$$\{(t,\mathbf{x})\in\mathbb{R}^3 | t\geq s,\ |\mathbf{y}-\mathbf{x}|\leq t-s\}.$$

We observe that K is homogeneous in the sense there exists $\tilde{K}:\mathbb{R}^3\to\mathbb{R}$ such that

$$K(x,y)=\tilde{K}(x-y), \qquad \forall x,y\in T\times\mathbb{R}^2.$$

In fact, we have $\tilde{K}(y) = 1_{C(0,\mathbf{0})}(s,\mathbf{y})$, where $y=(s,\mathbf{y})$.

Let $\eta\in\mathcal{G}(\overline{T})$, f be a smooth function whose derivatives of each order are bounded.

Proposition 3.1 *There is a unique solution $G\in\mathcal{G}(\overline{T})$ of the equation*

$$\left\{\begin{array}{rcl} LG &=& f(G,x)+\eta \\ G(0,\cdot) &\equiv& \dfrac{\partial G}{\partial t}(0,\cdot)\equiv 0. \end{array}\right. \tag{15}$$

Remark 3.2 *i) The proof is in Proposition 3.2 of [2]*

ii) If η where a continuous function, the unique solution would be associated (but not more) with the solution of the corresponding classical wave equation.

Sketch of the Proof.

Using (8), Proposition 3.1 of [2] prove that G is solution to (15) if and only if there are representatives R_G, $R_\eta\in\mathcal{E}_M(\mathbb{R}^3)$ such that every $\phi\in\mathcal{A}_0$,

$$\left\{\begin{array}{rcl} LR_G(\phi,x) &=& f(R_G(\phi,x),\,x)+R_\eta(\phi,x) \\ R_G(\phi,\cdot) &\equiv& \dfrac{\partial}{\partial t}R_G(\phi,\cdot)\equiv 0 \end{array}\right. \tag{16}$$

(16) is equivalent to the following integral form

$$R_G(\phi, x) = \int_T \{f(R_G(\phi, y), y) + R_\eta(\phi, y)\}\, dy. \tag{17}$$

If we fix $\phi \in \mathcal{A}_0$, (17) is an integral wave equation with smooth coefficients. Therefore it has a unique solution. The solution is moderate because R_η is moderate. Given R_{η_1}, R_{η_2} such that $R_{\eta_1} - R_{\eta_2}$ is null we consider the solution R_{G_i}, $i = 1, 2$ of (17) with $\eta = \eta_i$, $i = 1, 2$. It is possible to prove that $R_{G_1} - R_{G_2}$ is null. For this we need Gronwall type lemma but we also use the fact that all the partial derivatives DR_{G_i}, $i = 1, 2$, solve similar equations as (16) and (17).

Remark 3.3 *The case of heat equations on a bounded subset of $\mathbb{R}^d$ can also be treated. The case of elliptic equations is different because the derivatives of solutions to elliptic equations do not fulfill the same type of equation.*

The consequence for stochastic analysis of previous proposition is the following.

Proposition 3.4 *Let η be a random Schwartz distribution which is the samplewise version of a B-valued distribution.*

Then there is a unique random generalized function U which also belongs to $\mathcal{G}(\overline{T}, B)$ such that

$$\begin{cases} LU &= f(U(x), x) + \eta(x) \quad \text{on } T \\ U(0, \cdot) &\equiv \dfrac{\partial}{\partial t} U(0, \cdot) \equiv 0 \end{cases} \tag{18}$$

Moreover U is bounded by a deterministic constant.

Remark 3.5 *If f depends in a smooth way on a parameter $\lambda \in \Lambda$ where Λ is a domain of $\mathbb{C}$, then U may be samplewise seen as an element of $\mathcal{G}_\Omega(\overline{T} \times \mathbb{C})$.*

Remark 3.6 *The interesting case produces when η is a white noise W.* □

The aim of this section could be the characterization of some cases when the solution provided by propostion 3.4 is a random Schwartz distribution. Unfortunately, if the perturbation η is a white noise, the non–linearity f can be shown to act trivially for a large class of f.

Let $f : \mathbb{C} \to \mathbb{C}$ be such there exists a complex measure μ with finite total variation $|\mu|$ having all the moments

$$f(x) = \int_{\mathbb{R}} \exp(iax)\, d\mu(a). \tag{19}$$

We observe in particular that $f \in C_b^\infty(\mathbb{C})$.

(19) is verified for instance if $f : \mathbb{R} \to \mathbb{C}$ is such that Ref and Imf are difference of two smooth non–negative defined function (in the sense of characteristic function). Let $\lambda \in \mathbb{C}$ and consider the unique random generalized function $U = U(\lambda)$ such that

$$(20) \qquad \lambda) \left\{ \begin{array}{rcl} LU & = & \lambda f(U) + \dot{\overline{W}} \text{ on } T \\ U(0,\cdot) & \equiv & \dfrac{\partial U}{\partial t}(0,\cdot) \equiv 0 \end{array} \right.$$

Moreover $U \in \mathcal{G}(\overline{T}, B)$. If $\lambda = 0$ $U = \overline{X}$ where $\overline{X}$ is the "free solution" which is obtained in linear case. In fact $\overline{X}$ is associated with the random distribution

$$< \overline{X}, \alpha >= \int dW(y) \quad \int_T K(x,y)\alpha(x)\, dx.$$

The following result is theorem 4.1 of [2].

Theorem 3.7 *Let $\lambda \in \mathbb{C}$. The unique solution U of ((20); λ) is associated of ((20); 0) (free case), in the sense of B–valued generalized function.*

Sketch of the proof. For $\phi \in \mathcal{A}_0$, let us consider the unique solution of the integral equation

$$(21) \qquad R_U(\phi,\, x\,;\, \lambda) = \lambda \int_T K(x,y)\, f(R_U(\phi,\, y\,;\, \lambda)) + R_{\overline{X}}(\phi, x)$$

where

$$R_{\overline{X}}(\phi, x) = \int_T dy\, K(x,y)\, R_{\overline{W}}(\phi, y), \quad x \in \overline{T}.$$

$R_U(\phi,\cdot)$ is obviously a representative of U. Proposition 2.4 of [2] says that $\lambda \longrightarrow R_U(\phi, x\,;\,\lambda)$ is analytical, therefore it coincides with the asymptotic expansion

$$\sum_{n=0}^{\infty} R_n(\phi, x)\frac{\lambda^n}{n!}$$

where

$$R_n(\phi, x) = \frac{\partial^n}{\partial \lambda^n} R_U(\phi, x\,;\,\lambda)|_{\lambda=0}.$$

R_0 coincides with $R_{\overline{X}}$. The difficulty is to show that R_n is B–associated with zero. The basic argument is that for almost all $x_1, x_2 \in T$

$$\int_T (K(x_1,y) - K(x_2,y))^2\, dy = \infty,$$

see lemma 4.1 of [2].

The main ingredient to prove this will be the following:

$$(22) \qquad \|R_n(\phi^\varepsilon, x)\|_{L^p} \longrightarrow 0, \quad \forall x \in C, \forall p \geq 1$$

when $\varepsilon \to 0$. In order to check (22), we first treat the case $f(x) \equiv \exp(iax)$ and then by linearity we extend the result to the general case.

(22) implies the final result. □

In theorem 3.7, we have observed that the unique solution to the non–linear problem (20) is B–associated with the solution of the free case. There is a sort of converse formulation which is much easier.

Proposition 3.8 *The free solution $\overline{X}$ is such that $\square\overline{X} - \overline{W}$ is B–associated with the zero B–valued distribution.*

Remark 3.9 1. *Proposition 3.8 is less interesting than theorem 3.7 because no uniqueness result is known for problem (20) up to a B–generalized function which is associated.*

2. *The triviality results for f being difference of non–negative definite functions can be extended to equations where other partial differential operators L occur. In particular, L can be the heat operator or a D'Alembert operator of higher order.*

3. *The estimates have been done without the study of the covariance behaviour. In our case, the covariance behaves as the logarithm of the distance. If L is the heat operator with space order grater than 2, then the divergence is proportional to an inverse power of the distance. In this case the triviality result becomes even stronger : there is identity between U and $\overline{X}$ in the strong sense $\mathcal{G}_\Omega(\mathbb{R}^{d+1})$.*

4. *The condition for f being a difference of non–negative defined functions is fundamental for the triviality. Take for instance $f \equiv 1$. It has a Fourier decomposition $f(t) = \Sigma_n c_n \exp\left(\frac{in\pi t}{2}\right)$. But the measure $\mu = \Sigma_{n=0}^{\infty} c_n \delta(n\pi/2)$ is not a finite measure. Now the problem $LU = f(U) + \dot{W}$ becomes $(L+1)U = \dot{W}$ and the solution is given by a random Schwartz distribution which is different from the one obtained if $f \equiv 0$.*

5. *Theorem 3.7 opens the way to the construction of non–trivial solutions of modified PDE's. This has been partially studied for the case of a heat equation perturbed by a white noise.*

   In order to eliminate the trivialisation effect, we will replace $f(U)$ by a Wick renormalization type. The first step consists in the construction of weak solutions, see [1]. The construction of strong solution is the object of a paper in preparation ([3]). As for theorem 3.7 we expand the solution in λ ; we are able to prove that each single term of the expansion is associated with a random Schwartz distribution. It remains to check that the whole series is associated with a random distribution.

   Concerning the parabolic case with polynomial interaction, [20] and more recently [15], have obtained a weak solution (in measure) when the heat operator L is suitably regularized; there the sense of solution is not precised ; [4] and [5] have proved some weak existence and uniqueness using the framework of Dirichlet forms.

4 Stochastic integrals and stochastic differential equations

Let us consider $X = (X(t), t \geq 0)$ and $Y = (Y(t), t \geq 0)$ two continuous stochastic processes. They will be prolongated to $\mathbb{R}$ by setting

$$X(t) = X(0) \quad if\, t \leq 0.$$

We first recall the basic definitions coming from [36], [37] and [38]. For $\varepsilon > 0$, we set

$$\begin{aligned} \tilde{I}^{-}(\varepsilon, t, X, dY) &= \int_0^t X(s)\frac{Y(s+\varepsilon)-Y(s)}{\varepsilon}\, ds \\ \tilde{I}^{+}(\varepsilon, t, X, dY) &= \int_0^t X(s)\frac{Y(s)-Y(s-\varepsilon)}{\varepsilon}\, ds. \end{aligned} \tag{23}$$

If $\lim_{\varepsilon\to 0+} \tilde{I}^{\Delta}(\varepsilon, t, X, dY)$, $\Delta = -$ (resp. $+$), exist according to the ucp topology, we denote it $\int_0^t X\, d^- Y$ (resp. $\int_0^t X d^+ Y$) and it is called forward (resp. backward) integral of X with respect to Y. When these two integrals exist, we set

$$\int_0^t X d^0 Y = \frac{1}{2}\left\{\int_0^t X d^- Y + \int_0^t X d^+ Y\right\}.$$

It is the symmetric integral of X with respect to Y. We set

$$C^{\varepsilon}(t, X, Y) = \frac{1}{\varepsilon}\int_0^t (X(s+\varepsilon) - X(s))(Y(s+\varepsilon) - Y(s))\, ds$$

If $\lim_{\varepsilon\to 0+} C^{\varepsilon}(\cdot, X, Y)$ exists ucp, we will denote the limit by $[X, Y]$: it is the generalized brachet of X and Y.

In [36], [37], [38], [38], [39] we have developed part of a stochastic calculus based on the definition above. [44] has exploited some regularization techniques for establishing anticipative Ito formulas for two-parameter processes which are Skorohod integrators.

We recall here some basic facts.

1. Integration by parts, see [36].

$$\int_0^t X d^0 Y = XY(t) - XY(0) - \int_0^t Y d^0 X$$

2. Ito formula.

   Let us suppose that $[X, X]$ exists. For any $f \in C^2$

$$f(X(t)) = f(X(0)) + \int_0^t f'(X(s))\, d^- X(s) + \frac{1}{2}\int_0^t f''(X(s))\, d[X, X](s) \tag{24}$$

$$f(X(t)) = f(X(0)) + \int_0^t f'(X(s)) d^0 X(s) \tag{25}$$

   Ito formula can be of course generalised to the case of a multi–dimensional process.

3. Extended Ito formula.

   Let us suppose that X is a continuous and locally square integrable semimartingale such that for every $\tau > 0$, the process $\widehat{X}_\tau(t) = X(\tau - t)$, is a semimartingale. Then (25) holds for any $f \in C^1$. The basic example is of course constituted by Brownian motion. Other significant extensions are in [39].

4. In the classical adapted case, the forward integral coincides with the Ito integral, the symmetric integral corresponds to the Fisk–Stratonovich integral.

   If Y is a Brownian motion B, $\int_0^t X\, d^\Delta B$, $\Delta \in \{-,+,0\}$, is the Skorohod integral plus a trace term of DX, where DX is the derivative of X on Wiener space (sometimes called Malliavin derivative). The trace term is of superior, inferior or symmetric type. By the way we recall that the definition of [6] for the forward integral with respect to a Brownian motion is Skorohod integral plus trace term of inferior type.

5. If in (23) we replace the ordinary product with the Wick product, the limit of

$$\widetilde{I}(\varepsilon, t, X, dB)$$

   is the Skorohod integral $\int_0^t X\delta B$.

6. An important class of anticipating processes which can be studied with the help of an regularizing techniques is given by $\psi(\cdot, z)$, where z is any random variable and $\psi(\cdot, x)$ is a good stochastic flow as the solution of a stochastic differential equation

7. If $X(t) = \int_0^t Z(s)\delta B(s)$ admits some strong integral representation, then the bracket is well–defined :

$$[X, X](t) = \int_0^t Z^2(s)\, ds$$

   See [38], theorem 1.1.

8. We have studied in [36] an existence theorem for stochastic differential equation of forward type with anticipate initial conditions. In the one–dimensional case, a simplified version sounds like this.

Proposition 4.1 *Let σ, b some Lipschitz real functions. Let us consider the solution $X(t) = X(t, x)$ of the following Ito equation*

$$X(t) = x + \int_0^t \{\sigma(X(s))\, dB(s) + b(X(s))\, ds\} \tag{26}$$

Let ξ be any random variable. Then the process $Y(t) = X(t, \xi)$ is the solution of the following equation of forward type :

$$Y(t) = \xi + \int_0^t \left\{\sigma(Y(s))\, d^- B(s) + b(Y(s))\, ds\right\} \tag{27}$$

9. In [40], we have studied an existence theorem for the case of the equation of type (27) but with symmetric integral instead of forward integral.

Proposition 4.2 *Let σ, b some real function such that*
$\sigma \in C^1$, $\frac{1}{|\sigma|}$, σ', b *are bounded*
σ', b *are Lipschitz.*
Let $\Phi(t,x)$ be the continuous (Brownian) adapted solution of

$$\Phi(t,x) = x + \int_0^t \{\sigma(\Phi(s,x)) \circ dB(s) + b(\Phi(s,x))\, ds\} \tag{28}$$

Let ξ be any random variable.
Then $Y(t) = \Phi(t,\xi)$ is solution to

$$Y(t) = \xi + \int_0^t \{\sigma(Y(s))\, d^0 B(s) + b(Y(s))\, ds\}.\square \tag{29}$$

Remark 4.3 *The existence and uniqueness of continuous adapted solutions for (28) is obtained through Zvonkin method, see [35], Ex (2.11) Ch. IX.*

We remark that the Colombeau calculus is closely related with previous definitions of stochastic integration.

Let F, $G \in \mathcal{G}_\Omega(\mathbb{R}_+)$. We can define a stochastic integral of the symmetric Stratonovich type.

We define the following element of $\mathcal{G}_\Omega(\mathbb{R}_+)$

$$H(\tau) = \int_0^\tau F(t) \cdot \dot{G}(t)\, dt.$$

If R_F and R_G are representatives of F and G, then a representative of H will then be given by

$$R_H(\phi,\tau) = \int_0^\tau R_F(\phi,t)\dot{R}_G(\phi,t)\, dt.$$

If X and Y are stochastic processes as at the beginning of the section, we recall that classical representatives of X and $\dot{Y}$ are

$$\begin{aligned} R_X(\phi,t) &= \int ds \phi(t-s) X(s) \\ R_{\dot{Y}}(\phi,t) &= \int ds \dot{\phi}(t-s) Y(s) \end{aligned}$$

Proposition 4.4 *The random generalised function $H(\tau) = \int_0^\tau X(t)\dot{Y}(t)\, dt$ is ucp associated with the integral $\int_0^\tau X d^0 Y$ provided that this is well-defined.*

Proof : It is enough to check that for any $\tau > 0$

$$\sup_{t\leq\tau} \left| \int_0^t R_X(\phi^\varepsilon, s)\dot{R}_Y(\phi^\varepsilon, s)\, ds - \int_0^t X d^0 Y \right|$$

converges in probability to zero when ε goes to zero. Let $\varepsilon > 0$.

$$R_X(\phi^\varepsilon, s)\dot{R}_Y(\phi^\varepsilon, s) = \int dv \phi(v) X(s-\varepsilon v) \int \frac{du \dot{\phi}(u)}{\varepsilon} Y(s-\varepsilon u)$$

$$\int dv \phi(v) X(s-\varepsilon v) \int \frac{du}{\varepsilon} \dot{\phi}(u) Y(s-\varepsilon u).$$

Therefore

$$\begin{aligned} \int_0^t ds R_X(\phi^\varepsilon, s)\dot{R}_Y(\phi^\varepsilon, s) &= \\ &= \int dv \phi(v) \int du \frac{\dot{\phi}(u)}{\varepsilon} \int_0^t dv X(s-\varepsilon v) Y(s-\varepsilon u) \\ &= I_1 + I_2, \end{aligned}$$

where

$$\begin{aligned} I_1 &= \int_{-\infty}^{\infty} dv \int_{-\infty}^{v} du \phi(v)\dot{\phi}(u) \int_0^t ds X(s-\varepsilon v) \frac{Y(s-\varepsilon u) - Y(s-\varepsilon v)}{\varepsilon} \\ I_2 &= \int_{-\infty}^{\infty} dv \int_{v}^{\infty} du \phi(v)\dot{\phi}(u) \int_0^t ds X(s-\varepsilon v) \frac{Y(s-\varepsilon u) - Y(s-\varepsilon v)}{\varepsilon} \end{aligned}$$

Now

$$\begin{aligned} I_1 &= \int dv \int_{-\infty}^{v} du \phi(v)\dot{\phi}(u)(u-v) A(t,u,v) \\ I_2 &= \int dv \int_{v}^{\infty} dv \phi(v)\dot{\phi}(u)(u-v) A(t,u,v) \end{aligned}$$

where

$$A(t,u,v) = \int_{-\varepsilon v}^{t-\varepsilon v} ds X(s) \frac{Y(s+\varepsilon(u-v)) - Y(s)}{\varepsilon(u-v)}.$$

But

$$A(t,u,v) 1_{\{u>v\}} \text{ converges ucp to } \int_0^t X d^- Y$$

$$A(t,u,v) 1_{\{u<v\}} \text{ converges ucp to } \int_0^t X d^+ Y$$

with respect to (t,u,v). Therefore, when $\varepsilon \to 0$, ucp we have

$$\begin{aligned} I_1 + I_2 \longrightarrow \quad & \int dv \int_{-\infty}^{v} du \phi(v)\dot{\phi}(u)(u-v) \int_0^t X d^- Y \\ & + \int dv \int_{v}^{\infty} du \phi(v)\dot{\phi}(u)(u-v) \int_0^t X d^+ Y \end{aligned}$$

The conclusion is obtained after having observed that

$$\begin{aligned}\int dv \int_{-\infty}^{v} du\phi(v)\dot{\phi}(u)(u-v) &= \int dv \int_{-\infty}^{0} du\phi(v)\phi(u) = \frac{1}{2}\\ \int dv \int_{v}^{\infty} du\phi(v)\dot{\phi}(u)(u-v) &= \int dv \int_{0}^{\infty} du\phi(v)\phi(u) = \frac{1}{2}.\Box\end{aligned}$$

Remark 4.5 *In [37] we have defined analogous concept only by regularisation of the integration. In this case, the result depends on the support of the mollifier.*

$$\int_0^{\tau} dtX(t)\dot{R}_Y(\phi^{\varepsilon}, t) \longrightarrow$$
$$\int_{-\infty}^{0} dt\phi(t)\int Xd^{-}Y + \int_0^{\infty} dt\phi(t)\int Xd^{+}Y.$$

where the convergence holds in ucp. So if $supp\,\phi \subset \mathbb{R}_-$ *(resp.* $supp\,\phi \subset \mathbb{R}_+$*) the limit is the forward (resp. backward) integral. We only get the symmetric integral if* ϕ *is even.* □

What happens to 4.1, if we replace the ordinary product with the Wick product?

Proposition 4.6 *Let* Y *be a Brownian motion* B *and suppose* $X \in \mathcal{D}_{p,1}(L^2(0,\tau))$ *for some* $\tau > 0$ *for any* $p \geq 1$. *Then* $H^w(s) = \int_0^s X(t) : \dot{B}(t)\,dt$ *is* $L^{\infty,-}$*-associated with the Skorohod integral* $\int_0^s X(t)\,\delta B(t)$.

Proof : According to (12),

$$\begin{aligned}X(t) : \dot{R}_B(\phi^{\varepsilon},t) &= X(t) : \int \delta B(u)\phi^{\varepsilon}(t-u)\\ &= \int \delta B(u)(X(t)\,\phi^{\varepsilon}(t-u)).\end{aligned}$$

Since Skorohod and Lebesgue integration commute,

$$\begin{aligned}\int_0^s X(t) : \dot{R}_B(\phi^{\varepsilon},t)\,dt &= \int \delta B(u)\int_0^s X(t)\phi^{\varepsilon}(t-u)\,dt\\ &= \int \delta B(u)\int_{\frac{-u}{\varepsilon}}^{\frac{s-u}{\varepsilon}} dsX(s\varepsilon+u)\phi(s)\\ &\to \int_0^s \delta B(u)X(u)\end{aligned}$$

uniformly in $\mathcal{D}_{p,1}$ since the integrand in the second line converges in every $\mathcal{D}_{p,1}(L^2(0,\tau))$. □

Let us discuss some applications to stochastic differential equations. The aim should be the comprehension in Colombeau sense of anticipative stochastic differential equation.

Let f, g be C_b^{∞} functions, B a Brownian motion and Z any random variable.

Let us consider the following equation in Colombeau sense.

$$\left\{\begin{array}{lcl} \dot{X}(t) & = & f(X(t))\dot{B}(t) + g(X(t)) \\ X(0) & = & Z \end{array}\right. \tag{30}$$

The existence and uniqueness of a solution $X \in \mathcal{G}_\Omega(\mathbb{R}_+)$ can be reduced to the study of a deterministic problem

$$\left\{\begin{array}{lcl} \dot{X}(t) & = & f(X(t))\eta + g(X(t)) \\ X(0) & = & Z \end{array}\right. \tag{31}$$

where η is a given element in $\mathcal{G}(\mathbb{R}_+)$. This can be done by classical techniques of Colombeau calculus, see [12], [2].

Proposition 4.7 *Let us consider the following stochastic flow $(\psi_t(x),\ t \geq 0)$ solving the following SDE :*

$$\psi_t(x) = x + \int_0^t \{f(\psi_s(x)) \circ dB(s) + g(\psi_s(x))\, ds\} \tag{32}$$

in the Stratonovich sense. The unique solution of (30), is ucp associated with the process $\psi_t(Z)$.

Sketch of the proof. Let us replace in (30) Z with a deterministic initial condition y. The unique random generalized function X solving (30) has a representative $R_X^y(\phi, t)$ so that

$$R_X(\phi, t) = y + \int_0^t \{f(R_X(\phi, s))R_{\dot{B}}(\phi, s) + g(R_X(\phi, s))\}\, ds \tag{33}$$

[19] has studied the convergence of the solution to (32) with $\phi = \phi^\varepsilon$, when ε goes to 0, in the case supp $\phi \subset \mathbb{R}_+$.

It is possible to generalize this study to the general case, when supp $\phi \subset \mathbb{R}$. Adaptation of theorem 7.2 in [19] says that for any $\tau > 0$

$$\lim_{\varepsilon \to 0} E\left(\sup_{0 \leq t \leq \tau} |R_X^y(\phi^\varepsilon, t) - \psi_t(y)|^2\right) = 0.$$

Let C be a compact of $\mathbb{R}$, $\tau > 0$. There is a constant c such that

$$E\left(\sup_{t \leq \tau} |R_X^y(\phi^\varepsilon, t) - R_X^z(\phi^\varepsilon, t)|^2\right) \leq c|y - z|^2$$

for every $x, y \in C$.

By a classical application of Garsia–Rudemich–Rumey lemma as in [39], there is a suitable modification of R_X denoted by $\tilde{R}_X$ such that

$$\tilde{R}_X^Z(\phi^\varepsilon, \cdot) \to \psi_\cdot(Z)$$

in ucp. □

We remark that [32], [27], [39] have studied the existence of equation

$$Y(t) = Z + \int_0^t \{f(Y(s))\, d^0B(s) + g(Y(s))\, ds\} \tag{34}$$

In the three cases, the symmetric stochastic integral has been defined through regularization of the integrator. The existence is provided by $(\psi_t(Z), t \geq 0)$. Proposition 4.7 provides a sort of uniqueness result for equation (34).

[23] have studied a SDE with jumps in the sense of symmetric-Stratonovich integral. In fact they give a sense to the equation but they are not able to define the appropriate integral; in any case it does not correspond to the classical Stratonovich integral appearing in the case of jump processes. The question we could ask is the following. Does the Colombeau framework produce this kind of equation?

We conclude with a remark concerning Wick products. Let us suppose we replace the ordinary product with Wick product in (30). We have already recalled that Wick products are strongly related with Skorohod integrals. We recall that [7], [8], but also [33],[29] and [25] have examined anticipative Skorohod equations. In particular [7] and [8] have provided conditions for existence and uniqueness.

It is reasonnable to think that under some conditions of the type [7], [8], the solutions of (30) Wick product is $L^{\infty,-}(\Omega)$ - associated with the solution of the corresponding SDE of Skorohod type. □

ACKNOWLEDGEMENTS

We are particularly grateful to Prof. Ludwig and Margarida Streit for the invitation to Madeira and the encouragement to write this contribution. Part of this research has been financially supported by a Human Capital and Mobility fellowship. We remark that a great part of this work is consequence of joint works with S. Albeverio, Z. Haba, P. Vallois.

References

[1] S. Albeverio, Z. Haba, F. Russo: *Stationary solutions of stochastic parabolic and hyperbolic Sine-Gordon equations.* J. Phys. A, Vol. 26 (1993)

[2] S. Albeverio, Z. Haba, F. Russo: *Trivial solutions for a non-linear two-space dimensional wave equation perturbed by a space-time white noise.* Prépublication 93-17, LAPT Marseille.

[3] S. Albeverio, Z. Haba, F. Russo: In preparation.

[4] S. Albeverio, M. Roeckner: *Stochastic differential equations in infinite dimensions: solutions via Dirichlet forms.* J. of Funct. Analysis, 88, 395-436 (1990).

[5] S. Albeverio, M. Roeckner, T.S. Zhang: *Markov uniqueness and its applications to martingale problems, stochastic differential equatins and stochastic quantization* C. R. Math. Rep. Acad. Sci. Canada XV 1-6 (1993)

[6] J. Asch, J. Potthoff: *Ito lemma without anticipatory conditions.* Prob. Th. Rel. Fields 88, 17-46 (1991)

[7] R. Buckdahn: *Linear Skorohod stochastic differential equations.* Prob. Th. Rel. Fields 90, 211-238 (1991)

[8] R. Buckdahn: *Skorohod stochastic differential equations of diffusion type.* Prob. Th. Rel. Fields (1992)

[9] H.A. Biagioni: *A non-linear theory of generalized functions.* Lect. Notes in Math. 1421 (1990)

[10] N. Bouleau, F. Hirsch: *Dirichlet forms and Wiener analysis on Wiener space.* Walter de Gruyter (1991).

[11] R. Carmona, D. Nualart: *Random non-linear wave equations: smoothness of the solutions.* Prob. Th. Rel. Fields 79, 469-508 (1988)

[12] J.F. Colombeau: *Elementary introduction to new generalized functions.* North-Holland 113 (1985)

[13] J.F. Colombeau: *Multiplication of Distributions.* Lect. Notes in Math. 1532, Springer-Verlag (1992)

[14] C. Dellacherie, P.A. Meyer: *Probabilités et Potentiel.* Ch. I à IV, Hermann (1975)

[15] G. Gatarek, G. Goldys: *A simple proof of the stochastic quantization volume.* Preprint (1993).

[16] J. Glimm, A. Jaffe: *Quantum Physics: a functional integral point of view.* Springer-Verlag (1986)

[17] T. Hida, H. Kuo, J. Potthoff, L. Streit: *White noise, an infinite dimensional calculus.* Kluwer Dordrecht (1993)

[18] H. Holden,T. Lindstrom, B. Oksendal, J.Uboe, T.S. Zhang: *Stochastic boundary value problems. A white noise functional approach.* Prob. Th. Rel. Fields, Vol. 95, Nr. 3, 391-420 (1993).

[19] N. Ikeda, S. Watanabe: *Stochastic differential equations and diffusion processes.* North Holland (1981).

[20] P. Jona-Lasinio, P.K. Mitter: *On the stochastic quantization of field theory.* Comm. Math. Phys. 101, 409-436 (1985)

[21] H. Koenig: *Neue Begruendungen der Theorie der "Distributionen" von L. Schwartz..* Math. Nachr. 9, 1953, 129-148.

[22] H. Koenig: *Multiplikation von Distributionen.* J. Math. Ann. 128 (1955), 420-452.

[23] T.G. Kurtz, E. Pardoux, P. Protter: *Stratonovich stochastic differential equations driven by general semimartingales.* To appear: Ann. Inst. H. Poincaré. Série: Probabilités et statistiques.

[24] R. Léandre, F. Russo: *Small stochastic perturbations of a non-linear wave equation.* Proc. of the Silivri conference on Sochastic Analysis (1990), H. Korezlioglu, A.S. Ustunel eds. Progress in Probability (1992).

[25] T. Lindstrom, B. Oksendal, J.Uboe: *Wick multiplication and Ito Skorohod SDE* S. Albeverio et al. : "Ideas and methods in mathematical analysis". Cambridge Univ. Press 1992.

[26] C. Martias: *Colombeau generalized functions on Wiener space.* Preprint.

[27] A. Millet, D. Nualart, M. Sanz: *Large deviations for a class of anticipating stochastic differential equations.* Preprint.

[28] D. Nualart: *Non-causal stochastic integrals and calculus.* H. Korezlioglu, A.S. Ustunel eds. Lect. Notes in Math. 1316, 80-129 (1988).

[29] D. Nualart, M. Thieullen: *Skorohod stochastic differential equations on random intervals.* Preprint.

[30] M. Oberguggenberger: *Multiplication of distributions and applications to PDE's.* Longman scientific and technical, Wiley (1992). Stochastic analysis and related topics,

[31] M. Oberguggenberger: *Generalized functions and stochastic processes.* To appear: Proceedings of the Seminar on Stochastic Analysis, Random fields and Applications, eds. E. Bolthausen, M.Dozzi, F. Russo, june 1993.

[32] D. Ocone, E.Pardoux: *A generalized Itô-Ventsell formula. Application to a class of anticipating SDE's.* Ann. IHP 25, 39-71 (1989)

[33] E. Pardoux: *Applications of anticipating stochastic calculus to stochastic differential equations.* Lect. Notes in Math. 1444, Springer-Verlag (1988)

[34] E. Rosinger: *Generalized solutions of nonlinear partial differential equations.* North-Holland (1987).

[35] D. Revuz, M. Yor: *Continuous martingales and Brownian motion.* Springer Verlag, 1991.

[36] F. Russo, P. Vallois: *Forward, backward and symmetric stochastic integration.* Prob. Th. Rel. Fields 97, 403-421 (1993).

[37] F. Russo, P. Vallois: *Non-causal stochastic integration for ladlag processes.* Proceedings of the Oslo-Silivri Conference 1992. T. Lindstrom, B. Oksendal, A.S. Ustunel eds. Gordon and Breach, p. 227-263 (1993).

[38] F. Russo, P. Vallois: *The generalized quadratic variation process and Ito formula.* Prépublication LAPT-Marseille 94-4.

[39] F. Russo, P. Vallois: *Ito formula for C^1 functions of semimartingales.* Preprint.

[40] F. Russo, P. Vallois: *Anticipative Stratonovich equation via Zvonkin method.* In preparation.

[41] L. Schwartz: *Sur l'impossibilité de la multiplication des distributions.* CR Acad. Sci.Paris 239 (1954), 847-848

[42] L. Schwartz: *Distributions à valeurs vectorielles.* Annales de l'Institut Fourier, t. 7 (1957).

[43] B. Simon: *The $P(\phi_2)$ Eucleadian (quantum) field theory.* Princeton University Press (1974).

[44] M. Thieullen: *Calcul stochastique non-adapté pour des processus à deux paramètres: formule de changement de variables de type Stratonovich et de type Skorohod.* Prob. Th. Rel. Fields 89, 457-485 (1991)

[45] T.D. Todorov: *Colombeau's generalized functions and non-standard analysis.* Generalized functions, converge structures and its applications. Plenum press, New York (1988). Ed. B. Stankovic, E. Pap., S. Pilipovic, V.S. Vladimirov.

[46] J.B. Walsh: *An introduction to stochastic partial differential equations.*, Ecole d' été de Saint-Flour XIV, Lect. Notes Math. 1180 (1986).

[47] S. Watanabe: *Lectures on stochastic differential equations and Malliavin calculus.* Tate institute of fundamental research Bombay 1984.

[48] S. Watanabe: *Analysis of Wiener functionals (Malliavin calculus) and its applications to heat kernels.* Ann. Prob. 15, 1-39 (1987)

CONSTRUCTIVE FIELD THEORETIC METHODS AND STOCHASTIC PARTIAL DIFFERENTIAL EQUATIONS

R. SENEOR
Centre de Physique Théorique - UPR 14 -
Ecole Polytechnique
91128 Palaiseau Cedex - France

Contents

A. I. Cardoso et al. (eds.), Stochastic Analysis and Applications in Physics, 351–387.

The aim of these lectures is to enlarge the bridge between field theory methods and stochastic analysis. Such a bridge already exists: field theory can be reconstructed in principle from the study of a parabolic stochastic partial differential equation as this was explained 15 years ago by Parisi and Wu [P] and applied in full rigor 7 years ago by Jona-Lasinio and Mitter [J1] in the case of $P(\Phi)_2$ theories. On the other hand, probabilistic methods (not stochastic) were applied in the early 70's by Nelson [N] in the study of global existence of field theories.

The motivation to go deeper in the connection between these two fields is the study of dynamical systems perturbed by some random force whose prototype can be for example the Navier-Stokes equation:

$$\partial_t \overrightarrow{v} = -(\overrightarrow{v}.\overrightarrow{\nabla})\overrightarrow{v} - \overrightarrow{\nabla} p + \nu \nabla^2 \overrightarrow{v} + \overrightarrow{f} \tag{1}$$

for an uncompressible flow

$$\overrightarrow{\nabla}.\overrightarrow{v} = 0$$

where $\overrightarrow{v} = \overrightarrow{v}(\overrightarrow{x}, t)$ is the speed and $\overrightarrow{f} = \overrightarrow{f}(\overrightarrow{x}, t)$ is the random force depending on space and time, $\overrightarrow{x}$ being in R^3.

In a first part we present the principal methods of constructive field theories (C.F.T), we then review the to day status of the link between field theory and stochastic analysis, then we outline some strategy to extend the connection between these two fields and give two results in this direction.

1 Review of Euclidean Constructive Field Theory Methods

1.1 The statement of the problem

A relativistic quantum field theory is usually assumed to obey a set of axioms, the Wightman axioms. These axioms are shown to be equivalent to another set of axioms, the Osterwalder-Schrader axioms [O] which are the axioms for euclidean field theories for which Lorentz invariance is replaced by rotational invariance. The hyperbolic nature, in the relativistic framework, of the equations of motion is then replaced by a less singular elliptic behaviour in the euclidean framework where field theories are similar in nature with the statistical mechanics of continuous systems and therefore can be described in term of a probabilistic language.

For simplicity, we will choose to restrict our study to scalar massive field, $\phi(x), x \in R^d$. Usually, euclidean field theories are given by an action

$$S(\phi) = S_0(\phi) + S_I(\phi) \tag{2}$$

where

$$S_0(\phi) = \frac{1}{2}\int_{R^d} \phi(x)\Big((-\Delta + m^2)\,\phi(x)\Big)\, d^d x = \frac{1}{2}(\phi, (-\Delta + m^2)\,\phi) \tag{3}$$

$$S_I(\phi) = \lambda \int_{R^d} P(\phi(x))\, d^d x \tag{4}$$

m being the mass and λ the coupling constant. Here, $P(\phi(x))$ is a local polynomial that we choose of the form

$$P(\phi(x)) = \frac{1}{n}\phi(x)^n + \text{lower order terms}$$

By a variational principle, there corresponds the equation of motion

$$(-\Delta + m^2)\,\phi + \frac{\delta S_I(\phi)}{\delta\phi} = 0 \tag{5}$$

Formally, a quantum field is an operator valued (in some Hilbert space) distribution solution of this equation satisfying some initial conditions. For reasonnable models, there exists a normalized ground state $|\Omega>$, the vacuum state, $<\Omega|\Omega>= 1$. All objects of interest of the theory can be recovered from the vacuum expectation values (v.e.v.) of fields

$$<\Omega, \phi(x_1)\cdots\phi(x_n)\Omega>$$

which are symmetrical real functions of the arguments (this is a basic property of v.e.v. of euclidean fields). One then considers these v.e.v. as the moments of a measure μ

$$<\Omega,\, \phi(x_1)\cdots\phi(x_n)\,\Omega>= \int \phi(x_1)\cdots\phi(x_n)\, d\mu(\phi) \tag{6}$$

Formally, one expects (this will be justified later)

$$\begin{aligned} d\mu(\phi)_\lambda &= \frac{1}{Z}\prod_{x\in R^d} d\phi(x)\, e^{-S(\phi)} \\ Z &= \int \prod_{x\in R^d} d\phi(x)\, e^{-S(\phi)} \end{aligned} \tag{7}$$

It is natural for the existence of this measure to require the exponential to be bounded for large ϕ. Since the S_0 part is positive the only requirement for avoiding an arbitrary large value of the exponent is that the interaction polynomial P has its highest term of even order:

$$P(\phi) = \frac{1}{2n}\phi^{2n} + \cdots$$

from which follows that

$$P(\phi) > Cst$$

This bound expresses the boundedness from below of the energy, a usual physical statement.

To be more explicit, unless otherwise stated, we will suppose from now on that

$$P(\phi) = \frac{1}{4}\phi^4$$

but all results presented in the sequel extend to arbitrary polynomials with positive highest coefficient. The question we want now to discuss is how to make sense to the measure defined formally in 7. We first examine the case when the interaction is switched off, the free case, then the one of the complete theory.

1.2 The free case: gaussian measures

The basic mathematical tool is the Minlos theorem:

Theorem 1 *(Minlos)*

Let E be a topological space and C be a positive continuous bilinear form on $E\times E$, then there exists a measure μ_C such that for all $f \in E$

$$e^{-\frac{1}{2}(f,Cf)} = \int_{E^*} e^{i\phi(f)}\, d\mu_C(\phi) \tag{8}$$

For nearly all the applications made in field theory, it is usual to choose E to be the Schwartz space $\mathcal{S}(R^d)$ of C^∞ functions with rapid decrease at infinity and $E^* = \mathcal{S}'(R^d)$. In the present case, we will take

$$C = (-\Delta + m^2)^{-1}$$

for which one can take $E = H_{-1}(R^d)$.

With this choice, one can give a rigorous meaning to the measure μ_λ in the case $\lambda = 0$ and set

$$d\mu(\phi)_0 = \frac{1}{Z_0} \prod_{x \in R^d} d\phi(x)\, e^{-S_0(\phi)} = d\mu_C \tag{9}$$

We now give the main properties of gaussian measures.

From the definition of the measure given by 8, one can compute the moments

$$\int \phi(f_1) \cdots \phi(f_n)\, d\mu_C(\phi) = \frac{d^n}{d\lambda_1 \cdots d\lambda_n} e^{-\frac{1}{2}((\lambda_1 f_1 + \cdots + \lambda_n f_n), C(\lambda_1 f_1 + \cdots + \lambda_n f_n))}|_{\lambda_1 = \cdots = \lambda_n = 0}$$

Applying this formula in the case $n = 2$ gives

$$\int \phi(x)\, \phi(y)\, d\mu_C = C(x,y) = \frac{1}{(2\pi)^d} \int_{R^d} e^{ip.(x-y)} \frac{1}{p^2 + m^2}\, d^d p \tag{10}$$

thus

$$\int \phi(x)^2\, d\mu_C = C(x - x) = c(0) = \infty \qquad \text{if } d \geq 2$$

This result expresses the fact that for $d \geq 2$, powers of the fields are not measurable. This difficulty is solved by introducing Wick powers by which $\phi(x)^n$ is replaced by $: \phi(x)^n :$.

The Wick powers are relative to a measure i.e. to a covariance. Abstractly they can be defined as objects satisfying

For $n \in Z_+$

1. $: \phi^n :$ is measurable
2. $: a\phi^n := a : \phi^n :$ for $a \in C$
3. $\frac{d}{d\phi} : \phi^{n+1} := (n+1) : \phi^n :$
4. $: \phi := \phi$

We will see in the next section that it can be given a more intuitive reason to this definition.

In fact, one of the basic tool in the calculus with respect to gaussian measures is the formula of integration by part:

Lemma 1 (Integration by part formula)

Let μ_C be a gaussian measure of mean 0 and covariance C and F be a differentiable function of ϕ then

$$\int \phi(x) F(\phi)\, d\mu_C(\phi) = \int d^d y\, C(x,y) \int \frac{\delta F(\phi)}{\delta \phi(y)}\, d\mu_C(\phi) \tag{11}$$

This formula can be proved easily by going through finite approximations (cylindrical functions) of the functional integral.

The process of replacing in the integrant the field $\phi(x)$ and another field $\phi(y)$ in F by the covariance $C(x,y)$ is called a *contraction* or equivalently the contraction of $\phi(y)$ by $\phi(x)$. With this terminology, Wick powers can be defined by the fact that none of the fields of a Wick power can contract with the other ones. This leads to the following extension of the integration by part formula:

Lemma 2

$$\int :\phi(x)^n: F(\phi)\, d\mu_C(\phi) = \int d^d y\, C(x,y) \int :\phi^{n-1}(x): \frac{\delta F(\phi)}{\delta \phi(y)}\, d\mu_C(\phi) \tag{12}$$

It is then easy to prove the orthogonality condition

$$\int :\phi(x)^n::\phi(y)^m: \, d\mu_C(\phi) = \begin{cases} 0 & m \neq n \\ n! C(x,y)^n & m = n \end{cases}$$

1.3 The interacting case

How to define a measure in the interacting case. Since the total lagrangien is the sum of a quadratic one (the free part) and of an interacting part, the simplest idea is to suppose that the interacting measure has a density with respect to the free measure. Thus one expects that

$$d\mu_\lambda = \frac{1}{Z_\lambda} e^{-S_I}\, d\mu_C(\phi)$$

where the division by

$$Z_\lambda = \int e^{-S_I}\, d\mu_C(\phi)$$

is to ensure the measure to be a probability measure.

Moreover, one expects Z_λ to be regular in λ, or in another way

$$\int S_I\, d\mu_C(\phi), \quad \int S_I^2\, d\mu_C(\phi), \quad \cdots$$

to exist.

For $S_I = \frac{\lambda}{4} \int_{R^d} \phi(x)^4\, d^d x$

$$\int S_I\, d\mu_C(\phi) = \frac{\lambda}{4} \int\int \phi(x)^4\, d^d x\, d\mu_C(\phi) = \frac{3\lambda}{4} \int c(0)^2\, d^d x = \frac{3\lambda}{4} c(0)^2 \int d^d x$$

but

$$c(0) = \frac{1}{(2\pi)^d} \int_{R^d} \frac{1}{p^2 + m^2}\, d^d p$$

which is infinite for $d \geq 2$, thus one has to replace the usual powers by Wick powers and from now on one takes

$$S_I = \frac{\lambda}{4} \int_{R^d} : \phi(x)^4 : d^d x$$

Therefore, from the definition of the Wick powers

$$\int S_I \, d\mu_C(\phi) = 0$$

The term quadratic in S_I is

$$\begin{aligned} \int S_I^2 \, d\mu_C(\phi) &= \frac{\lambda^2}{16} \int \int \int : \phi(x)^4 :: \phi(y)^4 : d^d x \, d^d y \, d\mu_C(\phi) \\ &= \frac{\lambda^2 4!}{16} \int d^d y \int [C(x,y)]^4 \, d^d x = \frac{3\lambda^2}{2} \int d^d y \int [C(x)]^4 \, d^d x \end{aligned} \tag{13}$$

where in the last equality we used the translation invariance property of the laplacian.

This result shows two difficulties. First, the integral over y diverges as the volume of space. Second, the integral over x may diverge depending on the local behaviour and on the behaviour at infinity of $C(x)$.

Since

$$C(x) \simeq \frac{1}{|x|^{d-2}} e^{-m|x|} \tag{14}$$

thus

$$C(x)^4 \simeq \frac{1}{|x|^{4(d-2)}} e^{-4m|x|}$$

This function is therefore integrable at infinity but presents a local non integrable singularity for $4(d-2) > d$, i.e. $d > 2$.

These difficulties are solved by introducing two regularizations: a space cutoff Λ and, in dimension bigger than 2, a momentum cutoff κ. Since all the expressions are functions of fields, whose correlations are expressed in term of the covariance, we introduce a momentum cutoff on the covariance by replacing $C(x)$ by

$$\frac{1}{(2\pi)^d} \int \xi_\kappa(p) \, e^{ip.x} \frac{1}{p^2 + m^2} \, d^d p$$

where ξ_κ is some cutoff function.

A convenient cutoff function can be to take the cutoff value κ of the form M^ρ, for $\rho \in Z_+$ and $M > 1$, and to define

$$C_\rho(x,y) = \frac{1}{(2\pi)^d} \int e^{ip.(x-y)} \frac{1}{p^2 + m^2} \, e^{-M^{-2\rho}(p^2+m^2)} \, d^d p \tag{15}$$

For later convenience, we note ϕ_ρ the field of covariance C_ρ.

We now set

$$C^i = C_i - C_{i-1} \quad i = 1, 2, \cdots \rho, \qquad C^0 = C_0 \tag{16}$$

thus,

$$C_\rho = \sum_0^\rho C^i$$

and since each C^i is obviously of positive type,

$$d\mu_{C_\rho} = \otimes_i d\mu_i \tag{17}$$

where μ_i is the measure of covariance C_i. It follows from this formula that

$$\phi_\rho = \sum_0^\rho \phi^i \tag{18}$$

where the ϕ^i's are independent variables.

One can then easily show that there exists a constant K such that

$$C^i(x, y) = C^i(x - y) \leq K M^{i(d-2)} e^{-M^i|x-y|} \tag{19}$$

Moreover we can give analytic expressions for the Wick products of cutoff fields:

$$\begin{aligned} :\phi_\rho(x)^2: &= \phi_\rho(x)^2 - c_\rho \\ :\phi_\rho(x)^3: &= \phi_\rho(x)^3 - 3c_\rho\phi_\rho(x) \\ :\phi_\rho(x)^4: &= \phi_\rho(x)^4 - 6c_\rho\phi_\rho(x)^2 + 3\,c_\rho^2 && (20) \\ \cdots &= \cdots && (21) \end{aligned}$$

where $c_\rho = C_\rho(x, x)$.

From above, we see that to prove the existence of the interacting measures we have to prove:

1. start with a well defined expression, i.e. with space Λ and momentum M^ρ cutoff; denote the corresponding measure

$$d\mu^{(\Lambda,\rho)} = \frac{1}{Z^{(\Lambda,\rho)}}\, e^{-S_I^{(\Lambda,\rho)}} d\mu_{C_\rho} \tag{22}$$

2. try to remove, in the order, the momentum cutoff first, i.e. let $\rho \to \infty$, and then the space cutoff, i.e. let $\Lambda \to R^d$.

Question: do these limits exist?

The answer is *yes* for polynomial interactions (under the restriction mentionned above) in $d = 2$. The case of higher dimensions needs that we enter more deeply into the concept of renormalization, therefore we will limit our presentation to the two dimensional case and just comment on how to extend it to higher dimensions.

Thus, let us suppose from now on that we are in the case of a ϕ^4 interaction in $d = 2$.

We first bound the exponential, this is the so called **Wick bound**, and get:

$$e^{-S_I^{(\Lambda,\rho)}} \leq e^{K|\Lambda|(LnM^\rho)^2} \tag{23}$$

using

$$\int_\Lambda : \phi_\rho(x)^4 : d^2x = \int_\Lambda (\phi_\rho(x)^4 - 6c_\rho(\phi_\rho(x))^2 + 3c_\rho^2)\, d^2x \geq -6c_\rho^2 \int_\Lambda d^dx = -6|\Lambda|c_\rho^2$$

with $c_\rho \simeq \mathrm{Ln} M^{2\rho}$.

Since the Wick bound explodes when $\rho \to \infty$, there is no hope of bounding the measure in one step. On the other hand, the Wick bound is a natural bound for the exponential. Thus the idea is to apply the Wick bound in small regions of the space of integration and to produce some convergent factors which can beat this bound and at the same times ensure the control over the number of these small regions.

Let us therefore consider

$$< F(\phi) >_{(\Lambda,\rho)} = \int F(\phi)\, e^{-S_I^{(\Lambda,\rho)}}\, d\mu_{C_\rho}(\phi_\rho) \tag{24}$$

where $F(\phi)$ is an arbitrary polynomially bounded, not necessarly local, function of the field, for example $\phi(x)\phi(y)$.

Then, introduce an interpolating parameter τ, $0 \leq \tau \leq 1$ and replace everywhere ϕ_ρ by

$$\phi_\rho(\tau) = \tau\, \phi^\rho + \phi_{\rho-1}$$

thus $< F(\phi) >_{(\Lambda,\rho)}$ becomes a function of τ:

$$f(\tau) = < F(\phi) >_{(\Lambda,\rho)} (\tau)$$

The first step of the expansion is to set

$$\begin{aligned} < F(\phi) >_{(\Lambda,\rho)} &= f(1) = f(0) + f^{(1)}(0) + \cdots + \frac{1}{r_\rho!} f^{(r_\rho)}(0) \\ &+ \frac{1}{r_\rho!} \int_0^1 d\tau\, (1-\tau)^{r_\rho} f^{(r_\rho+1)}(\tau) \end{aligned} \tag{25}$$

where

$$f^{(n)}(\tau) = \frac{d^n}{d\tau^n} f(\tau)$$

When $\tau = 0$, $\phi_\rho(0) = \phi_{\rho-1}$, thus the terms for which $\tau = 0$ only depend on momenta of index strictly less than ρ and for example

$$< F(\phi) >_{(\Lambda,\rho)} (0) = < F(\phi) >_{(\Lambda,\rho-1)}$$

The remainder of the Taylor expansion contains a derivative of the field with respect to τ, i.e.

$$\frac{d}{d\tau}\phi_\rho(\tau) = \phi^\rho$$

and is estimated by bounding the exponential by the Wick bound and then by computing the remaining gaussian integral. In computing this term, there will be a contraction of the derived field ϕ^ρ giving a covariance C^ρ. The decrease of C^ρ for large values of the argument will produce a smaller coefficient than in the non derived terms and this will be used for convergence purpose. We illustrate this fact by looking at some explicit expression.

Consider the vacuum term

$$\int (S_I^{(\Lambda,\rho)})^2 \, d\mu_{C_\rho} \simeq |\Lambda| \int C_\rho(x)^4 \, d^2x \leq K_1|\Lambda|$$

for some constant K_1 since

$$C_\rho(x) \leq K e^{-|x|}$$

If now we look at

$$\int \frac{d}{d\tau}(S_I^{(\Lambda,\rho)})^2 \, d\mu_{C_\rho} \simeq 2|\Lambda| \int C_\rho(x)^3 C^\rho(x) \, d^2x \leq |\Lambda| K_2 M^{-2\rho}$$

since from $C^\rho(x) \leq K e^{-M^\rho|x|}$

$$2\int C_\rho(x)^3 C^\rho(x) \, d^2x \leq K_2 M^{-2\rho}$$

One can generalize this result and show that to each derivative $\dfrac{d}{d\tau}$ there corresponds a convergent factor $M^{-2\rho}$.

One can thus bound the remainder by

$$\frac{1}{(M^{2\rho})^{r_\rho}} \frac{1}{r_\rho!}(2r_\rho)! e^{K_3(\mathrm{Ln} M^{2\rho})}$$

where $(2r_\rho)!$ is an estimate on the number of contractions (since each derivative may hit the interaction term in ϕ^4, this factorial bounds the number of terms produced by computing $\int(\phi^4)^{r_\rho} \, d\mu_C$).

If now we choose $r_\rho = M^{2\rho}$, the remainder term is bounded by

$$K_4 \, M^{-2\rho}$$

The factor $M^{-2\rho}$ will be used for convergence purpose, for example to bound the expansion 25 and other expansions of this type since this Taylor expansion is only the first step of the expansion.

The next step of the expansion will be applied to the first r_ρ terms of the Taylor expansion with respect to τ since in each of these terms the cutoff index is now $\rho-1$ instead of ρ. Thus there will be ρ steps.

With all the estimates given above, one can prove the convergence of the complete expansion.

The removal of the space cutoff is obtained through a **cluster expansion**. One then shows

$$\begin{aligned} Z^{(\Lambda)} &\simeq e^{K|\Lambda|} \\ \int F(\phi)\, e^{-S_I^{(\Lambda)}}\, d\mu_C &\simeq K(F)\, e^{K|\Lambda|} \end{aligned}$$

and that

$$\lim_{\Lambda\to R^2} \frac{1}{Z^{(\Lambda)}} \int F(\phi)\, e^{-S_I^{(\Lambda)}}\, d\mu_C$$

exists.

In higher dimension, for example $d \geq 3$, contractions of Wick powers may produce non integrable singularities. For example

$$\int \phi(x) S_I^2\, \phi(y)\, d\mu_C \simeq \int\int C(x-z_1) \frac{e^{-3|z_1-z_2|}}{|z_1-z_2|^{3(d-2)}}\, C(z_2-y)\, d^d z_1\, d^d z_2$$

which represent a mass correction does not exist for $d \geq 3$.

For polynomial interactions, the perturbation expansion generates infinitely many divergent quantities and one is faced with the renormalization problem: is it possible to remove these divergences by modifying in a suitable way, in the cutoff theory, the initial interaction coefficients? If the answer is yes, the theory is said to be renormalizable. The proof of existence for renormalizable theories is in general very complicated. In fact, one can prove the existence of a limit only for theories for which the contribution of the interaction part decreases as the energy becomes larger and larger: these are the so called **asymptotically free** field theories. The proof is done by an expansion, the **phase space cell expansion**, which is much more complicated than the one we present above.

Summary

Following what we have seen above the rules for proving the existence of the complete measure μ can be detailed:

- regularize until all expressions which may be produced by perturbation expansions are finite, i.e. introduce cutoff in phase space (then the Radon-Nikodym theorem applies for the complete and free measures).
- expand in momentum slices (or in phase space cells depending on the dimension) and integrate by part to cancel local potential divergences and to produce convergent factors to get estimate independant of the cutoff.
- do a cluster expansion to control the infinite volume limit

2 The C.F.T. methods applied to some S.P.D.E.

The introduction of stochastic analysis to study field theory was advocated more then twenty years ago by K. Symanzik , but took its present expression only after the article of Parisi and Wu [P] who introduce an extra time with the idea that the initial theory appears as the stationary limit of some stochastically driven diffusion process. This idea was rigorously proved in the case of $P(\phi)_2$ field theories by G. Jona-Lasinio and P. K. Mitter [J1].

2.1 Statement of the problem

Within the notation introduced in the previous section, the measure μ is viewed as the **equilibrium measure** of an appropriate Markov process in $d+1$ dimensions

$$\begin{aligned} d\hat{\phi}_t(x) &= -\frac{1}{2}\frac{\delta S}{\delta\hat{\phi}_t(x)}dt + dW_t(x) \\ \hat{\phi}_0(x) &= \phi(x) \end{aligned} \tag{26}$$

$W_t(x)$ is a Wiener process indexed by $x \in R^d$ which expectations, since the process is gaussian, are defined through

$$\begin{aligned} E(dW_t(x)) &= 0 \\ E(dW_t(x)\,dW_s(y)) &= \delta(t-s)\delta^{(d)}(x-y) \end{aligned} \tag{27}$$

We will denote by $E_\phi(\cdots)$ the expectations of the stochastic process defined by 26. We want to show that

$$\lim_{T\to\infty} E_\phi(F(\hat{\phi}_T)) = \int d\mu F(\phi) \tag{28}$$

The method which is followed is similar to the one used in C.F.T. Suppose that

$$S = S_0 + S_I$$

where S_0 is quadratic in ϕ, more precisely (once for ever we fix the mass term to be $m=1$),

$$S_0(\phi) = \frac{1}{2}(\phi, (-\Delta+1)\phi)$$

and for later convenience we set

$$S_I(\phi) = V(\phi)$$

Then let us define a simpler process, the **Orstein-Ulhenbeck** process,

$$\begin{aligned} d\phi_t &= -\frac{1}{2}\frac{\delta S_0}{\delta\phi_t}dt + dW_t \\ &= -\frac{1}{2}(-\Delta+1)\phi_t dt + dW_t \end{aligned} \tag{29}$$

The generalized random stochastic variable $\phi(x)$ is a gaussian variable since a linear combination of Wiener processes, i.e. is completely defined by the knowledge of its mean and of its quadratic expectation. This S.P.D.E. can be integrated

$$\phi_t = e^{-\frac{t}{2}C^{-1}}\phi + \int_0^t e^{-\frac{t-s}{2}C^{-1}} dW_s \tag{30}$$

where $C = (-\Delta + 1)^{-1}$ and the left and right hand side of equation 30 have to be understood as function of x. For example

$$e^{-\frac{t}{2}C^{-1}}\phi(x) = \int e^{-\frac{t}{2}C^{-1}}(x,y)\phi(y)\, d^d y$$

the kernel of the exponential being defined by its power series or by its Fourier transform

$$e^{-\frac{t}{2}C^{-1}}(x,y) = \frac{1}{(2\pi)^d}\int e^{-\frac{t}{2}(p^2+1)} e^{ip(x-y)} d^d p$$

The first term in equation 30 is the mean value of ϕ_t and the second term is a gaussian variable of mean 0 and covariance

$$E_\phi\Big(\Big(\phi_t(x) - (\int_0^t e^{-\frac{t-u}{2}C^{-1}} dW_u)(x)\Big)\Big(\phi_s(y) - (\int_0^s e^{-\frac{s-v}{2}C^{-1}} dW_v)(y)\Big)\Big)$$

$$= Ce^{-\dfrac{|t-s|}{2}C^{-1}}(1 - e^{-inf(t,s)C^{-1}})(x,y) \tag{31}$$

Let $E_\phi^{(W)}$ be the expectation with respect to the Orstein-Ulhenberg process. Applying the Girsanov theorem (see Appendix) to the set of equations 26 and 30 one gets

$$E_\phi(\hat\phi_t(x_1)\cdots\hat\phi_t(x_n)) = E_\phi^{(W)}(\phi_t(x_1)\cdots\phi_t(x_n)\, e^{\xi_t}) \tag{32}$$

with the **Calmeron-Martin-Girsanov** exponent

$$\xi_t = -\frac{1}{2}\int_0^t (V'(\phi_s), dW_s) - \frac{1}{2}(\frac{1}{2})^2\int_0^t (V'(\phi_s))^2 ds$$

The integral

$$\int_0^t (V'(\phi_s), dW_s) = \int_0^t\int_\Lambda V'(\phi_s(x))\, dW_s(x)\, d^d x$$

is the Ito stochastic integral with respect to time.

Formula (32) is remarkable in two ways. First, since taking an expectation is related to a measure, it has the usual form one has encounter in C.F.T.: a free measure, the Orstein-Ulhenberg ones, times the exponential of an interaction, here given by ξ_t. Secondly, there is no need to normalize the measure to get a normalized probability measure since by the nature of ξ_t we have the key identity

$$E_\phi^{(W)}(e^{\xi_t}) = 1 \tag{33}$$

2.2 Functional calculus versus stochastic calculus

We have seen in the previous section that Wiener or Orstein-Ulhenbeck processes are gaussian. Formally for example, the Wiener process can be described by the White Noise measure (see L. Streit in this book or [H])

$$d\mu_\delta(W) \simeq \frac{1}{Z} e^{-\frac{1}{2}\int(\int \dot{W}_s^2(x)\, d^dx)ds} \prod_{s,x} dW_s(x)$$

More precisely, can we write

$$E^{(W)}(F(W)) = \int F(W)\, d\mu_\delta(W) \tag{34}$$

where the right hand side can be manipulated as a functional integral, i.e. can we, for example, do integration by part? The answer is that because of the singularity of the covariance, which is not a function but a distribution, one may expect the appearance of ill-defined expressions.

To see more explicitely the difficulties one may encounter in this identification and how to bypass them, let us consider a one dimensional process

$$dX_t = f_t dt + dW_t$$

where $f_t = f(t, X_t)$. From Girsanov formula we have formula 33 with

$$\xi_t = \int_0^t f_s\, dW_s - \frac{1}{2}\int_0^t f_s^2\, ds$$

provided f_s is regular enough (for example $\int f_s^2\, ds$ finite a.s.).

The stochastic way to prove 33 is by writing

$$E(e^{\xi_t}) = E(1) + E(\int_0^t de^{\xi_t})$$

and to use the Ito formula of differentiation

$$\begin{aligned} de^{\xi_t} &= \left(f_s\, dW_s - \frac{1}{2} f_s^2\, ds + \frac{1}{2} f_s^2\, ds\right) e^{\xi_t} \\ &= \left(f_s\, dW_s\right) e^{\xi_t} \end{aligned}$$

the first two terms of the r.h.s. in the first equality come from the usual differential calculus while the last one is the Ito one. Now it is a standard result to prove under the above regularity condition that

$$E(\int_0^t f_s e^{\xi_s}\, dW_s) = 0$$

If however one tries to prove 33 using functional and usual differential calculus

$$\begin{aligned}\int e^{\xi_t}\, d\mu_\delta(W) &= \int d\mu_\delta(W) + \int d\mu_\delta(W) \int_0^t d\xi_s e^{\xi_s} \\ &= 1 + \int d\mu_\delta \int_0^t (f_s dW_s - \frac{1}{2} f_s^2 ds) e^{\xi_s}\end{aligned}$$

one gets the wrong result.

What was wrong in the above derivation?

It was to suppose enough regularity of the integrands in the exponent to get derivable functions of the upper bound (this is what is corrected by the Ito term in the Ito differential formula). This obviously comes from the singular behaviour of the random variable due to the lack of regularity of the covariance. A way to overpass this difficulty is therefore to regularize the covariance.

So, let us replace the covariance $\delta(t-s)$ by (any other symmetrical regularization will suit)

$$\delta_a(t-s) = \frac{a}{2} e^{-a|t-s|}$$

Also replace for later convenience the Ito integral by the Stratonovich integral

$$\int_0^t f_s \circ dW_s = \int_0^t f_s\, dW_s + \frac{1}{2} \int_0^t f_s'\, ds \tag{35}$$

and replace ξ_t by

$$\xi_t(a) = \int_0^t f_s \circ\, dW_s - \frac{1}{2} \int_0^t f_s'\, ds - \frac{1}{2} \int_0^t \int_0^t f_s \delta_a(s-s') f_{s'} ds ds'$$

which formally converges to ξ_t when $a \to \infty$.

Then it is possible to write $dW_t = \dot{W}_t\, dt$ and

$$\begin{aligned}I &\equiv \int d\mu_{\delta_a}(W)\, e^{\xi_t(a)} - 1 \\ &= \int d\mu_{\delta_a}(W) \int_0^t e^{\xi_s(a)} \left(f_s \circ \dot{W}_s - \frac{1}{2} f_s' - f_s \int_0^t \delta_a(s-s') f_{s'} ds' \right) ds\end{aligned} \tag{36}$$

Integrating by part the Stratonovitch integral term

$$\begin{aligned}&\int d\mu_{\delta_a}(W) \int_0^t f_s \circ \dot{W}_s\, e^{\xi_s(a)}\, ds = \\ &= \int d\mu_{\delta_a}(W) \int_0^t f_s \circ \int \delta_a(s-s') \frac{\delta \xi_s(a)}{\delta \dot{W}_{s'}}\, ds'\, e^{\xi_s(a)} \\ &= \int d\mu_{\delta_a}(W) \int_0^t f_s \circ \int \delta_a(s-s') \Big(\int \frac{\delta f_u}{\delta \dot{W}_{s'}}\, dW_u \\ &\quad + \int f_u \delta(u-s')\, du - \int \frac{\delta f_u}{\delta \dot{W}_{s'}} \delta_a(u-v) f_v\, dv \Big)\, e^{\xi_s(a)}\end{aligned}$$

from which follows that

$$I = \int d\mu_{\delta_a}(W) \int_0^t e^{\xi_s(a)} f_s \circ \int \delta_a(s-s') \Big(\frac{\delta f_u}{\delta \dot{W}_{s'}} dW_u - \int \frac{\delta f_u}{\delta \dot{W}_{s'}} \delta_a(u-v) f_v \, dv \Big) \, ds'$$

Now it is easy to see that the r.h.s. of this last equality converges to 0 as $a \to \infty$.

A remark has to be made at this level. In the above formulas 36 and below, the Stratonovitch integrals can be replaced by limits of ordinary Riemann sums since the integrands are regular. Conversely, it is easy to show that the removal of the cutoff a leads to Stratonovitch integrals for expressions in which the derivative of the Wiener process enters explicitely. Both remarks show why it was convenient to replace the Ito integrals by Stratonovitch integrals.

3 The Jona-Lasinio - Mitter proof of existence of $P(\phi)_d$ for $d = 2$.

We now present some of the arguments used by Jona-Lasinio and Mitter to prove the existence of the stochastic dynamics generated by $P(\phi)_2$ field theories. We give the sketch of this proof in the case of a ϕ^4 theory.

The basic idea is to introduce some stochastic partial differential equation (SPDE) for ϕ, more exactly a Markov diffusion process, such that

$$d\mu = \prod_{x \in R^2} d\phi(x) e^{-S(\phi)} \tag{37}$$

is the invariant measure of the system.

We discuss the case of a finite dimensional system, a finite dimensional approximation of our problem, the extension to the infinite dimensional case being obvious.

We remember that a homogeneous Markov process is given by transition probabilities, for $t \in [0, T]$, ϕ n-dimensional, $\phi = (\phi^{(1)}, \cdots, \phi^{(n)})$, and B a Borel set,

$$P(\phi; t, B) = P(\phi_t \in B | \phi_0 = \phi) \tag{38}$$

satisfying the Chapman-Kolmogorov equation.

It is a homogeneous diffusion process if the transition probabilities satisfy for all $\phi \in R^n$ and for all $\epsilon > 0$

1.
$$\lim_{t \searrow 0} \frac{1}{t} \int_{|\psi - \phi| > \epsilon} P(\phi; t, d\psi) = 0$$

2. $\exists$ $f(\phi)$ such that
$$\lim_{t \searrow 0} \frac{1}{t} \int_{|\psi - \phi| > \epsilon} (\psi - \phi) P(\phi; t, d\psi) = f(\phi)$$

3. $\exists$ $B(\phi)$, a $n \times n$ matrix such that

$$\lim_{t \searrow 0} \frac{1}{t} \int_{|\psi-\phi|>\epsilon} (\psi - \phi)(\psi - \phi)^{tr} P(\phi; t, d\psi) = B(\phi)$$

From the transition probabilities one defines the **time semi-group** T_t on a measurable bounded function g by

$$T_t g(\phi) = E_\phi(g(\phi_t)) := \int g(\Phi)\, P(\phi; t, d\Phi) \tag{39}$$

The generator L of the semi-group is given by

$$Lg(\phi) = \lim_{t \to 0} \frac{1}{t} \left[T_t g(\phi) - g(\phi)\right] \tag{40}$$

and formally $T_t = e^{tL}$. Because of the conditions which caracterize the Markov diffusion process

$$L = \sum_i f_i(\phi) \frac{\delta}{\delta \phi^{(i)}} + \frac{1}{2} \sum_i \sum_j B_{ij}(\phi) \frac{\delta^2}{\delta \phi^{(i)} \delta \phi^{(j)}} \tag{41}$$

an associated stochastic differential equation is

$$d\phi_t^{(i)} = f_i(\phi_t) dt + \sum_i g_{ij}(\phi_t)\, dW_t^{(j)} \tag{42}$$

where $W_t^{(i)}$, $i = 1, \cdots n$, are n independant Wiener processes and g has to satisfy

$$g_{ik}(\phi) g_{kj}(\phi) = B_{ij}(\phi)$$

Thus, let us introduce a scalar product with respect to the measure 37

$$(f, g) = \int f(\phi) g(\phi)\, d\mu(\phi) \tag{43}$$

and suppose the infinitesimal generator has the form

$$L = \sum_i B_i \frac{\delta}{\delta \phi^{(i)}} + \sum_{i,j} A_{ij} \frac{\delta^2}{\delta \phi^{(i)} \delta \phi^{(j)}}$$

Now let us look at the constraints on the coefficients of L imposed by the fact that L is symmetrical

$$(Lf, g) = (f, Lg)$$

One gets that A has to be symmetrical and that

$$B_i = -\sum_j A_{ij} \frac{\delta S(\phi)}{\delta \phi^{(j)}}$$

If $A_{ij} = \frac{1}{2}\delta_{ij}$, then the associated S.P.D.E. is

$$d\phi_t = dW_t - \frac{1}{2}\frac{\delta S}{\delta\phi_t}\,dt$$

which, in the infinite dimensional space ($\phi^{(i)} \to \phi(x)$), becomes

$$\begin{aligned} d\phi_t(x) &= dW_t(x) - \frac{1}{2}\frac{\delta S}{\delta\phi_t(x)}\,dt \\ E(dW_t(x)\,dW_s(y)) &= \delta^{(d)}(x-y)\,\delta(t-s)\,dt\,ds \end{aligned}$$

But with our choice of S,

$$\frac{1}{2}\frac{\delta S_0}{\delta\phi} = (-\Delta+1)\phi$$

which demands too much regularity on ϕ and therefore leads to singular quantities.

One can regularize the theory by choosing $A = \frac{1}{2}C^{1-\rho} = \frac{1}{2}(-\Delta+1)^{1-\rho}$ with $0 < \rho < 1$. The corresponding equation is (in full notation)

$$d\phi_t(x) = \int C^{(1-\rho)/2}(x,y)\,dW_t(y)\,d^2y - \frac{1}{2}\int C^{1-\rho}(x-y)\frac{\delta S}{\delta\phi_t(y)}\,d^2y\,dt$$

Following Jona-Lasinio and Mitter, we rewrite the above system by introducing a modified Wiener process $\hat{W}$ and get

$$d\hat{\phi}_t(x) = d\hat{W}_t(x) - \frac{1}{2}\int C^{1-\rho}(x-y)\frac{\delta S}{\delta\hat{\phi}_t(y)}\,d^2y\,dt \tag{44}$$

$$\hat{\phi}_0 = \phi \tag{45}$$

where the modified Wiener process, of mean 0, is defined by its quadratic expectation

$$E(d\hat{W}_t(x)\,d\hat{W}_s(y)) = C^{1-\rho}(x-y)\,\delta(t-s)\,dt\,ds$$

We now introduce the cutoff: a space cutoff by restricting all volume integrations to a finite volume Λ and a momentum cutoff as follows. Let λ_n, e_n, $n = 1,\cdots\infty$, be an eigenbasis of $C^{-1} = -\Delta+1$, the cutoff is defined by expanding the Wiener process and the random variables with respect to this eigenbasis and by restricting the sums to the first N modes. Thus $\hat{W}_s$ becomes $\hat{W}_s^{(N)}$ with

$$\hat{W}_s^{(N)} = \sum_{n=1}^{N} \lambda_n^{-\frac{1}{2}(1-\rho)}\beta_s^{(n)} e_n$$

and $\beta_s^{(n)}$ is a normalized Brownien process such that

$$E(\beta_s^{(n)}\beta_t^{(m)}) = \delta_{mn} min(t,s)$$

similarly ϕ_s becomes

$$\phi_s^{(N)} = \sum_{n=1}^{N} (\phi_s, e_n)\, e_n$$

Finally the initial S.P.D.E. is

$$\begin{aligned} d\hat{\phi}_t &= d\hat{W}_t - \frac{1}{2}\Big(C^{-\rho}\hat{\phi}_t + \lambda C^{1-\rho} : \hat{\phi}_t^3 :\Big)\, dt \\ \hat{\phi}_0 &= \phi \end{aligned} \tag{46}$$

where the Wick ordering is with respect to the covariance C. It can be rewritten as an integral equation

$$\hat{\phi}_t = \phi_t - \frac{\lambda}{2}\int_0^t e^{-\frac{1}{2}(t-s)C^{-\rho}} C^{1-\rho} : \hat{\phi}_s^3 : \, ds \tag{47}$$

ϕ_t being the Ornstein-Ulhenbeck process satisfying

$$\begin{aligned} d\phi_t &= d\hat{W}_t - \frac{1}{2}C^{-\rho}\phi_t\, dt \\ \phi_0 &= \phi \end{aligned} \tag{48}$$

or in an integrated form

$$\phi_t = e^{-\frac{t}{2}C^{-\rho}}\phi + \int_0^t e^{-\frac{t-s}{2}C^{-\rho}}\, d\hat{W}_s$$

Comparing 46 and 48, Girsanov theorem says that there exists a density

$$e^{\xi_t} = e^{-\frac{\lambda}{2}\int_0^t (:\phi_s^3:, d\hat{W}_s) - \frac{\lambda^2}{8}\int_0^t (:\phi_s^3:, C^{1-\rho} :\phi_s^3:)} \tag{49}$$

such that equation 46 can be solved in a **weak sense** by

$$(e^{tL}F)(\phi) = E_\phi(F(\hat{\phi}_t)) = E_\phi^{(W)}(F(\phi_t)\, e^{\xi_t}) \tag{50}$$

Another way to say that this is a weak solution is by asserting that

$$Z_t = \hat{\phi}_t + \frac{\lambda}{2}\int_0^t e^{-\frac{1}{2}(t-s)C^{-\rho}} C^{1-\rho} : \hat{\phi}_s^3 : \, ds$$

is an Orstein-Ulhenbeck process.

Jona-Lasinio and Mitter proved the existence of a weak solution of 46 provided ρ is small enough, in fact for $\rho < 1/10$.

Remark that this is far from the initial equation which correspond to $\rho = 1$. It seems hard using the techniques used in two dimensional problems to be able to prove the existence for values of ρ close to 1.

To this extend, this seems to indicate that solving a field theory problem through a stochastic approach is harder than in usual field theory.

How do they prove this result?

First, they introduce space and momentum cutoff Λ and N, i.e. cutoff fields and Wiener processes $\phi^{(N)}$ and $W^{(N)}$, thus a cutoff Girsanov exponant $\xi_t^{(\Lambda,N)}$, in order to have a well defined initial expression.

Their key bound is that there exists p, $1 < p < \infty$ such that

$$E_\phi^{(W)}(e^{p\xi_t}) < \infty \tag{51}$$

Moreover, they showed that the process does not explode, i.e.

$$E_\phi^{(W)}(e^{\xi_t}) = 1 \tag{52}$$

This shows that

1. e^{tL} is well defined on bounded measurable functions and that
$$e^{tL}1 = 1$$

2. e^{tL} is a bounded self-adjoint strongly continuous contraction semi-group on $L^2(d\mu)$
$$d\mu = e^{-\frac{\lambda}{4}\int_\Lambda :\phi^4(x):\, d^2x}\, d\mu_C$$
and 1 is the unique ground state.
Also, for $f \in L^2(d\mu)$
$$\int e^{tL} f\, d\mu = \int f\, d\mu$$

3. From these results and standard C.F.T. analysis, one gets that e^{tL} is mixing,i.e.
$$\lim_{t\to\infty}(f, e^{tL}g)_{L^2(d\mu)} = \frac{1}{Z}\Big(\int f d\mu\Big)\Big(\int g d\mu\Big)$$
where $Z = \int d\mu$.

The main step of this proof go through an estimate of the stochastic integral, i.e. the Girsanov exponant ξ_t, in term of usual integrals of the random variable.

In fact, using the relationship between ϕ_t and $\hat{W}_t$

$$\int_0^t (\phi_s^3, d\hat{W}_s) = \int_0^t (\phi_s^3, d\phi_s) + \frac{1}{2}\int_0^t (\phi_s^3, C^{-\rho}\phi_s)\, ds \tag{53}$$

With the Wick ordering this relation becomes

$$\int_0^t (:\phi_s^3:, d\hat{W}_s) = \frac{1}{4}\int_\Lambda :\phi_t^4(x):\, d^2x - \frac{1}{4}\int_\Lambda :\phi^4(x):\, d^2x + \frac{1}{2}\int_0^t (\phi_s^3, C^{-\rho}\phi_s)\, ds \tag{54}$$

and the last integral can be estimated by the following inequality, true for any ρ, $0 < \rho < 1$,

$$\int_0^t (\phi_s^3, C^{-\rho}\phi_s)\, ds \geq \int_\Lambda \phi^4(x)\, d^2x$$

4 A strategy for more complicated cases

The purpose of this section is to define how these methods can be extended to cases where the stochastic integral cannot be exactly computed or when there are renormalization problems for which a detailed cancellation between various expressions have to be explicitely done.

4.1 The non gradient case

This is the case where the process is of the form

$$d\underline{\phi}_t = \underline{F}(\phi_t)\, dt + d\underline{W}_t \tag{55}$$

where $\underline{\phi}, \underline{F}$ and $\underline{W}$ are n-components objects and the non linear part of $\underline{F}$ is not the gradient of some action S.

Thus we cannot perform the stochastic integral. The question which arises is how to bound e^{ξ_t} in the Calmeron-Martin-Girsanov formula since the stochastic integrals whatever their forms

$$\int f\, dW, \qquad \int f\, d\phi \qquad \text{or} \qquad \int f \circ d\phi$$

depend on the whole history of the process.

4.2 The renormalization problems

The scalar cases (without derivatives) discussed in the preceeding section in $d \leq 2$ are very simple. In fact after suitable regularizations each term of ξ_t exists. This is no more the case in dimensions bigger than two. Existence of the theory go through subtle cancellations between contributions coming from the different terms in ξ_t.

These cancellations are obtained by local perturbation expansions which are generated by the integration by part formula. We have therefore to develop,as in field theory, a functional calculus.

4.3 The normalization problem

In field theory, in order to define a probability measure one has to explicitly divide the complete measure by the partition function. It appears that the control of the normalized measure cannot be done without a control on the dependance of the partition function with respect to the space cutoff. The existence of the normalized expression is therefore obtained by the way of a cluster expansion.

On the contrary, the stochastic approach gives measures, through for example the Calmeron-Martin-Girsanov formula, which are automatically normalized provided the process exists. In fact, if f is regular enough, the process

$$\begin{aligned} d\phi_t &= f(\phi_t)\,dt + dW_t \\ \phi_0 &= \phi \end{aligned}$$

exists and

$$E_\phi^{(W)}(e^{\xi_t}) = 1 \tag{56}$$

with

$$\xi_t = \lambda \int_0^t f(\phi_s)\,dW_s - \frac{\lambda^2}{2}\int_0^t f(\phi_s)^2\,ds$$

This identity results, as shown above, from the specific nature of the Ito differential calculus. Expanding in powers of λ, this identity, of the supersymmetric type, is equivalent to infinitely many identities as for example the well known

$$E_\phi^{(W)}\left(\left(\int_0^t f(\phi_s)\,dW_s\right)^2\right) = E_\phi^{(W)}\left(\int_0^t f(\phi_s)^2\,ds\right)$$

see [B-C] for a generalization of this formula.

4.4 The strategy

To study general S.P.D.E. with the help of field theory methods our strategy will be

1. regularize as far is necessary: space and mometum cutoff, white noise, ... until all expressions are well defined and enough regular
2. try to maintain as much as possible the normalization identity 56 since it expresses infinitely many identities
3. perform the stochastic integrals as often as possible
4. express everything, if possible, in term of more regular random variables than the ones of the Wiener process, for example these of an Orstein-Ulhenbeck process.
5. apply the methods of C.F.T., i.e. cancel potentially divergent expressions and do local perturbation expansions to prove the existence of Cauchy sequences allowing to remove all or some of the cutoff.

We will give some illustrations of these ideas in the next two sections.

5 The non gradient term problem

We present in this section some results towards the control of theories with non gradient contributions. We follow the content of [J2].

Problems of these types arise in the study of equations like

$$d\underline{\phi}_t = \nu\Delta\underline{\phi}_t\, dt + \check{\underline{F}}(\phi_t)\, dt + d\underline{W}_t \tag{57}$$

$$\underline{\phi}_0 = \underline{\phi} \tag{58}$$

where

$$\check{\underline{F}}(\phi_t) = -\frac{\delta V(\phi)}{\delta\underline{\phi_t}} + \underline{F}(\phi_t)$$

$\underline{F}(\phi)$ being non gradient.

Is it possible to control this stochastic process?

The answer is that we can control this process in a weak sense if we can bound e^{ξ} and this is possible if the restoring force provided by the gradient part V, $V > Cst.$, is strong enough to "dominate" the non gradient part $\underline{F}$. This can be expressed through an inequality that V and $\underline{F}$ have to satisfy. This in turn leads to a restriction on the class of theories which can be handle by these methods.

We give an example.

Let $\underline{\phi} = (\phi^{(1)}, \phi^{(2)})$, in $d = 2$, be a random variable and let $\check{\underline{F}}$ be a third order polynomial. Then consider the process

$$d\check{\underline{\phi}}_t = -(-\Delta+1)^{\rho}\check{\underline{\phi}}_t + (-\Delta+1)^{-1+\rho} : \underline{\underline{A}}(\hat{\phi}_t).\hat{\phi}_t : dt + d\underline{W}_t$$

with

$$E(dW_t^{(i)}(x)\, dW_s^{(j)}(y)) = \delta_{ij} C^{1-\rho}(x,y)\delta(t-s)\, dt\, ds$$

and

$$\underline{\underline{A}}(\phi) = \begin{pmatrix} \lambda_1(1-|\phi|^2) & -\lambda_2(1-|\phi|^2) \\ \lambda_2(1-|\phi|^2) & \lambda_1(1-|\phi|^2) \end{pmatrix}. \tag{59}$$

At the classical level, there exists a limiting cycle.

As in section 3 we expect the existence of a weak dynamics defined by

$$E_{\phi}(F(\hat{\phi}_t)) = E_{\phi}^{(W)}(F(\phi_t)e^{\xi_t})$$

with

$$\xi_t = \int_0^t (: A(\phi_s).\phi_s :, dW_s) - \frac{1}{2}\int_0^t ||C^{\frac{1-\rho}{2}} : A(\phi_s).\phi_s||^2\, ds \tag{60}$$

and ϕ_t such that

$$d\phi_t = -\frac{1}{2}C^{-\rho}\phi_t\, dt + dW_t$$

We know from the study of the gradient case in the previous section that to prove the existence of the weak dynamics one has to show that there exists $p > 1$ such that

$$E(e^{p\xi_t}) < \infty$$

We now prove this result for ρ small enough[1].

1. Write $A = A_D + A_{ND}$ where

$$\begin{aligned} A_D(\phi) &= \lambda_1(1-|\phi|^2)\begin{pmatrix} 1 & 0 \\ 0 & 1 \end{pmatrix} \\ A_{ND}(\phi) &= \lambda_2(1-|\phi|^2)\begin{pmatrix} 0 & -1 \\ 1 & 0 \end{pmatrix} \end{aligned} \tag{61}$$

   Since $A_D(\phi).\phi$ and $A_{ND}(\phi).\phi$ are orthogonal then with an obvious meaning

$$\xi_t = \xi_{D,t} + \xi_{ND,t}$$

   and

$$e^{p\xi_t} = e^{p\int A_D.\phi dW + p\int A_{ND}.\phi dW - \frac{p}{2}\int ||C^{\frac{1-\rho}{2}}A_D.\phi||^2 - \frac{p}{2}\int ||C^{\frac{1-\rho}{2}}A_{ND}.\phi||^2} \tag{62}$$

   We want to get rid of the second stochastic integral

2. Let us write

$$\begin{aligned} e^{p\xi_t} &= e^{p\int A_{ND}.\phi dW - \frac{p^3}{2}\int ||C^{\frac{1-\rho}{2}}A_{ND}.\phi||^2}\,. \\ &\quad \cdot e^{p\int A_D.\phi dW - \frac{p-p^3}{2}\int ||C^{\frac{1-\rho}{2}}A_{ND}.\phi||^2 - \frac{p}{2}\int ||C^{\frac{1-\rho}{2}}A_D.\phi||^2} \\ &= e^{K_1}e^{K_2} \end{aligned} \tag{63}$$

   Now we apply Hölder inequality

$$E(e^{K_1}e^{K_2}) \leq E(e^{pK_1})^{\frac{1}{p}}E(e^{\frac{p}{p-1}K_2})^{\frac{p-1}{p}}$$

   and use that

$$E(e^{pK_1}) = 1 \tag{64}$$

   since

$$e^{pK_1} = e^{p^2\int A_{ND}.\phi\, dW - \frac{p^4}{2}\int ||C^{\frac{1-\rho}{2}}A_{ND}.\phi||^2}$$

[1] All expressions are cutoff with N, but we do not explicitly write this cutoff dependence. Final estimates have to be uniform in N.

3. It remains to estimate $e^{\frac{p}{p-1}K_2}$. Since

$$\frac{p}{p-1}K_2 = \frac{p^2(p+1)}{2}\int_0^t ||C^{\frac{1-\rho}{2}}A_{ND}.\phi||^2 - \frac{p^2}{2(p-1)}\int_0^t ||C^{\frac{1-\rho}{2}}A_D.\phi||^2 + \frac{p^2}{p-1}\int A_D.\phi\, dW \tag{65}$$

we can rewrite the first two terms as

$$\frac{p^2}{2(p-1)}[\lambda_2^2(p^2-1)-\lambda_1^2]\int(||C^{\frac{1-\rho}{2}}(1-|\phi|^2)\phi^{(1)}||^2+||C^{\frac{1-\rho}{2}}(1-|\phi|^2)\phi^{(2)}||^2)$$

Thus if

$$\lambda_2^2(p^2-1) < \lambda_1^2 \tag{66}$$

$$E(e^{\frac{p}{p-1}K_2}) \le E(e^{\frac{p^2}{p-1}\int A_D.\phi\, dW})$$

and the last expectation is bounded uniformally in N. Condition 66 can be replaced by a more general one, i.e. it exists M, independant of N, such that

$$-\int_0^t ||C^{\frac{1-\rho}{2}}A_D.\phi||^2 + (p^2-1)\int_0^t ||C^{\frac{1-\rho}{2}}A_{ND}.\phi||^2 \le M$$

then

$$E(e^{\frac{p}{p-1}K_2}) \le e^{\frac{p^2}{p-1}M}E(e^{\frac{p^2}{p-1}\int A_D.\phi\, dW})$$

This result can be extended more generally to the case where $\underline{\phi}$ is a n-dimensional process in $d=2$ dimensions.

$$d\underline{\phi}_t = \{-(-\Delta+1)^\rho\underline{\phi}_t + (-\Delta+1)^{-1+\rho} : -\frac{\delta V}{\delta\underline{\phi}_t} + \underline{F} :\}\, dt + d\underline{W}_t$$

Following the same strategy, one finds the condition to be

$$-||:\frac{\delta V}{\delta\underline{\phi}_t} : C^{\frac{1-\rho}{2}}||^2 + (p^2-1)\left[||:\underline{F}(\phi_t) : C^{\frac{1-\rho}{2}}||^2 + 2(:\underline{F}:, C^{1-\rho} : \frac{\delta V}{\delta\underline{\phi}_t}:)\right] < M \tag{67}$$

This result is based on the fact that pK_1 has $p^2C^{\frac{1-\rho}{2}} : \underline{F} :$ for drift. Thus if

$$E(\int_9^t p^4||C^{\frac{1-\rho}{2}} : \underline{F} : ||^2\, ds) < \infty$$

it follows that

$$E(e^{pK_1}) \le 1$$

6 Existence of an invariant measure for some class of S.P.D.E.

The methods of C.F.T., in fact the cluster expansion, can be used to prove the existence and the properties of an invariant measure for some class of stochastic partial differential equations for which no analytical answer are known. This is for example the case of the special non gradient systems which were described in the preceding section.

For clarity reasons we will give a rigorous proof in the case of gradient drift and explain how it can be extended to the non gradient case. Also to simplify, we will consider the simplest gradient systems in looking at scalar one dimensional processes (i.e. no space dimension) [J3].

6.1 The model

The process is

$$d\phi_t = -\frac{1}{2}(\phi_t + \lambda : \phi_t^3 :)\, dt + dW_t \tag{68}$$

$$\phi_0 = \phi \tag{69}$$

According to what we said before, one is interested in the limit $T \to \infty$ of

$$E_\phi^T = E_\phi(F(\phi_T)\, e^{\xi_T}) \tag{70}$$

where F is a polynomial and

$$\xi_T = -\frac{\lambda}{2}\int_0^T : \phi_s^3 : dW_s - \frac{\lambda^2}{8}\int_0^T (: \phi_s^3 :)^2\, ds \tag{71}$$

We can show

Theorem 2 *Let ϕ be some finite number then for λ small enough independantly of ϕ, there exists*

$$< F(\phi) >= \lim_{TJ \to +\infty} E_\phi(F(\phi_T)e^{\xi_T}) \tag{72}$$

Moreover the limit is independant of the initial condition.

In this case, which is gradient, we can perform the stochastic integral and get

$$\xi_t = -\frac{\lambda}{2}\int_0^T : \phi_s^4 : ds - \frac{\lambda^2}{8}\int_0^T (: \phi_s^3 :)^2\, ds - \frac{\lambda}{4} : \phi_T^4 : + \frac{\lambda}{4} : \phi^4 :$$

Thus, since the exponent depends only on ϕ, $E(Fe^\xi)$ is the same as the expectation of Fe^ξ with respect to a gaussian measure of mean $\phi_s^0 = e^{-\frac{s}{2}}\phi$ and covariance

$$C(s,t) = e^{-\frac{1}{2}|t-s|}(1 - e^{-\inf(t,s)}) \tag{73}$$

We define now an expansion which shows that E_ϕ^T is

1. uniformally bounded

2. satisfies the Cauchy criterion in T

6.2 The expansion

To handle with E_ϕ^T we will exploid the fact that this expression can be either understood as a stochastic integral or as a functional integral.

The expansion is then based on the idea that using the exponential decrease of the covariance and the smallness of the coupling constant it is relevant to approximate E_ϕ^T by two independant factors, one corresponding to a finite length time interval containing T and the other one to its complementary interval in $[0,T]$. For this last one we can use the Girsanov normalization $E(e^\xi)=1$. By the way the expansion is defined the remainder is small by one order of magnitude. Then, applying another splitting step to this remainder will produce a small finite size factor and a remainder this time small by two orders of magnitude ... and so on and so forth.

6.2.1 The time intervals

Let us introduce the time interval characteristic functions

$$\chi_{k,j}(s) = \chi(s \in]j,k]) \qquad for \quad k > j \geq 1$$

$$\chi_j(s) = \chi_{j,j-1}(s) = \chi(s \in]j-1,j])$$

and

$$\chi_{(k)}(s) = \chi(s \in [0,k])$$

thus

$$\chi_{(T)}(s) = \chi(s \in [0,T]) = \chi_{(1)}(s) + \sum_2^T \chi_j(s)$$

6.2.2 The interpolation step

We introduce an interpolating parameter s_1, $0 \leq s_1 \leq 1$ and an interpolating covariance

$$C(s_1)(t,s) = [\chi_T(t)\chi_T(s) + (1-\chi_T)(t)(1-\chi_T)(s)]C(t,s)(1-s_1) + C(t,s)s_1 \quad (74)$$

Because

$$\frac{d}{ds_1}C(s_1)(t,s) = [1 - \chi_T(t)\chi_T(s) - (1-\chi_T)(t)(1-\chi_T)(s)]C(t,s)$$

the derivative of $C(s_1)$ with respect to s_1 couples a point in χ_T with a point in $(1-\chi_T)$.

We also introduce interpolated initial conditions by

$$\phi_T^4(s_1) = \phi_T^4 + (1 - s_1)\phi_{T-1}^4$$

Thus replacing C by $C(s_1)$ and ϕ_T^4 by $\phi_T^4(s_1)$, ξ_T and therefore E_ϕ^T become two functions of s_1, $\xi_T(s_1)$ and $E_\phi^T(s_1)$. Then applying the elementary formula of calculus

$$E_\phi^T = E_\phi^T(1) = E_\phi^T(0) + \int_0^1 \frac{d}{ds_1} E_\phi^T(s_1) = I_1 E_\phi^T + D_1 E_\phi^T \tag{75}$$

6.2.3 The term I_1

Since $s_1 = 0$, $C(0) = \chi C \chi + (1-\chi)C(1-\chi)$ is a direct sum of covariances thus the measure $d\mu_{C(0)}$ is a product of measures

$$d\mu_{C(0)} = d\mu_{\chi C \chi} d\mu_{(1-\chi)C(1-\chi)}$$

and

$$\xi_T = \xi_{T-1} + \xi_{T,T-1}$$

with

$$\xi_{T,T-1} = -\frac{\lambda}{2}\int_{T-1}^T : \phi_s^4 : ds - \frac{\lambda^2}{8}\int_{T-1}^T (: \phi_s^3 :)^2 \, ds - \frac{\lambda}{4} : \phi_T^4 :$$

Thus

$$I_1 E_\phi(F(\phi_T)\, e^{\xi_T}) = E_\phi(e^{\xi_{T-1}})\, E_\phi(F(\phi_T)\, e^{\xi_{T,T-1}}) = E_\phi(F(\phi_T)\, e^{\xi_{T,T-1}}) \tag{76}$$

where we used the normalization property of the expectation. Now using translation invariance

$$\lim_{T\to\infty} E_{\phi_0}(F(\phi_T)e^{\xi_{T,T-1}}) = E_0(F(\phi_1)e^{\xi_1}) \tag{77}$$

This equality comes from the fact that the expectation of $F(\phi_T)$ in the time interval $[T-1, T]$ with an initial condition at time $t = 0$ is equivalent to the expectation of $F(\phi_1)$ in the unit time interval with an exponentially small initial condition.

6.2.4 The term D_1

The action of the derivative with respect to s_1 on the expectation produces two terms depending on whether it acts on the measure or on the Girsanov exponent:

$$\begin{aligned}\frac{d}{ds_1} E_\phi^T(s_1) &= \int F(\phi_T)\, e^{\xi_T} \frac{d}{ds_1} d\mu_{C(s_1)} + \int F(\phi_T) \frac{d\xi_T}{ds_1}\, e^{\xi_T}\, d\mu_{C(s_1)} &(78)\\ &= A_1 + B_1 &(79)\end{aligned}$$

6.2.5 The term A_1

Since

$$\frac{d}{ds_1}d\mu_{C(s_1)} = \frac{1}{2}\int d\mu_{C(s_1)}\,du\,dv\frac{d}{ds_1}C(s_1)(u,v)\frac{\delta}{\delta\phi_u}\frac{\delta}{\delta\phi_v} \tag{80}$$

But $\frac{d}{ds_1}C(s_1)$ couples the intervals χ_T and $1-\chi_T$ and since $1-\chi_T = \sum_{j=1}^{T-1}\chi_j$

$$\begin{aligned}
A_1 &= \int\cdots\int \chi_T(u_1)C(u_1,v_1)(1-\chi_T)(v_1)\frac{\delta}{\delta\phi_{u_1}}(F(\phi_T)\frac{\delta\xi_T}{\delta\phi_{v_1}}e^{\xi_T})\,d\mu_{C(s_1)} \\
&= \sum_{j=0}^{T-2}\int\cdots\int \chi_T(u_1)C(u_1,v_1)\chi_{j+1}(v_1)\frac{\delta}{\delta\phi_{u_1}}(F(\phi_T)\frac{\delta\xi_T}{\delta\phi_{v_1}}e^{\xi_T})\,d\mu_{C(s_1)} \\
&= \sum_{j=0}^{T-2}A_1(j)
\end{aligned} \tag{81}$$

Since there is a derivative of ξ_T, A_1 is proportional to λ. Thus, if λ is small this term is small. Moreover, because

$$\chi_T(u_1)C(u_1,v_1)\chi_{j+1}(v_1) \simeq e^{-\frac{T-j}{2}}$$

each term $A_1(j)$ decreases exponentially in $T-j$. Thus

$$A_1(j) \simeq \lambda e^{-\frac{T-j}{2}}$$

6.2.6 The term B_1

B_1 is also small as λ since

$$B_1 = \frac{\lambda}{4}\int F(\phi_T) : \phi_{T-1}^4 : e^{\xi_T}\,d\mu_{C(s_1)}$$

6.3 The next step of the expansion

We have to decide how the expansion goes on for the I_1, A_1 and B_1 terms.

1). The expansion ends for the term I_1.

2). For the A_1 term we choose one of the $A_1(j)$. Let us consider $A_1(j_1)$, $0 \le j_1 < T-1$.

Let us introduce a new interpolating parameter s_2 to decouple $]j_1,T]$ from $[0,j_1]$ and

$$C(s_1,s_2)(t,s) = [\chi_{T,j_1}C\chi_{T,j_1} + (1-\chi_{T,j_1})C(1-\chi_{T,j_1})](t,s)(1-s_2) + C(t,s)s_2 \tag{82}$$

where here $C = C(s_1)$. Similarly

$$\phi_T^4(s_1,s_2) = \phi_T^4 + (1-s_1)\phi_{T-1}^4 + (1-s_2)\phi_{j_1}^4$$

and write

$$A_1(j_1) = (I_2 + D_2)A_1(j_1)$$

with $]j_1, T]$ playing the role of $]T-1, T]$ in the first step. Now

$$\begin{aligned} I_2 A_1(j_1) &= E_\phi(Polynomial \times e^{\xi_{T,j_1}})\, E_\phi(e^{\xi_{j_1}}) \\ &= E_\phi(Polynomial \times e^{\xi_{T,j_1}}) \end{aligned}$$

where

$$\xi_{T,j_1}(s_1, 0) = -\frac{\lambda}{2}\int_{j_1}^{T} :\phi_s^4: ds - \frac{\lambda^2}{8}\int_{j_1}^{T} (:\phi_s^3:)^2\, ds - \frac{\lambda}{4}(:\phi_T^4: + (1-s_1) :\phi_{T-1}^4:)$$

Now since the derivative of $d\mu_{C(s_1,s_2)}$ with respect to s_2 produces

$$\begin{aligned} \frac{d}{ds_2}C(s_1,s_2)(u_2,v_2) &= \chi_{T,j_1}(u_2)C(s_1)(u_2,v_2)(1-\chi_{T,j_1})(v_2) \\ &= \sum_{j_2=1}^{j_1-1} \chi_{T,j_1}(u_2)C(s_1)(u_2,v_2)\chi_{j_2+1})(v_2) \end{aligned} \tag{83}$$

one can write

$$D_2 A_1(j_1) = \sum_{j_2=1}^{j_1} A_{12}(j_1, j_2)$$

and

$$A_{12}(j_1, j_2) \simeq \lambda^2\, e^{-|T-j_1|-|j_1-j_2|} \simeq \lambda^2\, e^{-|T-j_2|} \tag{84}$$

3). The term B_1 is treated as $A_1(T-2)$
To these rules, one adds the fact that the expansion ends when the time interval $[0,1]$ is reached.

6.4 The generic step

Iterating the expansion steps we arrive at

$$E_\phi^T = I_1 E_\phi^T + I_2 D_1 E_\phi^T + \cdots + I_k D_{k-1}\cdots D_1 E_\phi^T + D_k \cdots D_1 E_\phi^T \tag{85}$$

with

$$D_k \cdots D_1 E_\phi^T = \int_0^1 ds_1 \cdots \int_0^1 ds_k \frac{d}{ds_k} \cdots \frac{d}{ds_1} \int d\mu_{C(s_1,\cdots,s_k)} F(\phi_T) e^{\xi_T(s_1\cdots s_k)}$$

and

$$I_k D_{k-1} \cdots D_1 E_\phi^T = \int_0^1 ds_1 \cdots \int_0^1 ds_{k-1} \frac{d}{ds_k} \cdots \frac{d}{ds_1} \int d\mu_{C(s_1,\cdots,s_k)} F(\phi_T)\, e^{\xi_T(s_1\cdots s_k)}\Big|_{s_k=0}$$

As shown above, each D_i is decomposed as a sum over unit time intervals orresponding to the localization of one of the variables, the variable v_i, of the derived covariance. Thus terms like $D_n D_{n-1} \cdots D_1 E^T_{\phi_0}$ can be written as sums over localization indices $j_1, \cdots, j_n$ with $j_m < j_l$ if $m < l$.

The expansion ends when $\chi_{T,j_k} = \chi_{(T)}$, i.e. $j_k = 0$, and the last term has the same form as the generic term.

It can be shown that for λ small enough, there exist $K > 0$ and $\mu < 1$ such that

$$|I_k D_{k-1} \ldots D_1 E^T_\phi| < K\mu^{k-1} \tag{86}$$

We now give some of the features which can justify bound

6.5 The estimates

6.5.1 The Wick bound

Since $: \phi^4_s : \geq -6$

$$: \phi^4_T(s_1, \cdots, s_j) : \geq -6[1 + \sum_1^j (1 - s_k)] \geq -6(j+1)$$

thus using $(: \phi^3 :)^2 > 0$

$$\xi_{T,j_k} \leq \frac{6\lambda}{4}(k+1) + \frac{6\lambda}{2}(T - j_k) \tag{87}$$

6.5.2 The basic estimates

We have shown in **6.4** that a generic term is a sum over time localizations $j_1, \cdots, j_n$. In fact each of these j_i corresponds to a time interval $[j_i, T]$, the support of $D_i \cdots D_1 E^T_\phi$, where by support we mean that this term can be written as the expectation of a Girsanov factor times a sum of product of random variables with their time dependance in this support. These expectations reduce to gaussian expectations after we bound the Girsanov exponent by the Wick bound. Therefore we need to have good enough estimates on gaussian expectations to be able, with the help of the expansion rules, to dominate the Wick bound and to sum over the localizations.

The whole expansion is based on two things, one is the smallness of the coupling constant and the other one is the exponential decrease of the derived covariances since, at step i, they exponentially couple the previous support $[j_{i-1}, T]$ with a unit interval in $[0, j_{i-1}]$. This decrease allows to sum over all possible values of j_i. More precisely suppose we know that $D_{i-1} \cdots D_1 E^T_\phi$ has a localization index j_{n-1}; this means that the next A term will give a derivative of the covariance with respect to s_i proportional to

$$C(u_i, v_i) \simeq e^{-|u_i - v_i|}$$

where $u_i \in [j_{i-1}, T]$ and $v_i \in [j_i, j_i + 1] \subset [0, j_{i-1}]$. Now the bound

$$\sum_{j_i < j_{i-1}} \chi_{j_i+1}(v_i) e^{-c|u_i - v_i|} \leq K(c) = \bar{K}_1$$

for some constant $K(c)$, shows that we can sum over all allowed values of j_i. We will choose $c = 1/4$ in order no to use the full exponential decrease for this sole purpose. Performing these sums starting from j_1, then j_2, etc. shows that the bound of a generic term reduces up to some constants to a bound over expressions with given support. Let us consider such a term.

After we bound e^{ξ} using the Wick bound, it remains to compute the gaussian expectation of a product of sums of monomials of random variables ϕ_s. Writing

$$\phi_s = \phi e^{-\frac{s}{2}} + \check{\phi}_s \tag{88}$$

we see that $\check{\phi}_s$ is now a gaussian variable of mean 0 and exponentially decreasing covariance. We have therefore to estimate the expectation of products of variables $\check{\phi}_s$.

This can be done using

Lemma 3 *Let $\tilde{C}$ be a covariance whose elements satisfy*

$$\tilde{C}(t, s) \leq \alpha \frac{1}{(1 + \beta|t - s|)^{1+\epsilon}}$$

for some $\alpha > 0$, $\beta > 0$ and $\epsilon > 0$.

Then for $s_j \in \chi_j$ and $n_j \in Z_+$, $j = 1 \ldots n$, there exits $\tilde{K} > 0$ such that

$$\int \prod_{j=1}^{n} \phi_{s_j}^{n_j} \, d\mu_{\tilde{C}} \leq \prod_j \sqrt{n_j!} \tilde{K}^n \tag{89}$$

Since the covariances which appear in our model decrease exponentially one fraction, c, of this decrease will be used, according to this lemma, to bound the gaussian expectations. However the application of the lemma show the appearance of factorials. To get rid off these factorials a careful analysis of how there can be so many variables in localized in a given time interval is necessary. We now explain the argument.

Let us choose a time interval χ_j. By formula 80 each derivative with respect to s may act on ξ, thus producing at most $(\phi_s)^5$ with $s \in [j - 1, j]$. If this happens k times, this means that the support has at least move to the left from k unit, each step giving rise to an exponential decrease. We thus have (since we look at product of monomials) an accumulation of exponential factors which is bounded by

$$\prod_{i=1}^{k} e^{-ci} \simeq e^{-ck^2}$$

Now the number of variables in the interval χ_j is at most $5k$, thus generating because of the lemma **3** $\sqrt{5k!}$, but

$$(ak)!e^{-ck^2} \leq K(a,c)$$

showing that up to a constant we can control the factorials.

Finally, the exponential decrease of the derived covariances produces a factor

$$e^{-c(T-j_n)}$$

which generalizes the bound 84 and dominate the factor in the Wick bound

$$e^{\frac{6\lambda}{2}(T-j_n)}$$

provided λ is small enough.

It remains from all these estimates a bound for a localized term of the form

$$\bar{K}_2^n$$

and thus for $n = 1, \ldots$

$$|I_n D_{n-1} \ldots D_1 E_\phi^T| \leq \lambda^{n-1} \bar{K}_1^n \bar{K}_2^n \, e^{\frac{3\lambda}{2}(n+1)} < K \mu^{n-1}$$

with

$$\mu = \lambda \bar{K}_1 \bar{K}_2 e^{\frac{3\lambda}{2}}$$

Thus, provided λ is small enough for μ to be less than 1

$$|E_\phi(F(\phi_T)\, e^{\xi_T})| < \frac{K}{1-\mu}$$

Remark. The sketch of the proof of the convergence does not show that λ can be taken independantly from the initial condition. This independance comes from the fact that at time s, the initial condition is dressed with an exponentially decreasing factor. The reason it does not change the proof is the following. For a given product of monomials of ϕ_s, in the splitting 88, the main contribution comes from a uniformally bounded number of choice of the initial condition, too many choices of initial conditions leading to exponentially small contributions.

The same methods which prove the uniform bound on the expansion can be used to show that

$$|E_\phi(F(\phi_T)\, e^{\xi_T}) - E_\phi(F(\phi_{T'})\, e^{\xi_{T'}})| \leq K\, e^{-c \inf(T,T')}$$

Moreover

$$\lim_{T\to\infty} \frac{d}{d\phi} E_\phi(F(\phi_T)\, e^{\xi_T}) = 0$$

The announced results follow from the last two estimates.

7 Extension to the non gradient case

We will give in this section an idea of what can be the proof in the non gradient case. We therefore suppose that we have some S.P.D.E. of the non gradient type as discussed in the previous section. We also suppose for simplicity that the initial condition $\phi_0 = \phi \equiv 0$.

The principle of the proof is the same as for the gradient case except that instead of interpolating the covariance it is the random variables themselves which are interpolated.

Thus let us consider $E(F(\phi_T)\, e^{\xi_T}$ with

$$\xi_T = \frac{\lambda}{2}\int_0^T \underline{f}(\phi_s)\, d\underline{W}_s - \frac{\lambda^2}{8}\int_0^T \underline{f}(\phi_s)^2\, ds$$

where

$$\underline{\phi}_s = \int_0^s e^{\frac{s-u}{2}}\, d\underline{W}_u$$

and

$$E(dW_u^{(i)}\, dW_v^{(j)}) = \delta_{ij}\delta(u-v)\, du\, dv$$

We just describe the first decoupling step, i.e. the decoupling of the time interval $]T-1,T]$ from $[0,T-1]$.

Let us set for simplicity $\chi^+ = \chi_T$ and $\chi^- = 1 - \chi_T$. Then $\chi^+\chi^- = 0$. Define

$$\underline{\phi}_s^+ = \int e^{-\frac{s-u}{2}}\chi^+(u)\, d\underline{W}_u \tag{90}$$

$$\underline{\phi}_s^- = \int e^{-\frac{s-u}{2}}\chi^-(u)\, d\underline{W}_u \tag{91}$$

Thus $\underline{\phi}_s = \underline{\phi}_s^+ + \underline{\phi}_s^-$. Define the interpolating random variables

$$\underline{\phi}_s(\tau) = \underline{\phi}_s^+[\chi^+(s) + \tau\chi^-(s)] + \underline{\phi}_s^-[\chi^-(s) + \tau\chi^+(s)]$$

then for any integer $n \geq 1$

$$[\underline{\phi}_s(0)]^n = [\underline{\phi}_s^+]^n\chi^+(s) + [\underline{\phi}_s^-]^n\chi^-(s)$$

from which follows that $\xi_T = \xi_T^+ + \xi_T^-$, ξ_T^+ and ξ_T^- being independant. We also set

$$F(\phi_T)(\tau) = F(\phi_T^+ + \tau\phi_T^-)$$

With an obvious definition for $\xi_T(\tau)$

$$E(F(\phi_T)(\tau)\, e^{\xi_T(\tau)})\Big|_{\tau=1} = E(F(\phi_T)(\tau)\, e^{\xi_T(\tau)})\Big|_{\tau=0} + \int_0^1 E(\frac{d}{d\tau}F(\phi_T)(\tau)\, e^{\xi_T(\tau)})d\tau \tag{92}$$

The first term of the r.h.s. of 92

$$E(F(\phi_T)(\tau)\, e^{\xi_T(\tau)})\Big|_{\tau=0} = E(e^{\xi_{T-1}})\, E(F(\phi_T^+)e^{\xi_T^+}) = E(F(\phi_T^+)e^{\xi_T^+})$$

since the expectation of $e^{\xi_{T-1}}$ is 1. In the second term of the r.h.s. of 92, through Leibnitz formula, the derivative with respect to τ finally acts on a random variable and

$$\frac{d}{d\tau}\underline{\phi}_s(\tau) = \underline{\phi}_s^+ \chi^-(s) + \underline{\phi}_s^- \chi^+(s) = \underline{\phi}_s^- \chi^+(s)$$

or

$$\frac{d}{d\tau}F(\phi_T(\tau)) = F'(\phi_T(\tau))\phi_T^-$$

Write

$$\underline{\phi}_s^- = \int e^{-\frac{s-u}{2}} d\underline{W}_u \chi^-(u) = \sum_{j=1}^{T-1} \int e^{-\frac{s-u}{2}} d\underline{W}_u \chi_j(u) \tag{93}$$

For j fixed (terms like $A_1(j)$) the rule is now to contract $d\underline{W}_u$, $u \in \chi_j$. The contraction will link the $u \in \chi_j$ to $s \in \chi^+$ by an exponential factor

$$e^{-\frac{s-u}{2}} \simeq e^{-\frac{1}{2}|T-j|}$$

thus generating the same small exponential factor than for the $A_1(j)$ term in the non gradient case. All other factors or steps are of the same type than for the non gradient case and we end with the same results.

A The Girsanov theorem

We set here the weak form of a theorem mainly due to Girsanov that we used during the lectures. We state it in its finite dimensional form.

Let for $t \in [0, T]$, the two systems of m-dimensional S.P.D.E.

$$\begin{aligned} d\phi_t^{(i)} &= a^{(i)}(t, \phi^{(i)})dt + \sum_{k=1}^{m} b^k(t, \phi^{(i)})dW_t^k \qquad i = 1, 2 \\ \phi_0^{(i)} &= \phi \end{aligned} \tag{94}$$

where $W_t^1, \cdots, W_t^m$ are m independant Wiener processes. Then one has the following theorem

Theorem 3 *Suppose that the coefficients of equations 94 satisfy*

1. There exists K such that

$$\begin{aligned}1)\quad & |a^{(1)}(t,\phi)-a^{(1)}(t,\psi)|+|a^{(2)}(t,\phi)-a^{(2)}(t,\psi)| \\ & +\sum_{k=1}^{m}|b^k(t,\phi)-b^k(t,\psi)|\le K|\phi-\psi| \\ 2)\quad & |a^{(1)}(t,\phi)|^2+|a^{(2)}(t,\phi)|^2+\sum_{k=1}^{m}|b^k(t,\phi)|^2\le K(1+|\phi|^2)\end{aligned}$$

2. There exist m continuous functions $\lambda^1(t,\phi),\cdots,\lambda^m(t,\phi)$ such that for $t\in[0,T]$

$$a^{(2)}(t,\phi)-a^{(1)}(t,\phi)=\sum_{k=1}^{m}\lambda^k(t,\phi)b^k(t,\phi)$$

then, with $E_\phi^{(i)}$ being the expectations with respect to the ith process, $i=1,2$, for any mesurable function one has

$$E_\phi^{(2)}(f(\phi_t^{(2)}))=E_\phi^{(1)}(f(\phi_t^{(1)})e^{\xi_t^{(1)}}) \tag{95}$$

where

$$\xi_t^{(1)}=\sum_{k=1}^{m}\int_0^t\lambda^k(s,\phi_s^{(1)})dW_s^k-\frac{1}{2}\sum_{k=1}^{m}\int_0^t(\lambda^k(s,\phi_s^{(1)}))^2\,ds \tag{96}$$

We will apply this theorem to equations 46 and 48 that we rewrite as

$$\begin{aligned}d\phi_t &= C^{\frac{1-\rho}{2}}dW_t-\frac{1}{2}\left(C^{-\rho}\phi_t+\lambda C^{1-\rho}:\phi_t^3:\right)dt \\ \phi_0 &= \phi\end{aligned} \tag{97}$$

and

$$\begin{aligned}d\phi_t &= C^{\frac{1-\rho}{2}}dW_t-\frac{1}{2}C^{-\rho}\phi_t\,dt \\ \phi_0 &= \phi\end{aligned} \tag{98}$$

Here we have, using m-dimensional approximations of the Riemann integrals,

$$\begin{aligned}a^{(1)}(\phi) &= -\frac{1}{2}C^{-\rho}\phi \\ a^{(2)}(\phi) &= -\frac{1}{2}(C^{-\rho}\phi+\lambda C^{1-\rho}:\phi^3:) \\ b(\phi) &= C^{\frac{1-\rho}{2}}\end{aligned} \tag{99}$$

then since

$$a^{(2)}(\phi)-a^{(1)}(\phi)=-\frac{1}{2}\lambda C^{1-\rho}:\phi^3:=C^{\frac{1-\rho}{2}}(-\frac{1}{2}\lambda C^{\frac{1-\rho}{2}}:\phi^3:)$$

condition 2 of the theorem means that

$$\lambda(\phi) = -\frac{1}{2}\lambda C^{\frac{1-\rho}{2}} : \phi^3 :$$

which gives the Calderon-Martin-Girsanov exponent of 49 provided one redefines dW as $C^{\frac{1-\rho}{2}} dW$. With suitable cutoff, the Lifschitz and the growth conditions 1 can be satisfied. Thus these formulas are shown provided they have meaningful limits when the cutoff are removed.

References

[B-C] Bertini L. and Cancrini N., [1992] Reduction Formula for Moments of Stochastic Integrals, Preprint.

[H] Hida T., Kuo H., Potthoff J. and Streit L., [1993] White Noise, Kluwer Academic Publish.

[J1] Jona-Lasinio G. and Mitter P.K., [1985] On the stochastic quantization of field theory *Commun. Math. Phys.* **101** 409

[J2] Jona-Lasinio G. and Seneor R., [1991] On a class of stochastic reaction-diffusion equations in two space dimensions *J. Phys. A:Math. Gen.***24** 4123

[J3] Jona-Lasinio G. and Seneor R., [1994] Study of non-gradient stochastic differential equations by field theoretic methods, to appear.

[N] Nelson E., [1973] Probability theory and Euclidian Field Theory *In: Constructive Quantum Field Theory* G. Velo and A.S. Wightman eds.,Springer

[O] Osterwalder K. and Schrader R.,[1973], Axioms for Euclidean Green's functions *Comm. Math. Phys.***31** 83

[P] Parisi G. and Wu Y., [1981] *Scienta Sinica* **24** 483

QUANTUM YANG-MILLS THEORY ON COMPACT SURFACES

AMBAR SENGUPTA
Department of Mathematics
Louisiana State University
Baton Rouge, Louisiana 70803-4918, U.S.A.

ABSTRACT. We describe the construction of the Yang-Mills measure for the (Euclidean) quantum theory of gauge fields associated to a bundle over a compact surface. This measure is obtained from the product of a Gaussian measure and a Haar measure by means of a conditioning process which encodes the topology of the bundle. We then describe expectation values of important random variables ('Wilson loops variables') associated to systems of loops on the surface. We also discuss results relating to the limit of the Yang-Mills measure as an associated parameter goes to zero, and the symplectic nature, in certain situations, of the set of minima of the Yang-Mills functional.

1. Introduction.

Quantum field theory has been a fundamental influence on the development of analysis on infinite dimensional spaces. In quantizing a classical theory, one replaces a classical phase space X by a Hilbert space $L^2(\Omega, \mu)$, where (Ω, μ) is a certain probability measure space associated to the system. The space Ω generally has an infinite dimensional differential structure. Construction of the measure μ is the basic problem in mathematically formulating a quatum field theory in the Euclidean framework. Once this measure is constructed, one may proceed to tackle a variety of questions which usually involve analysis on the space Ω. The following table illustrates some measure spaces of interest in this context :

	Space Ω	*Measure* μ
1.	∞−dimensional linear space	Gaussian Measure (White Noise)
2.	Path space (or loop space)	(pinned) Wiener measure
3.	Moduli space of connections	Yang-Mills (YM) measure.

A. I. Cardoso et al. (eds.), Stochastic Analysis and Applications in Physics, 389–403.

Our discussion here is concerned with Item 3. The YM measure describes the quantum theory of gauge fields. Unlike the case for Items 1 and 2, the study of the Yang-Mills measure is not well developed. In fact the measure has not even been constructed as a rigorously defined entity in most situations of interest (the work [MRS] is a recent breakthrough in the difficult problem of constructing the Yang-Mills measure in four dimensions). The problem of constructing the YM measure in two dimensions does not suffer from many of the technical difficulties that are the hallmarks of the situation in higher dimensions. The YM measure describing gauge fields over the plane R^2 is in fact Gaussian. Nevertheless, computation of 'Wilson loop expectation values' (i.e. $\int_\Omega f\,d\mu$ for certain interesting functions f) is a substantial problem even for R^2; this was solved rigorously in [GKS] and [Dr] (which contain references to previous work in this area). The construction of the YM measure for gauge fields on compact surfaces presents problems of a topological nature. This has been in the subject of investigations of D. Fine [F 1,2] and the author [Se 1,3,4]. Problems arising from quantum gauge theory on compact surfaces have been the subject of the works [Wi 1,2] by E. Witten, and the work [Fo] by R. Forman.

The pinned Brownian motion measure for loop spaces and the YM measure (for the case of compact surfaces) are not Gaussian measures in any natural sense; they are obtained from Gaussian measures by *conditioning.*

In the following, I will describe the construction of the YM measure for gauge fields associated to bundles of specified topology, possibly non-trivial, over compact surfaces. Then I will describe how to integrate interesting functions (Wilson loop variables) with respect to the YM measure. Most of the results are drawn from [Se 4] where detailed proofs are given. For the case where the gauge group is simply-connected (then all the bundles are trivial) the results and proofs are available in [Se 3].

Section 2 sets up the language and some of the basic problems. Section 3 describes the construction of the YM measure, and Section 4 (Theorem 2) contains the formula for integration of important random variables with respect to the YM measure. Some results relating to a limiting measure associated to the YM measure are described in Section 5, along with a description of the symplectic structure (which exists under certain hypotheses) of the space of Yang-Mills minima. In the concluding section, I mention some further avenues of investigation.

2. The Problems.

The YM measure is described heuristically by the formal expression :

$$d\mu_T(\omega) = \frac{1}{Z_T} e^{-S_{YM}(\omega)/T}[\mathcal{D}\omega]$$

(notation will be explained shortly).

Here :

* μ_T should live on an $\infty-$dimensional (usually) non-linear space $\mathcal{C}$ whose points represent gauge field configurations ("geometries");

* $\mu_T(\mathcal{C}) = 1$;

* T is a positive parameter.

Problems : (1) To construct the measure μ_T;

(2) To calculate $\int_{\mathcal{C}} f \, d\mu_T$ for interesting f;

(3) To determine the limiting measure $\lim_{T\to 0} \mu_T$.

Explanation of $\mathcal{C}$, $S_{YM}(\omega)$,... : The ingredients of the theory are :

(A) A *compact surface* (connected two dimensional Riemannian manifold, possibly with boundary) Σ. The area measure on Σ induced by the Riemannian metric will be denoted $d\sigma$.

(B) A *compact connected Lie group* G equipped with an $Ad-$invariant metric $\langle\cdot,\cdot\rangle_{\underline{g}}$ on its Lie algebra $\underline{g}$. For example, $G = SU(n)$ or some closed subgroup of $SU(n)$, with Lie algebra $\underline{g}$ equipped with the metric given by

$$< A, B >_{\underline{g}} = tr(A^* B).$$

(C) A *principal* $G-$*bundle* $\pi : P \to \Sigma$. Thus P is a manifold with a smooth free right action of $G : P \times G \to P : (p, g) \mapsto R_g p$, with quotient being (diffeomorphic to) Σ, making π the quotient map. For example the bundle of positively-oriented orthonormal frames of $T\Sigma$ (in case Σ is oriented) is a principal $SO(2)-$bundle over Σ.

(D) A *connection* on P is a $\underline{g}-$valued 1$-$form on P such that : (i) $\omega((R_g)_* X) = Ad(g^{-1})\omega(X)$, and (ii) $\omega(\frac{d}{dt}|_{t=0} R_{exp(tH)} p) = H$, for every $p \in P$, and $H \in \underline{g}$. We denote by $\mathcal{A}$ the set of all connections on P. The space $\mathcal{A}$ is : (a) an infinite dimensional affine space; (b) has an inner-product induced by the Riemannian metric on Σ and by the metric $\langle\cdot,\cdot\rangle_{\underline{g}}$ on $\underline{g}$. If $\omega_1, \omega_2 \in \mathcal{A}$ then the 'tangent vector' $\omega_2 - \omega_1$ is a $\underline{g}-$valued 1$-$form satisfying (i) and a modified version of (ii) : the right side of the equation in (ii) is replaced with 0. Let ζ and η be two such 1–forms. The inner-product $\langle \zeta, \eta \rangle$ is defined as follows. Let $x \in \Sigma$ and pick any $y \in \pi^{-1}(x)$ and vectors $e_1, e_2 \in T_y P$ such that $\pi_* e_1, \pi_* e_2$ form an orthonormal basis of $T_x \Sigma$; then $\langle \zeta(e_1), \eta(e_1) \rangle_{\underline{g}} + \langle \zeta(e_2), \eta(e_2) \rangle_{\underline{g}}$ does not depend on the choice of y and (e_1, e_2), and thus defines a function of $x \in \Sigma$. Integrating this over Σ with respect to $d\sigma$ we obtain $\langle \zeta, \eta \rangle$.

When Σ is orientable, the space $\mathcal{A}$ also has a *symplectic* structure. With ζ, η and x, y as above, let $X, Y \in T_x \Sigma$. Pick $X', Y' \in T_y P$ with $\pi_* X' = X$ and $\pi_* Y' = Y$. Then $\langle \zeta(X'), \eta(Y') \rangle_{\underline{g}}$ is independent of the choices of y, X' and Y'. Thus $\langle \zeta(X'), \eta(Y') \rangle_{\underline{g}} - \langle \zeta(Y'), \eta(X') \rangle_{\underline{g}}$ specifies a well-defined 2–form on Σ. Integrating this over Σ yields a number which we denote $\Theta(\zeta, \eta)$; this defines a 2–form Θ on $\mathcal{A}$ which is symplectic.

(E) An *automorphism* of P is a diffeomorphism $\phi : P \to P$ for which $\pi\phi = \pi$ and $\phi R_g = R_g \phi$ for every $g \in G$. The set $\mathcal{G}$ of all such automorphisms form a group under composition. We denote by $\mathcal{G}_m$ the subgroup consisting of those automorphisms $\phi : P \to P$ for which $\phi(y) = y$ for some (and hence every) y in the fiber $\pi^{-1}(m)$ over a point $m \in \Sigma$. *The group $\mathcal{G}$ acts on $\mathcal{A}$ by pullbacks* : $(\phi, \omega) \mapsto \phi^* \omega$.

(F) $\mathcal{C}_m = \mathcal{A}/\mathcal{G}_m$: *an infinite-dimensional space*, generally non-linear. We will write $\mathcal{C}$ for $\mathcal{C}_m$, when a fixed m is understood.

(G) *Curvature* of a connection ω is the $\underline{g}$–valued 2–form Ω^ω on P defined by :

$$\Omega^\omega(X, Y) = d\omega(X, Y) + [\omega(X), \omega(Y)].$$

(H) The **Yang-Mills functional**

$$S_{YM} : \mathcal{A} \to R : \omega \mapsto S_{YM}(\omega) = \frac{1}{2} \int_\Sigma ||\Omega^\omega||^2 \, d\sigma$$

where $d\sigma$ denotes the area measure on Σ induced by the Riemannian metric, and $||\Omega^\omega||^2$ is the function on Σ defined by :

$$||\Omega^\omega||^2(x) = ||\Omega^\omega(e_1, e_2)||^2_{\underline{g}}$$

where $e_1, e_2 \in T_y P$, for some $y \in \pi^{-1}(x)$ and $(\pi_* e_1, \pi_* e_2)$ is an orthonormal basis of $T_x \Sigma$.

Facts : (1) $||\Omega^\omega||^2(x)$ depends only on x (and ω); and (2) $S_{YM}(\phi^*\omega) = S_{YM}(\omega)$ for every $\phi \in \mathcal{G}$.

Thus S is well-defined on the quotient $\mathcal{A}/\mathcal{G}$, and, in particular, also on $\mathcal{C}_m$, for every $m \in \Sigma$.

We have explained all the terms on the right side of

$$d\mu_T(\omega) = \frac{1}{Z_T} e^{-S_{YM}(\omega)/T}[\mathcal{D}\omega]$$

except for $[\mathcal{D}\omega]$: this is the push-forward on $\mathcal{C}_m = \mathcal{A}/\mathcal{G}_m$ of the "Lebesgue measure" on $\mathcal{A}$, and Z_T is a 'normalizing constant' insuring that $\mu_T(\mathcal{C}_m) = 1$.

(I) *Parallel Transport* : Fix $m \in \Sigma$ and $u \in \pi^{-1}(m)$. Given a connection ω, a piecewise smooth path C in Σ from m to a point x can be 'lifted' to a unique path $\tilde{C}$ in P, starting at u, in a natural way : $\tilde{C}$ is the piecewise smooth solution of $\pi \circ \tilde{C} = C$ and $\omega(\tilde{C}'(t)) = 0$ (the derivative being the right/left derivative at the 'corners' of C). We shall denote the endpoint of $\tilde{C}$ by $\tau_\omega(C)u$; this is the parallel translate of u around C by ω.

(J) *Holonomy* : If C is a loop then $\tau_\omega(C)u = ug$ for a unique $g \in G$. This g is called the *holonomy* of ω around C, with initial point u, and we shall denote it by $g(C;\omega)$ (u being fixed.)

We can now state Problem 2 more precisely :

Problem 2. Given loops $C_1, ..., C_k$ in Σ based at m, and a bounded measurable function f on G^k, calculate :

$$\int_{\mathcal{C}} f(g(C_1;\omega), ..., g(C_k;\omega))\, d\mu_T(\omega).$$

3. Construction of μ_T :

The strategy is to cut up Σ, display Σ as quotient $q : D \to \Sigma$ (here D is the unit disk on the plane R^2 centered at the origin O), work on D, and finally project back to Σ. We describe this in some detail for the case when Σ is a torus. A similar formulation can be made for other surfaces. Let $t \mapsto x_t = (cos(2\pi t), sin(2\pi t))$ trace out ∂D. Divide ∂D into four arcs as follows. Let K_j be $[\frac{j-1}{4}, \frac{j}{4}] \to \partial D : t \mapsto x_t$. Thus K_1, K_2, K_3, K_4 split ∂D into congruent pieces. Form the quotient of D by identifying K_1 with reversed K_3, and K_2 with reversed K_4. To be precise, identify x_t with $x_{t'}$ whenever $t - \frac{j-1}{4} = \frac{j+2}{4} - t'$ with $j \in \{1, 2\}$, $t \in [\frac{j-1}{4}, \frac{j}{4}]$ and $t' \in [\frac{j+1}{4}, \frac{j+2}{4}]$.

The quotient space is a torus which we take to be Σ, and we denote the quotient map by $q : D \to \Sigma$.

For the rest of the discussion $m = q(O)$, wherein $O = (0,0)$ is the center of the disk D. If $x \in D$ then we write Ox to denote the radial path from O to x. The two basic loops $S_1 = q(Ox_0)q(K_1)q(x_0O)$ and $S_2 = q(Ox_0)q(K_2)q(x_0O)$ generate the fundamental group $\pi_1(\Sigma, m)$ and will be of great use to us.

Fix $u \in \pi^{-1}(m)$. Let $\omega \in \mathcal{A}$. Define the function $F^\omega : D \to \underline{g}$ as follows. Let $x \in D$. Consider the path $q(Ox)$ in Σ. Parallel translate u along $q(Ox)$ to obtain a point $s_\omega(x) \in P$. In this way we obtain a map $s_\omega : D \to P$.

Set

$$F^\omega(x) = \Omega^\omega(e_1, e_2)$$

where $e_1, e_2 \in T_{s_\omega(x)}P$ with $(\pi_* e_1, \pi_* e_2)$ forming an orthonormal basis in $T_{q(x)}\Sigma$, oriented positively with respect to the orientation on $T_{q(x)}\Sigma$ induced by q from the standard orientation of $D \subset R^2$.

So for each $\omega \in \mathcal{A}$, we have a map $F^\omega : D \to \underline{g}$, which describes the curvature of ω.

Moreover,

$$\int_\Sigma ||\Omega^\omega||^2 \, d\sigma = \int_D ||F^\omega||^2_{\underline{g}} \, d\sigma_D,$$

wherein $d\sigma_D$ is the area measure on D obtained by lifting $d\sigma$ from Σ, by q.

The following result characterizes connections on a principal bundle in terms of F^ω, the holonomies around the basic loops S_1 and S_2, and the topology of the bundle.

Theorem 1([Se 4]): Suppose Σ is a torus, and let $\tilde{G} \to G : \tilde{x} \mapsto x$ be the universal covering of G. Let ω be a connection on a principal $G-$bundle $P \to \Sigma$.

Let $a = g_u(S_1; \omega)$ and $b = g_u(S_2; \omega)$ (these are the holonomies around the two basic loops S_1 and S_2, and u is a fixed point on the fiber over m). Define $t \mapsto \tilde{g}_t \in \tilde{G}$ to be the solution of :

$$d\tilde{g}_t = -dM_t^\omega \tilde{g}_t$$

$$\text{with} \qquad \tilde{g}_0 = e \qquad (\text{the identity in } \tilde{G}),$$

where

$$M_t^\omega \stackrel{\text{def}}{=} \int_{D_t} F^\omega \, d\sigma_D$$

and $D_t = \{rx_s : (r,s) \in [0,1] \times [0,t]\} \subset D$.

Then :

$$\tilde{g}_1 = \tilde{b}^{-1}\tilde{a}^{-1}\tilde{b}\tilde{a}h \qquad (*)$$

wherein $\tilde{a}, \tilde{b} \in \tilde{G}$ cover $a, b \in G$, respectively, and $h \in ker(\tilde{G} \to G)$ is specified by (and specifies) the topology of the bundle P. (To be precise, specification of h as an invariant of P requires a choice of orientation on Σ; here it is the one induced by q from D.)

Conversely, given $F : D \to \underline{g}$, and $a, b \in G$, satisfying $(*)$ (with F for F^ω) and certain smoothness conditions, then there exists, uniquely up to the action of automorphsims in $\mathcal{G}_m$, a connection ω on the principal G−bundle P specified by h, with $F^\omega = F$ and with $g_u(S_1;\omega) = a$ and $g_u(S_2;\omega) = b$. ∎

Note : The torus is only an example; the theorem can be formulated and holds for all compact surfaces Σ. In the case where Σ is not orientable, $\tilde{G}$ is replaced by another covering of G, and then no 'choice of orientation' is involved in defining h as an invariant of P.

Recall the expression $d\mu_T = Z_T^{-1} e^{-S(\omega)/T}[\mathcal{D}\omega]$ for the YM measure. Recall also that $\int_\Sigma ||\Omega^\omega||^2 \, d\sigma = \int_D ||F^\omega||^2_{\underline{g}} \, d\sigma_D$. Consider the space $L^2(D;\underline{g})$ of $\underline{g}$−valued measurable functions on D whose $||\cdot||_{\underline{g}}$−norms are square integrable with respect to $d\sigma_D$. Thus :

$$e^{-S(\omega)/T} = e^{-\frac{1}{2T}\int_\Sigma ||\Omega^\omega||^2 \, d\sigma} = e^{-\frac{1}{2T}||F^\omega||^2_{L^2(D;\underline{g})}}.$$

Theorem 1 identifies $\mathcal{C}_m$ with $\{(F,a,b) : F : D \to \underline{g}, (a,b) \in G^2 \text{ satisfy}(*)\}$. Motivated by these observations, we realize the YM measure to be :

$\mu_T =$ Gauss measure on $L^2(D;\underline{g}) \times$ Haar measure on G^2 *conditioned* to satisfy $(*)$.

Here the Gauss measure "on" $L^2(D;\underline{g})$ is the one with formal density $e^{-||F||^2_{L^2(D;\underline{g})}/(2T)}$. We review some facts about this measure. As is well known, this measure actually lives on a larger Hilbert space $\overline{L^2(D;\underline{g})}$ which is the closure of $L^2(D;\underline{g})$ under a suitable norm. To each $f \in L^2(D)$ one can associate in a natural way a $\underline{g}$—valued, almost everywhere defined, Gaussian random variable on $\overline{L^2(D;\underline{g})}$ denoted $\omega \mapsto \int_D F^\omega(f)\, d\sigma_D$. In case $f = 1_S$ for some Borel $S \subset D$, we write $\int_S F^\omega\, d\sigma_D$ instead of $\int_D F^\omega(1_S)\, d\sigma_D$. If $E_1, ..., E_d$ form an orthonormal basis of $\underline{g}$, we can write $F^\omega = F_1^\omega E_1 + \cdots + F_d^\omega E_d$; then $\int_S F^\omega\, d\sigma_D = \sum_{i=1}^d \left(\int_S F_i^\omega\, d\sigma_D\right) E_i$, and the $\int_S F_i^\omega\, d\sigma_D$ are mutually independent real-valued Gaussian random variables, each with mean zero and variance $T|S|$ (here $|S|$ is the area of S with respect to the measure $d\sigma_D$).

Now μ_T itself is defined on a certain subspace $\overline{\mathcal{C}} \subset \overline{L^2(D;\underline{g})} \times G^2$: $(F, a, b) \in \overline{\mathcal{C}}$ if and only if (*) of Theorem 1 holds. However, since for almost every $\omega \in \overline{L^2(D;\underline{g})}$, F^ω is not a pointwise defined function on D, the conditions in Theorem 1 have to be interpreted in a stochastic sense. This is explained in the Note below.

Definition of YM measure ([Se 4]) : We equip $\overline{\mathcal{C}}$ with Gauss measure $\times$ Haar measure *conditioned* to satisfy $(*)$. *We take this to be the Yang-Mills measure* μ_T. ■

Note : Recall that $\tilde{g}_1 = \tilde{b}^{-1}\tilde{a}^{-1}\tilde{b}\tilde{a}h$, wherein $h \in ker(\tilde{G} \to G)$ specifies the topology of the bundle, $\tilde{a}, \tilde{b} \in \tilde{G}$ cover $a, b \in G$, respectively, and

$t \mapsto \tilde{g}_t \in \tilde{G}$ solves

$$d\tilde{g}_t = -dM_t \tilde{g}_t \qquad (**)$$

$$\tilde{g}_0 = e \qquad \text{(the identity in } \tilde{G}\text{)},$$

where

$$M_t : \omega \mapsto M_t^\omega \stackrel{\text{def}}{=} \int_{D_t} F^\omega\, d\sigma_D.$$

Under Gauss measure on the space of F's, $(**)$ has to be read as a *Fisk-Stratonovich stochastic differential equation*, and implies that the process $[0, |\Sigma|] \to \tilde{G} : |D_t| \mapsto \tilde{g}_t$ describes Brownian motion on $\tilde{G}$ (governed by the metric $\langle \cdot, \cdot \rangle_{\underline{g}}$).

Fact (*via Theorems*) ([Se 4]) : Such a conditioned measure μ exists. ■

Note : Once again, this works for all compact surfaces, not just for the torus.

In the approach of Fine [F1,2], carried out for orientable Σ without boundary and with simply connected G (in this situation the bundle P is trivial), $\mathcal{A}$ is viewed as a fiber bundle over a certain space of paths in G^{2g}, wherein g is the genus of Σ. The expression $e^{-S(\omega)}$ is used to understand the induced measures on the fibers and on the path space. Expectation values associated to holonomies around the basic loops (S_1, S_2 for the torus) are then determined and the results agree with Theorem 2 below (specialized to this situation).

The approach used by Witten [Wi 1], again for simply connected G and hence trivial bundles, is lattice theoretic. Thus in a sense, the rigorous starting point of this approach is the loop expectation value formula in Theorem 2 (specialized to $\tilde{G} = G$, $h = e$). In [Wi 2], in the setting involving non-trivial bundles, Witten used other arguments to identify the formal 'partition function' $\int_{\mathcal{A}} e^{-S(\omega)}[\mathcal{D}\omega]$, suitably normalized, with N_T (see Theorem 2).

4. Holonomy expectation values.

We have defined the holonomy $g_u(C;\omega)$ when ω is a smooth connection on P and C is a smooth closed curve on Σ, based at m. A glance at the definition shows that $g_u(C;\omega)$ is obtained after solving a diffierential equation (used in defining parallel transport) involving ω. Now a typical $\omega \in \overline{\mathcal{C}}$ does not correspond to a smooth 1−form on P; in fact, with μ_T−probability 1, ω does not even correpsond to a pointwise defined entity. Thus the differential equation defining parallel transport has to be reinterpreted in this setting. The proper interpretation of it is to understand it as a Fisk-Stratonovich stochastic differential equation. This idea was introduced in [GKS] and, in a related setting, in [AHH]. Using this idea it is possible to understand holonomy around closed curves on Σ which are projections by q of curves on D. For technical reasons, such a curve (in D) should be the composite of a finite number of pieces, each either radial or intersecting every radius at most once. Composing such loops with the basic loops S_1 and S_2 (and analogously for other compact surfaces), we obtain a class of loops C on Σ for which there is well-defined notion of holonomy; i.e. a G−valued random variable $\overline{\mathcal{C}} \to G : \omega \mapsto g(C;\omega)$ (we drop the fixed subscript u). Recall that this is defined using the solution of a stochastic differential equation. Loops for which this holonomy is defined will be called 'good' loops.

Let C be a closed contractible good loop on Σ, based at m, enclosing a region A (i.e. C forms the 'positive boundary' of A). Then (as proved in [Se 4]) for any bounded measurable function f on G,

$$\int_{\overline{\mathcal{C}}} f\Big(g(C;\omega)\Big)\, d\mu_T(\omega) = \frac{1}{N_T}\int_{G^2\times\tilde{G}} f(x)\tilde{Q}_{T|A|}(\tilde{x})\tilde{Q}_{T|\Sigma\setminus A|}(h\tilde{b}^{-1}\tilde{a}^{-1}\tilde{b}\tilde{a}\tilde{x}^{-1})\, da db d\tilde{x}$$

($\tilde{G} \to G : \tilde{x} \mapsto x$ is the universal cover). Here N_T is a normalizing constant, $|A|$ is the area of A, $d\tilde{x}$ is Haar measure on $\tilde{G}$, $\tilde{Q}_t(\tilde{x})$ is the density, with respect to the Haar measure $d\tilde{x}$, at time t of Brownian motion on G governed by the metric $\langle \cdot, \cdot \rangle_{\underline{g}}$, and da and db are the Haar measure on G of unit total mass. Since $ker(\tilde{G} \to G)$ is contained in the center of $\tilde{G}$, it follows that $\tilde{b}^{-1}\tilde{a}^{-1}\tilde{b}\tilde{a}$ is determined by $a, b \in G$, and is independent of the choices of $\tilde{a}, \tilde{b} \in \tilde{G}$ covering a and b, respectively. Recall that $h \in ker(\tilde{G} \to G)$ specifies the topology of the bundle.

More generally, if $C_1, ..., C_k$ are good loops on Σ, all based at m, and G is semi-simple or if Σ is not orientable, then it is possible to obtain an explicit formula for $\int_{\mathcal{C}} f(g(C_1; \omega), ..., g(C_k; \omega))\, d\mu_T(\omega)$. This expression involves the areas of the regions on Σ cut out by $\{C_1, ..., C_k\}$ and topological invariants of $\{C_1, ..., C_k\}$ on Σ, and the topology of the bundle (i.e. h). Theorem 2 below describes such a formula.

Let Σ be compact oriented of genus g. Let $\pi : P \to \Sigma$ be a principal $G-$bundle as before. Let $\{\Delta_1, ..., \Delta_n\}$ be the positively oriented two-simplices of a triangulation of Σ. Let $\mathcal{E} = \{e_1, \bar{e}_1, ..., e_M, \bar{e}_M\}$ be the set of oriented one-simplices of the triangulation. We shall use maps $\mathcal{E} \to \tilde{G} : b \mapsto \tilde{x}_b$ for which $\tilde{x}_{\bar{b}} = \tilde{x}_b^{-1}$, where $\bar{b}$ denotes the orientation-reverse of the oriented $1-$simplex b. If C is a closed curve which is the composite $b_r...b_1$ of $1-$simplices then we write $\tilde{x}(C)$ to mean $\tilde{x}_{b_r}...\tilde{x}_{b_1}$. Thus $\tilde{x}(\partial\Delta_i) \in \tilde{G}$, where $\partial\Delta_i$ is the boundary of Δ_i with some choice of initial point. We write x_b for the element of G covered by $\tilde{x}_b \in \tilde{G}$; so $x(C) \in G$ is covered by $\tilde{x}(C) \in \tilde{G}$.

Suppose now that $C_1, ..., C_k$ are loops, all based at $m \in \Sigma$ (which is assumed to be a $0-$simplex of the triangulation) are each composites of oriented $1-$simplices. Assume also that all curves involved form a good collection. Then :

Theorem 2 ([Se 4]) : For Σ compact oriented of genus g and without boundary, and semi-simple G, with $\tilde{G}$ equipped with Haar measure of unit total mass,

$$\int_{\overline{\mathcal{C}}} f(g(C_1; \omega), ..., g(C_k; \omega))\, d\mu_T(\omega) =$$

$$\frac{1}{N_T} \int_{\tilde{G}^M} f(x(C_1), ..., x(C_k)) \tilde{Q}_{T|\Delta_1|}(h\tilde{x}(\partial\Delta_1)) \prod_{i=2}^{n} \tilde{Q}_{T|\Delta_i|}(\tilde{x}(\partial\Delta_i))\, d\tilde{x}_{e_1}...d\tilde{x}_{e_M}$$

where

$$N_T = \int_{\tilde{G}^{2g}} \tilde{Q}_{T|\Sigma|}(h\tilde{b}_g^{-1}\tilde{a}_g^{-1}\tilde{b}_g\tilde{a}_g...\tilde{b}_1^{-1}\tilde{a}_1^{-1}\tilde{b}_1\tilde{a}_1)\, d\tilde{a}_1...d\tilde{b}_g. \quad \blacksquare$$

Note : If Σ is unorientable then $\tilde{G}$ is replaced by $\tilde{G}/K^2$ where $K^2 = \{x^2 : x \in ker(\tilde{G} \to G)\}$. If Σ has boundary then the bundle is trivial, $\tilde{G}$ is replaced by G and $h = e$. ■

Outline of Proof : The YM measure with gauge group $\tilde{G}$ for the disk D is Gaussian in F. The definition of parallel translation, with the equation interpreted stochastically, can be used to prove ([Dr], [GKS], [Se 1]) that the holonomy $g(C;\omega)$ has probability density function $\tilde{Q}_{TA}(\tilde{x})$ on $\tilde{G}$, where A is the area enclosed by the good closed contractible loop C, and is independent of $g(C';\omega)$ if C and C' enclose disjoint regions. Taking into account that every good loop in Σ, based at m, can be expressed as a composite of the basic loops $S_1, ..., S_k$ and loops of the type $q(C)$, for good loops C in D based at O, it can be proven by conditioning according to (*) that the expectation values are as stated above. The detailed argument is somewhat combinatorial in nature. ■

As a consequence we have the result (which holds even for non-semisimple G and all compact surfaces Σ) :

Theorem 3 ([Se 4]) : Let $\phi : \Sigma \to \Sigma$ be an area-preserving homeomorphism, and $C_1, ..., C_k$ be good loops on Σ, all based at m, with $\{\phi C_1, ..., \phi C_k\}$ also 'good'. Then

$$\int_{\overline{\mathcal{C}}} f(g(\phi C_1;\omega), ..., g(\phi C_k;\omega))\, d\mu_T(\omega) = \int_{\overline{\mathcal{C}}} f(g(C_1;\omega), ..., g(C_k;\omega))\, d\mu_T^{\phi}(\omega)$$

where we have written $d\mu_T^{\phi}$ to denote the YM measure (with the same parameter T) for the pullback bundle $\phi^* P$ over Σ. ■

5. The Limit $\lim_{T\to 0} \mu_T$ and Symplectic structure.

Recall the expression

$$d\mu_T = \frac{1}{Z_T} e^{-S_{YM}(\omega)/T} [\mathcal{D}\omega].$$

A natural question is : what happens to μ_T as $T \to 0$?

A simple situation is when the surface Σ is a sphere. For this case we have :

Theorem 4 ([Se 2]) : If Σ is a sphere then there is a probability measure μ_0 on

$$\mathcal{C}^0 = \{[\omega] \in \mathcal{C}_m : S_{YM}(\omega) = \inf_{\omega'} S_{YM}(\omega')\},$$

such that

$$\lim_{T\to 0}\int_{\overline{\mathcal{C}}} f\big(g(C_1;\omega),...,g(C_k;\omega)\big)\, d\mu_T(\omega) = \int_{\mathcal{C}^0} f\big(g(C_1;\omega),...,g(C_k;\omega)\big)\, d\mu_0(\omega)$$

for every collection $\{C_1,...,C_k\}$ of good loops on Σ all based at $m \in \Sigma$, and every continuous function f on G^k. ∎

When Σ is a sphere, $\mathcal{C}^0$ can be identified with the space of shortest length loops on G, based at the identity e, in a given homotopy class specified by, and specifying, the topology of P. The action on $\mathcal{C}^0$ of automorphisms of P which do not necessarily fix the fiber $\pi^{-1}(m)$ pointwise corresponds to the conjugation action of G on the space of loops. The corresponding quotient space contains just one point and so the content of the above result after passing to the full quotient under all automorphisms (i.e. under $\mathcal{G}$) is simply that, when the function f is invariant under the conjugation action of G, the limit on the left is actually equal to $f\big(g(C_1;\omega),...,g(C_k;\omega)\big)$ where ω is 'the' (up to bundle automorphism) point where S_{YM} is minimum. In this sense μ_0 has no essential significance for this case.

We assume from now on that Σ is a compact oriented surface without boundary and has genus $g > 0$. In this situation the space $\mathcal{C}^0$ of minima of S_{YM} has a more interesting structure. An automorphism of P induces, by pullbacks of connections, a map of $\mathcal{C}^0$ into itself; we denote the quotient space by $\mathcal{C}^0/G$; thus in the notation explained in (E) of Section 2, $\mathcal{C}^0/G$ is the set of minima of S_{YM} on $\mathcal{A}/\mathcal{G}$.We assume henceforth that G is semisimple (the connected component of the center of a more general G can be treated separately). Then the set of minima of S_{YM} consist of flat connections on P. By Theorem 1, these are specified completely up to bundle automorphisms in $\mathcal{G}_m$ by the holonomies $g_u(S_1;\omega)$,...,$g_u(S_{2g};\omega)$. Examination of this shows that the set $\mathcal{C}^0/G$ can be identified with $Hom\big(\pi_1(\Sigma;m),G\big)/G$, where $\pi_1(\Sigma,m)$ is the fundamental group of Σ with basepoint m, and the quotient is under the conjugation action of G on $Hom\big(\pi_1(\Sigma;m),G\big)$. This space has a natural symplectic structure which may be viewed either as arising from the canonical symplectic structure on the space $\mathcal{A}$ or from a certain cup product operation. It should be noted, however, that this space is not a smooth manifold, i.e. it has 'kinks', and the symplectic structure has to be treated with care. Theorem 5 below describes the symplectic form form on $\mathcal{C}^0/G$ in terms of variations of the holonomies around the basic loops $S_1,...,S_{2g}$.

Recall (Section 2, Item (D)) the symplectic form Θ on the space $\mathcal{A}$ of all connections on a principal G–bundle $\pi : P \to \Sigma$. Denote by $A_1, B_1, ..., A_g, B_g$ the $2g$ loops, all based at $m \in \Sigma$, generating the fundamental group $\pi_1(\Sigma; m)$ (thus $\overline{B}_g\overline{A}_g B_g A_g ... \overline{B}_1\overline{A}_1 B_1 A_1$ is contractible to m). Fix $u \in \pi^{-1}(m)$, and, for $\omega \in \mathcal{A}$, we write $\alpha_1 = g_u(A_1; \omega)$, $\alpha_2 = g_u(B_1; \omega)$, $\alpha_3 = \alpha_1^{-1}$, $\alpha_4 = \alpha_2^{-1}$, and, in general, $\alpha_{4i+1} = g_u(A_{i+1}; \omega)$, $\alpha_{4i+2} = g_u(B_{i+1}; \omega)$, $\alpha_{4i+3} = \alpha_{4i+1}^{-1}$, $\alpha_{4i+4} = \alpha_{4i+2}^{-1}$, with $i \in \{0, 1, ..., g-1\}$. Then $\omega \mapsto \alpha = (\alpha_1, ..., \alpha_{4g})$ sets up a bijection between $\mathcal{C}^0$ and the set $\mathcal{F} = \{\alpha \in G^{4g} : \tilde{\alpha}_{4g} ... \tilde{\alpha}_1 = h^{-1}, \alpha_{j+2} = \alpha_j^{-1}, j = 1, 2, 5, 6, ...\}$; here $\tilde{\alpha}_i \in \tilde{G}$, the universal cover of G, covers $\alpha_i \in G$, $\tilde{\alpha}_{j+2}$ is chosen to be $\tilde{\alpha}_j^{-1}$, for $j \in \{1, 2, 5, 6...\}$, and $h \in ker(\tilde{G} \to G)$ specifies the topology of the bundle P. A path $\epsilon \mapsto \alpha(\epsilon) \in \mathcal{F}$, smooth as a path in G^{4g}, where ϵ runs over a neighborhood of 0 and $\alpha(0) = \alpha$, corresponds to a path $\epsilon \mapsto \omega_\epsilon$ of flat connections on P, where $(\epsilon, x) \mapsto \omega_\epsilon|_x$ is smooth in ϵ and $x \in P$. We write $H_i = \alpha_i^{-1}\alpha_i'(0)$, and take $H = (H_1, ..., H_{4g}) \in \underline{g}^{4g}$ as the tangent to the path $\epsilon \mapsto \alpha(\epsilon)$. We set $\overline{\Theta}(H^{(1)}, H^{(2)}) = \Theta(\eta^{(1)}, \eta^{(2)})$, where $H^{(1)}, H^{(2)}$ are tangent vectors to $\mathcal{F}$ at α, and $\eta^{(1)}, \eta^{(2)}$ the corresponding vectors in $\mathcal{A}$ (tangent to paths of flat connections). The following result is joint work with C. King (a different approach, in terms of the cup-product description, is available in [Go]) :

Theorem 5 : With hypotheses and notation as above, $\overline{\Theta}$ is well-defined on the 'smooth part' of $\mathcal{F}$ and induces a well-defined closed 2–form on the smooth part of $\mathcal{F}/G$ (where the quotient is under the action of G on $\mathcal{F}$ which conjugates each component). If $H^{(1)}, H^{(2)}$ are tangent vectors to $\mathcal{F}$ at α then :

$$\overline{\Theta}(H^{(1)}, H^{(2)}) = \frac{1}{2} \sum_{1 \le i,k \le 4g} \epsilon_{ik} \langle f_{i-1}^{-1} H_i^{(1)}, f_{k-1}^{-1} H_k^{(2)} \rangle_{\underline{g}}$$

where ϵ_{ik} equals 1 if $i < k$, equals -1 if $i > k$, and equals 0 if $i = k$; and $f_i = Ad(\alpha_i ... \alpha_1)$. ■

For simply-connected G, Witten ([Wi 1]) has given arguments to show that :

$$\lim_{T \to 0} N_T = \text{ the symplectic volume of the space of minima of } \mathcal{C}^0/G.$$

Using this and another observation in [Wi 1], R. Forman ([Fo]) has shown :

Theorem 6 : Let Σ be a compact oriented surface without boundary of genus $g > 1$. Suppose G is simply connected (compact and connected). (In this case the bundle P is trivial.) Let f be a continuous function on G^k, invariant under the conjugation action of G on G^k; so $\omega \mapsto f\big(g(C_1;\omega),...,g(C_k;\omega)\big)$ induces a well-defined function on $\mathcal{C}^0/G$. Then :

$$\lim_{T\to 0}\int_{\overline{\mathcal{C}}} f\big(g(C_1;\omega),...,g(C_k;\omega)\big)\, d\mu_T(\omega) =$$

$$= \int_{\mathcal{C}^0/G} f\big(g(C_1;\omega),...,g(C_k;\omega)\big)\, d(\text{symp})(\omega)$$

wherein $d(\text{symp})(\omega)$ is the symplectic volume measure on $\mathcal{C}^0/G$, normalized to have volume 1. ■

The method used by Forman can probably be extended to cover non-trivial bundles as well. The case of genus 1 (torus) is singular. The author has proven that in case the gauge group is $SU(2)$ or $SO(3)$ and Σ is a torus then Theorem 6 holds. The proof is by direct calculation by expanding out the heat kernel in terms of characters of $\tilde{G}$, and by using the formula for the symplectic form given in Theorem 5.

6. Other Questions.

The following are some directions of interest : (1) construction and study of the YM measure for surfaces with boundary when the holonomies around the boundary components are specified; (2) determination and study of the limiting behavior of μ_T as $T \to 0$ (for non-trivial bundles, *etc.*); (3) the Duistermaat-Heckmann formula in the Yang-Mills setting (see [Wi 2]); (4) development of analysis, involving both differential structure and integration theory, on the moduli space of connections using the YM measure as the basic measure.

REFERENCES

[AHH] S. Albeverio, R. Hoegh-Krohn, and H. Holden, "Stochastic Multiplicative Measures, Generalized Markov Semigroups, and Group-Valued Stochastic Processes", *J. Funct. Anal.* **78** (1988), 154-184.

[Dr] B. K. Driver, "YM_2 : Continuum Expectations, Lattice Convergence, and Lassos", *Commun. Math. Phys.* **123**, 575-616 (1989).

[F 1] D. Fine, "Quantum Yang-Mills on the two-sphere", *Commun. *Math. Phys.* **134** 273-292 (1990).

[F 2] D. Fine, "Quantum Yang-Mills on a Riemann Surface", *Commun. Math. Phys.* **140**, 321-338 (1991).

[Fo] R. Forman, "Small volume limts of $2-d$ Yang-Mills", *Commun. Math. Phys.* **151**, 39-52 (1993).

[Go] W. Goldman, "The Symplectic Nature of Fundamental Groups of Surfaces", *Adv. Math.* **54** (1984) 200-225.

[GKS] L. Gross, C. King, and A. Sengupta, "Two Dimensional Yang-Mills Theory via Stochastic Differential Equations", *Ann. Phys.* **194**, 389-402 (1989).

[MRS] J. Magnen, V. Rivasseau, and R. Seneor, "Construction of YM_4 with Infrared Cutoff", *Commun. Math. Phys.* **155**, 325-383 (1993).

[Se 1] A. Sengupta, "The Yang-Mills Measure for S^2", *J. Funct. Anal.* **108**, 231-273 (1992).

[Se 2] A. Sengupta, "The Semi-Classical Limit of the Yang-Mills Measure on S^2", *Commun. Math. Phys.* 147, 191-197 (1992).

[Se 3] A. Sengupta, "Quantum Gauge Theory on Compact Surfaces "*Ann. Phys.* **221**, 17-52 (1993).

[Se 4] A. Sengupta, "Gauge Theory on Compact Surfaces", preprint (1993).

[Wi 1] E. Witten, "On Quantum Gauge Theories in Two Dimensions", *Commun. Math. Phys.* **141**, 153-209 (1991).

[Wi 2] E. Witten, "Two Dimensional Quantum Gauge Theory Revisited", *J. Geom. Phys.* **9**, 303-368 (1992).

Entropy and KMS norms

R.F.STREATER
Department of Mathematics, King's College, Strand, London WC2R 2LS

(Received:1/9/1993)

Abstract. We borrow the norm dual to that given by the Kubo-Martin-Schwinger construction and apply it to classical statistical mechanics, where it provides a much stronger sense of convergence than the ℓ_1-norm. We show that the "equivalence of ensembles" for the ideal gas holds in this sense, and obtain an estimate for the relative entropy in terms of this norm.

Key words: entropy, ensembles

1. Introduction

Equilibrium quantum statistical mechanics concerns the Gibbs state

$$\rho_\beta = e^{-\beta H}/Tre^{-\beta H} \tag{1}$$

which makes sense if H is of trace class. This defines the scalar product on the algebra of observables

$$\langle A, B\rangle = Tr\,(\rho_\beta A^* B)\,. \tag{2}$$

The norm of A given by this scalar product is

$$||A||_{\beta,2} = (Tr\,(\rho_\beta A^* A))^{1/2} \tag{3}$$

and this is just one of a scale of generalised Schatten norms

$$||A||_{\beta,\ell} = \left(Tr\left(\rho_\beta |A|^\ell\right)\right)^{1/\ell} \tag{4}$$

The norm $||\bullet||_{\beta,2}$ determines convergence in the KMS representation space. If $\beta = 0$ and $\ell = 2$ we get the Hilbert- Schmidt norm, while $\beta = 0$ and $\ell = 1$ gives the trace norm, often the one preferred for states (density matrices in quantum theory, measures in classical theory). The norm $||\bullet||_{\beta,\ell}$ is not a very strong norm for positive β, since it gives an exponentially small weight to the high energies. However, these norms are the natural norms only for the *observables* that is, the random variables or operators. For the states, we should use the dual norms

$$||\rho||_{-\beta,\ell'} = \sup_{A:||A||_{\beta,\ell}=1} |\langle A, \rho\rangle| \text{ where } 1/\ell + 1/\ell' = 1. \tag{5}$$

A. I. Cardoso et al. (eds.), Stochastic Analysis and Applications in Physics, 405–413.

This is a strong (large) norm, as it gives an exponentially increasing weight to the high-energy tail. In this paper we discuss the well-known theorem on the "equivalence of ensembles". We treat in detail the case of a collection of Boson oscillators, all of the same frequency, each oscillator labelled by a point $x \in \Lambda$. The microcanonical state ρ_E of total energy E is the uniform distribution over the corresponding energy shell. Let Λ_0 be a subset of Λ. Then the marginal distribution of ρ_E restricted to Λ_0 approximates the canonical distribution with mean energy $E/|\Lambda|$ per site if E and $|\Lambda|$, the size of Λ, is large compared with $|\Lambda_0|$. We show that the approximation is in the sense of the norm $\|\bullet\|_{-\beta,2}$.

Our technique makes use of estimates on the relative entropy, a simple case of which we give in the next section.

2. Estimates on the relative entropy

We consider for simplicity the classical case. Let Ω be a countable set, and let p and p' be probabilities on Ω. Then the *Kullback relative information* of p given p' is defined as

$$S(p|p') = \sum_{\omega\in\Omega} p(\omega)\left(\log p(\omega) - \log p'(\omega)\right). \tag{6}$$

If any one of the $p'(\omega)$ vanishes, we may put $S(p|p') = +\infty$, but this will not be a useful convention here. We shall always arrange that this does not happen. It is known that

$$S(p|p') \geq 0, \text{ with equality only if } p = p'. \tag{7}$$

We shall prove a stronger estimate. The Kullback relative information is sometimes called the relative entropy, but it has the wrong sign for this; it can be identified with $\beta(F - F')$, where F is the Helmholtz free energy, when p' is the canonical state of beta β. When p and p' have the same mean energy we find that $S(p|p') = S(p') - S(p)$, where

$$S(p) = -\sum_{\omega\in\Omega} p(\omega)\log\; p(\omega)) \tag{8}$$

is the Shannon entropy.

Because of (7) the relative information is often taken as an indication of how far apart the two measures p and p' are. But it has some disadvantages, as it is non-linear, and is not a metric. The following lemma will be used to show that if p' is the canonical state p_β, then with one extra condition, the relative information is related to the KMS norm $\|\bullet\|_{-\beta,2}$.

LEMMA 1. *Suppose p and q are positive functions in $\ell_1(\Omega)$ and there exists $K \geq 1$ such that*

$$p(\omega) \leq Kq(\omega) \text{ for all } \omega \in \Omega. \tag{9}$$

Then

$$S(q|p) \geq \frac{1}{2K^2} \sum_\omega q(\omega)^{-1} |p(\omega) - q(\omega)|^2 - \|p\|_1 + \|q\|_1. \tag{10}$$

Remark: if $q = p_\beta$, we say that p obeying (9) is close to equilibrium (K, β).
Proof
The Taylor series with Lagrange remainder gives

$$\begin{aligned} S(q|p) &= \sum q(\omega)(\log q(\omega) - \log p(\omega)) \\ &= \sum q(\omega)\{\log q(\omega) - [\log q(\omega) + \frac{p(\omega) - q(\omega)}{q(\omega)} - \frac{(q(\omega) - p(\omega))^2}{2\xi(\omega)^2}]\} \\ &= \sum (q(\omega) - p(\omega))^2 q(\omega)/(2\xi(\omega)^2) - \|p\|_1 + \|q\|_1. \end{aligned}$$

Here $\xi(\omega)$ is the intermediate value, obeying

$$q(\omega) < \xi(\omega) < p(\omega) \text{ or } p(\omega) < \xi(\omega) < q(\omega).$$

If ω is such that $\xi(\omega) < q(\omega)$ we get for each term in the sum a contribution greater than

$$\frac{1}{2} |q(\omega) - p(\omega)|^2 \frac{q(\omega)}{q(\omega)^2}.$$

If ω is such that $\xi(\omega) < p(\omega) < Kq(\omega)$, we get a term greater than

$$\frac{1}{2} |q(\omega) - p(\omega)|^2 \frac{q(\omega)}{K^2 q(\omega)^2}.$$

In either case, summing we get the lemma, which for simplicity was stated for the case $K \geq 1$.

We obtain a similar inequality, but in the other direction, for those p obeying $p(\omega) > kq(\omega)$.

Now let us use the lemma to get a result.

3. Equivalence of Ensembles

We consider a finite number of oscillators, one for each point $x \in \Lambda$, all having the same frequency. Thus the sample space is $\Omega(\Lambda) = \prod_{x\in\Lambda} \Omega_x$, where

$$\Omega_x = \{0, 1, 2, \ldots\}. \tag{11}$$

A sample point $\omega \in \Omega$ is a collection of non-negative integers $\{j_x\}_{x\in\Lambda}$. The ideal neoclassical Bose gas is described by the energy

$$H_0 = \sum_x H_0(x) \text{ where } H_0(x)(\omega) = j_x. \tag{12}$$

The energy is a random variable, that is, a function of ω. The microcanonical state p_Λ of energy E is the uniform distribution over the energy shell Ω_E:

$$\Omega_E = \{\omega : H_0(\omega) = E\}. \tag{13}$$

That is,

$$\begin{aligned} p_\Lambda(\omega) &= 1/|\Omega_E| \text{ if } \omega \in \Omega_E \\ &= 0 \text{ otherwise} \end{aligned} \tag{14}$$

We define the marginal state of this for a region $\Lambda_0 \subseteq \Lambda$ to be the probability:

$$p_{\Lambda,\Lambda_0}(\omega_0) = \sum_{\omega:\omega|\Lambda_0=\omega_0} p_\Lambda(\omega). \tag{15}$$

The theorem on equivalence of ensembles says that if Λ_0 is fixed and $|\Lambda|$ becomes large, such that the energy/site is constant, then p_{Λ,Λ_0} converges to the canonical distribution

$$p_{\Lambda_0,\beta}(\omega_0) = Z(\Lambda_0)^{-1} \exp\{-\beta H_0(\Lambda_0)\}. \tag{16}$$

Here, β is fixed by

$$\sum_{\omega_0} p_{\Lambda_0,\beta}(\omega_0) H_0(\omega_0) = E|\Lambda_0|/|\Lambda|. \tag{17}$$

We shall show that the convergence is in the sense of $\| \bullet \|_{-\beta,2}$, which is stronger than the usual statement [1]. Spitzer shows that the expectations of bounded observables coincide in the limit, which is the "vague" topology for measures, and the weak*-topology for C^*-algebras. Our method is to prove the bound (9) and to show that the relative information $S(p_{\Lambda_0,\beta}|p_{\Lambda,\Lambda_0})$ converges to zero as $|\Lambda| \to \infty$. The lemma then completes the proof.

Dobrushin and Tirozzi [2] obtained the equivalence of ensembles in the sense of the equality of correlation functions and also in the sense of pointwise convergence of probabilities, both for a lattice model with interactions of short range. Lewis et al. [3] have obtained equivalence, again for a class of interacting models, in the sense of the metric given by the total variation. This is the norm, in our notation, given by $\beta = 0$ and $\ell = 1$. It remains an open question whether interacting theories converge in the stronger sense we use here.

To find p_Λ for the energy-shell Ω_E we need to know the number of points in Ω_E, which is the number of ways of getting E as the sum of $|\Lambda|$ non-negative integers. This is the coefficient of z^E in

$$(1 + z + z^2 + \ldots)^{|\Lambda|} = (1 - z)^{-|\Lambda|}. \tag{18}$$

It follows that

$$|\Omega_E| = \frac{1}{E!}\frac{\partial^E}{\partial z^E}(1-z)^{-|\Lambda|}|_{z=0},$$

so the microcanonical state for $\omega \in \Omega_E$ is

$$p_{|\Lambda|}(\omega) = \frac{1}{|\Omega_E|} = \frac{E!}{|\Lambda|(|\Lambda|+1)\dots(|\Lambda|+E-1)}. \tag{19}$$

This depends only on $|\Lambda|$, as indicated. Let $\Lambda_0 \subseteq \Lambda$ and consider the marginal distribution of $p_{|\Lambda|}$ given the configuration ω_0 in Λ_0. This will depend on ω_0, but only through the total energy in Λ_0, say

$$J = H_0(\Lambda_0) = \sum_{\omega_0 \in \Omega(\Lambda_0)} H_0(\omega_0).$$

Then the canonical state, (16), becomes

$$p_{\Lambda_0,\beta} = Z_\beta^{-|\Lambda_0|}e^{-\beta J} \text{ where } Z_\beta^{-1} = 1-e^{-\beta}. \tag{20}$$

Clearly, on the energy shell Ω_E, we must have $J \leq E$. The number of terms in Eqn.(15) is the number of configurations *outside* Λ_0 that have total energy $E-J$. But this is the same calculation as before, with $|\Lambda|-|\Lambda_0|$ replacing $|\Lambda|$ and $E-J$ replacing E. Hence

$$p_{\Lambda,\Lambda_0}(\omega_0) = \frac{E!}{(E-J)!}\frac{(|\Lambda|-|\Lambda_0|)(|\Lambda|-|\Lambda_0|+1)\dots(|\Lambda|-|\Lambda_0|+E-J-1)}{|\Lambda|(|\Lambda|+1)\dots(|\Lambda|+E-1)}$$

which can be rearranged to give

$$\begin{aligned} p_{\Lambda,\Lambda_0}(J) &= \underbrace{\frac{E(E-1)\dots(E-J+1)}{(|\Lambda|+E-1)(|\Lambda|+E-2)\dots(|\Lambda|+E-J)}}_{J\text{ factors}} \\ &\quad \underbrace{\frac{(|\Lambda|-|\Lambda_0|)}{|\Lambda|}\frac{(|\Lambda|-|\Lambda_0|+1)}{(|\Lambda|+1)}\dots\frac{(|\Lambda|-|\Lambda_0|+E-J-1)}{(|\Lambda|+E-J-1)}}_{E-J\text{ factors }\leq 1} \qquad (21) \\ &\leq \frac{E-1}{|\Lambda|+E-1}\frac{E-2}{|\Lambda|+E-2}\dots\frac{E-J+1}{E-J+|\Lambda|+1}\frac{E}{E+|\Lambda|-J}.1^{E-J} \\ &\leq \left(\frac{1}{1+|\Lambda|/E}\right)^{J-1}.\frac{E}{|\Lambda|}. \qquad (22) \end{aligned}$$

Now $E/|\Lambda|$, the energy per site, determines a unique beta by

$$E/|\Lambda| = e^{-\beta}/(1-e^{-\beta}) \tag{23}$$

so $(1+|\Lambda|/E) = e^\beta$. Thus we have

$$p_{\Lambda,\Lambda_0} < (1-e^{-\beta})^{-1}e^{-J\beta},$$

and we have proved (9) with $K = (1-e^{-\beta})^{-(|\Lambda_0|+1)}$. Since K is bounded in any compact subset of $0 < \beta < \infty$, our estimate is uniform on any such subset.

We are going to use Eqn.(10) with $q = p_{\Lambda_0,\beta}$ and $p = p_{\Lambda,\Lambda_0}$. Since the latter is zero for those ω_0 with $J > E$, we must restrict the sum to $J \leq E$. In that case we have $\|p\|_1 = 1$ but $\|q\|_1 < 1$, since the tail is missing. While it is obvious that $\|q\|_1 \to 0$ as $E \to \infty$, it is not so obvious that the convergence is uniform in beta. To prove this, we compute the multiplicity of states in $\Omega(\Lambda_0)$ with energy J, often denoted $g(J)$. We have already found this to be

$$g(J) = \frac{|\Lambda_0|(|\Lambda_0+1)\ldots(|\Lambda_0|+J-1)}{1.2\ldots(J-1)J}. \tag{24}$$

We are interested in the case of large J, at least equal to $|\Lambda_0|$. In that case there is some cancellation, and

$$\begin{aligned} g(J) &= \frac{(J+1)(J+2)\ldots(J+|\Lambda_0|-1)}{(|\Lambda_0|-1)!} \\ &\leq \frac{(2J)^{|\Lambda_0|}}{(|\Lambda_0|-1)!}. \end{aligned} \tag{25}$$

So the tail of the canonical distribution $p_{\Lambda_0,\beta}$ sums to

$$\begin{aligned} \sum_{J_0}^{\infty} g(J)e^{-\beta J}(1-e^{-\beta})^{|\Lambda_0|} &\leq (1-e^{-\beta})^{|\Lambda_0|}e^{-\beta J_0}\sum_{j=0}^{\infty} g(j+J_0)e^{-\beta j} \\ &\leq C(\beta,|\Lambda_0|)e^{-\beta J_0}\sum (j+J_0)^{|\Lambda_0|}e^{-\beta j} \\ &\leq K(\beta_1,|\Lambda_0|)J_0^{|\Lambda_0|}e^{-\beta J_0} \end{aligned} \tag{26}$$

if $\beta \geq \beta_1$. This shows that letting $E \to \infty$, $\|q\|_1 \to 1$ exponentially and uniformly in beta, when we choose q to be equal to the canonical state $p_{\Lambda_0.\beta}$ for energy $\leq E$, and zero otherwise. A similar argument shows that the missing tail on the right-hand side of Eqn.(10), in the norm $\|p-q\|_{-\beta,2}$, is uniformly exponentially small. We complete the proof by showing that the relative information

$$S(p_{\Lambda_0,\beta}|p_{\Lambda,\Lambda_0}) = O(|\Lambda|^{-1})$$

uniformly for a compact set of positive beta. In the sum, only ω_0 with energy $\leq E$ are included.

We note that there is more cancellation in (21) and we get

$$p_{\Lambda,\Lambda_0}(J) = \underbrace{\frac{E(E-1)\dots(E-J+1)}{(|\Lambda|+E-|\Lambda_0|-1)(|\Lambda|+E-|\Lambda_0|-2)\dots(|\Lambda|+E-|\Lambda_0|-J)}}_{J\text{ factors}}$$
$$\underbrace{\frac{(|\Lambda|-|\Lambda_0|)(|\Lambda|-|\Lambda_0|+1)\dots(|\Lambda|-1)}{(|\Lambda|+E-1)(|\Lambda|+E-2)\dots(|\Lambda|+E-|\Lambda_0|)}}_{|\Lambda_0|\text{ factors}}. \tag{27}$$

We take out J factors $E/(|\Lambda|+E) = e^{-\beta}$. Each of the $|\Lambda_0|$ factors

$$\frac{|\Lambda|-|\Lambda_0|+k-1}{|\Lambda|+E-k}, \ ,k = 1,2,\dots|\Lambda_0|$$

converges to

$$|\Lambda|/(|\Lambda|+E) = 1-e^{-\beta} = Z_\beta^{-1}.$$

So these factors we have taken out converge to $p_{\Lambda_0,\beta}$ given by Eqn.(20). Taking logs we have, uniformly in a compact set of positive β, the estimate

$$\sum_{\omega_0} p_{\Lambda_0,\beta}(\omega_0)\left(\sum_{k=1}^{|\Lambda_0|}\left(\log\frac{|\Lambda|-|\Lambda_0|+k-1}{|\Lambda|+E-k}\right) - \beta J - \log p_{\Lambda_0,\beta}(\omega_0)\right) = O\left(\frac{1}{|\Lambda|}\right).$$

We have therefore proved the theorem if we can show the same for the remaining terms

$$\sum_{\omega_0} p_{\Lambda_0,\beta}(\omega_0)\sum_{k=0}^{J-1}\log(1-k/E) \tag{28}$$

and

$$\sum_{\omega_0} p_{\Lambda_0,\beta}(\omega_0)\sum_{k=1}^{J}\log\left(1-\frac{|\Lambda_0|+k}{|\Lambda|+E}\right). \tag{29}$$

In (28), k can reach $J-1$ and $J-1$ can reach $E-1$, so the logarithm is not small. But $p_{\Lambda_0,\beta}$ is exponentially small for this region of k and J. To exploit this we divide the sum (28) into ω_0 such that $J \le E/2$ and $E/2 < J \le E$. The first half of the range is estimated by

$$|\sum_{J\le E/2} p_{\Lambda_0,\beta}\sum_{k=0}^{J-1}\log(1-k/E)| \le \sum_{J\le E/2} p_{\Lambda_0,\beta}(J)\sum_{k=0}^{J-1}(k/E+k^2/(2E^2\xi^2)). \tag{30}$$

Here, the intermediate value ξ obeys $1-k/E \leq \xi \leq 1$, and as $k < J \leq E/2$, we see that $\xi \geq 1/2$. Now the sum over k of the quadratic polynomial in k gives a polynomial of degree 3 in J, and the third and lower moments of the energy J in the canonical state $p_{\Lambda_0,\beta}$ give a certain cubic function of the mean energy $|\Lambda_0|E/|\Lambda|$. The sum in (30) is only up to $E/2$, and so is less than this, and is therefore bounded by $O(|\Lambda_0|^2/E) + O(|\Lambda_0|^3/E^2)$.

In the second half of the sum (28) we have $J \leq E$ terms $\log(1-k/E)$. In fact, $|\log(1-k/E)| \leq \log E$. So the second half is bounded by

$$\sum_{E/2<J\leq E} p_{\Lambda_0,\beta} E \log E,$$

which is part of the tail of the canonical state, starting at $E/2$. Since the exponential decrease of the tail (26) easily beats the extra $E\log E$, we get uniform convergence to zero of this part too.

The sum (29) does not need to be split; in a similar way to (30)

$$|\sum p_{\Lambda_0,\beta} \sum_{k=1}^{J} \log\left(1 - \frac{|\Lambda_0|+k}{|\Lambda|+E}\right)| \leq \sum p_{\Lambda_0,\beta}\left(\frac{|\Lambda_0|+O(J^2)}{|\Lambda|+E} + \frac{|\Lambda_0|+O(J^3)}{2\xi^2(|\Lambda|+E)^2}\right)$$
$$= O(1/|\Lambda|).$$

This concludes the proof that $p_{\Lambda,\Lambda_0,E} \to p_{\Lambda_0,\beta}$ as $O(1/|\Lambda|)$ in the $\|\bullet\|_{-\beta,2}$-norm uniformly in β in $0 < \beta_1 \leq \beta \leq \beta_2 < \infty$.

4. Further Remarks

There is a similar theorem for the case of a neoclassical Fermion gas. The proof is easier, as the configuration space is finite; we just have to note that the number of states in Λ with energy E is the coefficient of z^E in $f(z) = (1+z)^{|\Lambda|}$. This case also covers the one-dimensional Ising model. We can also easily prove the equivalence of ensembles for an assembly of non-interacting N-level atoms with equal spacing between the levels. In that case the number of states with energy E is the coefficient of z^E in $f(z) = (1+z+z^2+\ldots+z^{N-1})^{|\Lambda|}$.

It can be shown that if T is a stochastic map such that p_β is a fixed point of T^*, then T^* is a contraction in $\|\bullet\|_{-\beta,2}$ and that if p is close to equilibrium (K,β) then T^* reduces the free energy by at least $\frac{1}{2K}(\|p\|^2_{-\beta,2} - \|T^*p\|^2_{-\beta,2})$. This improves Lindblad's theorem that a Markov matrix reduces the relative entropy [4], Theorem 1, at least for the classical case. We also show that if the Markov chain is irreducible, then the convergence of $T^{*n}p$ to the unique fixed point p_β is in the norm $\|\bullet\|_{-\beta,2}$, provided that p is close to equilibrium (K,β). The latter condition holds for all states in the dense set in the KMS Hilbert space obtained by acting on the state p_β by an element f of the algebra of bounded random variables. In this case, the constant K is obviously $\|f\|_\infty$. The details will appear in [5].

References

[1] F.Spitzer, Introduction aux processus de Markov à paramètres dans Z_ν. Lecture Notes in Mathematics, **390** Springer-Verlag, Berlin (1974).p.161.

[2] R.L.Dobrushin and B.Tirozzi, The Central Limit Theorem and the Problem of Equivalence of Ensembles, Communications in Mathematical Physics, **54**, 173-192 (1977).

[3] J.T.Lewis, C.-E. Pfister and W.G. Sullivan, Large Deviations and the Thermodynamic Formalism: A New Proof of the Equivalence of Ensembles. Dublin Institute for Advanced Study- STP-93-24, to be published.

[4] G.Lindblad, Entropy, Information and Quantum Measurements, Commun. Mathematical Phys.,**33**, 305-322 (1973).

[5] R.F.Streater, Detailed Balance and Free Energy, Proceedings, Symposium on Quantum Information Theory and Open Systems, Sept.6-8,1993. To be published in the RIMS Kokyuroku Series, Kyoto, Ed. M. Ohya.

AN INTRODUCTION TO WHITE NOISE ANALYSIS

LUDWIG STREIT
CCM, Universidade da Madeira
BiBoS, Univ. Bielefeld

Contents

1. Fundamentals of White Noise Analysis

1.1. WHITE NOISE FUNCTIONALS

The Gaussian white noise measure μ (on the Borel algebra over cylinder sets of real, tempered distributions $\omega \in S^*(R^d)$) is conveniently described by its characteristic function:

$$C(f) = E(e^{i<\omega,f>}) = \int_{S^*} d\mu\,[\omega]\, e^{i<\omega,f>} = e^{-\frac{1}{2}\int f^2(t)dt},\ f \in S(R^d). \tag{1.1}$$

A. I. Cardoso et al. (eds.), Stochastic Analysis and Applications in Physics, 415–439.

It is intimately connected with Brownian motion. Wiener's Brownian motion B(t) has a representation in terms of White Noise as follows. For $d = 1$

$$B(\tau)(\omega) = <\omega, 1_{[0,\tau]}> = \int_0^\tau \omega(t)dt \text{ , resp. } \omega(t) = \frac{d}{dt}B(t)(\omega), \tag{1.2}$$

i.e. in this sense White Noise is the velocity of Brownian motion, and functionals of Brownian motion can be seen as functionals of White Noise.

To obtain a large class of the latter we consider the Hilbert space:

$$(L^2) \equiv L^2(S^*, d\mu).$$

Polynomials in $< \omega, f >$, $f \in S$, span a dense subspace P of (L^2). In particular, orthogonalization leads to a basis of Hermite polynomials:

$$:< \omega, f >^n : = <: \omega^{\otimes n} :, f^{\otimes n} > .$$

By polarization and closure these extend to

$$\phi_n(\omega) = <: \omega^{\otimes n} :, F_n >, \qquad F_n \in Sy\, S_C(R^n)$$

and one calculates their scalar product as

$$(\phi_n, \phi_m) = \delta_{m,n} n! \int F_n^* F_m d^m t.$$

As a consequence of this one then obtains the canonical isomorphism between (L^2) and Fock space:

$$(L^2) \simeq \bigoplus_n Sy L^2(R^n, n! d^n t), \tag{1.3}$$

generated by

$$\phi_n \simeq F_n.$$

hence the general element of (L^2) has a representation

$$\phi(\omega) = \sum_{n=0}^{\infty} \phi_n(\omega) = \sum_{n=0}^{\infty} <: \omega^{\otimes n} :, F_n >$$

with kernel functions F in Fock space.

Another extremely useful description of White Noise functionals is through their **T-** and **S-Transforms**. For these we first introduce "normal ordered exponentials"

$$: \exp(< \omega, f >) :\equiv \sum \frac{1}{n!} <: \omega^{\otimes n} :, f^{\otimes n} > = C(f) \exp(< \omega, f >) \tag{1.4}$$

for $f \in S(R)$ and make the

Definition 1. *The transformation defined for all test functions $f \in S(R)$ and (L^2) functions ϕ by the expectation*

$$(S\phi)(f) \equiv E(\phi \cdot : \exp(< \cdot, f >) :). \tag{1.5}$$

is called the S-transform of ϕ at f. Likewise we define the T-transform

$$(T\phi)(f) \equiv E(\phi \cdot \exp(i < \cdot, f >). \tag{1.6}$$

For functionals represented in terms of their kernel functions

$$\phi(\omega) = \sum_{n=0}^{\infty} <: \omega^{\otimes n} :, F_n >$$

one calculates the S-transform as

$$(S\phi)\ (f) = \sum_{n=0}^{\infty} \int_{R^n} F_n(t_1, ..., t_n)\ f(t_1)...f(t_n)\ d^n t \tag{1.7}$$

i.e. knowledge of the S-transform in the form (2) reduces the question of the (L^2) property of ϕ to that of whether $\oplus F_n$ is in Fock space. This is best illustrated by some counterexamples:

(1) "local Wick powers"

$$\phi(\omega) =: \omega^m(t) : \qquad (S\phi)(f) = f^m(t) \tag{1.8}$$

Comparison of this S-transform with (2) shows that F_n are all zero, except for n=m. F_m is a distribution valued non-L^2 kernel, hence ϕ cannot be in (L^2) .

(2) "**Donsker's δ - function**":

$$\phi = \delta(B(t) - a) \qquad (S\phi)(f) = (2\pi t)^{-\frac{1}{2}} e^{-\frac{1}{2t}(F(t)-a)^2}, \tag{1.9}$$

with $F(t) = \int_0^t f(s) ds$.

- **Exercise 1:** Calculate the S-Transform of Donsker's δ-function equation (1.9)

  a) by using a Fourier representation of the δ-function

  b) using an L^2 basis $\{e_n\}$, so that $\omega(t) = \sum \omega_n e_n(t)$ and

$$d\mu = \prod \frac{d\omega_n}{\sqrt{2\pi}} e^{-\frac{\omega_n^2}{2}}.$$

  (Hint: Use $e_1 = const \cdot 1_{[0,t]}$)

- **Exercise 2:** Verify that the kernel functions F_n of Donsker's δ-function are square integrable, but that $\sum_n \| F_n \|^2 = \infty$.

These two counterexamples indicate at the same time two desirable extensions for the space of White Noise functionals beyond the square integrable ones. To include local powers of White Noise we should admit generalized kernel functions , to include Donsker's δ-function we must relax the summability. of norms. These generalizations are the topic of the next section.

1.2. TEST AND GENERALIZED FUNCTIONALS

1.2.1. Constructing the Space of Test Functionals

A standard procedure for the enlarging of function spaces is through the construction of a Gel'fand triple. We shall first define in (L^2) a suitable, smaller space (S) of test functions. The space $(S)^*$ of continuous linear functionals on (S) is then an extension of (L^2) :

$$(S) \subset (L^2) \subset (S)^* \tag{1.10}$$

We shall speak of "Hida test functions" and of "Hida distributions".

The construction will proceed in analogy of that of the triple of Schwartz test functions and the corresponding space S^* of tempered distributions

$$S \subset L^2 \subset S^*. \tag{1.11}$$

Recall that Schwartz space $S(R^d)$ may be defined as intersection (projective limit)

$$S = \bigcap_p H_p$$

of Hilbert spaces with increasing norms $|\cdot|_p$, such as e.g. in Example 1 below.

For the construction of the corresponding spaces of White Noise functionals we use the concept of "second quantization" Γ, operating on Hilbert spaces H:

$$\Gamma(H) = \bigoplus_{n=0}^{\infty} Sy\, H^{\otimes n}$$

a direct sum over symmetrized direct products with scalar product $(\cdot,\cdot)_{\Gamma(H)}$ and norm $\|\cdot\|_0$ generated by

$$\left(\varphi^{\otimes n}, \psi^{\otimes n}\right)_{\Gamma(H)} = n!\,(\varphi,\psi)_H^n\,.$$

Likewise we introduce, for q=0,1,2,..., Hilbert spaces $\Gamma_q(H_p)$ as closures of the space of finite direct sums

$$\bigoplus_n Sy\, H_p^{\otimes n}$$

with respect to the scalar product generated by

$$\left(\varphi^{\otimes n}, \psi^{\otimes n}\right)_{\Gamma_q(H_p)} = 2^{nq} n! \, (\varphi, \psi)^n_{H_p} \tag{1.12}$$

and denote the corresponding norms by $\|\cdot\|_{p,q}$.

Returning to White Noise Analysis we note that

$$\Gamma(L^2(R)) = \bigoplus_n SyL^2(R^n, n! d^n t)$$

which appears on the rhs of the isomorphism (1.3) as space of kernels (F_n) in the chaos expansion of White Noise functionals. Admitting only those kernel functions that lie in all the $\Gamma_q(H_p)$ we obtain our test function space

$$(S) \equiv \bigcap_{p,q} (S)_{p,q}$$

with

$$(S)_{p,q} \equiv \left\{ \sum_{n=0}^{\infty} <: \omega^{\otimes n} :, F_n > |(F_n) \in \Gamma_q(H_p) \right\}. \tag{1.13}$$

It can be shown that the construction of (S) does not depend on the particular family of norms that is employed in the construction of S.

Example 1. *Let* $|f|_p \equiv |A^p f|_{L^2}$ *, where*

$$A \equiv -d^2/dt^2 + t^2 + 1. \tag{1.14}$$

Then

$$S(R) = \lim_{\overleftarrow{p}} S_p, \text{ with } S_p \equiv D(A^p) \subset L^2(R).$$

$$\left(\varphi^{\otimes n}, \psi^{\otimes n}\right)_{\Gamma_q(S_p)} = 2^{nq} n! \, (\varphi, \psi)^n_{S_p}$$

In this case

$$= 2^{nq} n! \, (A^p\varphi, A^p\psi)^n_{L^2} = n! \, (B^p\varphi, B^p\psi)^n_{L^2}$$

where $B = B_q = 2^{q/2p} A$. *In other terms*

$$\Gamma_q(S_p) = D\left(\Gamma(B^P_q)\right)$$

with the "second quantization" $\Gamma(B)$ *of an operator* B *on* L^2 *defined as follows:*

$$\Gamma(B) = \bigoplus_{n=0}^{\infty} B^{\otimes n}$$

It is not hard to see that in particular

$$(S) \equiv \lim_{\overleftarrow{p}} (S)_p \tag{1.15}$$

where, using the canonical isomorphism (1.3),

$$(S)_p \simeq D(\Gamma(A^p)).$$

Among the useful properties we list:

- (S) is *nuclear*
- (S) is an *algebra* under pointwise multiplication.
- *Complex scaling* given by

$$\sigma_z : \phi \to \phi_z, \text{with } \phi_z(\omega) = \phi(z\omega), \tag{1.16}$$

  extends to a continuous linear transformation on (S) for all complex z.

- The representation of Hida test functions in terms of their kernel functions

$$\phi(\omega) = \sum_{n=0}^{\infty} <: \omega^{\otimes n} :, F_n >$$

  exists not just in an (L^2) sense, but pointwise, *as a continuous function* for all $\omega \in \; S^*(R)$; this is so since the kernel functions F_n are in fact Schwartz test functions.

1.2.2. The dual space $(S)^*$, "Hida Distributions"

For $\Phi \in (S)^*$ and $\phi \in (S)$ we denote the bilinear dual product $(S)^* \times (S) \to C$ by $<<\Phi, \phi>>$.

- Since (S) is nuclear, all distributions are of *finite order*:

$$(S)^* = \bigcup_p (S)_{-p} = \bigcup_{p,q} (S)_{-p,-q}$$

- Expectations extend continuously to distributions $\Phi \in (S)^*$:

$$E(\Phi) \equiv << \Phi, 1 >> \tag{1.17}$$

- Positive Hida distributions are **measures.**
- **T**$-$ and **S-transform** extend to $(S)^*$ since

$$(S\Phi)(f) = E(\Phi \cdot : \exp(< ., f >) :) \tag{1.18}$$

$$(T\Phi)(f) = E(\Phi \cdot \exp(i < ., f >)).$$

  are nothing but the application of the distribution Φ to the exponentials which are test functionals; this is the case not only for real test fuctions $f \in S(R)$, but extends to complex $f \in S_C(R)$.

These transforms provide a handy characterization of the space of generalized White Noise functionals. In particular they can easily be shown to fall within the class described in the following

Definition 2. : *A mapping $G : S(R) \to C$ is called a "***U-functional***" if $G(zf_1 + f_2)$ is "ray-analytic", i.e. entire in z for any pair of test functions $f_i \in S(R)$, and*

$$| G(zf) | \leq C_1 \exp(C_2 |z|^2 |f|_p^2) \; for \; some \;\; C_i \, , \, p > 0. \qquad (1.19)$$

Actually the U-functionals furnish a complete description of generalized white noise functionals:

Theorem 1. *The following are equivalent:*
(1) G is a U-functional
(2) G is the S-transform of a (unique) Hida distribution $\Phi \in (S)^*$
(3) G is the T-transform of a (unique) Hida distribution $\hat{\Phi} \in (S)^*$.

Example 2. **"Donsker's δ - function"**:

$$\phi = \delta(B(t) - a)$$

has

$$(S\phi)(f) = (2\pi t)^{-\frac{1}{2}} e^{-\frac{1}{2t}(F(t)-a)^2}$$

which evidently is a U-functional.

Example 3. **"Gauss Kernels"**

$$\phi(\omega) \equiv N \exp < \omega, K\omega > .$$

The normalization is fixed such that

$$(S\phi)(f) = e^{<f, \frac{K}{1-2K} f>} \qquad (1.20)$$

which amounts to

$$\phi(\omega)" = " \frac{e^{<\omega, K\omega>}}{E(e^{<\omega, K\omega>})}. \qquad (1.21)$$

However the point here is that in fact the existence (separately) of numerator and denominator in (1.21) is not needed for the above S-transform (1.20) to exist, i.e. the latter defines the Gauss kernel as a Hida distribution for a much larger class of K.

Example 4. "Evaluation Map" : *For* $\omega \in S^*(R)$, *define* δ_ω *by*

$$(T\delta_\omega)(f) = e^{i<\omega,f>}$$

Again the rhs is a U-functional and thus defines a Hida distribution δ_ω *by the characterization theorem, with the property*

$$<< \delta_\omega, \phi >>= \phi(\omega).$$

Various generalizations of the characterization theorem have been studied.

1. One may generalize the Gel'fand triplet and its characterization by using - instead of White Noise - more general Gaussian spaces as a starting point. Examples of interest are, among others, vector valued and multi-parameter White Noise.

2. Also one may relax or omit the growth condition (1.19) to obtain other triples [13], [20].

We further note that the characterization theorem implies the existence of two algebraic structures on $(S)^*$ since the U-functionals form an algebra under pointwise multiplication, namely a convolution:

$$T(\Phi) \cdot T(\Psi) = T(\Phi * \Psi)$$

and the "normal ordered product":

$$S(\Phi) \cdot S(\Psi) = S(\Phi \diamond \Psi)$$

with e.g. .

$$< \omega, f > \diamond < \omega, f >=:< \omega, f >^2: .$$

1.3. CALCULUS

As one might expect, White Noise test functions are arbitrarily often differentiable: the **Gateaux derivative**

$$D_h\phi(\omega) = \lim_{e\to 0} \frac{1}{e}(\phi(\omega + h) - \phi(\omega)) \tag{1.22}$$

maps (S) into (S) for all, even distribution valued directions $h \in S^*(R)$ and obeys the **chain rule**: if g$\in C^1(R)$ such that g$o\phi \in (S)$, then

$$D_h(g \circ \phi) = g' \circ \phi \cdot D_h\phi.$$

Likewise the **product rule** holds:

$$D_h(\psi \cdot \phi) = \psi \cdot D_h\phi + D_h\psi \cdot \phi.$$

We remark that the equivalent of the Gateaux derivatives under the isomorphism (1.3) are Fock space annihilation operators

$$D_h \simeq a(h).$$

For the particular case where h is the Dirac distribution δ_t, write $D = \partial_t$ (**"Hida derivative"**).
Furthermore, (S) is **Fréchet differentiable**, and

$$\nabla : (S) \to S(R) \otimes (S)$$

is given by

$$(\nabla\phi)(t,\omega) = \partial_t\phi(\omega)\ .$$

and we have

$$D_h\phi = < h, \nabla\phi > .$$

For later use we also note that

$$|\nabla\phi|^2 = \int dt\, |\partial_t\phi|^2 \in (S)\ for\ all\ \phi \in (S).$$

Infinite dimensional **Laplacians** have been defined in various ways -
Gross Laplacian:

$$\Delta_G = \int \partial_t\partial_t dt$$

Laplace Beltrami operator (Number operator in Fock space):

$$\Delta = -\int \partial_t^\dagger \partial_t dt = -N$$

For the **Lévy Laplacian** see the lecture by T. Hida in this volume.

So much for calculus on test function space. On the distribution space there is the adjoint of the derivative as a continuous operator

$$D_h^\dagger : (S)^* \to (S)^*.$$

It is related to the derivative D_h by

$$D_h^\dagger \supset -D_h + < \omega, h > \tag{1.23}$$

and corresponds in Fock space to a creation operator. One sees directly from (1.23) that the derivative and its adjoint obey canonical commutation relations

$$[D_g, D_h^\dagger] \subset < g, h > .$$

To control the **convergence** of sequences of Hida distributions we can employ their characterization in terms of S- or T-transform:

Theorem 2. *Let $\{G_n\}_{n\in N}$ denote a sequence of U-functionals with the following properties:*

1. *For all $f \in S(R)$, $\{G_n(f)\}_{n\in N}$ is a Cauchy sequence,*
2. *There exist C_i, p such that*

$$|G_n(zf)| \leq C_1 \exp(C_2 |z|^2 |f|_p^2)$$

uniformly in n.

Then there are unique $\Phi, \hat{\Phi} \in (S)^*$ such that $S^{-1}G_n$ and $T^{-1}F_n$ converges strongly to Φ and $\hat{\Phi}$.

Similarly we can control the integration of families of Hida distributions with respect to a parameter as follows:

Theorem 3. *Let (Ω, B, m) be a measure space, and Φ_λ a mapping defined on Ω with values in $(S)^*$. We assume that the T-transform of Φ_λ*
(1) is an m-measurable function of λ for any test function f
(2) obeys a U-functional estimate

$$| (T\Phi_\lambda)(zf) | \leq C_1(\lambda) \exp(C_2(\lambda) | z |^2 \| f \|_p^2)$$

for some fixed p and for $C_1 \in L^1(m), C_2 \in L^\infty(m)$

Then Φ_λ is (Bochner-) integrable in the Hilbert space $(S)_{-q}$ for q large enough,

$$\int_\Omega dm(\lambda) \Phi_\lambda \in (S)^* \tag{1.24}$$

and

$$T(\int_\Omega dm(\lambda) \Phi_\lambda)(f) = \int_\Omega dm(\lambda)(T(\Phi_\lambda)(f)). \tag{1.25}$$

Example 5. *A Fourier representation for Donsker's δ-function:*

$$\delta(< \omega, g > -a) \equiv \frac{1}{2\pi} \int d\lambda e^{i\lambda(<\omega,g>-a)} \tag{1.26}$$

Exercise: Verify that the rhs is a well-defined Bochner integral and calculate its T-transform.

Stochastic integration with respect to Brownian motion $\int dB(t)$ can be introduced as follows. Formally

$$\begin{aligned}\int dB(t)\Phi(t) &= \int dt\, \dot{B}(t)\Phi(t)\\ &= \int dt\omega(t)\Phi(t)\\ &= \int dt(\partial_t + \partial_t^\dagger)\Phi(t).\end{aligned}$$

For (Bochner-) integrable $\partial_t^\dagger\Phi(t)$ e.g., one can define the **Hitsuda-Skorochod Integral** :

$$I \equiv \int dt\partial_t^\dagger\Phi(t).$$

The Itô integral is a special case resulting when we insist that Φ be non-anticipating.

There is no immediate generalization of the Fourier transform

$$F : f(t) \to (2\pi)^{-n/2}\int\cdots\int d^n t\, e^{i(s,t)} f(t)$$

to the case of infinite dimensional analysis where n=∞. However an **infinite dimensional Fourier transform** can be introduced [16] with the help of S- and T-transform

$$F \equiv T^{-1}S.$$

Clearly

$$F : (S)^* \to (S)^*,$$

more importantly this definition is - up to multiples - the unique continuous linear operation on (S)* that intertwines coordinate multiplication by $<\omega, h>$ and differentiation D_h as a Fourier transform should.

Exercise: The evaluation map at zero is given as

$$\delta_0 : \phi \to \phi(0).$$

Calculate its Fourier transform.
Exercise: Show that

$$F^2\Phi(\omega) = \Phi(-\omega). \tag{1.27}$$

Hint: Use

$$T\Phi(f) = e^{-\frac{1}{2}\int f^2 dt}\cdot S\Phi(if).$$

2. Applications of White Noise Analysis

2.1. MORE ON DONSKER'S δ-FUNCTION

Recalling the S-transform of Donsker's δ–function

$$\phi = \delta(B(t) - a) \;\rightarrow\; (S\phi)(f) = (2\pi t)^{-\frac{1}{2}} e^{-\frac{1}{2t}(F(t)-a)^2}$$

one notes that the rhs is a U-functional not only for real argument but for all *complex* a.

Theorem 1. *For any g∈ $L^2(R)$ and any complex z there exists a Hida distribution $\delta_{g,z}$ given by*

$$S : \delta(<\omega, g> - z) \rightarrow (2\pi|g|^2)^{-\frac{1}{2}} e^{-\frac{1}{2|g|^2}((f,g)-z)^2}$$

Theorem 2. *1. Donsker's δ-function is homogeneous of order -1:*

$$For \quad \lambda > 0, \qquad \delta_{\lambda g,z} = \frac{1}{\lambda}\delta_{g,z/\lambda}$$

2. $\delta_{\lambda g,z}$ is obtained from $\delta_{g,z}$ by complex scaling :
There are δ-sequences $\delta^{(n)} \in (S)$ such that for $|\arg\lambda| < \pi/2$

$$\delta^{(n)}_{g,z} \rightarrow \delta_{g,z} \textbf{and } \sigma_\lambda \delta^{(n)}_{g,z} \rightarrow \delta_{\lambda g,z} \equiv \frac{1}{\lambda}\delta_{g,z/\lambda}.$$

3. $\lambda \rightarrow \delta_{\lambda g,z}$ is continuous from $C - \{0\}$ to $(S)^*$.

Proof: 1. follows from the characterization theorem. 2. can be read off directly from the S-transform. For 3. use

$$\frac{1}{2\pi}\int_{-ne^{-i\alpha}}^{ne^{-i\alpha}} e^{i\lambda(\langle\omega,\eta_n\rangle - a)} d\lambda \;=\; \frac{1}{\pi(\langle\omega,\eta_n\rangle - a)} \sin\left(ne^{-i\alpha}\left(\langle\omega,\eta_n\rangle - a\right)\right) = \delta^{(n)}_{g,a}(\omega)\,.$$

and control this by its S-transform:

$$S\sigma_\lambda\delta^{(n)}_{g,a}(f) = \frac{1}{2\pi}\int_{-n}^{n} \exp\left(-\frac{1}{2}(\lambda e^{i\alpha}s)^2\,|g|^2 + ie^{i\alpha}s\,(\lambda\,(f,g) - a)\right) ds\;.$$

For z=1 and $|\alpha| < \pi/4$ this

1. is well defined and

2. S-transform of a test function.

3. The same is true $\forall\lambda$ such that $|arg\lambda e^{i\alpha}| < \pi/4$.

4. As $n \to \infty$, have for such λ, α :

$$\sigma_\lambda \delta_{g,a}^{(n)} \to \delta_{\lambda g,a}.$$

Products of Donsker δ–functions appear in various contexts. For example, in quantum theory studies of the **Dyson series** (below) use

$$\prod_{l=1}^{n} \delta(B(t_l) - x_l).$$

The moments of **intersection local times** can be expressed by Tanaka's formula as

$$\int dt E(\prod_{l=1}^{n} \delta(B(t_{l_1}) - B(t_{l_2})))$$

and play a role in polymer models.

In either case the principal object to study is

$$\Phi = \prod_{j=1}^{n} \delta\left(\langle\cdot, g_j\rangle - a_j\right)$$

Theorem 3. *Let $g_i \in L^2$ be linearly independent, with Gram matrix $(G)_{k,l} = (g_k, g_l)$. Then*

$$\Phi = \prod_{j=1}^{n} \delta\left(\langle\cdot, g_j\rangle - a_j\right), \quad a_i \in C$$

has the T-transform

$$T\Phi(f) = ((2\pi)^n \det G)^{-\frac{1}{2}} \exp\left(\left(\overrightarrow{g}, f\right) + i\overrightarrow{a}, G^{-1}\left(\left(\overrightarrow{g}, f\right) + i\overrightarrow{a}\right)\right).$$

In particular

$$\Phi \in (S)^*.$$

Proof: Use

$$\prod_{j=1}^{n} \delta\left(\langle\cdot, g_j\rangle - a_j\right) = \frac{1}{(2\pi)^n} \int \prod_{j=1}^{n} \exp\left(i\lambda\left(\langle\cdot, g_j\rangle - a_j\right)\right) d^n\lambda.$$

2.2.

THE FEYNMAN INTEGRAL.

Feynman's well known ansatz

$$< F >= \int d^{\infty}x(\tau)\exp(\frac{i}{h}\int L(\dot{x},x,\tau)d\tau)F[x] \qquad (2.1)$$

in terms of the classical action functional

$$S[x] = \int L(\dot{x},x,\tau)d\tau = S_0[x] + S_1[x] = \frac{m}{2}\int \dot{x}^2(\tau)d\tau - \int V(x)d\tau \qquad (2.2)$$

has been as inspiring for quantum physics as it has been elusive under the scrutiny of mathematicians. For lack of a well defined notion of integrals $\int d^{\infty}x(\tau)$, one has thought of it as infinite limit of well defined integrals, as analytic continuation of well defined ones, etc.

Here we shall argue essentially three things:

1. The Feynman integral - for a large class of models - is the action of a Hida distribution on F.
2. This Hida distribution can be constructed explicitly along the lines suggested originally by Feynman, as a sum over histories.
3. Once constructed within this framework, the Feynman integral is then rigorously amenable to the manipulations available in White Noise Analysis which can serve to replace the artful but often tricky and ambiguous arguments in earlier treatments.

Let us now undertake this construction. An averaging over Brownian paths:

$$x(t) \equiv x_0 + \sqrt{\frac{\hbar}{m}}\int_0^t \omega(\tau)\, d\tau$$

would formally produce

$$e^{\frac{i}{\hbar}S_0} = e^{\frac{i}{2}\int \omega^2(\tau)d\tau}.$$

Averaging over such paths would however amount to integration with respect to the Gaussian White Noise measure $d\mu$

$$d\mu = N^{-1}e^{\frac{-1}{2}\int \omega^2(\tau)d\tau}d^{\infty}\omega.$$

instead of the (nonexisting) "flat" measure $d^\infty\omega$ proposed by Feynman.

Hence we shall attempt to cancel this Gaussian "density" by including its inverse in the Feynman integrand

$$I_0 \equiv N\exp(\frac{i+1}{2}\int \omega^2(\tau)d\tau) \tag{2.3}$$

with T-transform

$$TI_0(f) \equiv \exp(\frac{-i}{2}\int f^2(\tau)d\tau),$$

where we have used the normalized exponential as defined in Example 3.

To describe the propagation of a quantum particle from x_0 to the point x we fix the final point x of our paths through a Donsker δ-function, i.e. consider $I_0\delta(x(t)-x)$. Using e.g. the Fourier representation (1.26) of the δ-function it is easy to calculate

$$T(I_0\delta)(f) = \frac{1}{\sqrt{2\pi it}}\exp(\frac{-i}{2}\int f^2(s)ds + \frac{i}{2t}(\int_0^t f(s)ds + x - x_o)^2) \tag{2.4}$$

By the characterization theorem this clearly is the T-transform of a Hida distribution, hence well defined mathematically. Moreover it has the right physical properties

The expectation $E(I_0\delta) \equiv T(I_0\delta)(0)$ is the propagator for a free quantum mechanical particle,

$$K_0(x,t;x_0,0) = \frac{1}{\sqrt{2\pi it}}\exp(\frac{i}{2t}(x-x_o)^2)$$

solving

$$i\partial_t K_0(x,t;x_0,o) = H_0 K_0(x,t;x_0,o)$$

with boundary condition

$$K_0(x,t;x_0,o)\mid_{t=+0} = \delta(x-x_0).$$

In conclusion: admitting an averaging not with respect to a measure but instead with respect to a distribution, we were able to rigorously construct the latter along the lines suggested by Feynman as a sum over histories. An alternative definition would have been what is now stated as

Proposition 1. *Let $z^2 = i$. Then*

$$I_0 = \sigma_z^\dagger 1.$$

Proof: Comparison of T-transforms.

In other words, Feynman integrals formally given by

$$< F >= \int d^{\infty}x(\tau)\exp(\frac{i}{h}\int \dot{x}^{2}(\tau)d\tau)F[x]$$

allow the following mathematically rigorous

Definition 3. *Let $F \in (S)$, $z^2 = i$. Then the Feynman integral of F is defined as*

$$< F >= E(\sigma_z^{\dagger} 1 \cdot F) = E(\sigma_z F)$$

Note:

1. Heuristically, the role of σ_z is to pass from the real correlations of the heat equations to the complex ones of quantum mechanics.
2. Mathematically the question of admissible interactions and matrix elements is that of *extension of integrands beyond $F \in (S)$.*
3. Complex scaling not new to mathematical theory of Feynman integrals; see e.g. Doss [6]

   $$\psi(x,t) = E(g(x + \sqrt{i}B(t))\exp(-i\int_0^t V(x + \sqrt{i}B(s))ds)),$$

   if f and V satisfy suitable analyticity and growth conditions.

The point of invoking White Noise Analysis here is that we may, instead of analytic functions, consider *limits of analytic functions,* which may well fail to be random variables, as long as they are Hida distributions. We expose this in the following

Theorem 4. *Let F_n be a sequence of Hida test functions. Then the following are equivalent:*

1. *the sequence $I_o F_n$ converges to a distribution $I \in (S)^*$*
1. *the sequence $\sigma_{\sqrt{i}} F_n$ converges to a distribution in $(S)^*$*
1. *the sequence of expectations $E(\sigma_{\sqrt{i}} F_n \cdot \psi)$ converges for all $\psi \in (S)$.*

If one of the conditions holds, the action of I is given by

$$< I, \psi >= \lim_n E(\sigma_{\sqrt{i}}(F_n \cdot \psi)).$$

Example 6. *As stated above, the quantum mechanical propagator is obtained as expectation of I_0 times a Donsker δ, this multiplication is justified since by Theorem 1 there are $\delta-$sequences that remain convergent after complex scaling:*

$$\sigma_z \delta^{(n)}_{g,a} \to \delta_{zg,a} \in (S)^*.$$

Next we turn to a discussion of admissible interaction terms

$$\exp(-i \int_0^t V(x_0 + B(s))ds).$$

The following class of potentials is amenable to a perturbative treatment.

Theorem 5. *[17] Consider a potential given by a signed measure $v(x,t)$, let v_x denote the marginal measure*

$$v_x \left(A \subset B \left(R \right) \right) \equiv v \left(A \times \Delta \right)$$

and similarly

$$v_t \left(B \subset B \left(\Delta \right) \right) \equiv v \left(R \times B \right).$$

Assume that $|\nu|$ is finite and that v_x and v_t are such that
i) $\exists R > 0 \; \forall r > R : |v|_x (\{x : \; |x| > r\}) < e^{-\beta r^2}$ for some $\beta > 0$,
ii) $|v|_t$ has a L^∞ density.
Then

$$\begin{aligned} I = I_0 \delta \left(x \left(t \right) - x \right) + \sum_{n=1}^{\infty} \left(-i \right)^n \textstyle\int_{R^n} \int_{\Delta_n} \prod_{i=1}^{n} v \left(dx_i, dt_i \right) \\ \times I_0 \delta \left(x \left(t \right) - x \right) \prod_{j=1}^{n} \delta \left(x \left(t_j \right) - x_j \right) \in \left(S \right)^* \end{aligned} \tag{2.5}$$

and $E(I) = K\left(x, t | x_0, t_0\right)$ is the quantum mechanical propagator for the potential v:

$$i \, \partial_t \, K \left(x, t | x_0, t_0 \right) = \left(-\frac{1}{2} \partial_x^2 + v \right) K \left(x, t | x_0, t_0 \right), \; for \; t > t_0.$$

Proof: Use the characterization theorem and its corollaries to control the integrations and infinite sum in (2.5).

2.3. QUANTUM FIELD THEORY: EQUILIBRIUM STATES, DIRICHLET FORMS, τ-FUNCTIONALS

2.3.1. *Ground State Measures*

Constructive quantum field theory, in essence, is the construction of measures suitable to describe the expectations of quantum fields A(f):

$$< e^{iA(f)} > \stackrel{!}{=} \int e^{i\omega(f)} d\nu(\omega)$$

An obvious example is furnished by the expectation of free time zero fields

$$< e^{iA(f)} >= e^{-\frac{1}{2}(f,(-\Delta+m^2)^{-1/2}f)}$$

and of Euclidean free fields:

$$< e^{iA(f)} >= e^{-\frac{1}{2}(f,(-\Delta+m^2)^{-1}f)}.$$

In either case the rhs are U-functionals, hence the corresponding measures are Hida distributions: $\nu \in (S)^*$.

The same is true for the commonly studied bosonic models for **interacting fields** - $P(\varphi)_2$, Sine-Gordon, Hoegh-Krohn models.

To prove this one uses moment inequalities (Fröhlich bounds) established in constructive quantum field theory which are sufficient to produce the analyticity and bounds of the characterization theorem [1] [22].

2.3.2. *Quantum (field) dynamics in terms of the ground state*

Schrödinger Hamiltonians $H = -\frac{1}{2}\Delta + V$ are well known to have an alternate definition

$$\varepsilon(\psi, \psi) \equiv < \psi, H\psi >= \int (\nabla\psi)^2 \rho(x) dx \tag{2.6}$$

where ρ the ground state density; with states ψ in the Hilbert space $L^2(\rho(x)dx)$ (**ground state representation**). Mathematically, the rhs of (2.6) defines the Hamiltonian as a **local Dirichlet form.**

This Dirichlet form dynamics is well defined even for interactions which are too singular to be handled perturbatively, hence should be particularly suitable for canonical QFT, see e.g. [2]. In terms of ground state measures $\nu \in (S)^*$ one is led to the ansatz

$$\varepsilon(\psi, \psi) \equiv << v, |\nabla\psi|^2 >>$$

with $D(\varepsilon) = (S) \subset L^2(d\nu)$.

Closability of this form is as in the finite dimensional case, a condition on the admissible measures. the characteristic contraction property of Dirichlet forms (Markov property) holds whenever $v > 0$ on all open sets of S* [21].

2.3.3. *Time Ordered Expectation Values*

In quantum field theory there corresponds to the Feynman integral the "τ-functional", generating functional of the time-ordered vacuum expectation values. For a scalar field A this would take the form

$$\tau\{j\} =< 0|Te^{i\int A(x)j(x)dx}|0 >= \sum_n \frac{i^n}{n!} \int dx_1...dx_n j(x_1)...j(x_n)\tau(x_1...x_n)$$

with

$$\tau(x_1...x_n) =< 0|TA(x_1)...A(x_n)|0 >=< 0|A(x_{\sigma(1)})...A(x_{\sigma(n)})|0 > .$$

Here $x = (x^0...x^s) \in R^{s+1}$, with scalar product xy=$x^0y^0 - \sum_1^s x^iy^i$, and x_0 the time component. σ is the permutation which provides time-ordering, i.e. for which $x^0_{\sigma(k)} \geq x^0_{\sigma(k-1)}$ for all k. Correspondingly we shall need to introduce s+1-parameter White Noise $\omega = \omega(x)$. Concepts such as e.g. the Hida derivative $\partial = \partial_x$, and results such as the characterization theorem 1 remain essentially unaltered [23] [14], and one has e.g.

Theorem 6. *If A is a relativistic free field of mass m> 0 then $\tau\{\cdot\}$is a U-functional. I. e. there is a Hida distribution Θ such that$< \Theta, e^{i<\omega,j>} >= \tau\{j\}$*

Proof: From the explicit expression

$$\tau\{j\} = \exp(-\frac{i}{2}\int dx \int dy j(x)\Delta_g(x-y)j(y)) \tag{2.7}$$

with

$$\Delta_g(x) = \frac{1}{(2\pi)^4}\int d^{s+1}p \frac{e^{-ipx}}{p^2 - m^2 + i\epsilon}.$$

one sees that $\tau\{\lambda j_1 + j_2\}$ extends from real λ to the whole complex plane as an analytic function. An estimate for τ as in equation (1.19) follows from

$$|\int dx \int dy j(x)\Delta_g(x-y)j(y))| \leq |\langle \Delta_g, j * j\rangle| \leq c|A^p j|^2$$

for suitable constants c and p.□

For this simple case Schwinger's functional differential equation is of course linear. On the level of the τ-functional one obtains directly from (2.7) that

$$(i(\Box + m^2)\frac{\delta}{\delta j(x)} + j(x))\tau\{j\} = 0,$$

which translates into the following equation for the Hida distribution $\Theta = \Theta_0$:

$$(\partial_x^\dagger - i(\Box + m^2)\omega(x))\Theta_0 = 0 \tag{2.8}$$

since the T-transform intertwines multiplication and derivatives as follows:

$$\omega(x) \overset{T}{\leftrightarrow} -i\cdot\frac{\delta}{\delta j(x)} \text{ and } \partial_x^\dagger \overset{T}{\leftrightarrow} i\cdot j(x).$$

Formally, or invoking regularizations, interactions would be introduced through an additional action functional S_1 in

$$\Theta = \Theta_0 e^{iS_1}$$

so that

$$(\partial_x^\dagger - i(\Box + m^2)\omega(x) - i\cdot\partial_x S_1)\Theta = 0,$$

work is in progress regarding these models.

3. Supplements

(by Ju. G. Kondratiev and L. Streit)

3.1. AN EXTENSION OF $(S)^*$.

The space $(S)^*$ is too small for some interesting applications. We have already mentioned some possible extensions [13] [20]; here we want to sketch the construction of an even wider one which however still admits the useful tool of S-transform.

As in Section 1.2.1 we construct test functionals by restricting the space

$$\Gamma(H) = \bigoplus_{n=0}^{\infty} Sy\, H^{\otimes n}$$

of kernel functions.

Likewise we introduce, for q,p=0,1,2,..., Hilbert spaces $\Gamma_q^1(H_p)$ as closures of the space of finite Fock sequences in

$$\bigoplus_n Sy\, H_p^{\otimes n}$$

with respect to the scalar product generated by

$$\left(\varphi^{\otimes n}, \psi^{\otimes n}\right)_{\Gamma_q^1(H_p)} = 2^{nq}\,(n!)^2\,(\varphi,\psi)_{H_p}^n\,.$$

In comparison with equation (1.12) note the square of $n!$ which produces a smaller space of test functionals via

$$(S)^1 \equiv \bigcap_{p,q} (S)_{p,q}^1 \subset (S)$$

where

$$(S)_{p,q}^1 \simeq \Gamma_q^1(H_p)$$

analogously to (1.13). I. e.

$$(S)^1 \subset (S) \subset (L^2) \subset (S)^* \subset (S)^{-1}.$$

As in the case of (S) one shows that

- $(S)^1$ is a nuclear space
- $(S)^1$ is an algebra under pointwise addition and multiplication
- $(S)^1$ admits Gateaux and Fréchet derivatives of arbitrary order.

Contrary to the case of Hida test functions however, exponentials are test functions of finite order only. More precisely

$$: \exp\left(\langle \cdot, f\rangle\right) : \in (S)^1_{p,q} \text{ iff } 2^q \left|f\right|^2_{H_p} < 1, \tag{3.1}$$

and likewise for complex $f \in S_c(R)$.

Therefore the space $(S)^1$ does not contain nontrivial exponentials.

Let us denote the dual space of $(S)^1_{p,q}$ and $(S)^1$ by $(S)^{-1}_{-p,-q}$ and $(S)^{-1}$. Then we have again

$$(S)^{-1} = \bigcup_{p,q} (S)^{-1}_{-p,-q}.$$

Because of this we may still define the S-transform on $(S)^{-1}$. For a given $\Phi \in (S)^{-1}$ we choose p,q such that $\Phi \in (S)^{-1}_{-p,-q}$. Then, for any f as in (3.1) we may use the dual product to define

$$(S\Phi)(f) = \langle\langle \Phi, : \exp\left(\langle \cdot, f\rangle\right) : \rangle\rangle .$$

Let us denote by $Hol_0(S_c)$ the algebra of (germs of) functions which are analytic at the point $0 \in S_c(R)$ and locally bounded as follows

Definition 4. *$G \in Hol_0(S_c)$ iff there exist $\varepsilon > 0$ and $p \geq 0$ such that*

1. *$\forall \zeta_0 \in S_c(R)$ with $|\zeta_0|_p < \varepsilon$, and $\forall \zeta \in S_c(R)$, the function*

$$C \ni z \rightarrow G(z\zeta + \zeta_0)$$

is analytic at $z = 0$.

2. *$\exists c > 0$ such that for all $\zeta \in S_c(R)$ with $|\zeta|_p < \varepsilon$*

$$|G(\zeta)| < c.$$

This class of functions descibes the distributions in $(S)^{-1}$.

Theorem 1. *The following are equivalent:*

1. *$G \in Hol_0(S_c)$*

2. *G is the S-transform of a unique distribution $\Phi \in (S)^{-1}$.*

3.2. WICK CALCULUS

Since $Hol_0(S_c)$ is an algebra under pointwise addition and multiplication, Theorem 1 induces an algebraic structure on $(S)^{-1}$. For $\Phi, \Psi \in (S)^{-1}$ we denote their Wick product by

$$\Phi \diamond \Psi = S^{-1}\left(S(\Phi)S(\Psi)\right) \in (S)^{-1}.$$

It is not hard to rewrite the product in terms of kernel functions since in particular one has

$$<: \omega^{\otimes n} :, F_n > \diamond <: \omega^{\otimes m} :, G_m > = <: \omega^{\otimes(n+m)} :, F_n \hat{\otimes} G_m > .$$

Likewise we introduce Wick powers $\Phi^{\diamond n}$ by

$$S(\Phi^{\diamond n})(f) = (S(\Phi)(f))^n$$

More general Wick compositions are obtained from analytic functions

$$L(z) = \sum a_n z^n$$

by setting

$$L^{\diamond}(\Phi) = \sum a_n \Phi^{\diamond n}.$$

Regarding the existence of such Wick compositions we have the following

Theorem 2. *Let $\Phi \in (S)^{-1}$ and L be analytic at the point $z_0 = E(\Phi) \in C$. Then there exists the generalized functional $L^{\diamond}(\Phi) \in (S)^{-1}$ with S-transform*

$$S\left(L^{\diamond}(\Phi)\right) = L\left(S(\Phi)\right). \tag{3.2}$$

Let us remark that the definition of $L^{\diamond}(\Phi)$ via equation (3.2) is correct:

$$S\left(L^{\diamond}(\Phi)\right)(0) = L\left(S(\Phi)(0)\right) = L\left(E(\Phi)\right)$$

and, under the conditions stated in Theorem 2, $L\left(S(\Phi)\right) \in Hol_0(S_c)$.

Example 7. *Let $\Phi \in (S)^{-1}$ such that $E(\Phi) \neq 0$. Then we can introduce the distribution $\Phi^{\diamond(-1)}$ by its S-transform*

$$S\left(\Phi^{\diamond(-1)}\right) = 1/S(\Phi) \in Hol_0(S_c).$$

It allows us to solve equations

$$A \diamond X = B$$

if $A, B \in (S)^{-1}$ and $E(A) \neq 0$ by

$$X = B \diamond A^{\diamond(-1)} \in (S)^{-1}.$$

Exercise: Let $\Phi(\omega) =< \omega, g >^2$. Discuss its inverse $\Phi^{\diamond(-1)}$.

Example 8. *For any $\Phi \in (S)^{-1}$ there exists a distribution $\exp^\diamond(\Phi) \in (S)^{-1}$ which may be defined through its S-transform*

$$S(\exp^\diamond(\Phi))(f) = \exp(S(\Phi)(f))$$

or, equivalently, by

$$\exp^\diamond(\Phi) = \sum_{n=0}^{\infty} \frac{1}{n!} \Phi^{\diamond n}.$$

Example 9. *If $\Phi \in (S)^{-1}$ has a positive expectation $E(\Phi) > 0$, then we can introduce the "Wick logarithm"*

$$\log^\diamond(\Phi) = S^{-1}(\log S(\Phi)).$$

It solves the equation in $(S)^{-1}$

$$\exp^\diamond(X) = A$$

if $A \in (S)^{-1}$ with $E(\Phi) > 0$, by

$$X = \log^\diamond(\Phi) \in (S)^{-1}.$$

Exercise: Verify the functional equations

$$(\sin^\diamond(\Phi))^{\diamond 2} + (\cos^\diamond(\Phi))^{\diamond 2} = 1,$$

as well as

$$\exp^\diamond(\Phi) \diamond \exp^\diamond(\Psi) = \exp^\diamond(\Phi + \Psi)$$

and

$$\log^\diamond(\Phi) + \log^\diamond(\Psi) = \log^\diamond(\Phi \diamond \Psi).$$

For more on the Wick calculus in the general setting of Gaussian analysis see [15].

References

[1] ALBEVERIO, S., HIDA, T., POTTHOFF, J., RÖCKNER, M. and STREIT, L.: "Dirichlet forms in terms of White Noise Analysis:

I - Construction and QFT examples". *Rev. Math. Phys.* **1** (1990) 291 - 312.

"II - Construction of infinite dimensional diffusions." *Rev. Math. Phys.* **1** (1990) 313 - 312.

[2] ALBEVERIO, S., HØEGH-KROHN, R.: "Quasi-invariant measures, symmetric diffusion processes and quantum fields." In: *Les Méthodes Mathématiques de la Théorie Quantique des Champs, colloques Internacionaux du C.N.R.S.* **248**

[3] ALBEVERIO, S., KONDRATIEV, Ju. G. and STREIT, L.: "How to generalize White Noise Analysis to non-Gaussian measures." Bielefeld preprint (1992), to appear in Proc. IX. Bielefeld Encounters in Mathematics and Physics.

[4] CHOW, P. L.: "Generalized solution of some parabolic equations with a random drift." *J. Appl. Math. Optimization* **20** (1989) 81 - 96.

[5] COCHRAN, C. and POTTHOFF, J.: "Fixed point principles for stochastic partial differential equations." In "*Dynamics of Complex and Irregular Systems*" (Ph. Blanchard et al., eds.). World Scientific, 1993.

[6] DOSS, H.: Sur une resolution stochastique de l'équation de Schrödinger à coéfficients anaalytiques. *Comm. math. Phys.* **73** (1980) 247-264.

[7] de FARIA, M., POTTHOFF, J., and STREIT, L.: "The Feynman integrand as a Hida distribution." *J. Math. Phys.* **32** (1991) 2123 - 2127.

[8] HIDA, T.: "Infinite-dimensional rotation group and unitary group." Lecture Notes in Mathematics 1379. Springer. 1989. pp. 125-134.

[9] HIDA, T., KUO, H.-H., POTTHOFF, J. and STREIT, L.: *White Noise - an Infinite Dimensional Calculus*. Kluwer-Academic, 1993.

[10] HIDA, T., POTTHOFF, J. and STREIT, L.: "Dirichlet forms and white noise analysis." *Comm. math. Phys.* **116** (1988) 235 - 245.

[11] HIDA, T. and STREIT, L.: "Generalized Brownian functionals and the Feynman integral." *Stoch. Proc. Appl.* **16** (1983) 55 - 69.

[12] KHANDEKAR, D. C. and STREIT, L.: "Constructing the Feynman integrand." *Ann. Physik* **1** (1992) 49.

[13] KONDRATIEV, Ju. G. and STREIT, L.: "Spaces of white noise distributions. Constructions, descriptions, applications I." *Rep. Math. Phys.* **33** (1993) 341-366.

[14] KONDRATIEV, Ju. G., LEUKERT, P., POTTHOFF, J., STREIT, L., WESTERKAMP, W.: "Generalized Functionals in Gaussian Spaces - the Characterization Theorem Revisited". Univ. Mannheim preprint, 1994.

[15] KONDRATIEV, Ju. G., LEUKERT, P., and STREIT, L.: "Wick Calculus in Gaussian Analysis". Preprint 1994.

[16] KUO, H. - H.: "The Fourier transform in white noise calculus." *J. Multivariate Anal.* **31** (1989) 311 - 327.

[17] LASCHECK, A., LEUKERT, P., STREIT, L. and WESTERKAMP, W.: "Quantum mechanical propagators in terms of Hida distributions." *Rep. Math. Phys.* **33** (1993) 221-232.

[18] LASCHECK, A., LEUKERT, P., STREIT, L. and WESTERKAMP, W.: "More about Donsker's Delta Function". Preprint 1993, to appear in *Soochow Math. J.* (1994).

[19] LINDSTRØM, T., ØKSENDAL, B. and UBØE: "Dynamical systems in random media: a white noise functional approach." Preprint. Oslo. 1990.

[20] MEYER, P. A. and YAN, J. - A.: "Les fonctions caracteristiques" des distributions sur l'espace de Wiener." *Strasbourg preprint,* 1991.

[21] POTTHOFF, J. and RÖCKNER, M.: "On the contraction property of energy forms on infinite-dimensional space." *J. Funct. Anal.* **92 (1990)**

[22] POTTHOFF, J. and STREIT, L.: "Invariant states on random and quantum fields. ϕ-bounds and White Noise Analysis." *J. Funct. Anal.* **111** (1993) 259 - 277.

[23] STREIT, L. and WESTERKAMP, W.: "A generalization of the characterization theorem for generalized functionals of white noise." In "*Dynamics of Complex and Irregular Systems*" (Ph. Blanchard et al., eds.). World Scientific, 1993.

[24] WATANABE, H.: "The local time of self-interactions of Brownian motions as generalized Brownian functionals". Lett. Math.Phys. **23** (1991) 1-9.

EXCURSION MEASURES FOR ONE-DIMENSIONAL TIME-HOMOGENEOUS DIFFUSIONS WITH INACCESSIBLE AND ACCESSIBLE BOUNDARIES

A.TRUMAN
Department of Mathematics,
University College of Swansea,
Singleton Park,
Swansea SA2 8PP.

ABSTRACT. For one-dimensional time-homogeneous diffusions with drift b with accessible and inaccessible boundaries, let the expected number of $\substack{\text{upward}\\ \text{downward}}$ excursions from some point a, of duration exceeding t, per unit local time at a, be $\nu_a^{\pm}[t,\infty)$. We give some new expressions for $\nu_a^{\pm}[t,\infty)$ in terms of the potential $\frac{1}{2}(b^2+b')$ and the stickiness of the boundary. These include as a special case Lévy's results for Brownian motion.

1. INTRODUCTION.

We consider excursion theory for one-dimensional time-homogeneous diffusions with continuous sample paths with inaccessible and accessible boundaries. For these diffusions we give generalizations of Paul Lévy's results for Brownian motion.

1.1 Inaccessible Boundary Case (See Truman and Williams [1] and Truman, Williams and Yu [2]).

Let B be a $BM(\mathbb{R})$ process, normalised so that for each $s,t>0$

$$\mathbb{E}\{B(s)B(t)\} = \min(s,t),$$

with $B(0)=0$, where $\mathbb{E}$ is the expectation with respect to the Weiner measure $d\mu(\omega)$ on the probability space $\Omega = C_0 = \{\omega : [0,\infty) \to \mathbb{R}, \omega \text{ continuous with } \omega(0)=0\}$ and $B(t)(\omega)=\omega(t)$ each $\omega\in\Omega$, each $t\geqslant 0$.

Now let $X(\in\mathbb{R})$ be the one-dimensional time-homogeneous process, satisfying

$$dX(s) = b\big(X(s)\big)ds + dB(s), \qquad \text{with} \qquad X(0)=a,$$

b being a Lipschitz continuous drift, the boundary $\{-\infty,+\infty\}$ being inaccessible. To ensure the pathwise existence and uniqueness of X we demand that for some constant C, $b^2(x) < C(1+x^2)$.

The process X has generator

$$\mathscr{G} = 2^{-1}\frac{d^2}{dx^2} + b(x)\frac{d}{dx}$$

A. I. Cardoso et al. (eds.), Stochastic Analysis and Applications in Physics, 441–454.

and putative invariant density $\rho_0(x) = C \exp\left(2 \int^x b(u)\, du\right)$, C being a normalisation constant. Denote by H the quantum mechanical Hamiltonian

$$H = -2^{-1} \frac{d^2}{dx^2} + V(x),$$

where $V(x) = 2^{-1}\left(b^2(x) + b'(x)\right)$. Then, setting $D = \frac{d}{dx}$ and using

$$D + b(x) = e^{-\int^x b(u)du} D e^{\int^x b(u)du},$$

we obtain the formal operator identity

$$\mathscr{G} = -\rho_0^{-\frac{1}{2}} H \rho_0^{\frac{1}{2}},$$

for H the above Schrödinger operator.

Assumptions I.

(1) Potential V is continuously differentiable and bounded below, with V' bounded and continuous.

(2) Invariant density $\rho_0 \in L^1(\mathbb{R})$ and is bounded together with its first few derivatives.

The semiboundedness and continuity of V ensure that the differential operator H is limit point at $\pm\infty$. When equipped with domain $C_0^\infty(\mathbb{R})$, $H : L^2(\mathbb{R}) \to L^2(\mathbb{R})$ has a unique self-adjoint extension also denoted by H. (See e.g. Cycon, Froese, Kirsch and Simon [3], page 17 Corollary 2.2.) Moreover, the first assumption ensures that the semigroup e^{-tH} is given by a Feynman-Kac formula with a differentiable kernel. The second assumption implies that the boundary $\{-\infty, +\infty\}$ is inaccessible. For consider the Cauchy-Schwarz inequality :

$$\int_{x_0}^{x} \rho_0^{\frac{1}{2}}(y) \rho_0^{-\frac{1}{2}}(y) dy \leqslant \left\{\int_{x_0}^{x} \rho_0(y)\, dy\right\}^{\frac{1}{2}} \left\{\int_{x_0}^{x} \rho_0^{-1}(y)\, dy\right\}^{\frac{1}{2}}.$$

Then $\rho_0 \in L^1(\mathbb{R})$ and divergence of *l.h.s.* as $x \uparrow \infty$, ensures the divergence of $\int_{x_0}^x \rho_0^{-1}(y)\, dy$ as $x \uparrow \infty$, giving the result that $\int^x \rho_0^{-1}(y)\, dy \left\{\int^y \rho_0(z)\, dz\right\}$ diverges as $x \uparrow \infty$. The latter condition is the well-known Feller condition for the boundary ∞ to be inaccessible. (See e.g. Gihman and Skorohod [4].)

1.2 Excursions and Local Time.

Map $s \mapsto X(s)$ is almost surely continuous, so $\{s > 0 : X(s) \gtrless a\}$ is almost surely an open subset of $\mathbb{R}$. Thus, $\{s > 0 : X(s) \gtrless a\}$ can be decomposed into a countable number of open component intervals - the *excursion intervals.* Define

$$L^{\pm}(t) = \text{Leb}\left\{s \in [0, t] : X(s) \gtrless a\right\}$$

and define the *local time* at a

$$L^a(t) = \lim_{h\downarrow 0} h^{-1}\,\mathrm{Leb}\left\{s \in [0,t] : X(s) \in \left(a - \tfrac{h}{2}, a + \tfrac{h}{2}\right)\right\}.$$

$L^a(t)$ has inverse $\gamma^a(t) = \inf\{s > 0 : L^a(s) > t\}$. $\gamma^a(t)$ is a *stopping time* with $X(\gamma^a(t)) = a$. $\gamma^a(t)$ is how long you have to wait until (local time at a) L^a equals t. The all-important observation is that the jumps in $\gamma^a(t)$ are the excursions which X makes up to L^a equals t. (See Rogers and Williams [5].)

Lévy's Result.

For $b \equiv 0$ for each $\lambda, t > 0$

$$\mathbb{E}_a\{e^{-\lambda\gamma^a(t)}\} = \exp\left\{-t\int_0^\infty (1 - e^{-\lambda s})d\nu_0(s)\right\},$$

with

$$\nu_0[s,\infty) = \left(\frac{2}{\pi}\right)^{\frac{1}{2}} s^{-\frac{1}{2}},$$

each $s > 0$. It follows that, if $\sharp_s(t)$ denotes the number of excursions of duration exceeding s, up to L^a equals t,

$$\mathbb{P}(\sharp_s(t) = N) = e^{-t\nu_0[s,\infty)}\frac{(t\nu_0[s,\infty))^N}{N!},$$

$N = 0, 1, 2, \ldots$. (See Itô and McKean [6].)

Definition.

Let

$$V^+_{Sy}(x) = \begin{cases} V(x), & \text{for } x > a \\ V(2a - x), & \text{for } x < a \end{cases}$$

and define V^-_{Sy} similarly. Denote by $e^{-tH^\pm}$ the semigroup generated by the symmetrized Hamiltonians

$$H^\pm = -2^{-1}\frac{d^2}{dx^2} + V^\pm_{Sy}(x).$$

These Hamiltonians $H^\pm$ are also essentially self-adjoint on $C_0^\infty(\mathbb{R})$ and their semigroups are given by a Feynman-Kac formula, $V^\pm_{Sy}$ being semibounded and continuous. (See Cycon, Froese, Kirsch and Simon [3].)

Theorem 1.

Given the above assumptions on ρ_0 *and* V,

$$\mathbb{E}_a\{e^{-\lambda L^\pm(\gamma^a(t))}\} = \exp\left\{-t\int_0^\infty (1 - e^{-\lambda s})d\nu_a^\pm(s)\right\},$$

each $\lambda, t > 0$, where for $s > 0$

$$\nu_a^{\pm}[s,\infty) = \pm \int_{y \gtrless a} \rho_0^{\frac{1}{2}}(y)\rho_0^{-\frac{1}{2}}(a)\frac{\partial}{\partial x}\bigg|_{x=a} e^{-sH^{\pm}}(x,y)\,dy.$$

Remarks.

(1) $L^{\pm}(\gamma^a(t))$ are independent random variables, with

$$L^+(\gamma^a(t)) + L^-(\gamma^a(t)) = \gamma^a(t),$$

almost surely.

(2) Jumps in $L^{\pm}(\gamma^a(t))$ are the $\substack{\text{upward}\\ \text{downward}}$ excursions X makes from a up to L^a equals t.

(3) Defining $\sharp_s^{\pm}(t)$ as the number of $\substack{\text{upward}\\ \text{downward}}$ excursions as above,

$$\mathbb{P}\big(\sharp_s^{\pm}(t) = N\big) = e^{-t\nu_a^{\pm}[s,\infty)}\frac{\big(t\nu_a^{\pm}[s,\infty)\big)^N}{N!},$$

$N = 0, 1, 2, \ldots$.

Corollary 1.

Given the above assumptions and some mild additional smoothness and boundedness of the derivatives of ρ_0 and $V_{Sy}^{\pm}$, as $\epsilon \downarrow 0$,

$$\nu_a^{\pm}[\epsilon,\infty) = (2\pi\epsilon)^{-\frac{1}{2}} \pm 2^{-1}b(a) + \left(\frac{\epsilon}{2\pi}\right)^{\frac{1}{2}} V_{\pm} + \frac{\epsilon^{\frac{3}{2}}}{(2\pi)^{\frac{1}{2}}}\left[-\frac{1}{6}V_{\pm}^2 - \frac{1}{6}b(a)V_{\pm}' + \frac{3}{20}V_{\pm}''\right]$$
$$\pm \frac{\epsilon^2}{30}\left[b(a)V_{\pm}'' + V_{\pm}'''\right] + O(\epsilon^{\frac{5}{2}}),$$

where $V_{\pm} = V(a\pm) = V(a)$ etc..

Remarks.

(1) The first term gives the Lévy behaviour for Brownian motion.
(2) If V is symmetrical about $x = a$, ρ_0 is symmetrical about $x = a$ and $b(a) = 0$.
(3) Note absence of term in ϵ.
(3) In a very impressive application of Mathematica K.Y.Yu has taken this expansion to 10 terms. (See Yu [7].)

1.3 Accessible Boundary Case.

When the boundary is accessible there are some new features. We consider a persistent diffusion on $[0,\infty)$, with Lipschitz drift b and generator $\mathscr{G}$ above, with 0 accessible and ∞ inaccessible.

Define the speed measure m and scale function p by

$$m'(x) = 2e^{2\int^x b(u)du}, \qquad p'(x) = e^{-2\int^x b(u)du},$$

so that

$$m' = 2\rho_0 = 2(p')^{-1}.$$

Then, since

$$2^{-1}e^{-2\int^x b(u)du}De^{2\int^x b(u)du}D = 2^{-1}(D+2b)D,$$

we obtain the operator identity

$$\mathscr{G} = D_m D_p = \frac{1}{m'(x)}D\frac{1}{p'(x)}D,$$

where we assume $m', p' > 0$.

Assumptions A.

(1) $\int_{x_0}^{\infty} m'(x)dx < \infty$, so $\int_{x_0}^{\infty} \rho_0(x)dx < \infty$.

(2) $\int_0^{x_0} e^{-2\int^x b(u)du}dx < \infty$, $\int_0^{x_0} e^{2\int_0^x b(u)du}dx < \infty$ and

$$\int_0^{x_0} e^{2\int^y b(u)du}\left\{\int_0^y e^{-2\int^x b(u)du}dx\right\}dy < \infty.$$

Here the first assumption ensures ∞ is inaccessible. The second assumption ensures 0 is an accessible regular boundary point. (See e.g. Gihman and Skorohod [4].)

According to Feller [8] for the process X with generator $\mathscr{G} = D_m D_p$ to have continuous sample paths the only allowed boundary condition at 0 is of the form,

$$\chi_0 f(0) - \pi_0 D_p f(0) + \sigma_0 D_m D_p f(0) = 0,$$

$\chi_0, \pi_0, \sigma_0 \geqslant 0$. For an invariant measure to exist for this process it is necessary that $\chi_0 = 0$ (See e.g. Mandl [9].) This gives the accessible boundary condition,

$$\pi_0 D_p f(0) = \sigma_0 D_m D_p f(0),$$

$\sigma_0, \pi_0 > 0$. We call this the 'sticky boundary condition'.

Theorem 2.

For the above process X, with the above sticky boundary condition,

$$\tilde{\nu}_a^-(\lambda) = \int_0^\infty e^{-\lambda t} \nu_a^-[t,\infty) dt,$$

is given by

$$\tilde{\nu}_a^-(\lambda) = \frac{\pi_0 \tilde{p}_D'(0+,a)\tilde{p}_N'(a-,a) + \lambda\sigma_0\tilde{p}_N(0,a)\tilde{p}_D'(a-,a)p'(0)}{2\lambda\left(\pi_0\tilde{p}_D'(0+,a)\tilde{p}_N'(a,a) + \lambda\sigma_0\tilde{p}_N(0,a)\tilde{p}_D(a,a)p'(0)\right)},$$

where $\tilde{p}_N^D$ is the $\substack{Dirichlet \\ Neumann}$ resolvent kernel, $\tilde{p}^D(x,y) = \int_0^\infty e^{-\lambda t} p_t^D(x,y)dt$, being the partial derivative with respect to the first spatial argument, $p'(0) = \rho_0^{-1}(0)$.

Remark.

Needless to say there is a similar expression for $\tilde{\nu}_a^+(\lambda)$.

Corollary 2.

For $b \equiv 0$, for Brownian motion with the above sticky boundary condition,

$$\tilde{\nu}_a^-(\lambda) = \left(\frac{\sigma_0}{2} + \frac{\pi_0 \tanh\left(a\sqrt{2\lambda}\right)}{\sqrt{2\lambda}}\right)\left(\frac{\sigma_0}{2}\sqrt{2\lambda}\tanh\left(a\sqrt{2\lambda}\right) + \pi_0\right)^{-1}.$$

This gives correct to first order in $\frac{\sigma_0}{\pi_0}$

$$\begin{aligned}\nu_a^-[t,\infty) = {} & (2\pi t)^{-\frac{1}{2}} - 2e^{-\frac{4a^2}{2t}}(2\pi t)^{-\frac{1}{2}} - 2e^{-\frac{16a^2}{2t}}(2\pi t)^{-\frac{1}{2}} - \dots \\ & + \frac{2\sigma_0}{\pi_0}\left(\frac{2a}{t}e^{-\frac{4a^2}{2t}}(2\pi t)^{-\frac{1}{2}} - 2.\frac{4a}{t}e^{-\frac{16a^2}{2t}}(2\pi t)^{-\frac{1}{2}} + \dots\right)\end{aligned}$$

and correct to first order in $\frac{\pi_0}{\sigma_0}$

$$\begin{aligned}\nu_a^-[t,\infty) = {} & (2\pi t)^{-\frac{1}{2}} + 2e^{-\frac{4a^2}{2t}}(2\pi t)^{-\frac{1}{2}} + 2e^{-\frac{16a^2}{2t}}(2\pi t)^{-\frac{1}{2}} + \dots \\ & - \frac{4\pi_0}{\sigma_0}\left(\operatorname{Erfc}\left(\frac{2a}{2^{\frac{1}{2}}t^{\frac{1}{2}}}\right) + 2\operatorname{Erfc}\left(\frac{4a}{2^{\frac{1}{2}}t^{\frac{1}{2}}}\right) + 3\operatorname{Erfc}\left(\frac{6a}{2^{\frac{1}{2}}t^{\frac{1}{2}}}\right) + \dots\right).\end{aligned}$$

Again the first terms correspond to Lévy's result for Brownian motion.

2. PRELIMINARY RESULTS.

In this section we give some additional motivation and preliminary results needed in what follows.

2.1 Nelson's Stochastic Mechanics.

For the invariant density $\rho_0(x) = C\exp\left(2\int^x b(u)du\right)$, $\theta_0 = \rho_0^{\frac{1}{2}}$ satisfies

$$\frac{\theta_0'}{\theta_0} = b,$$

so differentiating

$$\frac{\theta_0''}{\theta_0} - \left(\frac{\theta_0'}{\theta_0}\right)^2 = b'$$

i.e.

$$(H - E)\theta_0(x) = 0,$$

where $H = -2^{-1}\frac{d^2}{dx^2} + V(x)$, $V(x) = 2^{-1}\left(b^2(x) + b'(x)\right)$ and $E = 0$. When $\rho_0 \in L^1(\mathbb{R})$, $\theta_0 \in L^2(\mathbb{R})$ is the ground-state eigenfunction of the Hamiltonian H. In this case the diffusion X is the ground-state Nelson diffusion process for the Hamiltonian H.

To see this, following Nelson, define

$$D_\pm f(X(t),t) = \lim_{h\downarrow 0}\mathbb{E}\left\{\frac{f(X(t\pm h), t\pm h) - f(X(t),t)}{\pm h}\middle| X(t)\right\}.$$

(See Nelson [10], [11].) It follows from Itô's formula that for sufficiently well-behaved f

$$D_\pm f(\mathbf{X}(t),t) = \left(\frac{\partial f}{\partial t} + D_\pm X(t)\frac{\partial f}{\partial x} \pm 2^{-1}\frac{\partial^2 f}{\partial x^2}\right)\Bigg|_{x=X(t)} (x,t),$$

$D_\pm X$ being the mean forward and backward drifts. From the original diffusion equation we have

$$D_+X(t) = b(X(t))$$

and Nelson showed that in the state with density ρ_0

$$D_-X(t) = b(X(t)) - \frac{\partial}{\partial x}\Bigg|_{x=X(t)} \ln\rho_0 = -b(X(t)).$$

If we then compute Nelson's stochastic acceleration, $2^{-1}(D_+D_- + D_-D_+)X$, we obtain

$$2^{-1}(D_+D_-X + D_-D_+X)(t) = -2^{-1}(D_+ - D_-)b = -\frac{d}{dx}\left(\frac{b^2(x) + b'(x)}{2}\right)\Bigg|_{x=X(t)}.$$

This is the stochastic analogue of Newton's second law of motion for the potential $V = 2^{-1}(b^2 + b')$. This suggests H should be important in determining the statistical properties of X.

2.2 First Hitting times and Excursions.

Here we give some elementary connections between excursion theory and first hitting times.

Definition.

The *first hitting time* of the level a for the process X with starting value x, is defined by

$$\tau_x(a) = \inf\{s > 0 : X(s) = a \mid X(0) = x\},$$

for each $x, a \in \mathbb{R}$.

We begin with a simple but powerful lemma. (See also Albeverio, Fukushima, Karwowski, and Streit [12].)

Lemma 1.

Let X be a one-dimensional process which is time-homogeneous and has continuous sample paths. Let $p_t(\cdot,\cdot)$ be its transition density so that

$$\mathbb{P}\big(X(t) \in dy | X(0) = x\big) = p_t(x, y)dy.$$

Then for a any point intermediate to x and y

$$\mathbb{E}\{e^{-\lambda\tau_x(a)}\} = \frac{\tilde{p}_\lambda(x,a)}{\tilde{p}_\lambda(a,y)},$$

for $\lambda > 0$, $\tilde{p}_\lambda(x,y) = \int_0^\infty e^{-\lambda s}p_s(x,y)ds$. In particular, if for each fixed x and λ, $y \mapsto \tilde{p}_\lambda(x,y)$ is continuous,

$$\mathbb{E}\{e^{-\lambda\tau_x(a)}\} = \frac{\tilde{p}_\lambda(x,a)}{\tilde{p}_\lambda(a,a)},$$

$\lambda > 0$, $x, y \in \mathbb{R}$.

Proof.

If a is any point intermediate to x and y and if X is a one-dimensional process with continuous sample paths for X to go from x to y in time t the first hitting time of a must be less than or equal to t. If the process starts afresh at each time, we must therefore have

$$p_t(x,y) = \int_0^t \mathbb{P}\big(\tau_x(a) \in du\big)p_{t-u}(a,y).$$

The desired result follows by taking a Laplace Transform. □

Now because $\gamma^a(t)$ is a stopping time with $X\big(\gamma^a(t)\big) = a$ there exists $\mathscr{C}(\lambda) > 0$ with

$$\mathbb{E}_a\{e^{-\lambda\gamma^a(t)}\} = e^{-t\mathscr{C}(\lambda)},$$

so $\mathscr{C}(\lambda)^{-1} = \int_0^\infty \mathbb{E}_a\{e^{-\lambda\gamma^a(t)}\}dt$. (See Rogers and Williams [5].) We now come to Lévy's beautiful result.

Lemma 2.

Given the above assumptions

$$\mathbb{E}_a\{e^{-\lambda\gamma^a(t)}\} = \exp\left(\frac{-t}{\tilde{p}_\lambda(a,a)}\right),$$

$\lambda, t > 0$, $a \in \mathbb{R}$, *where* $\tilde{p}_\lambda(a,a) = \int_0^\infty e^{-\lambda s} p_s(a,a)ds$.

Proof.

Because $\gamma^a(\cdot)$ is almost surely the inverse function of $L^a(\cdot)$,

$$\mathscr{C}(\lambda)^{-1} = \int_0^\infty \mathbb{E}_a\{e^{-\lambda\gamma^a(t)}\}dt = \mathbb{E}_a\left\{\int_0^\infty e^{-\lambda\gamma^a(t)}dt\right\} = \mathbb{E}_a\left\{\int_0^\infty e^{-\lambda s}dL^a(s)\right\}$$
$$= \int_0^\infty e^{-\lambda s}d\big(\mathbb{E}_a\{L^a(s)\}\big) = \tilde{p}_\lambda(a,a),$$

where in the last step we used $\mathbb{E}_a\{L^a(s)\} = \int_0^s p_u(a,a)du$. □

3. OUTLINE PROOFS.

3.1 Inaccessible Boundary Case.

First we give a simple consequence of the fact that $\theta_0 = \rho_0^{\frac{1}{2}}$ satisfies the time-independent Schrödinger equation $H\theta_0 = 0$, for the Hamiltonian $H = -2^{-1}\frac{d^2}{dx^2} + V(x)$, $V(x) = 2^{-1}\big(b^2(x) + b'(x)\big)$.

Lemma 3.

Assuming $\theta_0(x) = C\exp\big(\int^x b(u)du\big) \in L^2(\mathbb{R})$ *and that* $V = 2^{-1}(b^2 + b')$ *is continuous and bounded below, with* $b^2(x) < C(1+x^2)$*, the transition density for* X *is*

$$p_t(x,y) = \theta_0^{-1}(x)e^{-tH}(x,y)\theta_0(y),$$

for $x, y \in \mathbb{R}$*, the heat kernel* $e^{-tH}(x,y)$ *being given as an integral with respect to the Brownian bridge process on* $(0,t)$*,* $\alpha_t(s) \cong B(s) - \frac{s}{t}B(t)$*,*

$$e^{-tH}(x,y) = (2\pi t)^{-\frac{1}{2}} e^{-\frac{(x-y)^2}{2t}} \mathbb{E}_{\alpha_t}\left\{\exp\left(-\int_0^t V\left[\left(1-\frac{s}{t}\right)x + \frac{s}{t}y + \alpha_t(s)\right] ds\right)\right\},$$

$\mathbb{E}_{\alpha_t}$ *the corresponding integral.*

Proof.

The condition $b^2(x) < C(1+x^2)$ implies there exists $\delta > 0$ such that

$$\sup_{s<t} \mathbb{E}\{e^{\delta b^2(B_x(s))}\} < \infty.$$

The proof is then a simple consequence of the Girsanov-Cameron-Martin theorem. (See e.g. Gihman and Skorohod [4], page 90.) □

Remark.

The fact that ρ_0 is the invariant density for X translates here to the fact that θ_0 is the ground-state of H with eigenvalue zero.

Theorem 3.

Given the above assumptions,

$$\nu_a^{\pm}[t,\infty) = \rho_0^{-1}(a) \int_{x \gtrless a} \rho_0(x) \frac{d}{dt} \mathbb{P}\big(\tau_x(a) < t\big)\, dx,$$

for each $a \in \mathbb{R}$, $t > 0$.

Proof.

Firstly Lemma 3 proves that $p_t(x,y)$ is jointly continuous in x and y, V being continuous and bounded below. Thus, from Lemma 1

$$\mathbb{E}\{e^{-\lambda\tau_x(a)}\} = \frac{\tilde{p}_\lambda(x,a)}{\tilde{p}_\lambda(a,a)}.$$

Thus,

$$\int_{-\infty}^{\infty} \rho_0(x)\mathbb{E}\{e^{-\lambda\tau_x(a)}\}dx = \big(\tilde{p}_\lambda(a,a)\big)^{-1} \int_{-\infty}^{\infty} \rho_0(x)\tilde{p}_\lambda(x,a),$$

where $\tilde{p}_\lambda(x,a) = \int_0^\infty e^{-\lambda t} p_t(x,a)dt$. Lemma 3 and the semiboundedness of V gives the existence of a constant α so that

$$p_t(x,y) \leqslant e^{\alpha t}\theta_0^{-1}(x)e^{-tH_0}(x,y)\theta_0(y).$$

Using Fubini's theorem we get

$$\int \rho_0(x)\mathbb{E}\{e^{-\lambda\tau_x(a)}\}dx = \big(\tilde{p}_\lambda(a,a)\big)^{-1} \int_0^\infty e^{-\lambda t} dt \left(\int p_t(x,a)\rho_0(x)dx \right),$$

or

$$\int \rho_0(x)\mathbb{E}\{e^{-\lambda\tau_x(a)}\}dx = \lambda^{-1}\big(\tilde{p}_\lambda(a,a)\big)^{-1}\rho_0(a).$$

The desired result follows after an integration by parts. □

Theorem 4.

Let $H^{\pm}(a)$ be the Dirichlet Hamiltonians

$$H^{\pm}(a) = \lim_{\lambda\uparrow\infty}\left(-2^{-1}\frac{d^2}{dx^2} + V(x) + \lambda\chi_{\mp}(x)\right),$$

$\chi_{\mp}$ being the characteristic functions of $\{x : x \gtrless a\}$, respectively. Then, given the above assumptions on V and ρ_0, for $\lambda, t > 0$, $a \in \mathbb{R}$,

$$\mathbb{E}_a\{e^{-\lambda L^{\pm}(\gamma^a(t))}\} = \exp\left(-t\int_0^\infty (1-e^{-\lambda s})d\nu_a(s)\right),$$

with

$$\nu_a^{\pm}[s,\infty) = \left(\theta_0, H^{\pm}(a)e^{-sH^{\pm}(a)}\theta_0\right)_{L^2}\theta_0^{-2}(a).$$

Proof.

The point is that for $x \gtrless a$, $t > 0$,

$$\mathbb{P}(\tau_x(a) > t) = \theta_0^{-1}(x)\left\{e^{-tH^{\pm}}\theta_0(x)\right\},$$

respectively. To see this we note that for $x \gtrless a$, respectively,

$$\mathbb{P}(\tau_x(a) > t) = \lim_{\lambda\uparrow\infty}\mathbb{E}_x\left\{\exp\left(-\lambda\int_0^t \chi_{\mp}(X(s))ds\right)\right\},$$

(See Simon [13].). The desired result follows from the Girsanov-Cameron-Martin theorem and Theorem 3. □

To complete the proof we need the following version of the reflection principle.

Lemma 4.

For any potential U which is continuous, bounded below and symmetrical about $x = a$, for $x, y > a$,

$$\mathbb{E}\left\{\exp\left(-\textstyle\int_0^t U(x+B(s))ds\right)\Big|\,\tau_x(a) < t,\, x+B(t)\in dy\right\}$$
$$= \mathbb{E}\left\{\exp\left(-\textstyle\int_0^t U(x+B(s))ds\right)\Big|\, x+B(t)\in d(2a-y)\right\},$$

Proof.

A simple application of the reflection principle. □

This gives

$$e^{-tH^{\pm}}(x,y) = e^{-tH^{\pm}_{Sy}}(x,y) - e^{-tH^{\pm}_{Sy}}(2a-x,y),$$

for $x, y > a$, $H^{\pm}_{Sy}$ being the symmetrized Hamiltonians $-2^{-1}\frac{d^2}{dx^2}+V^{\pm}_{Sy}(x)$. To complete the proof of Theorem 1 is now a simple exercise in partial integration. To prove Corollary 1 we merely expand in powers of t the resulting expressions for $\nu = \nu_1 + \nu_2$, taking $a = 0$,

$$\nu_1[t,\infty) = \pm(2\pi t)^{-\frac{1}{2}} \int_{u \gtrless 0} \rho_0^{\frac{1}{2}}(t^{\frac{1}{2}}u)ue^{-\frac{u^2}{2}} \times$$
$$\times\, \mathbb{E}\left\{ \exp\left(-t\int_0^1 V\left[uvt^{\frac{1}{2}} + (v^{\frac{1}{2}} - v)B(t)\right]dv\right)\right\}du$$

and

$$\nu_2[t,\infty) = \mp(2\pi)^{-\frac{1}{2}} \int_{u \gtrless 0} \rho_0^{\frac{1}{2}}(t^{\frac{1}{2}}u)e^{-\frac{u^2}{2}}\mathbb{E}\left\{t\int_0^1 V'\left[uvt^{\frac{1}{2}} + (v^{\frac{1}{2}} - v)B(t)\right](1-v)dv \times\right.$$
$$\left.\times \exp\left(-t\int_0^1 V\left[uwt^{\frac{1}{2}} + (w^{\frac{1}{2}} - w)B(t)\right]dw\right)\right\}du,$$

where $\nu = \nu_a^{\pm}$ and $V = V^{\pm}_{Sy}$. (An example where this expansion seems to be correct up to arbitrary high orders is given in Truman, Williams and Yu [2].)

3.2 Accessible Boundary Case.

Let Assumptions A of section 1 be valid. The next theorem is the analogue of Theorem 3 in the accessible boundary case.

Theorem 5.

For the diffusion $X(\,\in[0,\infty))$ with ∞ inaccessible and the origin 0 an accessible boundary point (as in Assumptions A) and the sticky boundary condition $\pi_0 D_p f(0) = \sigma_0 D_m D_p f(0)$, $\sigma_0, \pi_0 > 0$, for each $a \in [0,\infty)$, $t > 0$,

$$\nu_a^-[t,\infty) = \frac{\sigma_0}{2\pi_0\rho_0(a)}\frac{d}{dt}\mathbb{P}(\tau_0(a) < t) + \int_{0+}^a \rho_0^{-1}(a)\rho_0(x)\frac{d}{dt}\mathbb{P}(\tau_x(a) < t)dx$$

and

$$\nu_a^+[t,\infty) = \int_a^\infty \rho_0^{-1}(a)\rho_0(x)\frac{d}{dt}\mathbb{P}(\tau_x(a) < t)dx.$$

Proof.

The point is that for this sticky boundary condition the invariant density is

$$\left(\sigma_0\delta_0(x)dx + 2\pi_0\rho_0(x)dx\right)\left(\sigma_0 + 2\pi_0\int_{0+}^{\infty}\rho_0(x)dx\right)^{-1}.$$

(See Mandl [9], page 82 Theorem 5.) We now repeat the argument in the proof of Theorem 3 with this modified expression for the invariant density. A careful analysis of the contributions which the terms in this sum make leads to the above result. □

To complete the proof we use Lemma 1 and the observation that for $\lambda > 0$

$$\tilde{p}_\lambda(x,y) = \frac{\pi_0\tilde{p}'_D(0+,y)\tilde{p}_N(x,y) + \sigma_0\lambda\tilde{p}_N(0,y)p'(0+)\tilde{p}_D(x,y)}{\pi_0\tilde{p}'_D(0+,y) + \sigma_0\lambda\tilde{p}_N(0,y)p'(0+)},$$

$\tilde{p}^D_N$ being the $\substack{\text{Dirichlet}\\ \text{Neumann}}$ resolvent kernels, $\tilde{p}_D(x,y) = \int_0^\infty e^{-\lambda t}\tilde{p}^D_t(x,y)dt$ etc.. We give as a final corollary :

Corollary 3.

We see that as $a \sim 0$

$$\nu_a^-[t,\infty) \sim \frac{1}{\rho_0(a)}\left\{\frac{\sigma_0}{2\pi_0}\frac{d}{dt}\mathbb{P}\big(\tau_0(a) < t\big) + \mathrm{O}(a)\right\}.$$

This shows the effect of the sticky boundary at 0.

4. ACKNOWLEDGEMENTS.

It is a pleasure to thank David Williams for permission to publish some of our joint work and the SERC for financial support through research grant number GR/J34095. It is also a pleasure to thank Professor Ludwig Streit for organising such a stimulating conference.

REFERENCES

1. A.Truman and D.Williams, *Images, Reflections and Excursions*, Proceedings of the International Conference on Stochastic Processes, Ascona-Locarno 1991 (S.Albeverio, D.Merlini, eds.), World Scientific Press, London, (to appear 1993).
2. A.Truman, D.Williams and K.Y.Yu, *Schrödinger Operators and Asymtotics for Poisson Lévy Excursion Measures for One-Dimensional time-Homogeneous Diffusions* (to appear 1993).
3. H.L.Cycon, R.G.Froese, W.Kirsch and B.Simon, *Schrödinger Operators*, Springer-Verlag, New York, 1987.
4. I.I.Gihman and A.V.Skorohod, *Stochastic Differential Equations*, Springer-Verlag, Berlin, 1972.
5. L.C.G.Rogers and D.Williams, *Diffusions, Markov Processes and Martingales, Volume 2, Itô Calculus*, John Wiley, Chichester, 1987.

6. K.Itô and H.P.McKean, *Diffusion Processes and their Sample Paths*, Springer-Verlag, Berlin, 1965.
7. K.Y.Yu, *Ph.D thesis*, University College of Swansea, 1993.
8. W.Feller, *The Parabolic Differential Equations and the Associated Semi-Groups of Transformations*, Annals of Mathematics **55** (1952), 468–519.
9. P.Mandl, *Analytical Treatment of One-Dimensional Markov Processes*, Springer-Verlag, Berlin, 1968.
10. E.Nelson, *Dynamical Theories of Brownian Motion*, Princeton University Press, Princeton, 1967.
11. ———, *Quantum Fluctuations*, Princeton University Press, Princeton, 1984.
12. S.Albeverio, M.Fukushima, W.Karwowski, and L.Streit, *Capacity and Quantum Mechanical Tunneling*, Communications in Mathematical Physics **81** (1981), 501–513.
13. B.Simon, *Functional Integration and Quantum Physics*, Academic Press, New York, 1979.

MIX
Papier aus verantwortungsvollen Quellen
Paper from responsible sources
FSC® C105338

If you have any concerns about our products,
you can contact us on
ProductSafety@springernature.com

In case Publisher is established outside the EU,
the EU authorized representative is:
Springer Nature Customer Service Center GmbH
Europaplatz 3, 69115 Heidelberg, Germany

Printed by Libri Plureos GmbH
in Hamburg, Germany